Energy Storage and Redistribution in Molecules

Energy Storage and Redistribution in Molecules

Edited by Juergen Hinze

University of Bielefeld
Bielefeld, Federal Republic of Germany

PLENUM PRESS · NEW YORK AND LONDON

CHEMISTRY

Library of Congress Cataloging in Publication Data

Main entry under title:

Energy storage and redistribution in molecules.

Proceedings of two workshops held June 23–July 1, 1980, at the Centre for Inter-
disciplinary Studies of the University Bielefeld, Germany.
Includes bibliographical references and index.
1. Molecules—Congresses. 2. Energy transfer—Congresses. I. Hinze, Juergen, 1937–
. II. Universität Bielefeld. Zentrum für Interdisziplinäre Forschung.
QD461.E524 1983   539′.6   83-2354
ISBN 0-306-41272-1

Proceedings of two workshops—on Molecular Structure, Rigidity, and Energy
Surfaces, and on Energy Scrambling in a Molecule; How Stationary Are Internal States?—
held June 23–July 1, 1980, at the Center for Interdisciplinary Studies of the University
Bielefeld, Bielefeld, Federal Republic of Germany

©1983 Plenum Press, New York
A Division of Plenum Publishing Corporation
233 Spring Street, New York, N.Y. 10013

Printed in the United States of America

PREFACE

We characterize an isolated molecule by its composition, i.e. the number and types of atoms forming the molecule, its structure, i.e. the geometrical arrangement of the composite atoms with respect to each other, and its possible, i.e. quantum mechanically allowed, stationary energy states. Conceptually we separate the latter, being aware that this is an approximation, into electronic, vibrational and rotational states, including fine and hyperfine structure splittings. To be sure, there is an intimate relation between molecular structure and molecular energy states, in fact it is this relation we use, when we obtain structural information through spectroscopy, where we determine transitions between various stationary states of the molecule.

The concepts above have proven extremely useful in chemistry and spectroscopy, however, the awareness of the limitations of these concepts has grown in recent years with the increasing recognition of

(i)   fluctional molecules,
(ii) multiphoton absorption processes and
(iii) influences due to the surroundings on "isolated" molecules.

Improved experimental techniques, advanced by the rapid developments in laser technology, have allowed extremely high resolution spectroscopy, including time resolved spectroscopy, and the possibility to "pump" several light quanta into a molecule with the ability to follow their fate as time evolves. With the accumulation of such new experimental findings, coupled with the apparently insurmountable difficulties to achieve a "mode selective" chemistry, we are led to ask the question: "How stationary are stationary (molecular) states?"

In a nutshell, this question addresses the by now convincing observations that energy deposited with several photons into one or a few modes of a molecule does not remain there, it is distributed, scrambled rapidly into various other modes of the system, just as if the molecule itself would be its own heat bath. This is puzzling at first sight, for the simple answer, "a vibrational mode

is not a pure state, it is coupled to other modes", is only a
pseudo answer, it applies only in a particular frame of approximate
reference states. The isolated molecule and the photons know
nothing about such artificial though extremely useful approximations
and concepts. In a frame of reference states in terms of which the
total Hamiltonian of the isolated molecule is diagonal there are no
internal coupling terms left and a state is a state is a state.
In such a representation of true molecular eigenstates the only
coupling terms left are due to external perturbations, perturbations
which cannot be eliminated; as a minimum the zero point radiation
field is always present. The questions remaining are those of state
preparation, sensitivity of internal states to external perturbations
and density of states relative to the natural, due to the zero point
field, state width.

The contributions to this volume, concerned with these questions
of energy storage and redistribution in molecules are the outgrowths
of two overlapping workshops "Molecular Structure, Rigidity and
Energy Surfaces" and "Energy Scrambling in a Molecule; How Stationary
are Internal States?" held consecutively in summer 1980 at the
Centre for Interdisciplinary Studies of the University Bielefeld.
These workshops brought experimentalists and theoreticians of widely
differing schools together to present their work and discuss their
ideas related to the general problems outlined above. The contributions here are collected such as to begin with nine experimental
papers followed by theoretical investigations focusing in turn on
a) the electronic structure and potential surfaces of molecules,
b) various approaches to separate different molecular modes approximately, c) characterising the couplings in these modes and
d) the description of the molecular dynamics, which leads to mixing
and chaos in individual molecules in reactive collisions and in
condensed phase chemistry.

To be sure, all the theoretical investigations are based on
various approximations and thus in the frame of conventional concepts
and a representation of approximate states. This is so, since we do
not know the true eigenstate of a molecule, and even if we knew
them and could formulate the problem in such a basis, chances are
we would gain but little insight and understanding.

Before closing this introduction with an appropriate rhyme,
written by B.T. Sutcliffe, with some assistance from R.S. Berry,
during the workshops, I would like to express my sincere gratitude
to the directorate and staff of the Centre for Interdisciplinary
Studies for their generous financial support, which made the workshops possible, and their assistance, which provided the atmosphere
for the success of the workshops. My special thanks go to
K. Mehandru for her tireless help in the compilation of this volume.

J. Hinze

How doth the little molecule scramble internal modes,
And pour out photons on the world, in loads and loads and loads?
Were we to listen to it through a coupled microphone,
Perhaps an answer would emerge, perhaps we'd hear it groan.

"If those experimentalists would stop tweaking me with light,
My coupled oscillations would not be in such a plight.
I would not need to scramble, I could lead a normal life,
And leave the population free from quasi-random strife.

I'd be in a pure quantum state (if you forget the field),
And guarantee, when breaking up, to give just unit yield.
So abandon all experiments that torture me, I beg,
And study me theoretically, don't scramble me, like egg.

Let gentle theoreticians caress me to sensual moans,
And join in soft sweet coupling to excite my overtones,
The ecstasy will supervene and in one exquisite moment,
I'll yield up all my secrets and tell you what the show meant."

B.T.S.
Bielefeld, July 1980
(with apologies to Isaac Watts
& Lewis Carroll)

CONTENTS

CONTENTS

CHEMISTRY AND SPECTROSCOPY OF MOLECULES AT HIGH LEVELS OF EXCITATION

Jeffrey I. Steinfeld

Department of Chemistry
Massachusetts Institute of Technology
Cambridge, Massachusetts   02139   USA

INTRODUCTION

The behavior of molecules possessing large amounts of internal vibrational energy is the subject of a great deal of current attention, particularly at this workshop. Such highly excited molecular species are encountered in a variety of situations: in addition to the pre-eminent case of multiple infrared-photon excitation with high-intensity lasers[1,2], these systems can be prepared by stimulated-emission pumping[3], by chemical excitation[4], and even by thermal excitation, for example, in flames[5]. High vibrational levels may also be accessed by one-photon transitions, as in overtone spectroscopy[6].

The problem of understanding the spectroscopy of these systems becomes progressively more difficult as either the level of excitation or the dimensionality of the molecule increases. For purposes of discussion, it may be useful to distinguish the various regimes in the following manner:

1. <u>Low Excitation (system in its ground or first excited vibrational state)</u>

   (a) <u>Low Dimensionality (diatomic or linear triatomic)</u>: simple rigid-rotor, harmonic-oscillator models can be applied with confidence.

   (b) <u>High Dimensionality (polyatomic molecules)</u>: the normal-mode approximation is usually adequate, although vibration-rotation interaction can be important, as the recent detailed infrared analysis of SF_6 has demonstrated[7]. In

1

either of these regimes, conventional infrared, microwave, and Raman spectroscopy provide the necessary information for characterizing the structure of the molecule.

2. High Excitation (system containing 1 eV or more of vibrational energy, i.e., >20-30% of a bond dissociation energy).

(a) Low Dimensionality: For these systems, typically diatomic molecules in high vibrational levels, RKR potentials and JWKB wavefunctions in these potentials generally provide an adequate description. Experimental techniques such as laser-induced fluorescence and optical-optical double resonance are available.

(b) High Dimensionality: It is in this regime that the principal difficulties arise, since the nature of the interaction of a radiation field with a polyatomic molecule at high internal energy is neither simple nor intuitive. The system probably does not behave as a collection of isolated two-level systems, so that the fraction of molecules interacting directly with a monochromatic field will depend on excitation level and intensity of the radiation. If a "hole" is "burned" into the absorption profile, relaxation processes (collisional and/or intramolecular) tending to "fill" that hole must be considered. While spectroscopic probes can, in principle, be used to determine the canonical distribution of energy in an ensemble of chemically or optically pumped molecules, there remains the question of the localization of the energy in a directly pumped mode vs. internal redistribution ("scrambling") at constant energy. Furthermore, if a coherent superposition state is produced by the existing radiation, there remains the question of intramolecular dephasing. In other words, the entire concept of "taking a spectrum" of a polyatomic molecule at high excitation needs to be thoroughly examined.

These questions are especially pertinent in reference to excitation produced by multiple infrared photon absorption (MIRPA). While this is a very popular and efficient method of producing high excitation levels in molecules, along with concomitant phenomena such as dissociation and optical emission, it possesses several drawbacks. The principal problem with MIRPA is the extremely complicated phenomenology of the excitation process, which produces a broad distribution of internal energies in the excited molecule, making quantitative interpretation difficult. Most measurements performed on MIRPA as a function of integrated pulse energy, or fluence ($\Phi \equiv \int I(t)dt$), yield results averaged over the fluence-dependent energy distribution $f(E;\Phi)$. The dissociation yield samples f only for energies greater than the dissociation threshold energy

E_o, and is insensitive to the low-energy region. The mean energy
or photon deposition, $\langle E;\Phi \rangle$ or $\langle n;\Phi \rangle$, with $\langle E \rangle = \langle n \rangle \hbar\omega_o$, samples
all energies, but weights the high energy end of f more strongly.
In any case, measurements of only such first-moment quantities as
yield or energy deposition cannot, by themselves, uniquely specify
f. Furthermore, spatial averaging over a focussed laser beam eliminates much of the specific information on $f(E;\Phi)$[8]; deconvolution
techniques, such as those suggested by Yablonovitch et al.[9] require data having higher precision than are usually available.

Lee and coworkers[10] have measured product translational energy
distributions following MIRPA-induced dissociation under molecular-
beam conditions, and have inferred a statistical, or Boltzmann-like
form for $f(E;\Phi)$ from their results. However, Goodman et al.[11] have
argued that essentially identical translational energy distributions
can result from widely differing internal vibrational relaxation
(IVR) models, so that Lee's results are not unambiguous indicators of
internal energy distributions either.

We are thus led to seek spectroscopic methods for the characterization of energy distributions resulting from MIRPA, and in general for the investigation of polyatomic molecules at high levels of
excitation. In the following discussion, we shall examine in turn
the principles, limitations, and representative accomplishments of
these methods, including infrared-infrared double resonance (IRDR),
optical-ultraviolet absorption double resonance, laser-induced fluorescence excitation/emission spectroscopy ($LIFE^2S$), and Raman scattering. While an exhaustive reveiw of all available methods will
not be attempted, we shall try to point out the important features
of several of these important techniques.

EXPERIMENTAL METHODS

Infrared Double-Resonance Spectroscopy (IRDR)

In IRDR, a low-intensity, c.w., tunable probe is used to measure the transient infrared absorption of vibrationally excited
molecules produced by a moderate-intensity infrared pump source.
The probe intensity should ideally be kept low, in order to avoid
saturation and other nonlinear responses to the observing radiation.
In early experiments on SF_6[12] and BCl_3[13], a line-tunable c.w. CO_2
laser was used as the probe; more recently, a semiconductor diode
laser has been used to provide nearly continuously tunable cover-
age[14].

The interaction of pump and probe fields with an isolated two-
level resonance is well understood. If the detuning $\Delta = \omega(\text{pump}) -$
ω_o (transition) is less than the Rabi frequency $\Omega = \mu_{12}\varepsilon(\omega)/\hbar$, the
absorption line is saturated; if $\Delta > \Omega$, an a.c. Stark modulation of
the line occurs. A new, coherently modulated lineshape is observed,

but no population is transferred between upper and lower states.
Even if the levels involved are highly degenerate (as in SF_6 pumped
by the CO_2 P_1 (18) laser line at 945.9802 cm^{-1}; the resonant transition is the A_2^1 component of the $J' = 32 \leftarrow J'' = 33$, or P(33)
"line"), we have shown   that a two-level-system model still applies providing the following conditions are met:

(1) the $M_J' \leftarrow M_J''$ transition moments are all independent of M_J;

(2) the state-to-state relaxation rates are also independent
of M_J, so that the phenomenological relaxation parameters $\kappa = 1/T_1$
and $\kappa' = 1/T_2'$ can be used; and

(3) the probe field is weak, i.e., Ω (probe) $\ll \Omega$ (pump). Under
these conditions, the two-level and three-level double-resonance
lineshapes derived by Mollow[17] and by Hansch and Toschek[18] can be
employed. The CO_2-laser-pumped IRDR spectrum in SF_6 has been modelled in this way [19], and all the observed features can be accounted
for in terms of a series of noninteracting 2-level and 3-level systems.

The IRDR method should prove to be particularly informative at
high pump intensities, at which true MIRPA into the quasicontinuum
may occur. Some initial attempts at such experiments have indicated the problems which may be encountered in doing so. The high-
power pump pulses are less reproducible, giving rise to variations
in the IRDR signal. Many experiments have used a pulsed CO_2 laser
or a high-intensity (>1 W/cm^2) c.w. laser as the probe, giving rise
to nonlinear absorber response; furthermore, the relatively simple
2- and 3-level models cannot be expected to hold true for such high
probe intensities[16]. Nevertheless, the reports of IRDR absorption
at pump fluences between 0.1 and 1.0J/cm^2 which have appeared in
the literature[20-25] do show some interesting results. Most of these
studies find an induced absorption extending to frequencies below
the 948 cm^{-1} ν_3 fundamental, although this absorption sometimes appears to increase monotonically[23,24], and sometimes to decrease
monotonically[22,25] over the range 920-940 cm^{-1}. Since measurements
are made only at discrete CO_2 laser line frequencies, it is impossible to say whether there is structure in this absorption, or whether
a true "quasicontinuum" has been reached in the activated molecule.
The problem in interpreting such spectra has been well stated by
Bagratashvili et al.[22]: "... neither the absorption spectrum $\sigma(\nu)$
nor the initial distribution of absorbers $N(\nu_i, J_i)$ are known",
making any unique assignment of these transient spectra extremely
difficult.

The time-dependence of the high-intensity IRDR signals is complicated by numerous contributing effects, including intramolecular
relaxation and dephasing, pressure-dependent vibrational and rotational energy transfer, depolarization, and field-dependent effects

such as saturation, a.c. Stark effects, free-induction decay, and
self-focussing. Estimates of the collision – free, "intramolecular"
vibrational relaxation time extracted from IRDR experiments have
differed by many orders of nagnitude. It is to be hoped that the
current generation of experiments, carried out with low-power, tunable diode laser probes under carefully controlled excitation conditions, will be successful in unravelling this multiplicity of
phenomena.

Optical/Ultraviolet Absorption Double Resonance

Information on internal vibrational energy content and relaxation can also be obtained from measuring the changes in the optical or u.v. photoabsorption cross section following MIRPA, even
when the u.v. absorption is to a dissociating state, leading to a
continuous spectrum. Of course, when a fluorescent transition is
available, it is generally advantageous to utilize emission spectroscopy [following section]. We will consider here the results of recent absorption double resonance experiments on ozone[26] and CF_3I[27,28].

The transmission of u.v. laser pulses through CF_3I at 308 nm
(XeCl) or 337 nm (N_2) has been monitored following excitation by an
intense I.R. pulse[27]. These wavelengths lie considerably to the red
of the peak of the CF_3I absorption continuum, as does the wavelength
of the cw Kr^+ion laser (350 nm) employed by Pummer et at.[29]. At all
these wavelengths, there is an instantaneous increase in absorbance,
essentially coterminous with the CO_2 laser pulse, followed by a
slower, pressure-dependent increase. Kudryavtsev and Lethokhov[27]
ascribed the latter process to equilibration between laser-pumped
and "cold" molecules, described by a single relaxation time $p\tau_{VV} \simeq$
5×10^5 $Torr^{-1}$ sec^{-1}. Padrick et al.[28] examined the wavelength dependence of the absorbance change by flash-kinetic spectroscopy;
however, since they could observe only between 250 and 300 nm, no
direct correlation with the single-frequency measurements could be
made. They found, however, that the absorbance curves for vibrationally "hot" and "cold" CF_3I molecules appeared to cross at 285
nm, in remarkable agreement with the wavelength dependence predicted
in Ref. 27.

A more thorough analysis of a simpler system, ozone, has been
carried out by Adler-Golden in our laboratory[26]. The change in u.v.
absorption in the 250-nm Hartley band, following excitation by a CO_2
laser at 9.5 μm, can be resolved into contributions from molecules
in the v_1, v_2, and v_3 vibrational modes. Three separate vibrational
relaxation times are required to describe the absorbance changes
following infrared excitation: a very rapid (1-2 collisions) equilibration between the v_1 and v_3 stretching modes, a V-V relaxation
between these modes and the v_2 bending mode, and a V-R, T relaxation
out of v_2. In addition, the amplitude of the absorbance change can
be directly related to the mean number of infrared photons absorbed

by the ozone molecules[30], and can thus be used to measure this quantity when direct transmission and opto-acoustic methods both present difficulties.

Laser-Induced Fluorescence Excitation and Emission Spectroscopy (LIFE^{2}S)

Just as fluorescence spectroscopy, with its high sensitivity, energy resolution, and time resolution, is the method of choice for analytical detection[31], combustion diagnostics[32], and probes for chemical dynamics[33], so it is for probing energy distributions following MIRPA. The limitation on this method is the relatively small number of systems for which the simultaneous requirements of a mid-infrared absorption susceptible to MIRPA and an easily accessible transition having a high quantum yield of fluorescence can be met; when it can be used, though, the LIFE^{2}S method is an extremely powerful one. A few representative examples of results obtained using this technique will now be given.

The initial, discrete-excitation steps in the MIRPA of D_2CO[34,35] and HDCO have been investigated by Orr and co-workers; the v'' and $J''_{K''}$ levels populated by the infrared laser pulse are probed by exciting fluorescence from these levels via the $\widetilde{A}$ $^1A''$ electronic state using a tunable ultraviolet dye laser. The initial excitation steps can be unambiguously identified, and rotational relaxation pathways and rate constants are measured. Rate constants of (50 ± 10) μsec^{-1} $Torr^{-1}$ and (25 ± 10) μsec^{-1} $Torr^{-1}$ are for $(\Delta J = \pm 1)$ and $(\Delta J = 0,$ $\Delta K_c = \pm 1)$ transferring collisions, respectively; these are several times greater than the gas-kinetic molecular collision rate.

A similar experiment has been carried out in propynal (HC $\equiv$ CCHO) by Guillory and co-workers[38]. The interpretation given for this experiment, however, is open to serious question. The assertion is made in Ref. 38 that "collisionless intramolecular" rotational and vibrational(mode-to-mode) relaxation occurs following excitation in v_6 or v_{10}. This conclusion cannot be accepted however, since the former process would violate the conservation of rotational angular momentum, and the latter would violate the conservation of energy (at the level of excitation being studied, the propynal molecule is still in a "sparse" regime - that is, the density of background vibrational levels is insufficient to function as a quasi-continuum). More careful experiments on the same system under molecular-beam conditions by Brenner et al. [39] have shown that, in fact, the initially populated vibrational and rotational states remain unrelaxed. The problem with the interpretation of Guillory's results may result from a totally inadequate curve-fitting procedure used instead of proper kinetic modelling; the very large rotational relaxation rate constants found by Orr et al. [36] may be one source of difficulty, if they are not taken into account properly. This particular system illustrates in a very dramatic way the necessity for the care that

must be taken in both the execution and the interpretation of experiments of this type.

The LIFE^{2}S technique has recently[40] been applied to thiophosgene
($Cl_2C = S$), with rather different and somewhat surprising results.
In this molecule, any one of a number of CO_2 laser lines in the 10-μm
region is coincident with an overtone ($2\nu_4 \leftarrow 0$) transition. Observation of vibrational and rotational populations following collision-
free MIRPA by exciting visible-wavelength fluorescence from the $\tilde{A} \, ^1A_2$
state shows that all rotational levels of the ground vibrational state
appear to be homogeneously depleted; furthermore, population is not
found in the vibrational levels $2\nu_4$, $4\nu_4$, etc., but in much higher-
lying levels with $v_4 \geq 10$. Since a single infrared laser frequency
cannot possibly interact with all the rotational lines of the $2\nu_4 \leftarrow 0$
band, even with power-broadening taken into account, the most likely
interpretation of these results is that coherent multiphoton excitation processes are playing an important role; this would allow a
whole range of rotational compensation schemes, each giving rise to
a separate multiphoton excitation ladder, to be operative[41]. The
variety of behavior which has been observed in just these three molecular systems demonstrates the importance of having a thorough knowledge of the spectroscopy of a particular molecule in order to understand its multiple infrared-photon excitation behavior[42].

In this connection, it is worth mentioning a variation of the
LIFE^{2}S technique, in which a single u.v. photon serves simultaneously
to prepare and to probe the state. These are the experiments of
Smalley and co-workers, in which fluorescence spectra from dynamically
cooled alkylbenzenes[43,44] and alkylanilines[45] are observed as a function of molecular complexity and excitation wavelength. In the larger
of these molecules, and at higher vibronic energies, the spectra appear to be increasingly broad and congested; this is taken to be evidence for intramolecular vibrational mode-to-mode coupling, or IVR,
occurring as the density of vibrational background states reaches an
appropriately high value. This approach is similar to that taken in
vibrational overtone spectroscopy[6], and in common with that method,
the interpretation of the results suffers from a certain amount of
ambiguity relating to the extent to which the observed linewidth is
due simply to inhomogeneous (e.g., rotational) broadening or sequence
congestion, rather than to a true homogeneous relaxation process.

MIRPA in Electronically Excited States. A phenomenon which was discovered during the course of the double-resonance experiments described above is that moderately long-lived electronically excited molecular states can undergo MIRPA, even when the ground electronic state
is unaffected by the infrared laser radiation. The classic example
is that of biacetyl ($(CH_3CO)_2$)[46]. The lowest triplet state of
biacetyl, T_1, can be populated by blue light such as the 4579Å-line
of an argon-ion laser, and emits a strong green vapor-phase phosphorescence having a lifetime of approximately 1 msec. When the triplet-

state molecules are irradiated with a CO_2 laser, the yield and lifetime of the phosphorescence drops sharply, and in addition some of
the emission is shifted back into the blue region of the spectrum.
This is interpreted as strong singlet-triplet mixing when the vibrational energy acquired by the triplet state exceeds the (S_1 - T_1)
splitting, about 2100 cm^{-1}. Similar results have been reported independently by Borisevich in the Soviet Union[47]. A blue shift of
the emission spectrum has also been observed in NO_2[48]; in addition,
NO is formed by multiple infrared photon dissociation out of the
electronically excited $\tilde{A}$ ($^2A_2$, $^2B_1$) state of NO_2. Experiments of
this kind can be very useful in studying the spectroscopy and dynamics of electronically excited states of molecules.

LIFE[2]S of Thermally Excited Molecules. Some polyatomic molecules
having unusually high chemical stability can achieve substantial
levels of vibrational excitation by purely thermal means, as in a
shock tube or even in a flame. Even though it is rarely possible to
resolve individual vibrational states under such conditions, some
useful information can still be obtained. An example of this is a
recent study[49] of polycyclic aromatic hydrocarbons, such as pyrene
or fluoranthene, entrained in an atmospheric (air + ethylene) flame
at $\approx$ 1500 K. Fluorescence from these extremely hot molecules could
be excited with a frequency-doubled flash-pumped dye laser. Normally, a molecule such as pyrene is excited in the $S_2 \leftarrow S_0$ band at 300-
330 nm, and emits in the $S_1 \rightarrow S_0$ band at 340-370 nm. At the high
temperatures of the flame, however, there is a substantial amount of
emission in $S_2 \rightarrow S_0$ as well, indicating rapid internal conversion of
the excess internal energy between electronic and vibrational degrees
of freedom. Aside from this feature, however, spectra taken over a
1000° temperature range are very nearly superimposable, even though
the pyrene molecule in the flame has a mean vibrational energy content 20-30,000 cm^{-1} greater than in the low-temperature vapor. This
implies that most of the energy resides in vibrational modes which
are not strongly coupled to the electronic transition, and that the
same Franck-Condon intensity arguments which apply in the normal-
mode regime can still be used at these very high levels of excitation.

Raman Probes for High Vibrational Excitation

As a final experimental technique, we turn to Raman spectroscopy as
a probe for molecules in highly excited vibrational levels. The advantage of this technique is that it is applicable to all molecules,
without requiring specific resonant transitions, and that is state-
specific; its obvious and main disadvantage is the extremely small
magnitude of the Raman scattering cross-section.

A Raman spectroscopic measurement of multiple infrared photon
excitation in SF_6 and CF_3I has been reported by the group at the
USSR Institute of Spectroscopy[50-52]. A triple monochromator and
optical multichannel analyzer were used in the experiments; anti-

Stokes signals were observed at the Raman-active modes ν_1 (775 cm^{-1}) in SF_6 and ν_2(741 cm^{-1}) and ν_5(540 cm^{-1}) in CF_3I. Of course, none of these vibrational levels is directly pumped by the CO_2 laser. A variation of anti-Stokes intensity with infrared laser fluence and gas pressure could be measured, as well as a weak time-dependence, but it was not possible to derive individual vibrational level populations from these data, although the appearance of the signals was interpreted to show two separate sub-populations of "hot and "cold" molecules, as suggested by the u.v. double resonance experiments[27].

A more promising approach would seem to be to make use of the increased scattering intensity available in a stimulated Raman process such as Coherent Anti-Stokes Raman Spectroscopy (CARS). The intensity of the CARS signal, which may be several orders of magnitude greater than that of spontaneous Raman scattering, depends on the square of the nonlinear third-order susceptibility given by[53]

$$(3)\,\text{CARS} \quad \chi_{(\omega_1,\ \omega_1,-\omega_2)} = \chi_{NR} + \frac{Nc^4(\rho_{aa}-\rho_{bb})}{\hbar\omega_1\omega_2}\frac{d\sigma_{Raman}}{d\Omega}\frac{1}{\omega_{ba}-\omega_1+\omega_2-i\Gamma_{ba}}$$

There are several obvious ways in which information about excess vibrational energy can influence the magnitude of this susceptibility. First, as population is driven by the infrared laser from the ground state $|a\rangle$ to some excited vibrational $|b\rangle$, the inversion density ($\rho_{aa}-\rho_{bb}$) will tend toward zero, decreasing the intensity of the normal CARS signal. New CARS signals appearing at the anharmonically shifted frequencies $\omega_{b'a'}$ in the resonance denominator will indicate population in higher vibrational levels. It is even conceivable that, at sufficiently high excitation levels, anomalously large linewidths (Γ_{ba}), due in part to anharmonic coupling to background states, would be observed. Experiments of this nature are currently under way at the Regional Laser Facility in the MIT Spectroscopy Laboratory.

RELATIONSHIP WITH THEORY

We have now considered infrared and ultraviolet absorption, laser-induced fluorescence, and Raman scattering as possible methods for probing energy distributions in highly vibrationally excited molecules. All of these techniques can yield significant new information about such systems; we must consider further, though, the interpretation which can be placed on the results of such experiments.

In the usual practice of spectroscopy, one derives spectroscopic constants from measured transition frequencies and term values and uses these to determine structural parameters such as equilibrium bond lengths and angles, curvatures of potential surfaces, magnetic coupling coefficients, etc. It is not at all obvious that the language of transitions between stationary energy levels is appropriate

for systems at high levels of excitation. A more suitable model for
such experiments may be constructed in terms of dynamic, rather than
structural quantities. While the theory of such excited-state dynamics is treated in many papers at this workshop, it may nevertheless
be useful at this point to consider their connection with experiments
such as we have been considering.

    A useful way of expressing the absorption intensity $I(\omega)$ in terms
of dynamical quantities is as the Fourier Transform of a dipole correlation function[54],

$$I(\omega) = \frac{1}{2\pi} \int_{-\infty}^{\infty} e^{-i\omega t} \langle \vec{\mu}(o) \cdot \vec{\mu}(t) \rangle \, dt$$

Taking the usual normal-coordinate expansion of the dipole moment

$$\vec{\mu}(Q) = \vec{\mu}(o) + \sum_{j=1}^{N} \left(\frac{\partial \vec{\mu}}{\partial Q_j} \right) Q_j + \cdots$$

gives

$$I(\omega) = \frac{1}{2\pi} \sum_{j,k=1}^{N} \left(\frac{\partial \vec{\mu}}{\partial Q_j} \right) \cdot \left(\frac{\partial \vec{\mu}}{\partial Q_k} \right) \int_{-\infty}^{\infty} e^{-i\omega t} \langle Q_j(t) Q_k(o) \rangle \, dt$$

Thus, if the classical trajectory $\{Q(t)\}$ of the molecular system can
be found, the spectral density $I(\omega)$ can be calculated. The fluorescence or Raman scattering intensity $F(\omega)$ can be similarly defined.

    Calculations of this type have been carried out by Marcus and
his co-workers[55-58]. In this work, they used a two-dimensional model,
the Henon-Heiles potential, for which the Hamiltonian is

$$\mathcal{H} = \frac{1}{2} (p_x^2 + p_y^2) + \frac{1}{2} (x^2 + y^2) + \lambda(x^2 y - \frac{1}{3} x^3)$$

While this model bears no resemblance to any actual molecular system,
it possesses attractive mathematical properties. In work reported at
this workshop, Marcus has extended this method to treat a model Oscillator which could represent a linear triatomic molecule such as OCS.
Hänsel has also presented model calculations of this type[59], and has
recently used this model to calculate spectra for an ozone-like molecule containing about 0.7 eV of vibrational energy. Since spectra
of ozone at such excitation levels are beginning to be available[26],
these calculations may be amenable to direct comparison with experiment.

    As a final comment, we take note of some results recently obtain-
ed[60,61] for the spectral response of a system subjected to specified
initial conditions. Although the particular system is chosen to represent an "IVR" process as reflected in the linewidths of "overtone"

vibrational transitions[6], the conclusion reached should be a general
one. It is found that the nature of the expected spectral density
function, $I(\omega)$, depends strongly on the nature of the initial excitation given to the system, and far less so on the internal dynamics of the molecular model. Put into more picturesque language[60],
the result is "what you pluck is what you see" (or, perhaps, "where
you twang is where you hang"). Viewed pessimistically, this would
imply that it may never be possible to define a "spectrum" for a
highly excited molecule, in the way that is done for a system near
its equilibrium configuration, which is independent of the details
of the excitation process. To conclude on an optimistic note, however, we can take this result to mean that by selection of the right
excitation conditions, it may be possible to induce a molecule to
behave very much as we might wish it to do.

ACKNOWLEDGMENTS

The author would like to thank Professors Jürgen Hinze and Karl
Welge, the University of Bielefeld, and the Zentrum für Interdiszipl-
inare Forschung for their generous hospitality during the period of
the workshops. The manuscript itself was prepared while the author
was a Visiting Professor in the Department of Chemistry at the University of Southern California. Work from our own laboratory cited
in this report reflects research sponsored by the National Science
Foundation, the Air Force Office of Scientific Research and the Air
Force Geophysics Laboratory.

REFERENCES

1.   R.V. Ambartzumyan and V.S. Letokhov, in "Chemical and Biochemical
      Applications of Lasers", C.B. Moore, ed., Vol. 3, pp. 167-316,
      Academic Press, New York (1977).

2.   J.I. Steinfeld, ed., "Laser-Induced Chemical Processes", Plenum
      Publishing Corp, New York (1981).

3.   This method has been used to prepare high v" levels of I_2 (J.B.
      Koffend and R.W. Field, J. Appl. Phys. <u>48</u>, 4468 (1977)).

4.   For a discussion of classical chemical activation methods, see
      D.C. Tardy and B.S. Rabinovitch, Chem. Revs. <u>77</u>, 369 (1977).

5.   For a review of spectroscopic probes for molecular species in
      combustion environments, see D.R. Crosley, ed., "Laser Probes
      for Combustion Chemistry", American Chemical Society Symposium
      Series No. 134, Washington, D.C. 1980.

6.   See, for example, R.G. Bray and M.J. Berry, J. Chem. Phys. <u>71</u>,
      4909 (1979).

7.   H.W. Galbraith, C.W. Patterson, B.J. Krohn, and W.G. Harter, J.
     Mol. Spectroscopy $\underline{73}$, 475 (1978); E.G. Brock, B.J. Krohn, R.S.
     McDowell, C.W. Patterson, and D.F. Smith, J. Mol. Spectroscopy
     $\underline{76}$, 301 (1979); B.J. Krohn, J. Mol. Spectroscopy $\underline{73}$, 462 (1978).

8.   C. Reiser and J.I. Steinfeld, Opt. Eng. $\underline{19}$, 2 (1980).

9.   P. Kolodner, C. Winterfeld, and E. Yablonovitch, Opt. Commun.
     $\underline{20}$, 119 (1977).

10. P.A. Schulz, Aa. S. Sudbø, P.J. Krajnovich, H.S. Kwok, Y.R.
     Shen, and Y.T. Lee, Ann. Rev. Phys. Chem. $\underline{30}$, 379 (1979).

11. E. Thiele, M.F. Goodman, and J. Stone, Opt. Eng. $\underline{19}$, 10 (1980).

12. J.I. Steinfeld, I. Burak, A.V. Nowak, and D.G. Sutton, J. Chem.
     Phys. $\underline{52}$, 5421 (1970).

13. P.L. Houston, A.V. Nowak, and J.I. Steinfeld, J. Chem. Phys.
     $\underline{58}$, 3373 (1973).

14. C.C. Jensen, T.G. Anderson, C. Reiser, and J.I. Steinfeld, J.
     Chem. Phys. $\underline{71}$, 3648 (1979).

15. R.S. McDowell, H.W. Galbraith, B.J. Krohn, C.D. Cantrell, and
     E.D. Hinkley, Opt. Commun. $\underline{17}$, 178 (1976); R.S. McDowell, H.W.
     Galbraith, C.D. Cantrell, N.G. Nereson, P.F. Moulton, and E.D.
     Hinkley, Optics Letts. $\underline{2}$, 97 (1978).

16. H.W. Galbraith and J.I. Steinfeld, Opt. Commun. (to be pub-
     lished).

17. B.R. Mollow, Phys. Rev. $\underline{A5}$, 2217 (1972); ibid $\underline{A8}$, 1949 (1973).

18. Th. Hänsch and P. Toschek, Z. Physik. $\underline{236}$, 213 (1970).

19. C. Reiser, J.I. Steinfeld, and H.W. Galbraith, J. Chem. Phys.
     $\underline{74}$, 2189 (1981).

20. H.S. Kwok and E. Yablonovitch, Phys. Rev. Letts. $\underline{41}$, 745 (1978).

21. T.F. Deutsch and S.R.J. Brueck, J. Chem. Phys. $\underline{70}$, 2063 (1979).

22. V.N. Bagratashvili, V.S. Dolzhikov, and V.S. Letokhov, Soviet
     Physics JETP $\underline{49}$, 8 (1979).

23. W. Fuss and J. Hartmann, J. Chem. Phys. $\underline{70}$, 5468 (1979).

24. W. Fuss, Chem. Phys. Letts. $\underline{71}$, 77 (1980).

25. J.L. Lyman, L.J. Radziemski, Jr., and A.C. Nilsson, I.E.E.E. J. Quantum Electronics QE-16, 1174 (1980).

26. S. Adler-Golden and J.I. Steinfeld, Chem. Phys. Letts. 76, 479 (1980).

27. Yu. A. Kudriavtsev and V.S. Letokhov, Chem. Phys. 50, 353 (1980).

28. T.D. Padrick, A.K. Hays, and M.A. Palmer, Chem. Phys. Letts 70, 63 (1980).

29. H. Pummer, J. Eggleston, W.R. Bischel and C.K. Rhodes, Appl. Phys. Letts. 32, 427 (1978).

30. S. Adler-Golden and E. Schweitzer (to be published).

31. J.I. Steinfeld, C.R.C. Crit. Revs. Anal. Chem. 5, 225 (1975).

32. D.R. Crosley, ed., "Laser Probes for Combustion Chemistry", A.C.S. Symposium Series No. 134, Washington, D.C. 1980.

33. J.L. Kinsey, Ann. Rev. Phys. Chem. 28, 349 (1977).

34. B.J. Orr and G.F. Nutt, Optics Letts. 5, 12 (1980).

35. B.J. Orr and G.F. Nutt, J. Mol. Spectroscopy 84, 272 (1980).

36. B.J. Orr and J.G. Haub, Optics Letts. (to be published).

37. B.J. Orr, J.G. Haub, G.F. Nutt, J.L. Steward and O. Vozzo, Chem. Phys. Letts. (to be published).

38. M.L. Lesiecki, G.R. Smith, J.A. Steward and W.A. Guillory, Chem. Phys. 46, 321 (1980).

39. D. Brenner, K. Brezinsky, and P.M. Curtis, Chem. Phys. Letts. 72, 202 (1980).

40. D.M. Brenner, J. Chem. Phys. 74, 2293 (1981); D.M. Brenner and M. Spencer (to be published).

41. Similar to suggestions put forward by M.V. Kuzmin [Optics Comms. 33, 26 (1980)], and others.

42. A recent review having just this title [W. Fuss and K.L. Kompa, Report PLF-30, Max-Planck-Gesellschaft Projektgruppe für Laser-forschung, Garching-bei-München, July 1980] cites a number of additional examples.

43. J.B. Hopkins, D.E. Powers and R.E. Smalley, J. Chem. Phys. $\underline{72}$, 5039 (1980).

44. J.B. Hopkins, D.E. Powers, S. Mukamel, and R.E. Smalley, J. Chem. Phys. $\underline{72}$, 5039 (1980).

45. D.E. Powers, J.B. Hopkins and R.E. Smalley, J. Chem. Phys. $\underline{72}$, 5721 (1980).

46. I. Burak, T.J. Quelly, and J.I. Steinfeld, J. Chem. Phys. $\underline{70}$, 334 (1979).

47. N.A. Borisevich, Izvest. Akak. Nauk. SSSR, Ser. Fiz. $\underline{44}$, 681 (1980).

48. D. Feldmann, H. Zacharias and K.H. Welge, Chem. Phys. Letts. $\underline{69}$, 466 (1980).

49. D. Coe and J.I. Steinfeld, Chem. Phys. Letts. $\underline{76}$, 485 (1980).

50. V.N. Bagratashvili, Yu. G. Vainer, V.S. Dolzhikov, S.F. Kol'-yakov, A.A. Makarov, L.P. Malyavkin, E.A. Ryabov, E.G. Silkis, and V.D. Titov, Appl. Phys. $\underline{22}$, 101 (1980).

51. V.N. Bagratashvili, Yu. G. Vainer, V.S. Dolzhikov, S.F. Kol'-yakov, A.A. Makarov, L.P. Malyavkin, E.A. Ryabov, E.G. Silkis and V.D. Titov, JETP Letts $\underline{30}$, 471 (1979).

52. V.N. Bagratashvili, Yu. G. Vainer, V.S. Dolzhikov, V.S. Letokhov, A.A. Makarov, L.P. Malyavkin, E.A. Ryabov, and E.G. Silkis, Optics Letts. $\underline{6}$, 148 (1981).

53. Definitions of the quantities in the expression for $\chi^{(3)}$, as well as further theoretical and experimental elaborations, may be found in J.P. Taran, "Coherent Anti-Stokes Raman Spectroscopy", Chemical and Biochemical Applications of Lasers, Vol. 4 (C.B. Moore, ed.), Academic Press, New York (1980).

54. This expression is derived in J.I. Steinfeld, "Molecules and Radiation: An Introduction to Modern Molecular Spectroscopy", M.I.T. Press, Cambridge, 1978; and originally by R.G. Gordon, Adv. Magn. Resonance $\underline{3}$, 1 (1968).

55. D.W. Noid and R.A. Marcus, J. Chem. Phys. $\underline{62}$, 2119 (1975).

56. D.W. Noid and R.A. Marcus, J. Chem. Phys. $\underline{67}$, 559 (1979).

57. D.W. Noid, M.L. Koszykowski and R.A. Marcus, J. Chem. Phys. $\underline{71}$, 2864 (1979).

58. D.W. Noid, M.C. Koszykowski, M. Tabor, R.A. Marcus, J. Phys. Chem. $\underline{72}$, 6169 (1980).

59. K.D. Hänsel, Chem. Phys. $\underline{33}$, 35 (1978).

60. E.J. Heller and W.M. Gelbart, J. Chem. Phys. $\underline{73}$, 626 (1980).

61. K.F. Freed and A. Nitzan, J. Chem. Phys. $\underline{73}$, 4765 (1980).

ENERGY AND PHASE RANDOMIZATION IN LARGE MOLECULES AS PROBED BY

LASER SPECTROSCOPY

Ahmed H. Zewail*

Science Centre for Advancement of Post Graduate Studies
University of Alexandria, Alexandria, Egypt†

and

Arthur Amos Noyes Laboratory of Chemical Physics‡
California Institute of Technology, Pasadena
California 91125 USA

INTRODUCTION

In this paper, which is based on the materials presented at the Bielefeld workshop on Molecular Structure and Energy Scrambling, I would like to focus on the problems pertaining to energy and phase randomization in large molecules. Specifically, I would like to discuss the following major problems:

a) electronic dephasing as manifested in the interactions between large-molecules in different electronic states and a bath of dense low-frequency modes.

b) vibrational dephasing as manifested in the dynamics of localized vibrational energy in real large molecules.

c) population and phase changes in multi-level systems (optical T_1 and T_2).

---

* Alfred P. Sloan Fellow and Camille and Henry Dreyfus Teacher-Scholar.

† A major portion of this paper was written while the author was participating in the Winter School on laser spectroscopy at the University of Alexandria, UNARC, Alexandria, Egypt.

‡ Permanent address.

First, let me start by asking the question: Why are large molecules interesting? In these molecules, several intriguing phenomena may occur when the molecule is optically excited. One such phonomenon is the multiphoton absorption of low-energy photons, a topic that will not be address here. Another phenomenon associated with large molecules is the randomization of energy and phase among the large number of vibrational states at high energies. On the time scale of the experiment, vibrational energy redistribution may or may not be complete. If it is complete, one has a statistical distribution and the statistical theories, like RRKM, may be used to describe the phenomenon. If, on the other hand, the vibrational energy redistribution is nonstatistical, then we may be able to selectively localize energy in certain modes of large molecules.

With the advances made recently in the development of ultra-short time laser pulses (picosecond to subpicosecond) and ultrahigh frequency resolution (MHz and KHz) some of the following dynamical questions pertinent to the above mentioned problems can perhaps now be answered:

   a) What is the nature of the state that we excite with light?

   b) How can we separate intra- and intermolecular effects?

   c) What determines the dephasing of large molecules?

In the following sections, I shall describe how laser spectroscopy can be used to probe dynamics that are relevant to the above questions. First, however, a brief description of the origin of dephasing will be given.

THE QUANTUM ORIGIN OF OPTICAL DEPHASING

To describe the origin of optical dephasing we shall consider only the semiclassical approach--the molecule is treated quantum mechanically and the laser field classically. Semiclassically, we describe the process as follows: the laser field E interacts with an ensemble of molecules to produce a time-dependent polarization, $P(t)$, which in turn changes as the molecules dephase. So, our task now is to find how the polarization is related to dephasing and what dephasing means on the molecular level.

Before going into details it is instructive to examine criti-cally the origin of dephasing in molecules with only two levels (see Fig. 1). We shall consider two vibronic states; a ground state $\psi_a(\underset{\sim}{r})$ and an excited state $\psi_b(\underset{\sim}{r})$. The laser field is simply a wave (propagation direction, z) of the form

$$E(z,t) = \varepsilon \cos(\omega t - kz) = \tfrac{1}{2}[\varepsilon \, e^{i(\omega t - kz)} + \varepsilon \, e^{-i(\omega t - kz)}], \quad (1)$$

where ε is the amplitude and ω is the frequency of the radiation.

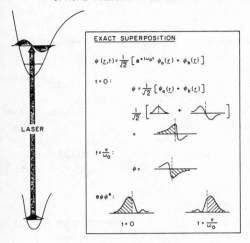

OPTICAL MOLECULAR COHERENCE

Fig. 1. A schematic for optical coherence in two level systems.

The state of the molecule (driven by the laser) at time t may be represented as

$$\psi(\underset{\sim}{r},t) = a(t)e^{-i\omega_a t} \psi_a(\underset{\sim}{r}) + b(t)e^{-i\omega_b t} \psi_b(\underset{\sim}{r}) \tag{2}$$

We now can calculate the time dependent molecular polarization:

$$P_m(t) = < \psi(\underset{\sim}{r},t)|\hat{\mu}|\psi(\underset{\sim}{r},t) >$$

$$= ab^* \mu_{ba} e^{-i(\omega_a-\omega_b)t} + a^*b\mu_{ab} e^{-i(\omega_b-\omega_a)t} \tag{3}$$

where $\hat{\mu}$ is the dipole-moment operator, and μ_{ba} and μ_{ab} are the transition moment matrix elements. Taking these matrix elements to be equal ($\equiv \mu$) and setting $\omega_b - \omega_a = \omega_0$, the transition frequency, we obtain

$$P_m(t) = \mu[ab^* e^{+i\omega_0 t} + a^*b e^{-i\omega_0 t}] . \tag{4}$$

Hence, the polarization, which is related to the radiation power, is zero if there is <u>no</u> coherent superposition or, in other words, if the molecule is certain to be in the state a or b. From Eq.(4) the total polarization for N molecules in the sample (assuming equal contribution and ignoring propagation effects) is therefore:

$$P(t) \equiv \tfrac{1}{2}[\overline{P} e^{i\omega_0 t} + \overline{P}^* e^{-i\omega_0 t}] = N \mu[\rho_{ab} + \rho_{ba}] \tag{5}$$

where $P(t) \equiv NP_m(t)$ and $\overline{P}$ is its complex amplitude, i.e., $\overline{P} = \overline{P}_{real} +$ $i\overline{P}_{imag}$. We chose the notation ρ_{ab} and ρ_{ba} for the cross terms $ab* e^{+i\omega_0 t}$ and $a*b e^{-i\omega_0 t}$ because they are indeed the off-diagonal elements of the ensemble density matrix $\underset{\sim}{\rho}$.

Equation (5) shows that in order to create a polarization or optical coherence we need a nonvanishing interference term or equivalently the off-diagonal elements of ρ must be nonzero in the zero-order basis set. Quantum mechanically, one can calculate these coherence terms and, for the most experimental description, one can perform a rotating (coordinate) frame analysis for the components of P. From knowledge of the polarization, the resultant coherent field in the sample can be found using the following self-consistent prescription:

$$\text{Laser field + Molecules} \longrightarrow \text{Polarization,P} \longrightarrow \text{Sample field} \tag{6}$$

Basically, P can be calculated using Eq.(5), and Maxwell's equations can be used with $\overline{P}$ being the source term to calculate the resultant sample field amplitude $\overline{\varepsilon}$. It is through $\overline{P}$ (or $\overline{\varepsilon}$) that we can monitor the nonlinear optical behavior of the sample and hence the changes in the rate of optical dephasing.

OPTICAL T_1 AND T_2: DEPHASING BY HOMOGENEOUS AND INHOMOGENEOUS BROADENINGS

From Eq.(2) we see that the probability of finding the system in the excited (ground) state is simple $|b|^2$ $(|a|^2)$. These probabilities decay by time constants, say T_{1b} and T_{1a}, respectively. Such phenomenological decay is the result of the Wigner-Weisskopf approximation, i.e., an exponential decay of the amplitudes a and b; a or b $\propto e^{-t/2T_1}$. Also, the cross terms of Eq.(2) will decay possibly by a different rate from the diagonal terms. Hence the ensemble density matrix can now be written as

$$|a> \equiv \psi_a(\underset{\sim}{r}) \qquad\qquad |b> \equiv \psi_b(\underset{\sim}{r})$$

$$
\begin{array}{c} <a| \\ \\ <b| \end{array}
\left[
\begin{array}{cc}
|a_0|^2 e^{-t/T_{1a}} & (a_0 b_0* e^{+i\omega_0 t}) e^{-t/T_2} \\
& \\
c.c. & |b_0|^2 e^{-t/T_{1b}}
\end{array}
\right] , \tag{7}
$$

where

$$\frac{1}{T_2} = \frac{1}{T'_2} + \frac{1}{2}\left(\frac{1}{T_{1a}} + \frac{1}{T_{1b}}\right)$$ (8)

The T_1-term in Eq.(8) comes from the diagonal elements and represents an average rate for the loss of population in the ab-levels. Physically, the T'_2-term represents the additional decay caused by phase changes in the cross terms. In other words, the <u>random</u> and rapid variation in $\omega_0(t)$, the transition frequency, causes the off-diagonal elements to decay faster than the diagonal ones. One can show that the linewidth of the transition a — b is $\frac{1}{\pi T_2}$ if the band profile is Lorentzian. The <u>total</u> dephasing rate is $\frac{1}{\pi T_2}$ therefore $\frac{1}{T_2}$: It contains $T'_2{}^{-1}$ -- the rate for phase coherence loss (pure dephasing) and T_1^{-1} -- the rate for irreversible loss of population in the two levels. The phenomenology described here is the optical analogue of magnetic resonance T_1 and T_2 of Bloch's equations, but the physics is different.

If we monitor emission say from b, it is clear now that all we can measure is T_{1b}. However, if the ensemble is homogeneous (i.e., consists of only those molecules that follow the uncertainty relationship $\Delta\nu T_2 = \pi^{-1}$) then an absorption experiment will give T_2 and an emission experiment will give T_1 and we have T'_2 at hand. Unfortunately, there are two problems. First, the optical transitions are inhomogeneously broadened (IB). This implies that the homogeneous ensemble is a subensemble of a grand ensemble as, e.g., in the case of Doppler broadening in gases (see Fig. 2). Because of IB, the homogeneous resonance is hidden under the usually broader IB transition. Similar to homogeneous widths, the width of IB transition can be related to a dephasing time, T_2^* using the uncertainty relationship.

The second problem concerns the measurement of T_2. If the homogeneous broadening (HB) is very small, it is difficult to measure HB through absorption methods using state-of-the-art lasers. By time-resolved experiments, however, the narrower the resonance, the longer the decay time, and it is relatively easy to measure T_2. It is this ability to separate T_2 (T_1 and T'_2) and T_2^* with ease that makes coherent laser spectroscopy a useful technique for unravelling dynamical optical processes in molecules. For example, most large molecules in low temperature solids show IB of $\sim 2cm^{-1}$, which implies that $T_2^* \sim 5$ picoseconds. If T_2^* were interpreted simply in terms of intrinsic HB, it would be said that all these molecules in solids dephase on the picosecond time scale. As we shall see later, this is not true and, in fact, the HB is orders of magnitude smaller than the IB, especially at low temperatures.

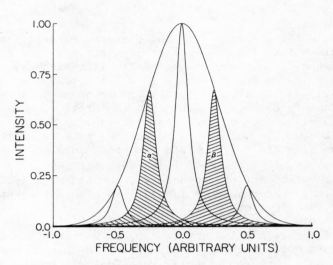

Fig. 2. Homogeneous and inhomogeneous broadenings: α and β are
homogeneous packets.

EXPERIMENTAL PROBING OF DEPHASING

        In Fig. 3, we depict the coherent transients obtained by
using a CW laser and a switch. A single mode laser (< 5MHz width)
is used to excite coherently a homogeneous subgroup of molecules.
The laser is then either diffracted acoustically so it will no
longer "see" the sample or switched into another frequency within
the IB line. The $\pi/2$ and π pulses are made by controlling the dif-
fraction duration time or the frequency switching time. Details can
be found in Refs. 1 and 2.

        When a group of molecules are excited coherently the super-
position of states discussed before will be established. As a result, a polarization is induced and if, e.g., the laser is immediately turned off, a coherent burst of light, sometimes called superradiance,[3] can be detected. Several transients can be observed
depending on the pulse sequence. These include photon echo, optical
nutation, incoherent resonance decay and optical free induction decay (OFID). Using the incoherent spontaneous emission, all these
coherent transients have been observed. (For a review see Refs.
1,2, and 4.) The important point is that from the transients we

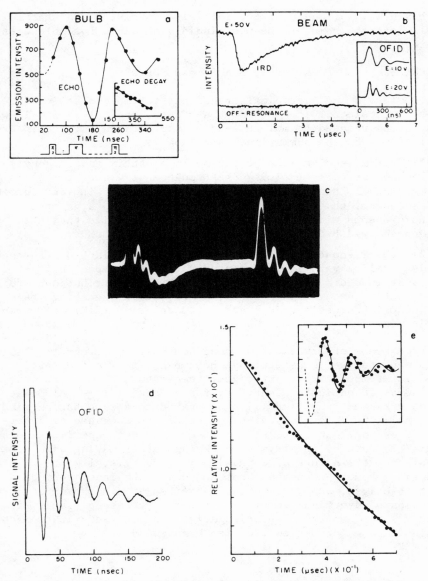

Fig. 3. The different coherent transients observed in gases and
        beams of iodine:[1] (a) the photon echo observed on the spon-
        taneous emission; (b) the IRD and OFID observed in an iodine
        beam; (c) OFID and nutation of iodine gas observed by fre-
        quancy switching a ring dye laser (Wm. Lambert, M. Burns,
        and A. H. Zewaill, to be published); (d) the OFID of iodine
        gas observed by frequency switching a conventional dye
        laser; (e) the transient nutation of iodine gas detected on
        the spontaneous emission.

can obtain T_1, T_2, and μ of the molecules involved. In what follows I shall describe some of the new findings obtained for molecules in different phases.

COHERENT TRANSIENTS IN GASES, BEAMS, AND SOLIDS

In Figs. 3 and 4, typical transients for molecules in different phases are depicted. From these transients, optical T_1 and T_2 were measured and related to molecule-bath interactions. The main experimental points can be summarized as follows:

1) for molecules like pentacene in a matrix (p-terphenyl) at low temperatures ($\sim$ 1.7K) the homogeneous width ($1/\pi T_2$) is very small, being $2.4 \times 10^{-4} cm^{-1}$ in contrast with the inhomogeneous width ($\sim$ 2cm^{-1}). Similarly for I_2 in a gas the homogeneous width at $\sim$ 10 mtorr is 579KHz while the inhomogeneous width due to Doppler broadening is 400MHz.

2) the T_1-contribution to the homogeneous width of pentacene is total at very low-temperatures (no pure dephasing). For I_2 at 10 mtorr the homogeneous width has contributions from radiative decay (128 ± 2KHz), nonradiative quenching (64Å^2 quenching cross section) and pure dephasing (71%), as shown in Fig. 5.

From these observations we conclude that in effect the large molecule pentacene in p-terphenyl host behaves as a two-level system at low-temperatures. At higher temperatures, however, this is not true as found independently by two groups.[1,2,4,5] Also, at low-temperatures and <u>long</u>-times (long compared to the dephasing time), pentacene is no longer a real two-level system because of intersystem crossing. In fact, using these transient experiments one deduces the rate of intersystem crossing, after making an approximation about the nature of dephasing in a multilevel system.[2] In conclusion, homogeneous and inhomogeneous dephasings are much different in origin, at least in the system we studied. Furthermore, extraction of dynamical information from spectroscopic analysis of lineshapes is "dangerous" particularly if phase randomization contributes in a major way to the line broadening.

VIBRATIONAL DEPHASING: HIGH-ENERGY VIBRATIONAL OVERTONES

A great deal of attention has been focused recently on the origin of relaxation of molecules in vibrationally-hot states, i.e., molecules with a high degree of excitation associated with "local" vibrational modes in the ground electronic state. These states presumably play an important role in the mechanistic description of processes such as multiphoton dissociation of molecules, laser isotope separation, and possibly selective laser-induced chemistry.

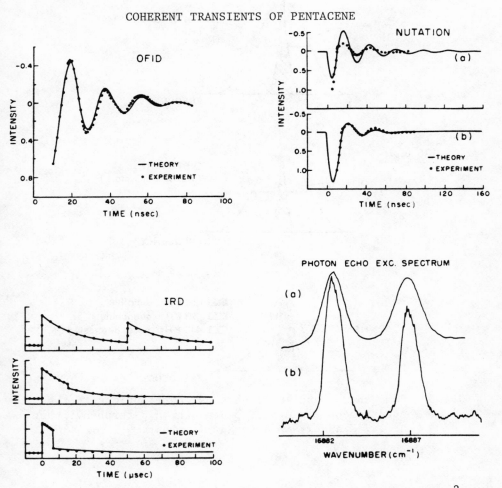

Fig. 4.   The different coherent transients observed in solids:[2] pen-
          tacene in p-terphenyl at ~ 1.7K. The OFID, the nutation,
          and the IRD at different pulse widths. The photon echo
          excitation spectrum is the courtesy of Dr. D. A. Wiersma.[5]

Studies of the energetics of these states have been done on
molecules in liquids and gases.[6] We have chosen to study the overtones in low-temperature matrices (1.3°K) in order to obtain the
dephasing after eliminating the contribution of thermal spectral
congestion. Frequency-resolved spectra were used, since the apparent dephasing time is very short (< 1 psec.).

For naphthalene (Figs. 6 and 7),Perry and Zewail[7] have found that
α and β-CH stretches of the fifth overtone have different apparent

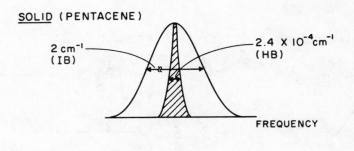

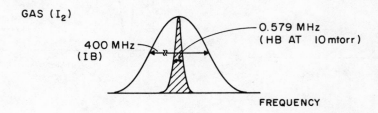

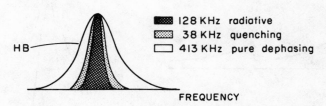

Fig. 5.   Homogeneous and inhomogeneous contributions to the
         spectroscopic width of pentacene in p-terphenyl host
         at 1.7°K and I_2 in a gas at 10 mtorr.

dephasing times (T_2^{α} = 73 fs and T_2^{β} = 0.11 psec). If this dephasing
is dominated by T_1 processes, then we must conclude that the vibrational energy of the five quanta CH modes is "flowing" very rapidly
to other modes.   Interestingly, when we compare the results on
naphthalene with those on benzene[6] we find that the β dephasing
time (and not α) appears to be comparable to the benzenic dephasing
time.   Furthermore, in molecules like durene (the spectra is shown
in Fig. 8), the relaxation time and the energies of methyl CH are
different from those of aromatic CH.   The result in Fig. 8 for

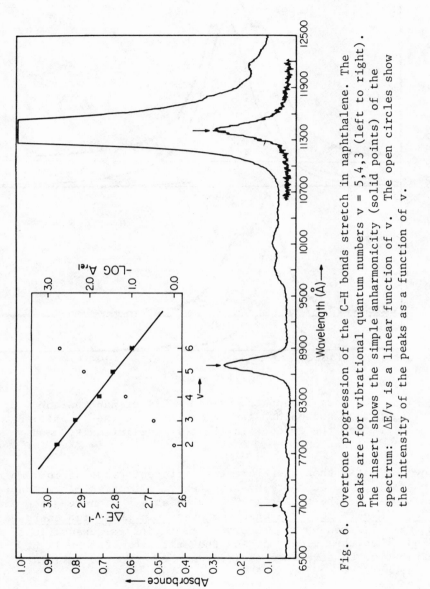

Fig. 6.   Overtone progression of the C-H bonds stretch in naphthalene. The peaks are for vibrational quantum numbers v = 5,4,3 (left to right). The insert shows the simple anharmonicity (solid points) of the spectrum: ΔE/v is a linear function of v. The open circles show the intensity of the peaks as a function of v.

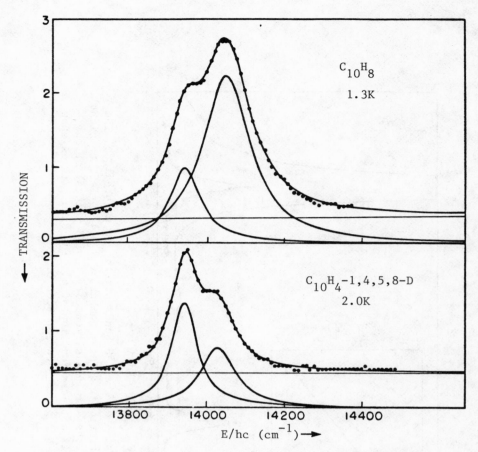

Fig. 7.   Transmission spectra at 1.3 - 2°K of naphthalene (upper
          graph) and naphthalene deuterated in the α-positions.
          Notice the changes in the relative intensity upon
          deuteration.

durene indicates that there are two different types of intramolecular baths for the CH to relax its energy. It shows clear, narrower
transitions for the aliphatic CH's even at v = 5 energy; the narrow-
est resonance (~ 20 cm⁻¹) observed so far in this high energy overtone region! Armed by the theory of radiationless transitions,[8] we
conclude that the relaxation and the states excited by light for
aromatic-type CH stretch are different from those of aliphatic-type
CH stretches.[9]

     Another technique for investigating overtone dephasing at low-
temperatures was reported by Smith and Zewail.[7] Smith observed
emission to the overtones from high-energy electronically excited
states.   In contrast to absorption measurements, very small

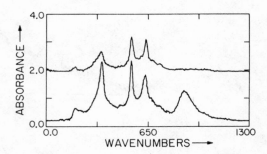

Fig. 8.   Spectra of durene and partially deuterated durene (in the
          aromatic CH's) obtained at low-temperatures. Notice the
          absence of the broad absorption of aromatic CH's in the
          spectrum of deuterated durene (upper graph). Also, note
          the sharpness of the lines due to methyl CH's.

concentrations can be used (1 part in 10^6) and the detection
sensitivity is quite high. In benzophenone (Fig. 9) we have located up to v = 6 in the ground state and measured the apparent
width as a function of excitation energy. These experiments are
quite relevant to a recent theoretical description of local and
normal modes made by Heller and Gelbart.[10] The apparent dephasing
rate in benzophenone appears to be linear in the v-energy. The
latter is a very important finding because it might add to our
understanding of the inter-mode coupling in the molecule, e.g.,
between C=O and C—C bonds. Smith has now completed the line
width studies at different temperatures and from these experiments
we have learned that overtone relaxation is efficient, possibly
due to the participation of the torsional modes of benzophenone.
This would be an example of vibrational relaxation and dephasing
due to forces imposed on the CO "oscillator" by neighboring
bonds.[7,11]

POPULATION RELAXATION AND PHASE RANDOMIZATION IN MULTI-LEVEL
SYSTEMS: THE LARGE MOLECULE-SMALL CRYSTAL LIMIT

     Before concluding this article, it is perhaps relevant to
comment on energy and phase randomization in multi-level systems.
The picture depicted earlier in the text using the two-level model
is adequate provided we have only two levels in contact with a
real bath, i.e., the correlation time for the bath is much shorter
than the dephasing time constants. In large molecules, with a
large number of levels (e.g., rotational or vibrational states) in

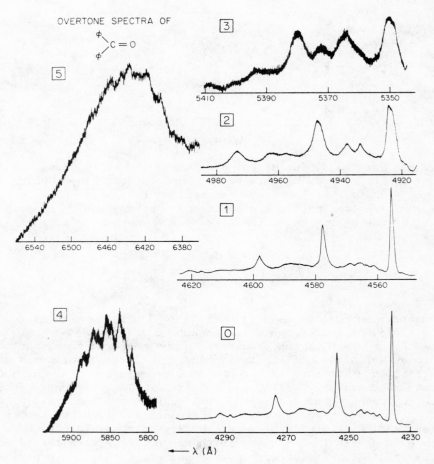

Fig. 9.   Overtone spectra of the C=O stretch of benzophenone
          obtained by emission from the first triplet state at
          low-temperature.

quasi-resonance with the optically-excited level one must be careful about the description of T_1 and T_2, especially if intense radiation is used.[12] To illustrate this point we shall make here a
connection between this problem and the problem of T_1 and T_2 for
the small-crystal limit.

    In a real one-dimensional solid, an excited molecule can
interact with a neighboring unexcited molecule by, e.g., intermolecular exchange couplings. This coupling leads to the formation
of an exciton band of finite width. The exciton states possess a
quasi-momentum, $\underset{\sim}{k}$, which varies in magnitude from 0 to π/c where
c is the stacking axis. To conserve momentum, light in the visible

or u.v region excites only the $k \simeq 0$ level. Hence all other $k \neq 0$
levels will not be excited directly by the light, but they are in
quasi-resonance with the optically-excited $k \simeq 0$ level. Our idea
was to measure T_1 and T_2 for such a system and to compare them with
the apparent relaxation time that one would obtain from a direct
absorption (or emission) experiment. The system we had in mind is
1,4-dibromonaphthalene (DBN) crystal simply because it is a quasi-
one-dimensional solid and because we know quite a bit about the
spectroscopy of the first triplet exciton.[13]

The absorption lineshape of the $k \simeq 0$ level at $2\,^{\circ}K$ and $25\,^{\circ}K$
taken by Burland is shown in Fig. 10. The shape is Lorentzian and

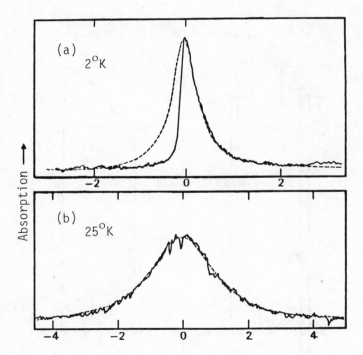

Fig. 10.   Absorption spectra of the $k \simeq 0$ region of the first
            triplet exciton of DBN at $2\,^{\circ}K$ and $25\,^{\circ}K$ (courtesy of
            Dr. D. Burland; see also D. Burland and A. H. Zewail,
            in Advances in Chemical Physics, eds. I. Prigogine and
            S. A. Rice (John Wiley & Sons, 1979) Vol. 40, p.369).
            Notice the Lorentzian fits to the lines.

implies a relaxation time constant of few ps. Does this relaxation
time reflect $k \simeq 0 \rightarrow k \neq 0$ population transfer or does it reflect
the occurrence of pure dephasing? To answer this question, Smith
and Zewail[13] have performed <u>laser-line-narrowing</u> experiments on
DBN at different temperatures. The idea of these experiments is
shown in Fig. 11. We simply tuned a narrow-band laser to the
$k \simeq 0$ absorption band. We then observed emission from the entire
band ($k \simeq 0$ and $k \neq 0$ levels) to a ground state vibrational exciton.
The emission was monitored at different times after the excitation
pulse was turned off. If the population of the $k \simeq 0$ level is not
transferred to other k states on the time scale of our experiment,
then a narrow emission line, similar to the absorption, will be
observed at all times. On the other hand, if the population is
transferred to $k \neq 0$ levels, then the emission characteristics will
be different for different delay times after the pulse is turned
off.

TRANSIENT BAND-TO-BAND SCATTERING

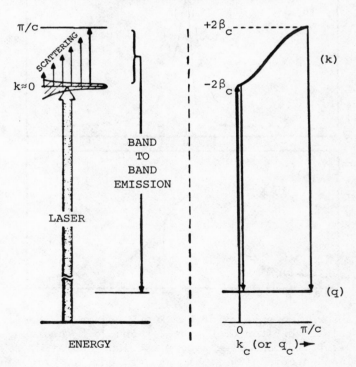

Fig. 11. A schematic for the idea behind the experiments of laser-
          line narrowing and population transfer in DBN.

Clearly, T_1 and T_2' are much different. By analogy one has to be very careful about extracting dynamical information from the spectra of _large_ molecules, especially at elevated temperatures.

CONCLUDING REMARKS

I have tried to give an overview of the problem of optical dephasing by energy and phase randomizations in three different cases: electronic excitation, high-energy vibrational excitation and multi-level excitation (the small crystal limit). The main conclusions are: a) energy and phase randomization can be separated experimentally by the aid of coherent optical spectroscopy; b) pure dephasing depends on the level structure and in some cases dominates the observed spectroscopic width of transitions; c) slower intramolecular vibrational dephasing can be achieved in some cases by weakening the active mode-bath coupling. The interplay between pure dephasing and energy relaxation is quite important for localization of energy in certain vibrational states as discussed in the recent article in _Physics Today_.[9]

ACKNOWLEDGMENTS

This work was supported by the National Science Foundation. I wish to thank several members of my research group whose contributions made the writing of this article possible. Finally, I wish to thank Professors Dr. A. El-Sadr, A. El-Bayoumi and S. Morsi for their generous hospitality during my stay at the Science Centre for Advancement of Post Graduate Studies, University of Alexandria, Alexandria, Egypt. This article is Contribution No. 6387 from the Division of Chemistry and Chemical Engineering, California Institute of Technology, Pasadena, California.

REFERENCES

1. A. H. Zewail, "Optical Molecular Dephasing: Principles of and Probings by Coherent Laser Spectroscopy," Acc. Chem. Res. _13_, 360 (1980), and references therein.
2. T. E. Orlowski and A. H. Zewail, J. Chem. Phys. _70_, 1390 (1979).
3. M. Sargent III, M. Scully, and W. Lamb, Jr., _Laser Physics_ (Addison-Wesley Publishing Co., Reading, Mass., 1979).
4. A. H. Zewail, D. Godar, K. Jones, T. Orlowski, R. Shah, and A. Nichols,in _Advances in Laser Spectroscopy_, ed. A. H. Zewail (SPIE Publishing Co., Bellingham, Washington, 1977) Vol.113,p.42.
5. T. J. Aartsma, J. Morsink, and D. A. Wiersma, Chem. Phys. Lett. _47_, 425 (1977); Phys. Rev. Lett. _36_, 1360 (1976).
6. B. R. Henry, Acc. Chem. Res. _10_, 207 (1977); H. L. Fang and R. L. Swofford, J. Chem. Phys. _73(6)_, 2607 (1980); R. Swofford, M. E. Long, and A. C. Albrecht, J. Chem. Phys. _65_, 179 (1976);

R. Bray and M. Berry, J. Chem. Phys. 71, 4909 (1979); C. K. N. Patel, A. C. Tam, and R. J. Karl, J. Chem. Phys. 71(3), 1470 (1979).

7. J. W. Perry and A. H. Zewail, J. Chem. Phys. 70, 582 (1979); Chem. Phys. Lett. 65, 31 (1979); D. Smith and A. Zewail, J. Chem. Phys. 71, 540 (1979).

8. M. Sage and J. Jortner, Chem. Phys. Lett. 62, 451 (1979); D.F. Heller, Chem. Phys. Lett. 61, 583 (1979); E.J. Heller, J. Chem. Phys. 72, 1337 (1980).

9. A.H. Zewail, "Laser Selective Chemistry--Is it Possible?", Physics Today, 33, 27 (1980); J. W. Perry and A. H. Zewail, J. Phys. Chem (to be published 1981).

10. E. Heller and W. Gelbart, J. Chem. Phys. 73, 626 (1980).

11. K. Shobatake, S. A. Rice, Y. T. Lee, J. Chem. Phys. 59, 2483 (1973); see also D. Micha, Chem. Phys. Lett. 46, 188 (1977).

12. J. Jortner and J. Kommandeur, Chem. Phys. 28, 273 (1978); S. Mukamel, Chem. Phys. 31, 327 (1978); I. Rabin and S. Mukamel, J. Phys. B (in press).

13. D. Smith and A. Zewail, J. Chem. Phys. 71, 3533 (1979); D. Smith, D. Millar, and A. Zewail, J. Chem. Phys. 72, 1187 (1980).

NOTE ADDED IN PROOF

This article was written in 1980 and since then many new relevant work has appeared in press. Here we mention some contributions from this laboratory:

(i)   For a recent review of work on electronic dephasing see the article by M.J. Burns, W.K. Liu, and A.H. Zewail, in Molecular Spectroscopy, series in Modern Problems in Solid State Physics, eds. V. Agranovich and R. Hochstrasser (North Holland Publishing Co., Amsterdam, 1983) Chap. 8.

(ii) For a recent work on dephasing and energy relaxation of high-energy vibrational states see J.W. Perry and A.H. Zewail, J. Phys. Chem. 86, 5197 (1982).

(iii) For a report on the new technique of Multiple Phase-Coherent Laser Pulses in Optical Spectroscopy see the papers by W.S. Warren and A.H. Zewail, J. Chem. Phys. 78, 2279 (1983); ibid p. 2298.

(iv) For a recent observation of dephasing in isolated large molecules see Wm. R. Lambert, P.M. Felker, and A.H. Zewail, J. Chem. Phys. 75, 5958 (1981); A. Zewail, et al., J. Phys. Chem. 86, 1184 (1982).

INTRAMOLECULAR PROCESSES IN ISOLATED POLYATOMIC MOLECULES

Martin R. Levy

Corporate Research Science Laboratories
Exxon Research and Engineering Company
P.O. Box 45
Linden, NJ   07036

and

Anita M. Renlund, Tom A. Watson, Metin S. Mangir,
Hanna Reisler, and Curt Wittig

Departments of E. E., Physics, and Chemistry
University of Southern California
Los Angeles, CA   90007

INTRODUCTION

This paper deals with certain rather well known unimolecular
processes which occur in isolated polyatomic molecules. These processes are (i) chemical reactions, forming fragments with identi-
fiable internal and translational energy distributions, and (ii) the
coupling of vibrational and electronic degrees of freedom, a phenomenon which is responsible for so-called radiationless transitions.
The work is mainly experimental, and capitalizes on the fact that
infrared lasers can be used to excite the vibrations of polyatomics
in the absence of collisions. Thus, parent translational and
rotational energies are near ambient, while vibrations can be
excited rather easily, allowing us to measure elementary
unimolecular processes of these species without interference from
collisions. Although molecular excitation cannot be adjusted to
yield monoenergetic species of excitation, still the "average
excitation level" can be controlled, enabling us to determine
effects qualitatively. Above dissociation threshold, the balance
between the optical pumping rate and the disssociation rate results
in a rather narrow range of energies from which dissociation occurs,
and by adjusting the laser intensity the mean unimolecular rate can
be controlled.

35

In the material that follows, two rather distinct experiments
are described which deal with vibration-electronic (V-E) coupling,
and with laser driven unimolecular reactions of polyatomics. In the
case of the former, it is non-trivial to determine unambiguously
the identity of the emitting species and much of our work has
centered on this issue. In the case of the latter, we are concerned
with energy disposal into the product degrees of freedom and how
this is influenced by excitation level.

ELECTRONIC EMISSION ACCOMPANYING IR MULTIPLE PHOTON EXCITATION

"Prompt" luminescence from electronically excited photofragments (e.g. C_2, CH) was a common feature of many of the earliest
experiments in which ir multiple photon excitation and decomposition
(MPE, MPD) were used to excite and dissociate polyatomic molecules
[1]. Direct production of such species via MPD required access to
highly excited potential surfaces, and therefore was a most
unattractive proposition. Fortunately (or unfortunately, depending
on one's point of view), most cases of luminescence disappeared when
the experiments were repeated under conditions wherein collisions
were suppressed. However, there remains a small but growing number
of molecules for which electronic emission appears concomitantly
with ir MPE under collision free conditions. Among these are the
(substituted) alkenes ($C_2H_3Cl[2,3]$, $C_2H_3CN[4,5]$, C_2H_3CHO, [2,3]),
tetramethyldioxetane[6,7], $F_2CO[8]$, $OsO_4[9,10]$, and $CrO_2Cl_2[11-13]$.
In most of these cases, the identity of the emitting species has
yet to be determined unambiguously. The emission spectra are usually
broad and structureless and the spontaneous emission lifetimes
rather long, thus eliminating diatomics and many of the polyatomics
whose spectra are known. Upon scrutinizing this phenomenon, it
becomes clear that V-E coupling causes the emission intensity to
depend on the rovibronic state densities of the ground and electronically excited states [3,14]. While useful, this concept does not
enable one to distinguish between electronically excited parent and
fragment species.

Several terms have been coined in connection with this phenomenon (e.g. "inverse electronic relaxation" [14]). These terms are
sometimes harder to justify than some of the experiments, and so we
will refer to the processes involved as simply "V-E coupling".

To answer the basic question of identity, it is clear that
further experiments are required. Since emission spectra alone are
of marginal value, we have measured the distribution of velocities
of the emitting species. At the very least, this allows us to
distinguish between parent and fragment, and with the proper choices
of parent, may enable us to identify the emitting species. In order

to carry out the measurements, we have used a pulsed molecular beam arrangement. Molecules in the molecular beam can be excited with a pulsed ir laser, and because of the long spontaneous emission lifetime, can be detected downstream from the intersection of the molecular and laser beam axes. Consequently, angularly resolved time-of-flight (TOF) data can be obtained for emitting species only, and the center-of-mass (c.m.) velocity distribution can be determined. Thus, in the case of fragment emission, the details of the dissociation process(es) can be determined for the emitting species. This is an important point, and we can distinguish minor dissociation channels with this technique since it is sensitive only to emitters.

To date, we have used this technique to investigate four molecules: C_2H_4, C_2HCl_3, C_2H_3CN, and CrO_2Cl_2 [15,16]. Of these, the ir MPD of C_2H_3CN has been studied most thoroughly. The luminescence peaks both in the ir and around 390 nm, and both spectral features have a radiative lifetime = 22 μs. Comparison of the fluence dependence for the production of the luminescence with that for the productions of $C_2(a^3\Pi_u)$ and $CN(X^2\Sigma^+)$ suggests that the emitter may be an excited precursor of C_2 and CN. The most likely sequential decomposition scheme is:

$$C_2H_3CN \rightarrow H_2 + C_2HCN \qquad \Delta H_1^o = 176 \text{ kJ mol}^{-1} \qquad (1)$$

$$\text{or} \begin{cases} C_2H_3CN \rightarrow H + C_2H_2CN & \Delta H_2^o = 450 \text{ kJ mol}^{-1} \qquad (2) \\ C_2H_2CN \rightarrow H + C_2HCN & \Delta H_3^o = 166 \text{ kJ mol}^{-1} \qquad (3) \end{cases}$$

$$C_2HCN \rightarrow H + C_2CN \qquad \Delta H_4^o = 551 \text{ kJ mol}^{-1} \qquad (4)$$

$$C_2HCN \rightarrow CN + C_2H \qquad \Delta H_5^o = 578 \text{ kJ mol}^{-1} \qquad (5)$$

$$C_2CN \rightarrow C_2 + CN \qquad\qquad\qquad\qquad\qquad (6)$$

Reaction (1) is expected to have a barrier of ~335 kJ mol^{-1} in excess of ΔH_1^o [17]; reaction (5), though not favored, is expected to make some contribution. Clearly, there are several possibilities for the emitter: C_2H_3CN, C_2H_2CN, C_2HCN, C_2CN, and C_2H.

The emission from C_2H_4 has a radiative lifetime $\cong$ 8 μs and the spectrum resembles the long wavelength portion of the emission from C_2H_3CN. In addition, both C_2 [18-21], and H atoms [22] have been detected in ir MPD, and the following decomposition scheme has been proposed [22].

$$C_2H_4 \rightarrow C_2H_2 + H_2 \qquad \Delta H_7^o = 174 \text{ kJ mol}^{-1} \qquad (7)$$

$$\text{or} \begin{cases} C_2H_4 \to C_2H_3 + H & \Delta H_8^\circ = 454 \text{ kJ mol}^{-1} & (8) \\ C_2H_3 \to C_2H_2 + H & \Delta H_9^\circ = 156 \text{ kJ mol}^{-1} & (9) \end{cases}$$

$$C_2H_2 \to C_2H + H \qquad \Delta H_{10}^\circ = 468 \text{ kJ mol}^{-1} \qquad (10)$$

$$C_2H \to C_2 + H \qquad\qquad\qquad\qquad\qquad\qquad (11)$$

Again, the elimination reaction (7) is expected to require ~335 kJ mol^{-1} in excess of ΔH_7 [17]. The formation of C_2 via ir MPD of C_2H_4 is truly remarkable. The experimental verifications of C_2 as a nascent photoproduct are almost certainly correct, and yet reaction (11) is rather hard to live with, given that there is no definitive independent evidence of a triatomic having been dissociated via collision free ir MPD. The reader should bear in mind that the amount of C_2 produced via the ir MPD of C_2H_4 is quite small, roughly two orders of magnitude less than the amount of C_2 produced via the ir MPD of C_2H_3CN. Thus, as yet undetermined secondary processes (which transpire without collisions) may be responsible for C_2 production in the ir MPD of C_2H_4.

The authors of ref. 22 found that the amount of C_2H_4 which is decomposed is about an order of magnitude larger than the number of H atoms produced. They were thus drawn to the conclusion that reaction (7) is the major first step in the sequential dissociation. As far as luminescence is concerned, the candidate species are C_2H_4, C_2H_3, C_2H_2, and C_2H.

In the case of C_2HCl_3, the emitting species has a radiative lifetime of > 40 μs, and the maximum in the spectrum is near 500 nm [2]. As with C_2H_3CN and C_2H_4, $C_2(a^3\Pi_u)$ has also been detected [2]. HCl production was observed at low fluences and under collisional conditions [23], but a molecular beam experiment [24] found only $C_2HCl_2^+$ in the mass spectrometer, suggesting Cl atom elimination. Recent ir chemiluminescence measurements at USC [25] indicate that at fluences ~40 J cm^{-1} both HCl and Cl are produced though possibly sequentially. A decomposition scheme such as those above can obviously be devised; possible emitters are C_2HCl_3, C_2HCl_2, C_2Cl_2, and C_2Cl.

In the first reports on CrO_2Cl_2 it was thought that the emission derived from electronically excited parent [11,12]. Identification was based on pressure and fluence dependences of the emission intensity, as well as the similarity of the observed emission spectrum to emission spectra of CrO_2Cl_2 reported in earlier studies. More recent work [13,16], however, shows that a very significant amount of the emission is due to fragments, with parent emission only occurring at the lowest fluences, if at all. The emission lacks any vibrational structure, in contrast to the reported CrO_2Cl_2 spectrum [26-29], and the radiative lifetime is 250 μs. Estimates

of the translational energy of the emitter indicate that only a very
small amount of recoil has been imparted to this species.

Experimental

     The apparatus we have designed and employed is shown schemati-
cally in fig. 1.   A detailed description can be found in ref. 15, so
only the general principles will be given here.   Briefly, a pulsed
molecular beam issues from a Gentry/Giese type nozzle source [30], and
is collimated to narrow angular width by one or two skimmers mounted
on large baffles.   As the gas pulse passes through the center of the
reaction chamber, it is intersected by the focused output from a CO_2
TEA laser (typical fluence $\sim 10^2$ J cm^{-2}) tuned to an absorption
feature of the molecule concerned.   Luminescing species are detected
with a photomultiplier which views the scattering center and rotates
about it ($\pm 45°$) within the plane of the molecular and laser beams.
Narrow apertures in front of the photomultiplier serve to define
both a narrow angle of acceptance, and a relatively uniform flight
path before detection by the photomultiplier or quenching at its
surface.   Since luminescence is into 4π steradians, the PMT
collection efficiency changes abruptly as emitting species pass
through the 7 mm aperture in front of the PMT (see fig. 2), thereby
discriminating severely against emission orginating from species
which have not passed through the 7 mm aperture.   TOF measurements
are obtained by connecting the PMT output to a transient
digitizer/signal averager combination and pulsing the beam source
and laser many times.   Figure 3 shows a typical TOF trace.   The
initial spike is due to light scattered from surfaces and from the

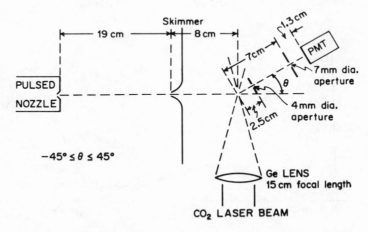

Fig 1.   Schematic view of the experimental arrangement.

scattering center; it rapidly falls to zero as emitting species
travelling at other than the detection angle move out of the line of
sight.   The broad secondary hump derives from molecules which emit
in the vicinity of the photomultiplier surface; compared to a "true"
TOF spectrum, it is somewhat foreshortened by the emitters' radiative decay.   When optically transparent material is placed over the

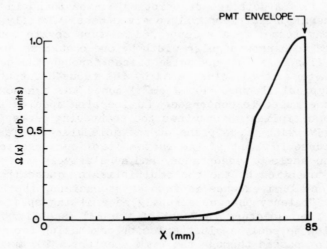

Fig. 2   Probabality of detection, $\Omega(x)$, for a species emitting
         uniformly into 4π steradians while travelling towards the
         PMT, as a function of distance x from the scattering center.
         $\Omega(x)$ is obtained by compounding the solid angle subtended by
         the PMT surface from point x with the probability of
         transmission through the window of a photon arriving from x,
         integrated over the surface of the window.

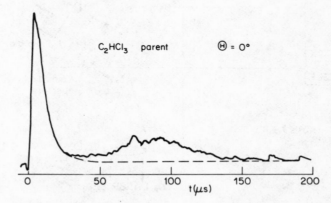

Fig. 3   Typical luminescence time profile (see text).   Dashed
         portion indicates baseline.

7 mm aperture, thereby quenching emission at this position, the
secondary hump disappears. Angular distributions are measured by
averaging, for a relatively small number of pulses, the total amount
of signal in the secondary hump (e.g. 40-200 μs in fig. 3) and
normalizing to 0°.

Experiments have been performed using both pure and seeded
beams. In the case of the former, beam velocity distributions can
be measured quite well with the Gentry/Giese type fast ionization
gauge [30]. For seeded beams, the distributions are known only
approximately, since mass analysis should be employed here.

Results

Results of the measurements for C_2H_3CN [15] are shown in figs.
4 and 5. Here, the data are most complete, consisting of TOF
traces at 5° intervals between 0° and 25°, as well as the time
integrated angular distribution. In addition, the time profile of
the C_2H_3CN beam pulse was measured rather carefully at several
points along the beam axis, allowing extraction of the number
density velocity distribution of the molecules in the scattering
center at the instant the laser fires. By convoluting this distribution with the viewing function of the photomultiplier (fig. 2),
the radiative decay, and various assumed forms for the photofragment
c.m. flux density velocity distribution, $U(u)$, it was possible to
simulate rather well the TOF traces and the time integrated angular
distributions (the scattering is assumed to be isotropic in the
c.m.).

A satisfactory fit to the data could not be obtained for the
distribution of translational energies, $P(E_t) = U(u)/u$ [31], thought
to be characteristic of statistical energy partitioning along the
reaction coordinate [32], for any reasonable average energy, number
of oscillators, and pair of fragments. Reasonable agreement, as
shown in figs. 4 and 5, could only be obtained when $P(E_t)$ peaked at
non-zero values of E_t. The following functional form was employed
for $U(u)$:

$$U(u) = \left|\frac{u}{u_0}\right|^n \left\{\exp\left[\alpha\left(1-\left(\frac{u}{u_0}\right)^m\right)\right] - \exp\left[\alpha\left(1+\left(\frac{u}{u_0}\right)^m\right)\right]\right\} \qquad (12)$$

Here, n and m are small integers and the distribution may be
regarded as sort of shifted Boltzmann. The fit shown in figs. 4 and
5 were achieved with $u_0 = 300\pm30$ m s^{-1}, n = m = 2, and $\alpha = 3.9$
(+0.5,-1.0). We found that a very reasonable fit could also be
obtained without the second exponential in eq. (12), and $U(u)$ and
$P(E_t)$ were not significantly affected. Figure 6 shows $P(E_t)$, obtained from eq. (12) with $u_0 = 300$ m s^{-1} and $\alpha = 3.9$, plotted vs.
u^2, since the magnitude of E_t depends on the (yet to be determined)
mass of the emitter.

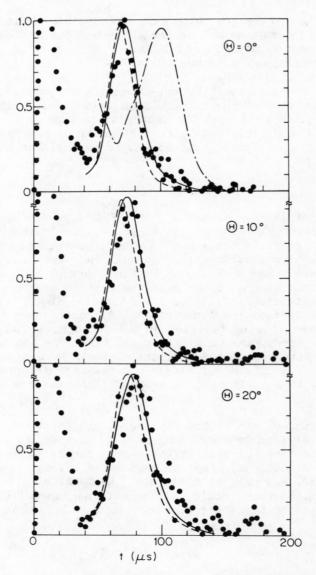

Fig. 4   Luminescence time profiles from ir MPE of C_2H_3CN at $\Theta = 0°$,
10°, and 20°. For convenience, the data are normalized to
the same peak height. The CO_2 laser is fired at t = 0, and
the points, at 2 μs intervals, were obtained by averaging
1024 traces. The solid and broken lines were calculated
using the c.m. velocity distribution of eq. (12) with u_0 =
300 m s^{-1} (solid line) and u_0 = 325 m s^{-1} (broken line).
The dot-dash line for $\Theta = 0°$ represents the profile that
would have been obtained if the emitting species were
electronically excited parent.

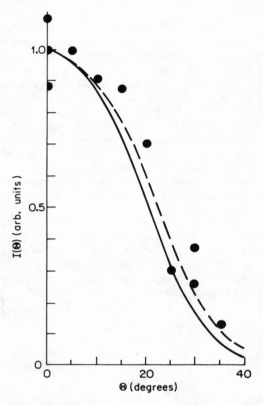

Fig. 5   Time integrated (40-200 µs) angular distribution of emitting
         species from the ir MPE of C_2H_3CN.

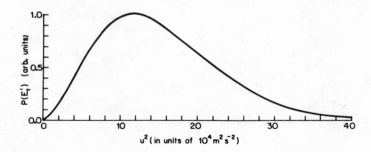

Fig. 6   Product translational energy distribution, $P(E_t')$, for emit-
         ting species from the ir MPE of C_2H_3CN. $P(E_t') = U(u)/u$,
         where $U(u)$ is given by eq. (12), with $u_0 = 300$ m s^{-1} and $\alpha =$
         3.9. The absolute value of E_t is not known a priori, but is
         obtained by mutliplying the ordinate, u^2, by m/2, were m is
         the mass of the emitting species.

Data for the other molecules which we studied are not so extensive as for C_2H_3CN. Figure 7 shows rather rough TOF profiles for
C_2HCl_3 at $0°$, $15°$, and $25°$. As with C_2H_3CN, intermediate angles
have been measured but are not shown. Here, the data points are
separated by 0.2 µs in contrast to 2 µs in fig. 4, so the time
profiles appear continuous. The same number of laser firings was
used at each angle, so, although the angular distribution has not
yet been measured, it must be rather broad by virtue of the
substantial scattering at $25°$. In these experiments, a seeded beam
was used, so the detailed shape of the C_2HCl_3 velocity distribution
has yet to be determined. Using an approximate form for the velocity distribution, we find that U(u), in fact, is qualitatively
similar to that for the case of C_2H_3CN. The observed TOF profiles,
of course, appear broader than for C_2H_3CN due to the longer radiative lifetime.

Time profiles for the luminescing species from C_2H_4 are shown
in fig 8. Here, the short radiative lifetime effectively quenches
much of the useful information, but substantial scattering is evident out to $25°$. In this case, simulations can only indicate
general features. It is, however, clear that $P(E_\perp)$ peaks at non-
zero values of $E_\perp$.

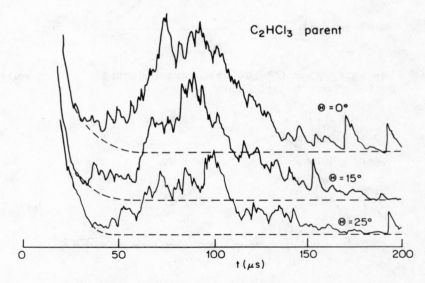

Fig. 7 Luminescence time profiles from the ir MPE of C_2HCl_3,
        showing relative magnitudes at $\Theta = 0°$, $15°$, and $25°$ as
        obtained from the same number of laser firings. The data
        appear continuous since points are seaparated by 0.2 µs.
        The (rough) angular distribuiton is obtained by integrating
        the area under each curve for $t > 40$ µs.

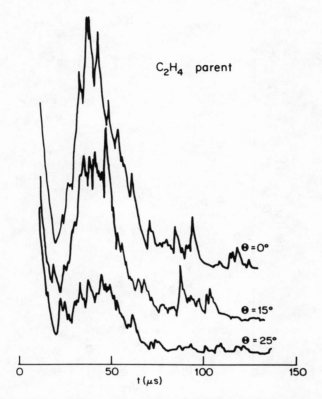

Fig. 8   Luminscence time profiles from the ir MPE of C_2H_4 at Θ = 0°, 15°, and 25°; details as per fig. 7.

In the case of CrO_2Cl_2, only the integrated angular distribution has been measured at this time [16], and this is shown in fig. 9. It is markedly narrower than for C_2H_3CN, although scattering can still be detected at 20°. These data are in excellent accord with simple bond fission and the data can be fit quite well with an average recoil energy of 40 kJ mol^{-1}. For the case of CrO_2Cl_2, there is no reason to believe that decomposition involves <u>other than</u> simple bond fission.

<u>Discussion</u>

The most obvious characteristics of the measurements is that the luminescence derives from scattered species (i.e. one or more photofragments) rather than the parent molecule. If there <u>is</u> any parent luminescence, it is relatively weak since the signal at 0° scattering angle shows no singularity. A second important feature is that in all cases expect CrO_2Cl_2 there is evidence for a

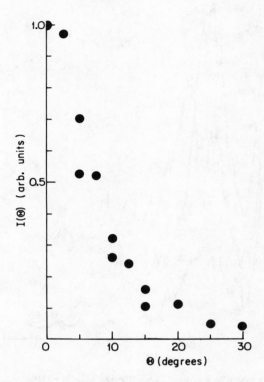

Fig. 9   Time integrated angular distribution of emitting species
         from the ir MPE of CrO_2Cl_2.

barrier in excess of the endoergicity. A barrier in excess of ΔH°
is expected for molecular elimination reactions, and molecular beam
studies of the ir MPD of halogenated organic molecules have found
$P(E_t')$ which are similar to that shown in fig. 6, for the cases of
four center hydrogen halide elimination [33,34]. In these studies,
mean <u>total</u> product translational energy was ~40 kJ mol^{-1}. In the
present case of C_2H_3CN, the emitter, with mass ~50 amu, has a mean
translational energy of ~3 kJ mol^{-1}. Thus, the mean <u>total</u>
translational energy release in reaction (1) is ~80 kJ mol^{-1} which
is the proper order of magnitude. There is no rational way to
explain our observations <u>other than</u> reaction (1), and we thus conclude that the emitting species is either C_2HCN or C_2CN.

Since there is also evidence for a barrier in the ir MPD of C_2HCl_3 and C_2H_4, the choice of emitters here is reduced to C_2Cl_2/C_2Cl and C_2H_2/C_2H respectively. It is questionable whether excited states of C_2H_2 are low enough to account for the observed emissions [35-37], and therefore C_2H must be taken as a serious candidate. The energy states of C_2H are elusive [38-43], and without going into detail over the various predictions and controversies, we should point out that in our laboratory we have been unable to find an absorption system of C_2H in the region 220-700 nm. We have, however, confirmed the presence of C_2H by monitoring the chemiluminescent reaction

$$C_2H + O_2 \rightarrow CH \ (A^2\Delta) + CO_2(X^1\Sigma_g^+) \tag{13}$$

which can be initiated by the (uv) photolysis of C_2H_2, C_2HBr, or propynal, as well as in the different ir photolysis experiments [44].

In the case of C_2H_3CN, we cannot make a clear choice between C_2HCN and C_2CN. For C_2HCN, the lowest known electronic transition is $\tilde{A}^1A' \leftarrow \tilde{X}^1\Sigma^+$, which has been observed in absorption in the range 230-270 nm [45]. Emission could clearly occur at longer wavelengths. With C_2CN, the ground state is $\tilde{X}^2\Sigma^+$ [46,47], and a spectrum analogous to that of C_2H might be expected. Similarly, in the case of C_2HCl_3 there appear to be no reports in the literature of the uv/visible spectrocopy of C_2Cl_2 and C_2Cl.

In the case of CrO_2Cl_2, the data suggest that only bond fission reactions have occurred. Thus, we feel that the emitting species is either CrO_2Cl or CrO_2. We see no evidence of parent emission in our experiments, but we should emphasize that laser fluences $\sim 10^2$ J cm^{-2} were used in our work, and it may be possible to detect parent emission at lower fluence. Our results with this species are of only marginal scientific interest, and were it not for previous papers which drew an unusual amount of attention [11,12], we probably would not have put CrO_2Cl_2 in the beam apparatus.

To summarize these experiments, we have shown that the collision free electronic emissions which accompany the ir MPE of C_2H_4, C_2HCl_3, C_2H_3CN, and CrO_2Cl_2 derive from photofragments, not parent molecules. For the first three cases, the first step in the reaction sequence is probably molecular eliminatin (e.g. $C_2H_3CN \rightarrow C_2HCN+H_2$), while for CrO_2Cl_2 simple bond fission occurs. In all cases, the photofragments are then pumped to levels where V-E coupling is present, and this coupling leads to the observed emissions. We find that all of our results are in very reasonable accord with the well established laws of physics, and nothing very magical occurs. The experimental technique is rather unique and will certainly find its way into other investigations.

LASER DRIVEN UNIMOLECULAR REACTIONS

The roles of translational and internal excitation in
elementary chemical processes have been of central interest to
kineticists and dynamicists for many years. Consequently, a wide
variety of experiments have been devised which focus on the translational and internal energies of both reactants and products. These
experiments have provided a means of elucidating pathways from
reactants to products, thereby contributing to the significant
advances in bimolecular reaction dynamics which have occurred during
the last two decades.

Detailed studies of unimolecular reactions are also of great
interest, and although such reactions have several features which
are in common with bimolecular reactions, there are also several
intriguing and important differences. The internal energy of a
reacting species is, of course, above dissociation threshold, and in
order to investigate the role(s) played by this internal energy, it
is most desirable to prepare reactants with a selected, but
variable, amount of excitation and observe how this excitation is
channeled into the product degrees of freedom. By monitoring nascent product state distributions at a fixed excitation of the
parent, one may obtain dynamical information about the reaction. As
the parent excitation is then varied, it may become clear how parent
excitation is transferred to product excitation via the dynamics and
the statistics of the reaction. Thus, the careful measurement of
nascent product state distributions provides us with a direct
measurement of the consequences of the forces which occur during the
unimolecular reaction.

Experimentally, it is best to energize isolated molecules, in
order to insure that nascent product distributions are unaffected by
collisions. A very direct and rather universal way to do this is
via ir MPE. This method also allows the experimenter to
systematically vary the amount of energy above dissociation threshold by simply changing the laser intensity. In the past, this
energy above dissociation threshold has often been provided by
thermal or chemical activation. These techniques are straightforward in principle, but can be augmented by using ir MPE to study
unimolecular processes.

Much of the early research concerning ir MPE was concerned with
finding ways to violate Schroedinger's and Maxwell's equations.
Presently, it is rather well accepted that the vast majority of
molecules which undergo unimolecular reaction following ir MPE
approach the transition state with a vibrational energy distribution
which is statistical. Above dissociation threshold, optical pumping
to higher energies is clearly in competition with unimolecular
reaction. Since the rate of unimolecular reaction varies rather

rapidly with excitation, those molecules which dissociate while the laser field is on do so from a rather narrow range of energies, determined by the condition that the unimolecular rate equals the net rate of optical pumping upward. The important consequence of this is that by varying the laser intensity, we can vary the rate of unimolecular reaction and the vibrational energy above dissociation threshold. Although non-trivial, it is within the limits of state-of-the-art technology that the intensity of the output from a CO_2 TEA laser can be varied by roughly 7 orders of magnitude, while maintaining adequate fluence to measure nascent product state distributions (30 ps, 1 J → 3 μs, 10 mJ).

There are various extensions and consequences of the above reasoning. For example, if ir MPE is done with high frequency photons, then the spread of parent energies from which dissociation occurs is determined only by the level occupancy of the parent prior to excitation. Reasons for using either large or small photons are typically "experiment dependent". In any event, there is considerable flexibility in the experimental technique and there is much more work ahead of us than behind us.

Experimental Arrangement

Several means are available by which one can vary the laser intensity. In order to avoid any confusion which may arise due to simultaneously changing the fluence and intensity, we have chosen, in our first experiments, to maintain constant fluence while changing intensity. This was achieved following the methods of Lyman et al. [49] and King and Stephenson [50] in their studies of the effects of intensity on dissociation yields. These authors compared dissociation yields obtained by using a high intensity modelocked laser and a lower intensity single mode laser. The fluence was the same for each laser. In a similar fashion, we have measured the effect of laser intensity on the vibration-rotation (V,R) energy distribution of CN which is generated via the ir MPD of propenenitrile (C_2H_3CN) while maintaining constant laser fluence [51]. CN is efficiently produced, despite the sequential nature of the photolyses indicated by reactions (1)-(6).

The experimental arrangement is similar to those used in several previous studies in our laboratory [51,52], and is shown schematically in fig. 10. Briefly, the vapor of a molecular parent is excited and dissociated in the center of a chamber with the focused output from a CO_2 TEA laser. Nascent CN fragments are monitored by LIF using a tunable dye laser which enters the chamber at right angles to the CO_2 laser beam. Fluorescence is monitored at right angles to both laser beams with a PMT which uses interference filters and/or a telescope to discriminate against scattered light.

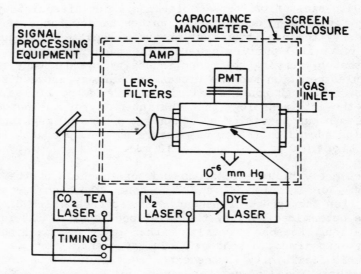

Fig. 10   Schematic view of the experimental arrangement.

        Both modelocked (ML) and single longitudinal mode (SLM) laser
pulses were obtained using a modified Tachisto CO_2 TEA laser, the
details of which are given elsewhere [51,53]. The temporal behavior
of the laser output was displayed on an oscilloscope. A typical ML
pulse had a roughly 200 ns train of narrow spikes separated by the
round trip time of the cavity, 17 ns. The width of the individual
spikes was ~1 ns (detector and amplifier response times distorted
the true time behavior), and the intensity of individual spikes
decreased after the first few. Such ML pulses have been described
in detail by Lyman et al. By contrast, SLM operation of the laser
resulted in a smooth pulse envelope of 60 ns fwhm. Laser energies
were very similar (typically 400 mJ) for both ML and SLM operation
of the laser, and the average intensity of the ML output was > 6
times the intensity of the SLM output.

        In later experiments, intensity and fluence were varied
simultaneously. This was achieved by using attenuators, and also by
taking spectra at different delays with respect to the onset of the
laser pulse. Here, we found it most convenient to use the output
from a Lumonics 103 CO_2 TEA laser, and no attempt was made to
control the temporal or spectral character of the laser output.
Despite the partially modelocked nature of the laser output, we
obtained reliable and repeatable results, although at the expense of

having to use "average intensities" which covered a broader range of
intensity than we would have preferred. Even with nicely tailored
pulses, the intensity near the focal region is anything but uniform
and well defined. Fortunately, the unimolecular rate is a strong
function of the internal excitation of the molecule, so large intensity fluctuations result in only a small change in the internal
excitation of the dissociating species.

In the case of C_2H_3CN, LIF spectra were obtained with the dye
laser tuned to the (0,0) band of the $B^2\Sigma^+ \leftarrow X^2\Sigma^+$ transition of CN,
while monitoring fluorescence through a narrow band interference
filter (centered at 420 nm) which passes $B \rightarrow X$ (0,1) emission. The
dye laser was delayed 1.3 μs from the onset of the CO_2 laser pulse,
and timing was initiated from the onset of the CO_2 laser pulse. The
LIF spectra are composites of data at 450 grating positions, where
data from 4 laser firings were summed at each grating position.

Results: $\underline{C_2CN}$ Precursor

Two LIF spectra of the CN fragment produced via ir MPD of
C_2H_3CN are shown in fig. 11. As mentioned above, the precursor to
CN in this case is C_2CN, as photolysis occurs via reactions (1)-(6).
The CO_2 laser fluence was 20 J cm^{-2}, and the dye laser was
delayed 1.3 μs following the beginning of the CO_2 laser pulse in
both experiments. Figure 11a is the spectrum obtained using the ML
laser output, while 11b is from the SLM laser. The only difference
in experimental conditions therefore is the   factor of 6 difference
in CO_2 laser intensity.

Qualitative information can be obtained by a visual comparison
of the two spectra. Both the more pronounced band head of the P-
branch, and the less sharply decreasing intensity of increasing
R(N") lines in fig. 11a as compared to 11b are indicators of the
"hotter" rotational distribution for those CN fragments produced
by the ML laser pulse. Also, note that there is negligible CN
population in v" = 1, whose LIF spectrum would be observed in the
same spectral region. The rotational temperature was determined in
the usual manner for a Boltzmann distribution by a plot of
$ln \{I/(2N"+2)\}$ vs. N"(N"+1), where N" is the rotational quantum number
of the lower state and I is the intensity of the R(N") line [55].
The data from fig. 11 were analyzed in this fashion, giving
Boltzmann rotational temperatures of 780±20 K and 570±15 K for the
CN radicals generated by the ML and SLM laser outputs respectively.
Spectra obtained at a slightly lower laser fluence, ~15 J cm^{-2}, gave
almost identical rotational temperatures (770±20 K and 590±10 K for
the ML and SLM laser outputs respectively.)

Other groups have done similar measurements and have obtained
qualitatively similar results [56,57]. In those experiments fluence

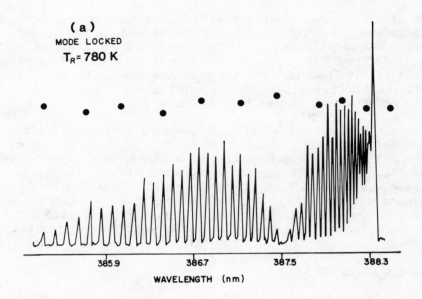

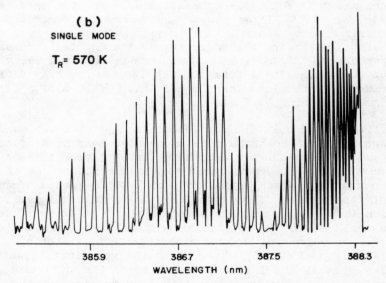

Fig. 11   LIF excitation spectra of CN produced via ir MPD of C_2H_3CN.
The spectrum in (a) was obtained when a ML laser was used
to dissociate C_2H_3CN; the spectrum in (b) was obtained
using a SLM laser.   The black dots indicate the dye laser
energy at different wavelengths.   Each spectrum is
normalized to its rotational maximum.   The rotational
temperature, T_R, was obtained for each spectrum by analysis
of R-branch lines.

and intensity were varied simultaneously. It is apparent from the
experiment described above that the observed differences in
rotational energy distributions were due solely to laser intensity,
and this was the only experimental parameter varied.

The difference of 200 K in the rotational temperatures of the
CN fragments formed by the ML and SLM laser pulses corresponds to a
difference in the CN rotational energy of 140 cm^{-1}. Since the CO_2
laser output cannot interact with nascent CN, this energy difference
must be due to vibrational excitation of the CN precursor prior to
dissociation. Thus, C_2CN was excited to higher levels by the ML
laser than by the SLM laser. Since excess parent vibrational energy
was also partitioned between C_2 rotation and relative translation of
the two fragments, it follows that C_2CN absorbed about one
additional laser photon (10^3 cm^{-1}) when the laser intensity was
increased by a factor of > 6.

Since the unimolecular reaction rate is "controlled" by the
optical pumping rate, and optical pumping rates cannot assume
arbitrarily large values, this limits the amount of parent
excitation that can be realized experimentally. We have calculated
reaction rates for the two species of interest to us using the RRKM
program of Bunker and Hase [59], and the results are shown in fig.
12. Notice that above dissociation threshold the unimolecular rate
rises quite rapidly for the case of C_2CN. Thus, for a small species
such as C_2CN, very large intensity changes are required in order to
bring about significant changes in product excitations. For larger
species such as CF_3CN parent excitation above dissociation

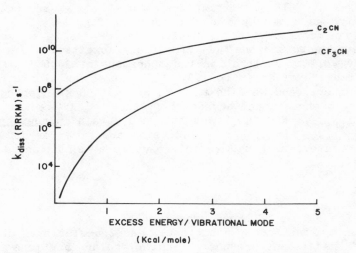

Fig. 12   A plot of the calculated unimolecular (RRKM) dissociation
          rates vs the excess energy/vibrational mode for C_2CN (6
          vibrational modes) and CF_3CN (12 vibrational modes).

threshold is more easily controlled, and in these cases we expect
to both achieve a greater amount of product excitation at high
fluence, and to be able to vary this excitation over a far larger
range than in the case of C_2CN.

Results: CF_3CN

The results here are in contrast to the case of C_2CN. For
C_2CN, only modest rotational excitation was observed and no
vibrational excitation was detected as the laser intensity was
varied by a factor of 6. This is sensible in light of the RRKM
calculations, and it is clear that the increase in CN rotational
excitation derives from additional parent vibrational excitation.
In separate experiments we verified that translational energies were
modest, and also varied weakly with laser intensity.

For the case of CF_3CN, the high laser intensity produced
considerable vibrational and rotational excitation, with vibrational
excitation clearly evident in the $\Delta v = 0$ sequence and with
rotational temperatures as high as 1200 K at laser intensities
comparable to those responsible for the 780 K distribution obtained
with a C_2CN precursor. With lower laser intensities, $T_R = 450$ K was
obtained, and vibrational excitation was no longer detectable. In
these experiments, the laser output was simply attenuated in order
to achieve low intensities, rather than using the ML and SLM
arrangement. This proved very convenient and gave reliable results.

At high intensity, rotational distributions for $v'' = 1$ and 2
were obtained via LIF by pumping the $\Delta v = -1$ and -2 sequences
respectively, as shown by the CN spectra in fig. 13. This resulted
in very clean rotational distributions for $v'' \leqslant 2$. The results of
these measurements are that (i) vibrational excitation can be described very well by a vibrational temperature of 2400 ± 150 K, and
(ii) rotational excitation for $v'' = 0,1$, and 2 can also be described
by a temperature, and we find that $T_R = 1200\pm100$ K independent of
v''. For the translational degree of freedom, a TOF technique was
used, and $T_{trans} = 850\pm150$ K was obtained, again independent of v''
for $v'' = 0, 1, 2,$ and 3! T_{trans} was also the same for different
rotational states, although we only checked a few J for each v''.

Discussion

From the outset of these experiments, it has been our
position that vibrational, rotational, and translational degrees of
freedom of the nascent products of unimolecular reactions are not
necessarily in equilibrium with one another. Since product V,R,T
excitations derive from parent excitations in a non-trivial manner,
there is no a priori reason to expect V,R,T excitations of nascent

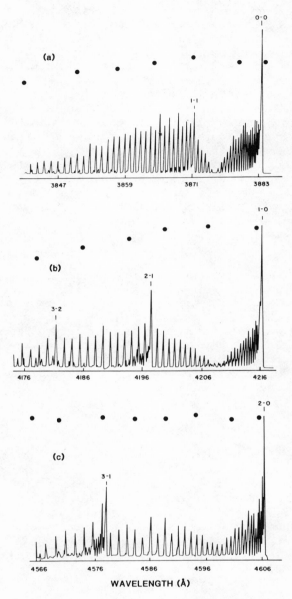

Fig. 13   LIF excitation spectra of CN produced via MPD of CF_3CN at
          high CO_2 laser fluence (150 J cm^{-2}).   The spectrum in (a)
          was obtained by exciting the $\Delta v=0$ sequence of the CN $(X \to B)$
          system; (b) was obtained by exciting the $\Delta v=-1$ sequence;
          (c) was obtained by exciting the $\Delta v=-2$ sequence.
          Fluorescence was observed in all cases by monitoring the CN
          $\Delta v=0$ emission.

products to be in any well defined equilibrium with one another.
The deposition of parent excitation into fragment V,R,T excitations
is influenced (a) in part by the (statistically distributed) parent
vibrational energy, (b) in part by the dynamics of the reaction, and
(c) in part by the coupling of (a) and (b). To see this in a very
simple case, consider the dissociation of C_2CN, where the molecule
is linear and bond rupture is instantaneous. Here, parent rotation
is transformed almost entirely to product translation. The same
kinds of arguments can be extended to other excitations for other
parent and product species. Clearly, experiments which measure only
translational energy distributions cannot infer the parent
excitation without a quantitative picture of the respective roles
played by (a), (b), and (c). It is our intention to develop a
quantitative basis for predicting product excitation in unimolecular
reactions using essentially classical mechanics to describe the
relevant physical processes. Calculations are in progress which
will provide insight into the factors which influence product state
distributions.

Obviously, the energy partitioning described above is sensitive
to the amount of parent excitation, and this is the next parameter
that will be dealt with in the experiments and calculations. Also,
other parent species will be studied in turn. It is not easy, and
often it is not wise, to attempt to predict future pathways in
experimental research. Nevertheless, we feel that the future is
bright for research in this area, and that within the next 5 years
unimolecular processes will be probed in far greater detail than was
possible during the last 30 years.

ACKNOWLEDGEMENTS

The authors are indebted to Trisha Woitena, who prepared the
manuscript, and who constantly suffers because of the authors'
inability to meet deadlines. We thank J. Caballero, D. M. Cox, M.
Duncan, R. C. Estler, T. A. Fischer, W. R. Gentry, H. Helvajian, D.
R. Herschbach, F. Kong, and Y. T. Lee for helpful discussions. MRL
thanks Exxon Research and Engineering for a postdoctoral fellowship.
Financial support for the research was provided by the U.S. Air
Force Office of Scientific Research, the U.S. Department of Energy,
and the National Science Foundation.

REFERENCES

1. See, for example, R.V. Ambartzumian and V.S. Letokhov, in
    "Chemical and Biochemical Applications of Lasers, II" (C.B. Moore
    Ed., Academic Press, N.Y., 1977), and references cited therein.
2. H. Reisler, M.S. Mangir, and C. Wittig, unpublished.

3. H. Reisler and C. Wittig, Adv. Chem. Phys., in press.
4. M.H. Yu, H. Reisler, M. Mangir, and C. Wittig, Chem. Phys. Lett. 62, 439 (1979).
5. M.H. Yu, M.R. Levy, and C. Wittig, J. Chem. Phys. 72, 3789 (1980).
6. Y. Haas and G. Yahav, Chem. Phys. Lett. 48, 63 (1977).
7. G. Yahav and Y. Haas, Chem. Phys. 35, 41 (1978).
8. J.W. Hudgens, J.L. Durant Jr., D.J. Bogan, and R.A. Coveleskie, a) Bull. Am. Phys. Soc. 24, 638 (1979). b) J. Chem. Phys. 70, 5906 (1979).
9. R.V. Ambartzumian, Yu. A. Gorokhov, G.N. Makarov, A.A. Puretzky, and N.P. Furzikov, Chem. Phys. Lett. 45, 231 (1977).
10. R.V. Ambartzumian, G.N. Makarov, and A.A. Puretzky, Appl. Phys. 22, 71 (1980).
11. Z. Karny, A. Gupta, R.N. Zare, S.T. Lin, J. Nieman, and A.M. Ronn, Chem. Phys. 37, 15 (1979).
12. J. Nieman and A.M. Ronn, Opt. Eng. 19, 39 (1980).
13. I. Burak and J.Y. Tsao, Chem. Phys. Lett. 77, 53 (1981).
14. A. Nitzan and J. Jortner, a) Chem Phys. Lett. 60, 1 (1978). b) J. Chem. Phys. 71, 3524 (1979)
15. T.A. Watson, M. Mangir, C. Wittig, and M.R. Levy, J. Chem Phys., in press.
16. T.A. Watson, M. Mangir, C. Wittig, and M.R. Levy, J. Phys. Chem., in press.
17. S.W. Benson, private communication.
18. N.V. Chekalin, V.S. Dolzhikov, V.S. Letokhov, V.N. Lokhman, and A.N. Shibanov, Appl. Phys. 12, 191 (1977).
19. J.H. Hall, Jr., M.L. Lesiecki, and W.A. Guillory, J. Chem. Phys. 68, 2247 (1978).
20. N.V. Chekalin, V.S. Letokhov, V.N. Lokhman, and A.N. Shibanov, Chem. Phys. 36, 415 (1979).
21. S.V. Filseth, J. Danon, D. Feldmann, J.D. Campbell, and K.H. Welge, Chem. Phys. Lett. 66, 329 (1979).
22. C.R. Quick, Jr., A.B. Horwitz, R.E. Weston, Jr., and G.W. Flynn, Chem. Phys. Lett. 72, 352 (1980); see also references cited therein.
23. C. Reiser, F.M. Lussier, C.G. Jensen, and J.I. Steinfeld, J. Am. Chem. Soc. 101, 350 (1979).
24. Aa.S. Sudbo, P.A. Schulz, E.R. Grant, Y.R. Shen, and Y.T. Lee, J. Chem. Phys. 68, 1306 (1978).
25. J. Caballero and C. Wittig, submitted.
26. M. Spoliti, J.H. Thirtle, and T.M. Dunn, J. Mol. Spectr. 52, 146 (1974).
27. V.E. Bondybey, Chem. Phys. 18, 293 (1974).
28. J.R.McDonald, Chem. Phys. 19, 423 (1975).
29. R.N. Dixon and C.R. Webster, J. Mol. Spectr. 62, 271 (1976).
30. W.R. Gentry and C.F. Giese, Rev. Sci. Instr. 49, 595 (1978).
31. E.A. Entemann and D.R. Herschbach, Disc. Farad. Soc. 44, 289 (1967).

32. S.A. Safron, N.D. Weinstein, D.R. Herschbach, and J.C. Tully, Chem. Phys. Lett. 12, 564 (1972),

33. Aa.S. Sudbo, P.A. Schulz, E.R. Grant, Y.R. Shen, and Y.T. Lee, J. Chem. Phys. 70, 912 (1979).

34. Aa.S. Sudbo, P.A. Schulz, Y.R. Shen, and Y.T. Lee, J. Chem. Phys. 69, 2312 (1978).

35. G. Herzberg, "Electronic Spectra of Polyatomic Molecules" (Van Nostrand, N.Y., 1966).

36. S. Trajmar, J.K. Rice, P.S.P. Wei, and A. Kuppermann, Chem. Phys. Lett. 1, 703 (1968).

37. D. Demoulin, Chem. Phys. 11, 329 (1975).

38. W.R.M. Graham, K.I. Dismuke, and W. Weltner, Jr., J. Chem. Phys. 60, 3817 (1974).

39. M. Okabe, J. Chem. Phys. 62, 2782 (1975).

40. K.H. Becker, D, Maaks, and M. Shurgers, Z. Naturforsch. 26a, 1770 (1977).

41. J.R. McDonald, A.P. Baronavski, and V.M. Donnelly, Chem. Phys. 33, 161 (1978).

42. S. Shingkuo, S.D. Peyerimhoff, and R.J. Buenker, J. Mol. Spectr. 74, 124 (1979).

43. M.E. Jacox, Chem. Phys. 7, 424 (1975).

44. H. Reisler, M. Mangir, and C. Wittig, Chem Phys. 47, 49 (1980).

45. V.A. Job and G.W. King, J. Mol. Spectr. 19, 155, 178 (1966).

46. M. Guelin and P. Thaddeus, Astrophys. J. 212, L81 (1977).

47. S. Wilson and S. Green, Astrophys. J. 212, L87 (1977).

48. R. Thomson and P.A. Warsop, Trans. Farad. Soc. 65, 2806 (1969).

49. J.L. Lyman, J.W. Hudson, and S.M. Freund, Opt. Commun. 21, 119 (1977).

50. D.S. King and J.C. Stephenson, Chem. Phys. Lett. 66, 33 (1979).

51. A.M. Renlund, H. Reisler, and C. Wittig, Chem Phys. Lett. 78, 40 (1981).

52. H. Reisler, M.S. Mangir, and C. Wittig, J. Chem. Phys. 71, 2109 (1979).

53. R.A. Dougal, C.R. Jones, M. Gundersen and L.Y. Nelson, Appl. Optics 18, 1311 (1979).

54. H. Reisler, F. Kong, A.M. Renlund and C. Wittig, submitted.

55. G. Herzberg, "Molecular Spectra and Molecular Structure, Vol. 1, Spectra of Diatomic Molecules" (Van Nostrand, Princeton, 1950)

56. C.M. Miller and R.N. Zare, Chem. Phys. Lett. 71, 376 (1980).

57. M.N.R. Ashfold, G. Hancock, and M.L. Hardaker, J. Photochem. 14, 85 (1980).

58. W.L. Hase and D.L. Bunker, Quantum Chemistry Program Exchange, No. 234.

59. See, e.g., P.J. Robinson and K.A. Holbrook, "Unimolecular Reactions", (Wiley Interscience, N.Y. 1972); W. Forst, "Theory of Unimolecular Reactions" (Academic Press, N.Y. 1973).

PULSE-PROBE MEASUREMENTS IN LOW-TEMPERATURE,

LOW-PRESSURE SF$_6$

John L. Lyman

University of California
Los Alamos Scientific Laboratory
Los Alamos, NM 87545

INTRODUCTION

The experimental determination of the intramolecular distribution of vibrational energy following infrared laser excitation
of polyatomic molecules has been a goal of many researchers in
recent years. One popular method of investigation has been the
pulse-probe technique with SF$_6$ and similar species.

This paper reviews some of our own work with SF$_6$ (Ref. 1
gives a more complete account) and gives some general conclusions about the kind of information these experiments can give.

We performed our experiments at low temperature (145 K)
and pressure (0.02 to 0.08 torr) and with a 30-fold laser fluence range (0.018 to 0.54 J/cm^2). The probe laser covered the
frequency range of the SF$_6$, ν_3 absorption band. The procedure
we used gave a fairly complete mapping of the induced spectrum
of this species for a variety of experimental conditions.

Our experiments extend and, in many respects, compliment
earlier work (see Refs. 2-7 and others cited in Ref. 1). Several
observations were common to all experiments such as the anharmonic shift of the induced spectrum to lower frequencies. However,
the investigators do not agree on several mechanistic conclusions
from their experiments, such as the time scale for collisionless
intramolecular vibrational-energy randomization and the role of
intermolecular vibrational-energy transfer.

EXPERIMENTAL

We show a diagram of the apparatus used in these experiments in Fig. 1. A sample of cold, low-pressure SF_6 was irradiated with a pulsed CO_2 laser while monitoring the intensity of the cw CO_2 probe laser after it passed through the irradiated region. A transient digitizer-minicomputer processed the detector signals and temporarily stored them.

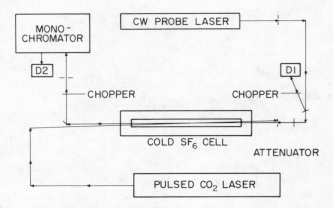

Fig. 1. Apparatus diagram for pulse-probe experiments.

The cold SF_6 absorption cell was a 1.00-m-long by 2.54-cm-id cylinder with CsI windows that contained the SF_6 gas sample. A cooling jacket surrounded this cylinder, and a cylindrical vacuum jacket with KCl windows enclosed the inner cylinder and cooling jacket. The cell temperature was 145 ± 3 K for all experiments.

We operated the pulsed CO_2 laser with a He-CO_2 gas mixture (no N_2) that gave multimode pulses (180 ns FWHM) with no long "tail". An intracavity iris and long focal-length mirrors controlled the spatial profile of the beam at the absorption cell. Pulse energies were lowered by attenuating the beam with a set of partially-reflecting germanium flats that had antireflection coatings on one side. The pulse energy was reproducible to ±5%. Table 1 gives average amount of energy absorbed per molecule at the fluences and pressures we used in our experiments.

The probe laser was a line-tunable cw CO_2 laser equipped with Invar stabilizing rods and piezo-electric control of the cavity length. The latter feature allowed us to tune the laser across the gain profile of a single CO_2 laser line (∼120 MHz).

Table 1.   Fluence (Φ) and Energy Absorbed per molecule (η),
           Averaged Along Probe Path

Φ (J/cm^2)	η		
	0.08 torr	0.04 torr	0.02 torr
0.54	3.7	2.7	2.2
0.19	1.4	1.1	0.89
0.048	0.42	0.33	0.29
0.016	0.15	0.13	0.11

The laser power was about 0.1 W at the absorption cell. The beam
was a few millimeters in diameter, yielding an intensity of about
1 W/cm^2 near the beam center.

    Detectors D1 and D2 were HgCdTe photovoltaic infrared detec-
tors (SAT) with the total circuit-response time estimated to be
20 ns. D1 was used to keep the laser power and frequency stable
and both detectors were employed to take absorption spectra over
the laser gain band. For the pulse-probe experiments a Tektronix
R7912 Transient Digitizer took the signal from detector D2 for
the time period shortly before, during, and after the laser
pulse, and digitized and stored the trace. The monochromator was
used with all experiments having delay times shorter than 5 ns,
to discriminate effectively against the pulsed laser radiation at
short time.

    The pulse and probe beams crossed within the absorption
cell. We attempted to keep the beam geometry the same for all
experiments. The centers of the two beams were 0.50 cm apart at
the entrance and exit windows of the inner cylinder of the absorption cell and the radial profile of the pulsed beam was very
nearly Gaussian with a radius parameter of 0.50 cm.

EXPERIMENTAL RESULTS

    Reference 1 gives the procedures for converting the primary
data to graphs of absorption cross section vs time or frequency.
In Figs. 2-6, we show examples of these graphs. The curves cover
the time period during and shortly after the pump-laser pulse for
a variety of experimental conditions (see Table 1). They show
that pump fluence, SF$_6$ pressure, and probe-laser frequency all
have a major effect on the probe signals.

    The "on line", "off line" labels in Figs. 3 and 4 refer to
the frequency of the probe laser relative to strong absorption

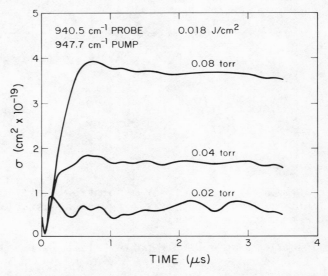

Fig. 2.   Absorption cross section vs time for SF₆ at 145 K.

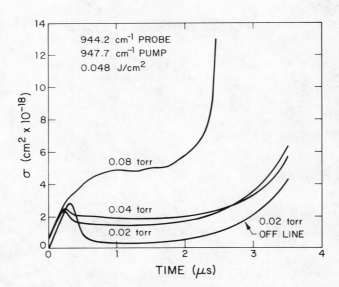

Fig. 3.   Absorption cross section vs time for SF₆ at 145 K.
The upper three curves are all "on line".

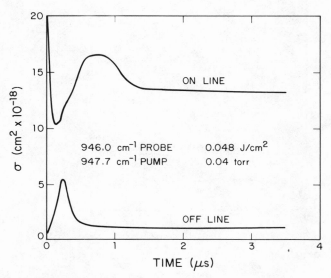

Fig. 4.   Absorption cross section vs time for SF$_6$ at 145 K.

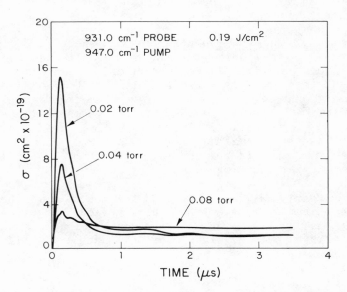

Fig. 5.   Absorption cross section vs time for SF$_6$ at 145 K.

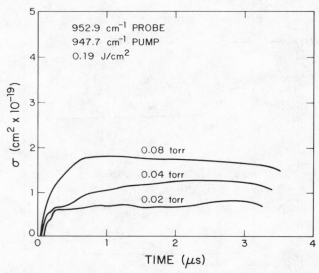

Fig. 6.   Absorption cross section vs time for SF$_6$ at 145 K.

features.   The frequency difference between the "on line" and
"off line" curves is about 25 MHz.   From the pulse-probe data,
one can also construct composite spectra at different times for
various experimental conditions.   Figures 7-9 give examples of
these composite spectra.

In Fig. 7, we see the evolution of the spectrum during and
shortly after the laser pulse for the highest fluence and pressure we employed.   Straight lines connect all data points in this
and subsequent figures.   The large feature in the unexcited spectrum (t = 0) is due to ^{32}SF$_6$, and the smaller feature is due to
^{34}SF$_6$.   The thermal spectrum is for the vibrational temperature
one obtains (670 K in this case) from a thermal distribution of
the absorbed energy among the vibrational degrees of freedom.
In Fig. 8 we show spectra for several fluences at 0.40 μs, which
is at the end of the exciting pulse.

The induced spectrum of this band relaxes back to the original 145 K spectrum over a very long time period.   Figure 9 shows
spectra at various times during that relaxation period after
irradiation with an average fluence of 0.54 J/cm^2.

DISCUSSION

We make some qualitative observations and conclusions about
these data and those data others have obtained.   Many phenomena

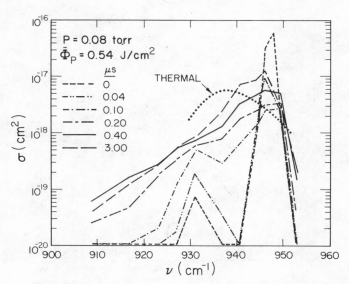

Fig. 7. Absorption cross section vs frequency at successive times after the beginning of the laser pulse. The thermal spectrum is from Ref. 8 (670 K).

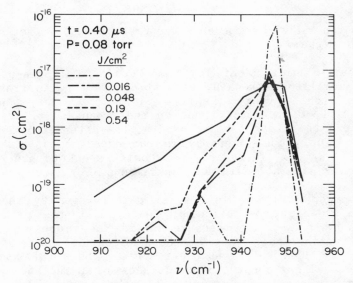

Fig. 8. Absorption cross section vs frequency near the end of the laser pulse for different fluences.

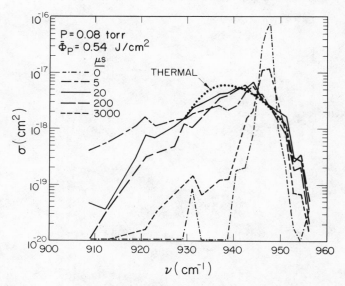

Fig. 9.   Absorption cross section vs frequency at successive
          times after the beginning of the laser pulse. The
          temperature for the thermal spectrum is 670 K.

can and do contribute to the observed probe signals. Let us consider some of them.

    For our conditions of low temperature, low pressure, and
high SF_6 oscillator strength, saturation of a given transition
probe laser could occur well below 1 mW/cm^2. The saturation
intensity will, of course, depend on the frequency of the probe
laser. Where the cross section is low, the saturation intensity
is higher. The probe intensity in our experiments was on the
order of 1.0 W/cm^2. Therefore, we should expect to see effects
of probe-laser saturation in our experiments.

    For times before or well after perturbation by the pulse
laser, the rate of absorption of energy from the probe beam
approaches a steady state. Because collisional processes such as
rotational relaxation, velocity changing collisions, and power
broadening all enhance that absorption rate, one would expect the
measured probe laser absorption cross section in the steady-state
regions to increase with pressure for the frequencies that do
saturate.

We see this evidence of probe laser saturation in the experiments. In Figs. 2, 3, and 6 we see a monotonic increase of absorption cross section with pressure at times beyond 1 μs. There are, of course, other pressure dependent effects in this time period. Saturation effects play some role in the absorption process, but we should not attribute all pressure effects to saturation.

The pump laser will perturb the probe laser's steady-state absorption rate. It may preferentially excite either the upper level or the lower level of the probe laser transition, and it may excite other species whose subsequent spectrum has a higher absorption cross section at the probe frequency. After perturbation by the pulsed laser, the time necessary to return the levels coupled by the probe laser to a steady-state rate could be shorter than the pulse length for some frequencies. We may see a transient absorption if the pump laser depletes the upper of two levels connected by the probe, or a transient transparency if the pump laser depletes the lower level. The net change in probe-laser absorption cross section before and after the laser pulse is a measure of the change of population of species that interact with the probe frequency.

The transient signals we see in the first 0.5 μs in Figs. 3, 4, and 5 could be due to the pump laser depletion of the levels connected by the probe laser. This effect should give larger signals at lower pressure as we see in Figs. 3 and 5; and it should give positive transients (i.e., increased absorption cross section) at the "off line" frequencies as we see in Figs. 3 and 4.

Another effect that probably contributes to the transient signals during the pump pulse is Rabi splitting of levels that are at or near the energy of the upper or lower probe levels.[6] If, for example, the pump laser is the right frequency to couple the upper probe level with a higher one, the resulting Rabi splitting will produce a transient absorption in the wings of a line and a transient transmission near the line center. This is precisely what we see in Fig. 4.

The effects we have discussed here are relevant to other experiments as well. The conditions for Deutsch and Bruecks's experiments[2,3] were similar to ours. They also saw transient absorption signals similar to Fig. 5. Kwok and Yablonowitch[5] observed rapid transient absorption during much shorter (30 ps) pulses. In both of these experiments the effects we discussed could have contributed to the observations.

Rotational relaxation processes could modify the induced spectrum in two ways: First, by contributing to the amount of

pulse laser energy absorbed by the sample; and second by modifying the rotational energy distribution (and hence, the induced
spectrum) during as well as after the laser pulse. Table 1 shows
that the amount of absorbed laser energy increases with pressure.
A major contribution to this effect is the feeding of molecules
by rotational relaxation processes into states that have high
absorption rates. This effect can be quite large, even for
pressures that are as low as the ones we used in these experiments.[9] The pressure trend we see in Figs. 2 and 3 may be influenced by this increase of absorbed energy with pressure as well
as by the saturation effect we discussed above. The relaxation
time for the process that Moulton, et al.[6] attribute to rotational energy redistribution is 36-ns-torr for SF_6 at room temperature. A correction for collision rate alters this time to 25-ns
torr at the temperature of our experiments, 145 K. This gives
rotational-relaxation times of 300 ns at 0.08 torr. The broadening of the spectrum to higher frequency with time during and
shortly after the pulse (Fig. 7) is probably due to collisional
excitation of higher energy-rotational states.

It will be difficult for us to determine the rate of collisionless, intramolecular V-V energy transfer. Many experiment-
ers[2,3,5,10] have attempted to measure the rate of this elusive
process in SF_6, and the times they have reported differ by many
decades. We would be able to observe changes in the spectrum
from this process only if they would occur on a time scale on the
order of the laser pulse length. If the time scale for the process were shorter than the pulse length, the changes in the spectrum would tend to follow the pulse, and if it were much longer
than the pulse length, collisional processes would dominate. A
detailed analysis of the induced spectrum could perhaps determine
whether or not the absorbed energy remained for any length of
time in the normal mode. Until recently the indications were
that vibrational excitation in the ν_3 normal mode would give a
much greater anharmonic frequency shift than the same amount of
vibrational energy statistically distributed among all normal
modes of the molecule. (See, for example, the analyses in Refs.
2 and 3). The recent analysis[11] of the $3\nu_3$ overtone band of SF_6
indicates that this is not so. The anharmonicity constant, X_{33},
is about half of the previous estimate.[12] The effects of anharmonic splitting within the ν_3 mode are greater than the small
differences between (1) the anharmonic shifts due to excitation
in the ν_3 mode and (2) those from a statistical distribution of
the same energy among all modes.[8,12] This reevaluation of X_{33}
clouds the interpretations others[3,4,16] have made about intramolecular vibrational energy transfer.

Because of these spectroscopic reasons it will also be
difficult to diagnose the effect of <u>collisional</u> intramolecular

V-V energy transfer. If the <u>collisionless</u> process were rapid, and if it produced a random intra-molecular energy distribution, the collisional process would play no role.

The case for extracting intermolecular V-V energy-transfer information from the pulse-probe experiments is better. It is clear from many experiments, including those we report here, that the vibrational-energy distribution that a pulsed-CO_2 laser produces in low pressure SF_6 is one that deviates significantly from a thermal distribution. The frequency shift of the spectrum is roughly proportional to the amount of vibrational energy.[8] Intermolecular V-V energy-transfer processes drive the population from the initial nonequilibrium distribution towards one describable by a single vibrational temperature. For example, we see in Fig. 7 that changes in the spectrum occur between 0.4 μs and 3.0 μs that are consistent with the redistribution of the available vibrational energy. We see similar changes between the 5 and 20 μs in Fig. 9. Over this time period the low-frequency tail and the band center portion of these spectra fall, while the intermediate portion rises and the peak of the spectra shift to lower frequencies. Thus, the 20 μs spectrum is much nearer the thermal spectrum. One would expect some changes on this time scale since the mean time between gas-kinetic collisions is about 1 μs at 0.08 torr. These changes are consistent with de-excitation of high-energy species along with excitation of lower-energy species, i.e., intermolecular V-V energy transfer. Bates, et. al.[13] give 1.5 μs-torr for an intermolecular V-V relaxation time in room temperature SF_6. This translates to 1.0 μs-torr for our temperature. This is consistent with the relaxation time we derived from our experiments of 1-2 μs-torr.

Deutsch and Brueck[2,3] observed similar changes in the induced spectrum of SF_6 at room temperature. They interpreted the changes to be due to collisionless intramolecular energy redistribution. We feel, however, that a more consistent interpretation of their data is the same intermolecular energy transfer processes that we are postulating for our experiments. The "collisionless" nature of their relaxation times is an artifact of the data reduction method they used.

Fuss and Hartmann[4] observed some structure in their pulse-probe spectra that they attributed to two- and three-photon resonances. We would expect to see less of this effect in our experiments because our pump frequency was at the peak of the one-photon spectrum. The major multiphoton resonances occur at lower frequencies.[13,14]

There are two other processes that could influence the induced spectra after long delay times. These are V-T transfer and diffusion. The latter dominates the spectra beyond a few hundred microseconds (Fig. 9) and the former is too slow[7] to have more than a minor effect.

CONCLUSIONS

In these experiments we found that for some frequencies saturation of the probe transitions alters the probe signals. These saturation effects, along with Rabi splitting of states by the pulsed laser, produce rapid transient signals for some probe frequencies.

The experiments indicate that the initial distribution of absorbed laser radiation in a low pressure, low temperature SF_6 sample is not thermal. The laser tends to produce more highly excited molecules and to leave more unexcited molecules than would be the case for a thermal distribution.

Rotational energy transfer also plays an important role in the absorption and redistribution of laser radiation. This process is particularly important during the laser pulse.

The preliminary analysis of the pulse-probe data gave little information about the intramolecular distribution of absorbed laser radiation or the (collisionless) rate of approach of that distribution to a statistical one. Understanding the extent and rate of intramolecular randomization of vibrational energy would require a more thorough analysis, and perhaps, independent experiments.

We feel that intermolecular vibration-to-vibration energy transfer contributes significantly to the relaxation of that energy for our conditions. We estimated this V-V relaxation time to by 1-2-µs torr in systems that do not deviate significantly from equilibrium.

REFERENCES

1.   J. L. Lyman, L. J. Radziemski, and A. C. Nilsson, "Pulse-Probe Measurements in the ν_3 Band of SF_6 at Low Temperature and Low Pressure," IEEE J. Quantum Electron. (in press).

2.   T. F. Deutsch and S. R. J. Brueck, "Collisionless Intramolecular Energy Transfer in Vibrationally Excited SF_6," Chem. Phys. Lett. 54:258 (1978).

3.   T. F. Deutsch and S. R. J. Brueck, "ν_3 Mode Absorption
      Behavior of CO_2 Laser Excited SF_6," J. Chem. Phys. 70:2063
      (1979).

4.   W. Fuss and J. Hartmann, , IR Absorption of SF_6 Excited up
      to the Dissociation Threshold," J. Chem. Phys. 70:5468
      (1979).

5.   H. S. Kwok and E. Yablonovitch, "Collisionless Intramo-
      lecular Vibrational Relaxation in SF_6," Phys. Rev. Lett.
      41:745 (1978).

6.   P. F. Moulton, D. M. Larsen, J. N. Walpole, and A. Mooradian,
      "High-Resolution Transient-Double-Resonance Spectroscopy in
      SF_6," Opt. Lett. 1:51 (1977).

7.   K. S. Rutkovskii and K. G. Tokhadze, "Study of Vibrational
      Relaxation of SF_6 at Low Temperatures by the Method of Double
      Infrared Resonance," Zh. Eksp. Teor. Fiz. 75:408 (1978).

8.   A. V. Nowak and J. L. Lyman, "The Temperature-Dependent Ab-
      sorption Spectrum of the ν_3 Band of SF_6 at 10.6 μm," J.
      Quant. Spectrosc. Radiat. Transfer 15:945 (1975).

9.   O. P. Judd, "A Quantitative Comparison of Multiple-Photon
      Absorption in Polyatomic Molecules," J. Chem. Phys. 71:4515
      (1979).

10.   D. G. Steel and J. F. Lam, "Two-Photon Coherent-Transient
      Measurement of the Nonradiative Collisionless Dephasing Rate
      in SF_6 via Doppler-Free Degenerate Four-Wave Mixing," Phys.
      Rev. Lett. 43:1588 (1979).

11.   C. W. Patterson, B. J. Krohn, and A. S. Pine, "Interacting
      Band Analysis of the High-Resolution Spectrum of the $^3\nu_3$
      Manifold of SF_6," J. Mol. Spectrosc. (submitted for publi-
      cation).

12.   R. S. McDowell, J. P. Aldridge, and R. F. Holland, "Vibra-
      tional Constants and Force Field of Sulfur Hexafluoride,"
      J. Phys. Chem. 80:1203 (1976).

13.   R. D. Bates, J. T. Knudtson, G. W. Flynn, and A. M. Ronn,
      "Energy Transfer Among Excited Vibrational States of SF_6,"
      Chem. Phys. Lett. 8: 103 (1971).

ENERGY SCRAMBLING IN MOLECULES: TWO EXAMPLES

FROM MOLECULAR BEAM SPECTROSCOPY

Donald H. Levy

The University of Chicago
The James Franck Institute and The Department of Chemistry
Chicago, Illinois 60637, U.S.A.

INTRODUCTION

In this article I would like to discuss two systems in which
we have observed intramolecular energy scrambling in some detail.
The two cases, free base phthalocyanine and rare gas-iodine van der
Waals molecules, would appear to be about as unrelated as it is
possible to get given the finite size of the periodic table, and
yet the phenomena that we observe in these two systems can be described using very similar langauge.

Of course the initial relationship between these two molecules
was the technique which we used to study them, laser induced fluorescence of a supersonic free jet of a trace amount of the molecule
seeded into a carrier gas. This technique has been described in
detail in a number of places,[1] but it may be useful to review it
briefly here before describing the experimental results and interpretation.

The supersonic free jet is used as a refrigerator to cool the
internal degrees of freedom of the molecule. The translational degrees of freedom of the carrier gas are directly cooled by the expansion, and collisions between the molecule and the translationally
cold carrier gas cool the molecular vibrations and rotations. Because the gas is expanding, the density decreases as the gas mixture
flows downstream. At some point the mixture is too rarified to
produce collisions, and from this point on downstream nothing
changes. Because the cooling is only effective for a finite period
of time, the system need not come to equilibrium, and the state of
the molecules in the collision free region is determined more by
kinetics than by thermodynamics. The degrees of freedom that

equilibrate most rapidly with the translationally cold bath are
cooled the most, those that take longer to equilibrate are cooled
less.

The equilibration between the translational degrees of freedom
of the carrier gas and both the translational and rotational degrees
of the seed molecule is fast. In the expansions which we use the
translational temperatures of the carrier gas and the molecule are
essentially identical, and the rotational temperature is only
slightly higher. Vibrational-translational equilibration is rather
slower although we are learning that the low energy collisions that
are dominant in the expansion are much more effective in cooling
molecular vibrations than are the higher energy collisions that
occur in a room temperature static gas.[2]

In a typical experiment a few ppm of the molecule is seeded
into up to 100 atmospheres of carrier gas, helium or a rare gas
mixture containing mostly helium. The gas is expanded through a
25μ pinhole and can reach downstream translational temperatures as
low as 0.05 K. Under these conditions the typical rotational temperature would be ∿0.5 K and the vibrational temperature would be
10-50 K. The unique feature of the supersonic expansion is that it
can produce these very low internal temperatures with little or no
condensation, and therefore we are able to prepare a sample of internally cold isolated gas phase molecules. If one wishes to examine intramolecular energy scrambling this is a very useful sample
indeed. Since the sample consists of isolated gas phase molecules,
intramolecular scrambling can be observed in the absence of external
collisions. Since the sample is internally cold, the molecular
spectrum is greatly simplified, and it is possible to use spectroscopy as a probe of the phenomena that we wish to study.

PHTHALOCYANINE

The structure of free base phthalocyanine is shown in Fig. 1.
The molecule contains 58 atoms, has 168 vibrational modes, and has
a molecular weight of 514. It has negligible vapor pressure at
room temperature but is stable at 400-450°C where it has a vapor
pressure of 10^{-2}-10^{-1} torr,[3] and this is sufficient for gas phase
spectroscopy. A molecule of this size is an excellent candidate
for the study of intramolecular energy scrambling, but under static
gas conditions its optical spectrum is broad and unresolved, and
it is difficult under these conditions to obtain much information
about the distribution of internal energy.

The fluorescence excitation spectrum of free base phthalocyanine which has been cooled in a supersonic expansion is shown in
Fig. 2.[4] Superimposed on the cold jet spectrum is the static gas
absorption spectrum taken from the data of Eastwood et al.[5] As seen
in this figure, the cooling produced in the supersonic free jet has

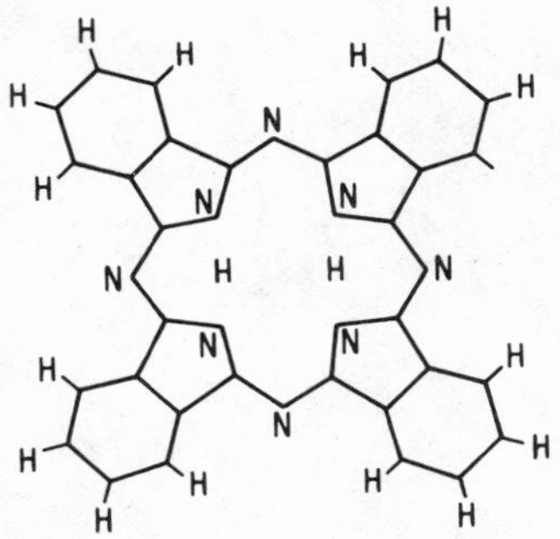

Fig. 1. Structure of free base phthalocyanine.   The positions of
the two inner hydrogens is unknown.

greatly simplified the spectrum and allowed resolution of individual
vibrational features.

     Free base phthalocyanine is almost but not quite fourfold sym-
metric. An excited state that would be orbitally degenerate in
fourfold symmetric metal bearing phthalocyanines is split into two
nearly degenerate excited electronic states in free base phthalocyanines, and this electronic structure may be seen in the fluorescence excitation spectrum. The strong feature at 15132 cm^{-1} is the
origin of the first excited singlet state and the weaker features
for the next several hundred wavenumbers to the blue are vibrational
progressions built upon this origin.* The strong congested structure
centered around 16600 cm^{-1} is due to the origin of the nearby second
excited singlet state which mixes with excited vibrational levels of
the first singlet.

     Since the resolution provided in the supersonic jet spectrum
allows us to excite individual vibrational features, it is possible
to probe the energy scrambling by examining the dispersed fluorescence spectrum. The question being addressed is the following:

---

     *The feature to the red is due to a chlorine-containing im-
purity.[4]

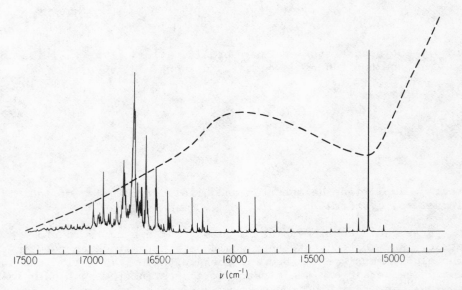

Fig. 2.   Fluorescence excitation spectrum of free base phthalocya-
         nine cooled in a supersonic free jet.   The dashed curve is
         the static gas absorption spectrum taken from Ref. 5.

what has happened to the initially localized energy between the
time of absorption and the time of emission, an interval of $\sim$10 nsec.

Figure 3 shows the dispersed fluorescence spectrum obtained following excitation of the 15132 cm^{-1} absorption feature (the S_1 origin).[6] The spectrum consists of one very strong feature at the excitation frequency plus several very much weaker features to the red
of the strong band. The strong feature in Fig. 3 has a small contribution from scattered laser light, but almost all of it is molecular fluorescence.

Since the state excited at 15132 cm^{-1} contains no excess vibrational energy, there is no question of vibrational energy scrambling
occurring. The concentration of the emission into a single strong
feature indicates that, as might be expected, the potential surface
is not changed very much by the electronic excitation, and the large
Franck-Condon factor in emission obeys the propensity rule $\Delta v=0$.
The weaker features to the red are due to progressions ($\Delta v \neq 0$) with
small but non-zero Franck-Condon factors.

Figure 4 shows the dispersed emission spectrum that results
when the 15697 cm^{-1} band is excited. This emission spectrum is almost but not quite identical with that of Fig. 3 (in Fig. 4 the
strong feature at the exciting frequency is all scattered laser
light). In Fig. 3 the frequency of the strong emission band is
15132 cm^{-1} while in Fig. 4 the strong feature is at 15127 cm^{-1}.

The qualitative features of the spectrum in Fig. 4 look very
much like vibrational relaxation has taken place, but the quantitative measurements imply that this is not the case. Qualitatively,
the emission spectrum is independent of the exciting wavelength,
and in solids this behavior is taken as evidence of intermolecular
vibrational relaxation between the vibrational modes of the molecule and the phonons of the lattice. However in our case the slight
frequency shift of the emission spectrum that occurs when the exciting frequency is changed implies that the emitting levels are not
the same in Fig. 3 and Fig. 4.

We believe that the similarity of the emission spectra in these
two cases is due to the Franck-Condon principle. If the potential
surface changes very little when the molecule is electronically excited, the vibrational level structure in the two electronic states
will be the same. Moreover the strong Franck-Condon factors will be
between corresponding levels in the two electronic states, i.e.
levels where $\Delta v=0$. For these transitions the emission frequency
will be given by $\nu_{eg}^i = \nu_0 + \nu_e^i - \nu_g^i = \nu_0 + \Delta \nu^i$. Here ν_{eg}^i is the
frequency of the strong emission feature when the ith vibrational
level is excited, ν_0 is the frequency difference between zero point
levels of the ground and excited electronic states, ν_e^i is the frequency difference between the ith vibrational level of the excited

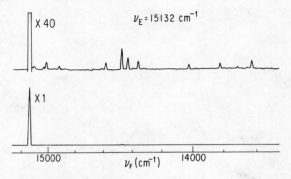

Fig. 3. The dispersed fluorescence spectrum of the free base phthalocyanine excited at 15132 cm^{-1}. The upper and lower
traces are taken at high and low sensitivity, respectively.
The spectral resolution was 9 cm^{-1}. The feature at the
exciting wavelength is mostly fluorescence with a small
amount of scattered laser light.

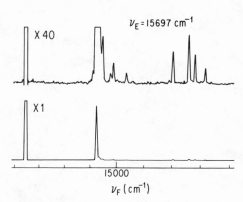

Fig. 4. The dispersed fluorescence spectrum of free base phthalo-
         cyanine excited at 15697 cm^{-1}. The upper and lower traces
         are taken at high and low sensitivity, respectively. The
         spectral resolution was 9 cm^{-1}. The strong feature at the
         exciting frequency is scattered laser light.

electronic state and the zero point level of that state, and ν_g^i is
the corresponding quantity for the ground electronic state. If the
two potential surfaces were very different, $\Delta\nu^i$ would be large and
different for each i, and the unrelaxed emission spectrum would be
a strong function of the state originally excited. If the two surfaces were identical, $\Delta\nu^i$ would be zero in all cases, and the emission spectrum would be the same for all excited levels. In the case
of phthalocyanine $\Delta\nu^i$ is small but non-zero, and the emission spectrum shows a slight frequency shift as the excited state is observed.

The spectrum in Fig. 5 results from excitation at 15925 cm^{-1},
this being 793 cm^{-1} above the origin. In Fig. 5 the spectral features are qualitatively in the same position as they were in Figs. 3
and 4 (again, there is a slight frequency shift), but now the emission bands are very much broader than they were previously. This

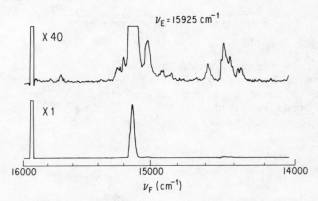

Fig. 5. The dispersed fluorescence spectrum of free base phthalocyanine excited at 15925 cm^{-1}. The upper and lower traces are
taken at high and low sensitivity, respectively. The spectral
resolution was 9 cm^{-1}. The strong feature at the exciting
wavelength is scattered laser light.

broadening may be more easily more seen in Fig. 6 where we show the strong emission feature that results when phthalocyanine was excited at 15808 cm^{-1}, 15852 cm^{-1}, and 16188 cm^{-1}. We have found that when the exciting frequency is less than or equal to 15808 cm^{-1}, the emission spectrum has only the instrumental linewidth. When the excitation frequency is greater than or equal to 16188 cm^{-1} the emission feature is broad and unresolved. For the single case where ν_E = 15852 cm^{-1}, the emission is broad but structured. It should be noted that in all cases no broadening is observed in the absorption band.

We believe that the broadening of the emission features is an indication of intramolecular energy scrambling, although it should be emphasized that the reciprocal linewidth should not be interpreted as a relaxation time. If the reciprocal emission width were the relaxation time, then the absorption band would have the same width, and this is not the case.

To understand the mechanism responsible for the emission broadening in phthalocyanine, one must consider the vibrational energy level structure in the molecule. As already mentioned, the largest Franck–Condon factors are for $\Delta v=0$ transitions. Some of the vibrational modes have small but finite $\Delta v \neq 0$ Franck–Condon factors and features involving excitation of these modes appear in the absorption spectrum. However many (perhaps most) of the vibrational levels of the excited electronic state have unobservably small Franck–Condon factors with the zero point level of the ground electronic state. Therefore only a small fraction of the total number of vibrational states is actually seen in the absorption spectrum. In the following discussion we will refer to observed and unobserved states.

Small interactions such as anharmonic coupling can mix the observed and unobserved states. These couplings are likely to be very small since the vibrational modes that are being mixed can be very different. Nonetheless, as the excitation energy is increased, the density of unobserved levels rapidly increases, and the probability of one or more unobserved levels being nearly degenerate with an observed level also increases. The amount of mixing depends on the ratio of the coupling strength to the energy difference between the levels that are being mixed, and when the levels become near degenerate even very weak couplings will produce substantial mixing.

We believe that the broadening in the emission spectrum of phthalocyanine occurs when the density of unobserved levels is large enough so that appreciable mixing occurs between the observed level that provides the Franck–Condon factor and the unobserved levels. When this happens the wavefunction of the emitting levels is a linear combination consisting of a small contribution from the observed level and a large contribution from the unobserved levels.

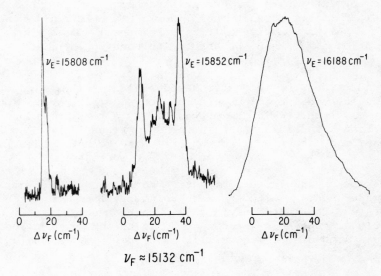

$$\nu_F \approx 15132 \ cm^{-1}$$

Fig. 6. The strongest features in the dispersed fluorescence
spectrum of free base phthalocyanine when excited at 15808,
15852, and 16188 cm^{-1}. Resolution was 2 cm^{-1} but increasing the resolution to 0.6 cm^{-1} did not produce any additional structure.

Therefore the character of the absorption spectrum will be determined by the wavefunction of the observed levels, but the character
of the emission spectrum (and any other physical properties) will
be determined by the wavefunction of the unobserved levels.

It must be remembered that the unobserved levels are only unobserved in absorption, that is to say, their Franck-Condon factors
with the populated zero point level of the ground electronic state
is very small. These levels are observable in emission since they
must have large Franck-Condon factors with their corresponding vibrational levels in the ground electronic state. This is shown in
Fig. 7.

The width of the absorption spectrum is determined by the
strength of the couplings. If the couplings are weak, the coupled
levels must be close together in order to mix. The width of the
emission spectrum is determined by the change in vibrational frequencies between the two electronic states. If the ground and excited state frequencies were identical, the emission spectrum would
be narrow even when interstate mixing occurred, since the strong
emission lines from all vibrational states would fall at the same
frequency. If the vibrational frequencies changed greatly, then
each excited state vibrational level would emit at a different frequency, and interstate mixing with a random collection of unobservable levels would lead to an emission spectrum with lines spread
over a wide frequency region. In phthalocyanine the frequencies
change only slightly, and the emission lines from the various
coupled levels fall at almost but not quite the same frequency
leading to a broadening of the emission band. Therefore we believe
that the width of the absorption lines is a measure of the strength
of the vibrational state couplings while the width of the emission
spectrum is a measure of the frequency changes between ground and
excited electronic states. The structure in the emission spectrum
produced by 15852 cm^{-1} excitation is the intermediate case where
only a few unobservable levels are coupled and their emission
spectra are partially resolved. Below this point the density of
unobservable levels is too small to allow any mixing, and above this
point the density of levels is high enough so that their emission
spectra are blended together at our best experimental resolution.

In the preceding discussion we have implicitly equated intramolecular energy scrambling with vibrational state mixing. Whether
this is a correct use of the term is debatable, but it seems to be
in the spirit of this conference. Whether or not one could actually
observe a time dependent intramolecular vibrational relaxation in
phthalocyanine would depend on how the mixed state was prepared.
If the preparation was coherent, then immediately after excitation
the time dependent wavefunction of the prepared state would be that
of the single observed level that carried the Franck-Condon factor.
This initial state would then dephase over some time period into

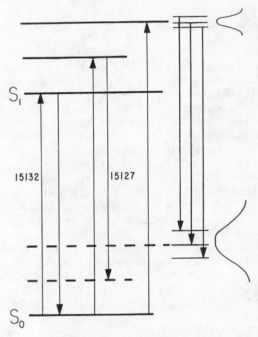

Fig. 7.   Proposed mechanism of broadening of fluorescence features
          in free base phthalocyanine. Excitation is from the zero
          point level of S_0 (heavy solid line) to certain vibrational
          levels of S_1 (heavy solid lines). These levels are mixed
          with nearby background levels (lighter solid lines) in a
          time shorter than the fluorescence lifetime. Emission is
          always from a given level of S_1 to the corresponding level
          of S_0.

the set of coupled unobservable levels. The dephasing time could
be considered a relaxation time since in principle the emission
spectrum would change on this time scale. On the other hand, if the
initial excitation were incoherent, there would be no time variation
of the emission spectrum, even in principle. In this meeting we
have tended to refer to intramolecular vibrational relaxation and
energy scrambling in those situations where systems could be prepared so as to show time dependent behavior. We have not been very
concerned about just how this coherent preparation could be carried
out in practice. In this spirit, we believe that gas phase phthalocyanine shows intramolecular energy scrambling when excited by more
than several hundred cm^{-1} above its zero point level.

RARE GAS-IODINE VAN DER WAALS MOLECULES

As the name implies, van der Waals molecules are molecules
that are bound in part by van der Waals forces.[7] In general these
forces are weak and produce binding energies much less than kT_R
where T_R is room temperature. Therefore van der Waals molecules are
unstable under ordinary laboratory conditions because they are dissociated by essentially every two body collision.

In the supersonic jets that we use to study these species, the
translational temperatures can be as low as 0.05 K, and at this
temperature even the weakest van der Waals force will produce a
binding energy that is large compared to kT. The translational
temperature is defined by the width of the velocity distribution in
the jet. Therefore the translational temperature is a measure of
the relative velocity between colliding partners and is the parameter of interest in comparing binding energy to relative collisional
kinetic energy.

At the translational temperature of the expanding jet, van der
Waals molecules are stable with respect to collisions, and the development of supersonic molecular beams has produced a very active
area of research into the properties of these molecules. Much of
the effort has been directed toward determining the structure of
these species, the structure being the key to an understanding of
the van der Waals force itself.

The relevance of van der Waals molecules to a conference on
molecular energy scrambling is somewhat different. In some sense
van der Waals molecules can be viewed as the simplest of photochemical systems, and if one wishes to understand the vibrational energy
scrambling that ultimately produces the rupture of a chemical bond,
van der Waals molecules seem to offer many advantages.[8]

Because van der Waals forces are so much weaker than ordinary
chemical binding forces, it is useful to separate the vibrational
modes of a van der Waals molecule into two groups: high frequency

chemical modes involving displacements of chemical bonds, and low frequency van der Waals modes involving displacements of the van der Waals bonds. Usually the chemical modes are only slightly different from the vibrational modes of the free chemically bound substrate that was used to form the van der Waals molecule. Therefore it is possible to excite vibrations in the van der Waals molecule and to have a very good description of the initially excited state.

Because van der Waals bonds are weak, it is usually the case that a very few quanta of a chemically bound vibration will contain more than the van der Waals binding energy. In the case of the weakest van der Waals bonds involving the lightest rare gases, a single quantum will be sufficient. In those cases where a chemically bound vibration is excited with more than the van der Waals binding energy, one gets a very good indication of intramolecular energy scrambling. After some period of time, an atom drops off.

In the experiments which I will describe here, we use a visible laser to both electronically and vibrationally excite the van der Waals molecule. This is an experimental convenience for two reasons. In the first place it provides a built-in probe to measure the extent of vibrational energy scrambling. After the electronically excited van der Waals molecule has dissociated, it frequently leaves behind an electronically excited fragment molecule. By analyzing the fluorescence from the fragment molecule we are able to measure the quantum state distribution produced by the photochemical reaction. Since we have a good description of the initially excited state, this provides a detailed understanding of the intramolecular vibrational energy scrambling. A second experimental advantage is the fact that tunable visible light is very much easier to produce than tunable infrared.

The experiments I want to discuss all involve van der Waals molecules consisting of an iodine molecule bound to one or more rare gas atoms. These van der Waals molecules are produced by expanding a mixture consisting of $\sim$1 ppm I_2 in a rare gas mixture of either pure helium or helium plus several percent of a second rare gas.[9] The van der Waals molecule is excited by a tunable dye laser to its B 0_u^+ electronic state. The laser excitation is always to some excited vibrational level, $v_i' \neq 0$, of the iodine stretching vibration. After some time energy flows from the initially excited vibrational mode, the iodine stretch, into the van der Waals modes leading to a rupture of the van der Waals bonds and the production of an electronically excited I_2 fragment in a lower final vibrational state, $v_b < v_i$.

There are two questions that we will consider. How long does it take for the energy to travel from the initially populated storage mode, the I_2 stretch, to the dissociating mode, the van der Waals stretch? What is the final disposition of the excess energy that was originally placed in the storage mode?

Our best approach to the measurement of the dissociation lifetime has been to measure the line broadening produced by the finite
lifetime. The fluorescence excitation spectrum of the van der Waals
molecule I_2He is shown in Fig. 8. In the upper trace the molecule
has been excited to the $v'=7$ vibrational level of the excited electronic state, and the linewidth is just slightly wider than the experimental width. In the lower trace the molecule has been excited
to $v'=27$, and the additional broadening produced by the shorter lifetime is clearly observable. We have measured the lifetimes of various levels of I_2He using the observed linewidth corrected for non-
lifetime broadening.[10] The results are displayed in Fig. 9 where
it may be seen that the lifetimes are in the range of tens to
hundreds of psec for $v'=12-26$. These lifetimes correspond to hundreds of vibrational periods, and in these weakly coupled molecules
vibrational energy scrambling is a relatively slow process.

We next consider the question of where the excess energy goes
between the time of excitation of the van der Waals molecule and
the time of fluorescence of the fragment I_2^* produced in the photochemical reaction. It is important to realize that although the
energy scrambling lifetime is long when compared to a vibrational
period, it is very short when compared to the radiative lifetime
of 1 μ sec. Therefore essentially all of the van der Waals molecules dissociate before they radiate, and essentially all of the
fluorescence that we observe comes from the fragment I_2^*, not from
the van der Waals molecule. We have discovered that the cross section for vibrational relaxation of the initial vibrational product
state distribution increases greatly as the translational temperature decreases.[2] Therefore we do observe collisional relaxation in
the jet at gas densities that would be far too small to produce
vibrational relaxation in a room temperature static gas. These low
energy relaxation processes alter the original product state distribution by roughly 10% and corrections for relaxation can be made.
The Franck-Condon factors for the B→X transition are well known and
the spectrum is well understood, and therefore measurement of the
relative fluorescence intensities of the fragment I_2^* provides a
very good measure of the initial product state distribution.

The dissociation of I_2He is a very efficient reaction in the
sense that it uses the minimum amount of energy necessary to break
the van der Waals bond and leaves the rest in the initially excited
storage mode.[11] If the state v' of the van der Waals molecule is
excited, roughly 97% of the light emitted comes from the $v'-1$ state
of the fragment I_2^*. In I_2He$_2$, roughly 95% of the light comes from
$v'-2$, as though the rupture of the two van der Waals bonds were
separate, uncoupled events each producing the product state distribution that would have been produced had the second helium atom not
been present.

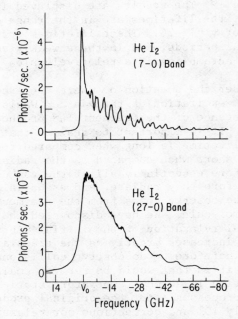

Fig. 8. The fluorescence spectrum of I_2He excited in the
$B(v'=7) \leftarrow X(v''=0)$ and $B(v'=27) \leftarrow X(v''=0)$ bands.

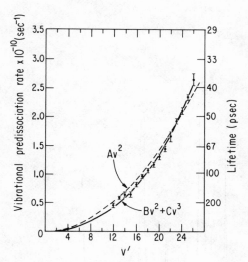

Fig. 9.   Vibrational predissociation rate (and lifetime) of I_2He
          as a function of the vibrational state of the I_2 stretching
          mode that was excited.   Points are experimental measure-
          ments, and dashed and solid curves are the best least
          squares fits of functional form A v^2 and B v^2 + C v^3.

        In the case of neon-containing van der Waals molecules, the
dissociation dynamics become more complicated.   Figure 10 shows
the fluorescence excitation spectrum of a mixture of I_2, Ne, and He.
As may be seen from the figure, this mixture produces a large number
of different chemical species, and we have observed spectra from
van der Waals complexes as large as I_2Ne_7.   The assignment of the
various species is largely based on the band shift rule of Kenny
et al.[12]

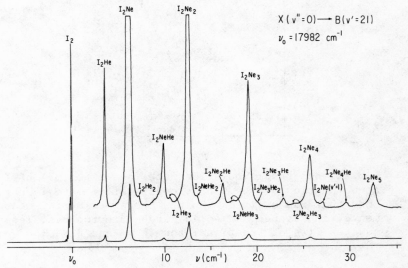

Fig. 10. The fluorescence excitation spectrum of a supersonic expansion of iodine in a mixture of neon and helium. Assignment of the various van der Waals species formed in the expansion is shown.

The dispersed fluorescence spectra that are produced by excitation of I_2Ne, I_2Ne_2, and I_2Ne_3 to their B(v'=22) levels are shown in Fig. 11. The product state distribution produced when I_2Ne dissociates is qualitatively similar to that produced when I_2He dissociates. That is, there is a strong propensity to take the minimum amount of energy out of the storage mode, and 93% of the light comes from the v'-1 level of the fragment.

In the case of I_2Ne_2 and I_2Ne_3, the reaction becomes less efficient and the product state distribution begins to broaden. In the case of I_2Ne_2, the one quantum per atom channel (v'=20 products produced by v'=22 reactants) is still the largest, but the three quantum channel is almost as large and the four and five quantum channels are observable. For I_2Ne_3 the largest cross section is for the four quantum channel, not the three quantum channel. This trend toward a broader product state distribution continues as the complex grows larger. In Fig. 12 the relative cross sections for the various dissociation channels are shown for the complexes I_2Ne-I_2Ne_6. In the case of I_2Ne_6 we do not observe a six quantum dissociation, and the largest cross section is for the eight quantum dissociation.

It is likely that the broader product state distribution in the fragment I_2^* that is produced when the larger complexes dissociate is due to the increasing number of non-dissociating van der Waals modes. In triatomic I_2Ne there are three vibrational modes: the I_2 stretch (the storage mode), the van der Waals stretch (the dissociating mode), and the van der Waals bend (a non-dissociating mode). The I_2 stretch is the only chemically bound vibration and is naturally the highest frequency mode. The vibrational frequency of the van der Waals bend is not known, but it is likely to be less than that of the van der Waals stretch (24.7 cm^{-1}). Theories of the dissociation of van der Waals molecules[13] predict that the strongest coupling and most rapid energy flow will be between modes most nearly matched in frequency. Therefore in triatomic van der Waals molecules one might imagine an efficient transfer of energy from the storage mode to the dissociating mode largely bypassing the single non-dissociating mode.

In the larger complexes the same strength of coupling arguments still hold, the modes closest in frequency will still be the storage mode and the several dissociating stretching modes. However, as the complex grows in size, the number of non-dissociating modes rapidly grows, and eventually statistical considerations will overwhelm any specific strength of coupling mechanisms. As in the case of phthalocyanine, energy flow will be produced by extremely weak coupling if only the density of accepting levels is high enough. In the limit, all modes must be heated statistically using energy originally deposited in the storage mode, and energy will not be preferentially directed into the dissociating modes.

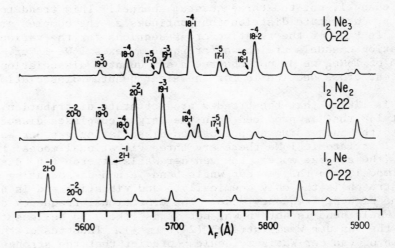

Fig. 11. Dispersed fluorescence spectra of the product I_2^* formed
upon dissociation of the van der Waals complexes I_2Ne,
I_2Ne_2, and I_2Ne_3. In all cases the complex was originally
excited to the $B(v'=22$ state.

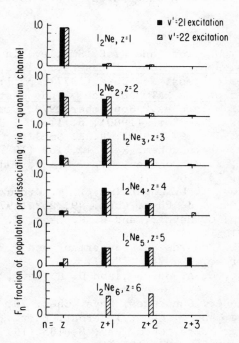

Fig. 12. The relative cross sections for the process $I_2Ne_z(v')* \rightarrow$ $I_2*(v'-n) + zNe$ for the complexes $I_2Ne-I_2Ne_6$ excited to $v'=21$ and $v'=22$.

We have discussed energy scrambling in two quite different systems: free base phthalocyanine and iodine-rare gas van der Waals molecules. The major conclusion to be drawn from this work is that it is possible to get very detailed experimental information about intramolecular energy scrambling by utilizing the greatly increased resolution provided by supersonic jet spectroscopy.

This work was supported by the U. S. Public Health Service under Grant 5-R01-GM25907, by the donors of The Petroleum Research Fund administered by the American Chemical Society, and by the National Science Foundation under Grant CHE78-25555.

REFERENCES

1. D. H. Levy, L. Wharton, and R. E. Smalley, in Chemical and Biochemical Applications of Lasers, Vol. 2, C. B. Moore, ed., Academic Press, New York (1977), p. 1; D. H. Levy, in Proceedings of the NATO Advanced Study Institute "Quantum Dynamics of Molecules," September 1979, Cambridge, U.K., Plenum Press, New York (1980), p. 115; D. H. Levy, Ann. Rev. Phys. Chem. 31, 197 (1980); R. E. Smalley, L. Wharton, and D. H. Levy, Accts. Chem. Res. 10, 139 (1977).

2. T. D. Russell, B. M. DeKoven, J. A. Blazy, and D. H. Levy, J. Chem. Phys. 72, 3001 (1980).

3. Yu. Kh. Shaulov, I. L. Lopatkina, I. A. Kiryukhin, and G. A. Krasulin, Russ. J. Phys. Chem. 49, 252 (1975).

4. P. S. H. Fitch, C. Haynam, and D. H. Levy, J. Chem. Phys. 73, 1064 (1980).

5. D. Eastwood, L. Edwards, M. Gouterman, and J. Steinfeld, J. Mol. Spectrosc. 20, 381 (1966).

6. P. S. H. Fitch, C. A. Haynam, and D. H. Levy, J. Chem. Phys., submitted.

7. W. Klemperer, Ber. Bunsenges. Phys. Chem. 78, 128 (1974); G. E. Ewing, Angew. Chem. Internat. Edit. 11, 486 (1972); G. E. Ewing, Accts. Chem. Res. 8, 185 (1975); G. E. Ewing, Can. J. Phys. 54, 487 (1976).

8. D. H. Levy, in Photoselective Chemistry, Advances in Chemical Physics, J. Jortner, R. D. Levine, and S. A. Rice, eds., Wiley-Interscience, in press.

9. R. E. Smalley, D. H. Levy, and L. Wharton, J. Chem. Phys. 64, 3266 (1976).

10. K. E. Johnson, L. Wharton, and D. H. Levy, J. Chem. Phys. 69, 2719 (1978).

11. W. Sharfin, K. E. Johnson, L. Wharton, and D. H. Levy, J. Chem. Phys. 71, 1292 (1979).

12. J. E. Kenny, K. E. Johnson, W. Sharfin, and D. H. Levy, J. Chem. Phys. 72, 1109 (1980).

13. G. E. Ewing, J. Chem. Phys. 71, 3143 (1979); G. Ewing, Chem. Phys. 29, 253 (1978); J. A. Beswick, G. Delgado-Barrio, and J. Jortner, J. Chem. Phys. 70, 3895 (1979); J. Beswick and

J. Jortner, Chem. Phys. Lett. <u>49</u>, 13 (1977); J. A. Beswick and J. Jortner, J. Chem. Phys. <u>68</u>, 2277 (1977); J. A. Beswick and J. Jortner, J. Chem. Phys. <u>68</u>, 2525 (1977); J. Beswick and J. Jortner, J. Chem. Phys. <u>71</u>, 4737 (1979).

THE PHOTODISSOCIATION DYNAMICS OF H_2S AND CF_3NO

Paul L. Houston

Department of Chemistry
Cornell University
Ithaca, New York 14853 USA

1. INTRODUCTION

It has been known since the earliest investigations of photochemistry that energy scrambling in a molecule can lead to dissociation if the total energy of a molecule exceeds the dissociation limit of one or more molecular bonds. But how complete is the scrambling, and what does it tell us about the nature of the potential energy surfaces or the mechanism of dissociation? A partial answer to these questions can be gained by a newly developed experimental technique that uses one laser to dissociate a parent molecule and a second to probe the energy distribution of the photofragments.

In this paper I will discuss two contrasting cases of photodissociation examined by this new experimental technique. Photolysis of CF_3NO[1,2] provides an example of a dissociation in which excess energy is scrambled nearly statistically in the product fragments. Photodissociation of H_2S,[3] on the other hand, provides an example of very specific energy disposal; nearly all the available energy is partitioned into the relative translation of the products.

2. 193-nm PHOTODISSOCIATION OF H_2S*

2.1 Introduction

The photodissociation of H_2S has been of interest to researchers since at least 1929[4-9]. The spectrum of this molecule between 190 and 230 nm is nearly structureless and presents an absorption coefficient of 0.23 cm^{-1} torr^{-1} at 193 nm.[5] Roughly 20,000 cm^{-1}

---

*In collaboration with W. G. Hawkins

of energy is available to the fragments following dissociation at
this wavelength.

2.2 Experimental

The experimental apparatus is of a design similar to that used
by Baronavski and McDonald.[10] Briefly, an ArF excimer laser was
used to photolyze the parent compound, while a tunable, doubled dye
laser was used to probe the energy content of the SH radicals by
the laser induced fluorescence technique. While photolysis of low
pressure H_2S samples contained in a bulb was used for the initial
investigations, later studies employed a pulsed nozzle source to
cool the parent compound before photolysis.

2.3 Results

Figure 1 displays the SH laser induced fluorescence spectrum
recorded near 30,800 cm^{-1} for an initial H_2S pressure of 5 mtorr
and a delay time of 1.0 μsec between the photolysis and probe laser
pulses. The parent compound was contained in a bulb at room temperature for this experiment. Assignments in the figure identify
the rotational transitions of the (v"=0 → v'=0) band of the
$^2\Pi_{3/2} \to {}^2\Sigma^+$ system. A similar spectrum was obtained for the corresponding $^2\Pi_{1/2} \to {}^2\Sigma^+$ band. Although several attempts were made to
observe any SH(v"=1) produced by the dissociation, none was detected. Our measurements place an upper limit of 0.005 on the SH(v"=1)
/SH(v"=0) ratio.[3] The laser induced fluorescence spectra of the
$^2\Pi_{3/2} \to {}^2\Sigma^+$ and $^2\Pi_{1/2} \to {}^2\Sigma^+$ systems were compared under identical conditions of starting pressure, delay time, and detection geometry.
The integrated intensity of the $^2\Pi_{3/2}$ system was found to be 3.75
± 0.20 times that of the $^2\Pi_{1/2}$ system.

2.4 Discussion

The preceding results show that the SH product of 193-nm

Figure 1. Portion
of the SH laser
induced fluorescence
spectrum.

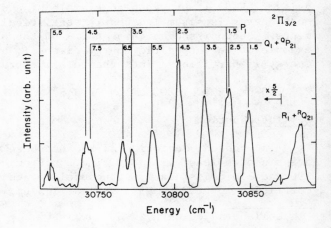

dissociation of H_2S is born with very little internal excitation.
The $^2\Pi_{3/2}/^2\Pi_{1/2}$ ratio of 3.75 corresponds to an electronic temperature of 410°K. The upper limit on the fraction of vibrationally
excited SH in either electronic state was only 0.005. The rotational distribution was also found to be unexcited. The intensities
of the rotational transitions in Fig. 1 were converted to relative
rotational populations using standard methods. When the logarithm
of the resulting populations was plotted against rotational energy,
straight line fits were obtained, as shown for the $^2\Pi_{3/2}$ case in
Fig. 2. The slopes of these lines gives rotational temperatures
of 375°K for the $^2\Pi_{3/2}$ state and 220°K for the $^2\Pi_{1/2}$ state. We
conclude that little of the 20,000 cm^{-1} in available energy appears
in the SH rotational, vibrational, or electronic degrees of freedom.
Most of the available energy must, therefore, appear in the relative
recoil coordinate between the H and SH fragments.

These conclusions are in good agreement with those of previous
studies[8,9] and with the predictions of a simple quasidiatomic
kinematic model for dissociation. The major assumptions of this
model have been reviewed elsewhere.[3,11] While detailed calculations
for the H_2S case have not been performed, the basic features of the
kinematics follow easily from the assumption that the dissociation
takes place along the original bond. Since the HSH bond angle is
92°, it is clear that recoil along the original bond is nearly
orthogonal to the SH vibration. It is not likely, therefore, that
the recoil will couple energy into product vibration. The absence
of SH rotational excitation also follows from the assumption that
the dissociation takes place along the original H-SH bond. By conservation of angular momentum, the initial rotation of the H_2S, $\vec{J}$,
must be partitioned into the internal angular momentum of the SH
fragment, $\vec{J}'$, and the orbital angular momentum, $\vec{L}'$, of the half
collision on the repulsive surface: $\vec{J}=\vec{J}' + \vec{L}'$. The maximum value

Figure 2. Relative
rotational populations
of the SH fragment
produced by dissociation
of room temperature H_2S.

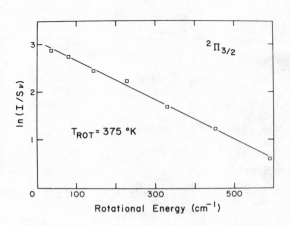

of L' is equal to $\mu v_{max} b$, where $\mu \approx 1$ amu is the H-SH reduced mass, $v_{max} \approx 2 \times 10^6$ cm/sec is the maximum relative recoil velocity, and b is the impact parameter for the half collision. If dissociation takes place along the original S-H bond, then $b \approx 4 \times 10^{-2}$ Å and $L_{max} \approx 1.2$ h. By contrast, the average values of J and J' are much higher; for example, $J \approx 5$ ℏ from the measured distribution for the $^2\Pi_{3/2}$ state. If we make the approximation that $L' << J, J'$, then $J \approx J'$, or the rotation momentum of the SH fragment is simply equal to the original rotational momentum of the H_2S parent. The energy corresponding to the angular momentum of the SH fragment, 320 ± 20 cm^{-1}, is indeed close to the original rotational energy of the H_2S parent, 3 RT/2 = 315 cm^{-1}.

The validity of the kinematic model proposed above may be tested by cooling the H_2S in a supersonic nozzle expansion prior to its dissociation. While a detailed description of the pulsed nozzle apparatus used in our laboratory will be provided elsewhere,[12] it is sufficient here to note that temperatures of a few degrees Kelvin are obtained by seeding the H_2S in helium. The rotational distribution of the SH fragment from dissociation of H_2S cooled to these temperatures is shown in Fig. 3, where the slopes of the stright line fits to the data give rotational temperatures of 170 ± 20 °K for the $^2\Pi_{3/2}$ state and 75 ± 15 °K for the $^2\Pi_{1/2}$ state. These temperatures are well below those obtained for dissociation of room temperature H_2S. As expected from the kinematic model, most of the SH rotational excitation from dissociation of room temperature H_2S is due to the initial rotational angular momentum of the H_2S parent molecule. The rotational distribution of SH($^2\Pi_{3/2}$) produced from cold H_2S is very nearly that which would be expected from an average angular momentum of $L = \mu v b = 1.2$ ℏ. We conclude that all of our observations concerning the energy distribution in the SH fragment are consistent with a simple quasidiatomic picture of the dissociation. Since nearly all of the available energy is channeled into

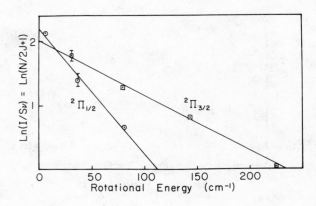

Figure 3. Relative rotational populations of the SH fragment produced by dissociation of H_2S cooled to a few degrees Kelvin.

relative fragment recoil, the scrambling of energy in the H_2S dissociation is far from statistical. Dissociation of CF_3NO, discussed
below, provides a contrasting example.

3.   600–680–nm PHOTODISSOCIATION OF CF_3NO^*

3.1 Introduction

The photochemistry of CF_3NO has come under increasingly detailed investigation during recent years. Early studies[13,14] of the
primary photochemical pathways have been followed by more recent
investigations of the CF_3NO lifetime.[15,16] Roellig and Houston[17]
have detected the NO product following visible photodissociation
of CF_3NO. While these studies have shown that the primary photolysis of CF_3NO yields CF_3 + NO, the current study was initiated to
investigate the dynamics of the dissociation.

3.2 Experimental

The experimental apparatus consisted of two pulsed dye lasers,
each pumped by a separate Nd:YAG laser. The first YAG laser pumped a "red" dye laser which was tuned between 600 and 680 nm to
dissociate CF_3NO via its lowest $n \to \Pi^*$ transition. The second YAG
laser pumped a "blue" dye laser which was tuned between 450 and
500 nm to excite NO to its $A^2\Sigma^+$ state via a two-photon transition.
The two dye lasers were propagated collinearly in opposite directions through a cell containing flowing CF_3NO and were each focused to the same spot by a 5-in. focal length lens. Fluorescence
from the NO A state was detected by a solar-blind photomultiplier.
The delay time between the two laser pulses was adjustable to
within a 5 nsec jitter. Different excited states of CF_3NO were
prepared by varying the frequency of the red laser, while different
internal states of the NO product were probed by varying the frequency of the blue laser. The temporal evolution of the NO product was monitored by scanning the time delay between the red and
blue lasers.

3.3 Results

Vibrational distributions for the NO product were obtained by
comparing the intensities of the $v'' \to v' = 0 \to 0$, $1 \to 0$, $2 \to 0$, and $3 \to 0$
transitions. For 600–nm dissociation the relative populations of
$v''=0,1,2$, and 3 were found to be 1.00 : 0.03 : 0.01 : 0.003, as
shown on a logarithmic scale in Fig. 4a. The dashed line in this
figure presents the populations which would be expected for a
statistical distribution of energy in the CF_3 and NO fragments,
while Fig. 4b plots the surprisal as a function of the fraction of

---

*In collaboration with M. P. Roellig, M. Asscher, and Y. Haas.

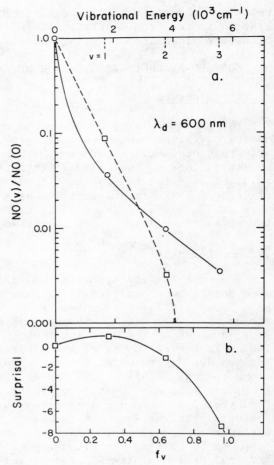

Figure 4. (a) Relative NO
vibrational populations observed following 600 nm photodissociation of CF_3NO. (b)
Vibrational surprisal as a
function of the fraction of
energy which appears in the NO
vibration.

available energy appearing in the NO vibration.

    As the dissociation wavelength was increased from 600 to 660 nm
there was a gradual decrease in the fraction of energy appearing in
the NO vibrational levels, as shown for NO(v"=0,1) in Fig. 5. The
dashed line again plots the variation in relative NO(v"=1) population expected for a statistical distribution.

    The rotational distribution of the NO product is highly excited,
as shown in Fig. 6 for NO($^2\Pi_{1/2}$, v"=0). Figure 6a presents the
0_{12}-branch excitation spectrum of an 8-mtorr sample of pure NO,
while Fig. 6b presents the excitation spectrum of NO produced by
photolysis of 0.1 torr of CF_3NO at 670 nm. While the former spectrum peaks at J=7 1/2, the latter peaks at J>12 1/2. The best fit
of a Boltzmann distribution to the spectrum of Fig. 6b gives a

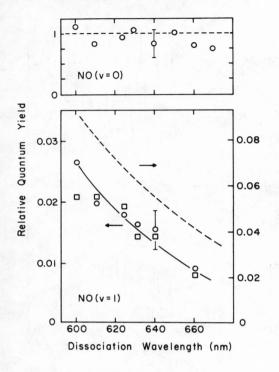

Figure 5. Relative populations of NO (v''=0) and NO (v''=1) as a function of dissociation wavelength.

rotational temperature of 900 ± 300 °K, whereas the predicted rotational temperature for a statistical energy distribution in the fragments is 1025 °K. For different wavelengths and NO vibrational states the measured (and statistical) rotational temperatures were as follows:

$$v''=0, \lambda=600 \text{ nm}: \quad 1100 \pm 200 \ (1400)$$

$$v''=0, \lambda=640 \text{ nm}: \quad 1000 \pm 200 \ (1200)$$

$$v''=0, \lambda=670 \text{ nm}: \quad 900 \pm 300 \ (1025)$$

$$v''=1, \lambda=600 \text{ nm}: \quad 900 \pm 200 \ (1000)$$

The temporal evolution of the NO product was obtained by scanning the delay of the blue laser with respect to the red laser. At a dissociation wavelength of 670 nm the appearance time of the NO(v''=0) product was 18.5 nsec, whereas at 640 and 600 nm the appearance times were 8.8 and 4.3 nsec, respectively. These values are in good agreement with the corresponding CF_3NO lifetimes measured by Spears and Hoffland.[15]

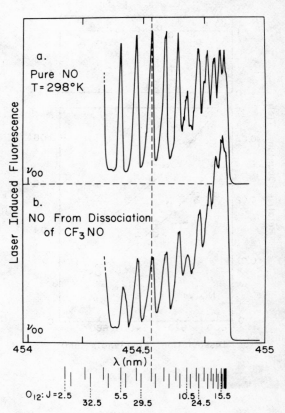

Figure 6. (a) Two-photon
laser induced fluorescence
spectrum of pure NO.
(b) Two-photon laser
induced fluorescence
spectrum of NO produced by
photodissociation of
CF_3NO.

3.4 Discussion

The lifetime of fluorescence from excited CF_3NO is several
orders of magnitude shorter than that calculated from the integrated absorption coefficient. Thus, it appears that a rapid radiationless process governs the lifetime. The facts that the appearance
time of the NO is comparable to the decay time of the excited CF_3NO
and that the yield of NO (Fig. 5) is basically constant as a function
of dissociation wavelength suggest that the rapid radiationless pro-
cell is predissociation to CF_3 + NO. The mechanism of the predissociation could be vibrational predissociation on the S_1 potential
surface, crossing from S_1 to a repulsive surface, or internal conversion to S_0 followed by vibrational predissociation. A detailed
analysis of these possibilities[2] suggests that the variation in
appearance rate of the NO fragment with wavelength is most consistent with the supposition that predissociation follows internal conversion to S_0.

4. CONCLUSION

Both the vibrational distributions shown in Figs. 4 and 5 and the rotational distributions presented in Section 3.3 indicate that energy is distributed nearly statistically among the available modes of the CF_3 and NO dissociation fragments. To be sure, somewhat more population in v''=1 and somewhat less in v''=2,3 is observed than expected, and the measured rotational temperatures seem consistently to be somewhat less than those predicted on a statistical basis alone. Nevertheless, the close adherence to a statistical energy distribution among the photodissociation fragments of CF_3NO stands in striking contrast to the highly specific energy disposal observed in photodissociation of H_2S. One is tempted to conclude that excited H_2S dissociates so rapidly along a repulsive surface that excess energy has little opportunity to sample a large region of phase space. By contrast, it is clear from the predissociative nature of the CF_3NO photolysis that ample time is available for energy scrambling in this case. Future experiments are planned to explore this speculation. The dependence of the polarization of the SH absorption on the polarization of the H_2S photolysis source should indicate whether the H_2S molecule dissociates faster or slower than it rotates. Measurement of the NO rotational distribution as a function of the amount of torsional energy deposited in the excited CF_3NO fragment should indicate whether the NO fragment retains any memory of the state in which it was initially prepared. Experiments such as these, made possible by new applications of lasers to molecular dynamics, should help in understanding how and why energy becomes scrambled in molecules.

ACKNOWLEDGMENTS

This work has been supported by the National Science Foundation under grants CHE-76-21991 and CHE-80-13691. Acknowledgment is also made to the donors of the Petroleum Research Fund, administered by the American Chemical Society, for partial support of this research.

REFERENCES

1. M. Asscher, Y. Haas, M. P. Roellig, and P. L. Houston, Two-photon Excitation as a Monitoring Technique for Photodissociation Dynamics: $CF_3NO \rightarrow CF_3 + NO$ (v,J), J. Chem. Phys. 72:768 (1980).
2. M. P. Roellig, P. L. Houston, M. Asscher, and Y. Haas, CF_3NO Photodissociation Dynamics, J. Chem. Phys. 73:5081 (1980).
3. W. G. Hawkins and P. L. Houston, 193-nm Photodissociation of H_2S: The SH Internal Energy Distribution, J. Chem. Phys. 73:297 (1980).
4. E. Warburg and W. Rump, Über die Photolyse der Lösungen von Schwefelwasserstoff in Hexan und in Wasser, Z. Physik, 58:291 (1929).

5. C. F. Goodeve and N. O. Stein, The Absorption Spectra and the Optical Dissociation of the Hydrides of the Oxygen Groups, Trans. Farad. Soc. 27:393 (1931).

6. G. Porter, The Absorption Spectroscopy of Substances of Short Life, Disc. Farad. Soc. 9:60 (1950).

7. D. A. Ramsay, Absorption Spectra of SH and SD Produced by Flash Photolysis of H_2S and D_2S, J. Chem. Phys. 20:1920 (1952).

8. R. G. Gann and J. Dubrin, Energy Partition in the Photodissociation of a Polyatomic System $H_2S + h\nu (2138 \text{ Å}) \rightarrow H + SH$, J. Chem. Phys. 47:1867 (1967).

9. G. P. Sturm and J. M. White, Photodissociation of Hydrogen Sulfide and Methanethiol. Wavelength Dependence of the Distribution of Energy in the Primary Products, J. Chem. Phys. 50:5035 (1969).

10. A. P. Baronavski and J. R. McDonald, Electronic, Vibrational and Rotational Energy Partitioning of CN Radicals from the Laser Photolysis of ICN at 266 nm, Chem. Phys. Lett. 45:172 (1977).

11. K. Holdy, L. Klotz, and K. R. Wilson, Molecular Dynamics of Photodissociation: Quasidiatomic Model for ICN, J. Chem. Phys. 52:4588 (1970).

12. W. G. Hawkins and P. L. Houston, in preparation.

13. J. Mason, Perfluoroalkyl Compounds of Nitrogen. Part VI. The Photolysis of Trifluoronitrosomethane, J. Chem. Soc. 1963: 4537.

14. J. Jander and R. N. Haszeldine, Addition of Free Radicals to Unsaturated Systems. Part VI. Free-radical Addition to the Nitroso-group, J. Chem. Soc. 1954:696.

15. K. G. Spears and L. Hoffland, Radiationless pathways in CF_3NO, J. Chem. Phys. 66:1755 (1977).

16. K. G. Spears, A Stochastic Description for Vibrational Relaxation in Electronically Excited CF_3NO, Chem. Phys. Lett. 54:139 (1978).

17. M. P. Roellig and P. L. Houston, Photodissociation of CF_3NO: Observation of NO(v=1), Chem. Phys. Lett. 57:75 (1978).

PHOTOFRAGMENT SPECTROSCOPY OF THE NO_2 DISSOCIATION

H. Zacharias, K. Meier and K.H. Welge

Fakultaet fuer Physik
Universitaet Bielefeld
D 48 Bielefeld 1, West Germany

INTRODUCTION

The photodissociation is an important elementary process in
the interaction of light with molecules. In the investigation of
the dynamics of the photodissociation special attention has been
directed to the question how the excess energy is distributed
among the dissociation products and among their different degrees
of freedom. The excess energy is the difference between the dissociating photon energy and the necessary dissociation energy of
the molecule. One of the most successful experimental attempts
for clarifying these questions is molecular beam photofragment
spectroscopy with mass spectrometric time-of-flight fragment detection [1,2]. This method is generally applicable and yields
inter-fragment recoil and momentum distributions. Quantum state
specific measurements are normally not possible because of limitation of the velocity resolution. With this method the dissociation of NO_2 into NO and O has been previously investigated by
Busch and Wilson [1] at $\lambda = 347$ nm. The translational energy resolution achieved was about 200 cm^{-1}. A theoretical interpretation
of their results has been given by Quack and Troe [3] assuming a
statistical distribution of the excess energy.

An alternative way of fragment detection is laser induced
fluorescence [4,5]. Provided of course spectroscopically suitable molecules are involved, this technique allows the complete
analysis, with respect to the energy and the angular momentum,
of the dissociation process on a state selective basis. Ideally,
the parent molecule can also be prepared in a definite quantum
state before the dissociation takes place. Experiments at that
degree of completeness have not yet been carried out with poly-

atomic molecules, however. In this paper we report experimental
results on the dissociation of the NO_2 molecule into NO and O
fragments under collision free conditions in bulk at three dissociation wavelengths, λ = 351 nm, 337 nm, and 308 nm. A complete
measurement and analysis of the internal energy distribution,
i. e., vibrational, rotational, and electronic energy, of the NO
fragment has been carried out, allowing also a conclusive deduc-
tion of the kinetic energy distribution in the center-of-mass
system of both fragments, NO and O, by applying the energy conservation law to the measured internal energy distribution.

EXPERIMENTAL

In the experiments a conventional photolysis apparatus has
been used, which was described in detail previously [6,7]. It
consists of an excimer laser as photolysis light source, a frequency tunable dye laser to excite NO fluorescence, and a high
vacuum photolysis cell equipped with a photomultiplier to monitor
the NO fluorescence. The excimer laser (Lambda Physik, EMG 500)
was operated with XeF, N_2 and XeCl to generate laser radiation
at 351 nm, 337 nm and 308 nm, with energies of 15 mJ, 3 mJ, and
25 mJ, respectively, at the center of the photolysis cell. The

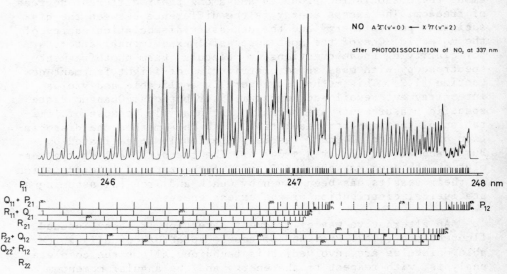

Fig. 1 Part of the excitation spectrum of NO (A $^2\Sigma^+$, v' = 0 $\leftarrow$
X $^2\Pi$, v" = 2) after NO_2 dissociation at λ = 337.1 nm. NO_2 pressure: 130 µbar. Delay between dissociation and probe laser: 5 ns.

pulse duration was about 10 ns, and the bandwidth about 10 cm^{-1}, varying slightly with the specific laser gas. The probe laser was a nitrogen laser pumped dye laser system with oscillator and two amplifier stages. Using different coumarin dyes the spectral range from 440 nm to 545 nm was covered. The frequency of this laser was doubled in different nonlinear crystals (KB5, LFM, and KDP) to excite the NO fragment molecules in the (A $^2\Sigma^+ \leftarrow$ X $^2\Pi$)-bands from the vibrational states v" = 0 to v" = 4 of the electronic ground state to the v' = 0 state of the electronically excited state by a one photon transition. The bandwidth of this laser was less than 1 cm^{-1} in the ultraviolet, sufficient to resolve single rotational transitions. The probe pulse energy varied from 5 μJ to 50 μJ per pulse in the ultraviolet, depending on the specific crystal used for frequency doubling. This probe laser pulse with about 5 ns duration was electronically delayed with respect to the dissocia-tion laser by about 30 to 40 ns. The time jitter of both laser pulses was about 10 ns with respect to each other.

Nitrogen dioxide was taken from a cylinder and purified by the usual methode of bubbling O_2 through the liquid at 0 °C. This re-duced the NO contamination to less than 0.1 %. During the experi-ment the sample was kept at - 48 °C, where the NO_2 pressure is about 4 mbar. The gas was introduced into the photolysis cell at a con-stant flow to avoid accumulation of photolysis products, particular-ly NO. Under normal conditions the pressure of nitrogen dioxide was about 65 to 130 μbar, which was low enough to preclude relaxa-tion in the initially produced rotational and vibrational popula-tion distribution of the NO fragment.

Fluorescence was observed by a solar blind photomultiplier (EMR, 541 - F) through a reflection interference filter (Schott, UV - R - 250) with a peak transmission of 0.9 at 250 nm and a band-width (FWHM) of 40 nm. The combination of both suppressed effec-tively the scattered light from the high power photolysis laser.

RESULTS

NO excitation spectra

Nitric oxide molecules were excited from the (X $^2\Pi$, v" = 0 - 4) ground states always to the (A $^2\Sigma^+$, v' = 0) state; thus, data which are to be compared do not need to be corrected for Franck-Condon factors for the fluorescence, filter transmission, detector sensitivity, etc..

Figure 1 shows the excitation spectrum of the (A $^2\Sigma^+$, v' = 0 ← X $^2\Pi$, v" = 2) transition of nitric oxide after photodissociation of NO_2 at 337 nm. A line identification is given at the bottom of the figure. A contribution to the observed spectrum from the natural

NO contamination can be neglected, because at room temperature the relative population of v" = 2 state is about 10^{-8}. This was confirmed since no signal was observable when the dissociation laser was blocked. The same was true for v" = 1,3, and 4. For the lowest vibrational state v" = 0, however, a contribution was observed for J" < 30 1/2 for the comparatively low power nitrogen dissociation laser (λ = 337 nm) and for J" < 20 1/2 for the other two dissociation wavelength (λ = 351 nm and λ = 308 nm). The probable level population in these levels are thus only indicated by broken lines.

Rotational distribution

The rotational distributions obtained are shown in Figures 2, 3 and 4 for the dissociation wavelengths λ = 351 nm, λ = 337 nm, and λ = 308 nm, respectively. The relative population density $N_{v''}(J'')$ is plotted versus J", separately for the two electronic states, $^2\Pi_{1/2}$ and $^2\Pi_{2/3}$, values of which are indicated by the open and filled symbols, respectively. The rotational branches are also distinguished by different symbols. As one can see the populations derived from different rotational branches deliver the same and consistent results. This consistency enhances the degree of confidence in the measurement since corresponding lines of different branches belong to various part of the spectra. Despite the small

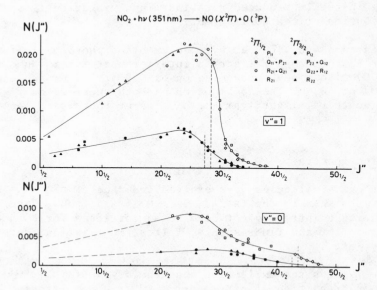

Fig. 2 NO(X $^2\Pi$) rotational population distribution after NO$_2$ dissociation at λ = 351 nm.

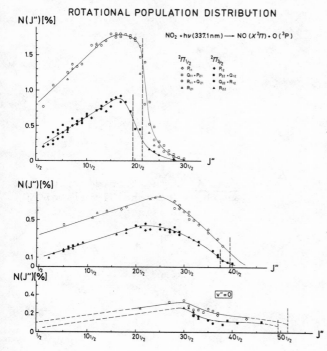

Fig. 3 NO(X $^2\Pi$) rotational population distribution after dissociation of NO_2 at λ = 337.1 nm. Open symbols: $^2\Pi_{1/2}$, filled symbols: $^2\Pi_{3/2}$. The denotion of the different rotational branches is the same as in Fig. 2 [7].

splitting the relative population of the two Λ-components e and f of each rotational level in the $^2\Pi_{1/2}$ and $^2\Pi_{3/2}$ states can be readily determined, since the lines from the f component belong to the branches (Q_{11} + P_{21}) and R_{21}, whereas the lines from the e component to the branches P_{11} and (R_{11} + Q_{21}) in the $^2\Pi_{1/2}$ state. For the $^2\Pi_{3/2}$ state the corresponding branches are (Q_{12} + P_{22}) and R_{22} for the f, and P_{12} and (R_{12} + Q_{22}) for the e component. The above mentioned equality of population derived from different rotational branches led to the conclusion that both Λ-components are equally populated.

The vertical dashed lines in the figures indicate the maximum rotational level which could be populated when dissociation of only the lowest rotational level in the NO_2 parent molecules is considered, i. e. dissociation of NO_2 at 0 K. According to this assumption, in the λ = 308 nm dissociation of NO_2 (Fig. 4) there should be no population in v" = 4. Any population in this state, and correspondingly in rotational levels with quantum numbers higher than indicated by the dashed lines in the other states, reflects therefore the dissociation of rotationally excited NO_2 molecules.

From the measured rotational distribution one can derive the relative state population ln $[N_{J''}/(2J'' + 1)]$ in each vibrational

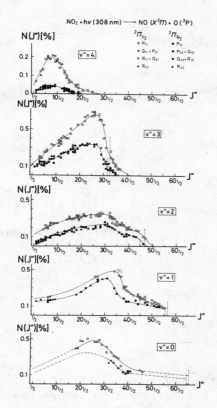

Fig. 4 Population distribution
in the rotational levels of NO
after photodissociation of NO₂
at λ = 308 nm.

level. Plotting this function versus the rotational energy E_{rot} (J")
allows a decision whether or to what degree the rotational distributions can be represented by an equilibrium temperature. For E_{rot} (J")
we used the general expression of Bennett [8] with rotational constants of Hallin et al. [9]. Typical results for λ = 351 nm and
λ = 308 nm dissociation are shown in Figures 5 and 6. We found that
at all three dissociation wavelengths and in all vibrational states
the rotational distribution deviates much from an equilibrium distribution, as indicated by the non-linearity of the dependence of
$\ln[N_{J"}/(2J" + 1)]$ on the rotational energy. This will be discussed
further below.

Vibrational distribution

The relative vibrational population distributions are obtained
by summing over the rotational distributions. Lines from different
bands have to be calibrated against each other, and laser powers
and Franck-Condon factors have to be taken into account. In Table I
are given the average band frequencies and Franck-Condon factors
used in this work. The rotational distributions were already nor-

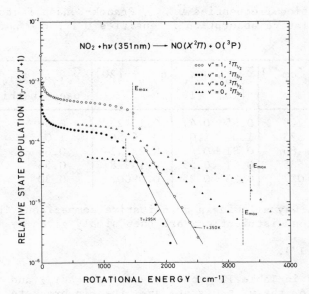

<u>Fig. 5</u> Relative state population $N_{J''}/(2 J'' + 1)$ versus the rotational energy for photodissociation of NO_2 at $\lambda = 351$ nm.

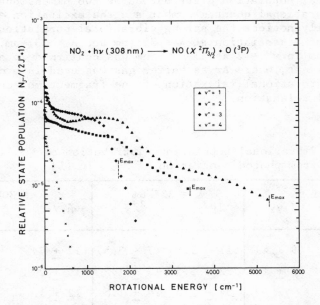

<u>Fig. 6</u> Relative state population $N_{J''}/(2 J'' + 1)$ in $NO(X\ ^2\Pi_{3/2})$ versus the rotational energy. Photodissociation of NO_2 at $\lambda = 308$ nm.

Table I Transition frequencies $\bar{\nu}_{v'v''}$, Franck-Condon factors $q_{v'v''}$, and relative absorption intensities $\bar{\nu}_{v'v''} q_{v'v''}$ of the NO γ-bands

$v' - v''$	$\bar{\nu}_{v'v''}[cm^{-1}]$	$\bar{\nu}_{v'v''} q_{v'v''}$ [18] [19]	$q_{v'v''}$ [20]	$\bar{\nu}_{v'v''} q_{v'v''}$ used in this work
0 - 0	44 248	0.65 0.47	0.16725	0.66
0 - 1	42 373	1.0 1.0	0.26456	1.0
0 - 2	40 486	0.81 0.83	0.23741	0.86
0 - 3	38 685	0.51 0.55	0.15978	0.56
0 - 4	37 037	0.23 0.35	0.09007	0.30

malized in this way allowing a quantitative comparison. The vibrational population distributions are then simply given by

$$N(v'') = \sum_{J''} N_{v''}(J'').$$

They are listed in Table II separately for the $^2\Pi_{1/2}$ and $^2\Pi_{3/2}$ state. Values for the $v'' = 0$ state are given in brackets, because due to the normal NO contamination only a small part of the population distribution could have been examined. In Figure 7 is shown this vibrational population distribution as two dimensional plot versus the vibrational quantum number and the excitation energy of NO_2. One readily notices the strong vibrational population inversion in the cases of dissociation at $\lambda = 351$ nm and $\lambda = 337$ nm. The population minimum in $v'' = 2$ at $\lambda = 308$ nm was confirmed in an independent experiment, where Ar as buffer gas was used in order to obtain only a rotational relaxation of the fragment molecules, but no vibrational relaxation.

Table II Vibrational population distribution of NO $(X\ ^2\Pi)$ from photodissociation of NO_2, in [%]

v''	$\lambda = 351$ nm $^2\Pi_{1/2}$	$^2\Pi_{3/2}$	$\lambda = 337$ nm $^2\Pi_{1/2}$	$^2\Pi_{3/2}$	$\lambda = 308$ nm $^2\Pi_{1/2}$	$^2\Pi_{3/2}$
0	(28 ± 10)	(9 ± 4)	(11 ± 5.5)	(7 ± 3.5)	(17 ± 7)	(12 ± 6)
1	48 ± 5	15 ± 1.5	22 ± 2.5	12 ± 1.5	16 ± 2	12 ± 1
2	-	-	34 ± 3.5	14 ± 1.5	12 ± 1	7 ± 0.7
3	-	-	-	-	14 ± 1.5	7.5 ± 0.7
4	-	-	-	-	2 ± 0.2	0.5 ± 0.1

Electronic energy distribution

As one easily notices the relative population in the $^2\Pi_{1/2}$ and $^2\Pi_{3/2}$ electronic substates were found to be different (see Fig. 2, 3 and 4). At all dissociation wavelengths the greatest difference appears in the highest vibrational level, becoming smaller as the vibrational quantum number decreases. In Fig. 8 is plotted the population ratio $N(\Omega'' = 3/2)/N(\Omega'' = 1/2)$ versus $(v''_{max} - v'')$ for the three dissociation wavelengths used. This quantity $(v''_{max} - v'')$ is proportional to the average excess energy of a particular vibrational state. This ratio shows a more or less monotonical increase as the excess energy increases for all three dissociation wavelengths.

Internal energy distribution

Knowing the vibrational and rotational population distributions, $N_{v''}(J'')$, one can easily convert them to the internal energy distribution of the NO fragment, $N[E_{int}(NO)]$:

$$E_{int}(NO) = E_{v''} + E_{J''} + E_{\Omega''}.$$

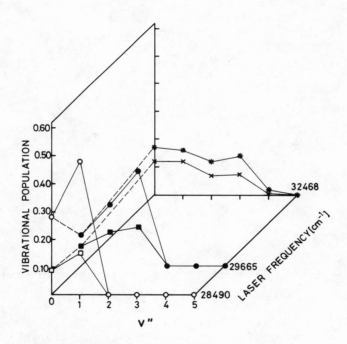

Fig. 7 Plot of the vibrational population of NO versus the vibrational quantum number and the frequency of the NO_2 photodissociation.

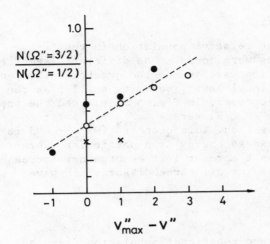

<u>Fig. 8</u> Plot of the population ratio $N(\Omega'' = 3/2)/N(\Omega'' = 1/2)$ versus $(v''_{max} - v'')$, which is proportional to the vibrational excess energy.

DISTRIBUTION OF INTERNAL ENERGY

$NO_2 + h\nu$ (308 nm) $\longrightarrow$ NO ($X^2\Pi_{1/2}$) + $O(^3P)$

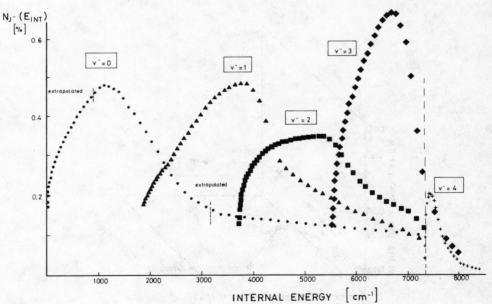

<u>Fig. 9</u> Population distribution of the NO($X^2\Pi_{1/2}$) fragments versus the internal energy of the specific rotational-vibrational levels. Photodissociation of NO_2 at λ = 308 nm.

For this derivation we used the smoothed curves of Figures 2, 3 and 4. In Fig. 9 is shown the distribution obtained for NO fragments from the dissociation of NO_2 at λ = 308 nm. For clarity only the $^2\Pi_{1/2}$ state is shown. The dashed line represents again the maximum possible excess energy of 7336 cm^{-1} for dissociation at λ = 308 nm. The population in a vibrational level at a certain internal energy decreases strongly when the next higher vibrational state becomes energetically accessible. This behaviour is observed at all three dissociation wavelengths for all vibrational states.

Total energy distribution

The complete energetics of the dissociation is given by

$$E(\nu, \hat{e}) + E_{int}(NO_2) - D_0^0$$

$$= E_{int}(NO) + E_{int}(O) + E_{kin}(NO_i, O_j; \delta, \varphi)$$

$E(\nu, \hat{e})$ denotes the energy and polarization of the dissociating light, D_0^0 is the dissociation energy, $E_{int}(NO_2)$ the internal energy of the NO_2 parent molecule at room temperature, $E_{int}(NO)$ the internal energy of the NO fragment in a specific (Ω'', v'', J'')-state, $E_{int}(O)$ the internal electronic energy of the oxygen atom in one of the states $^3P_0$, $^3P_1$, and $^3P_2$, and finally, E_{kin} is the total center-of-mass kinetic energy of the both fragments in one of their internal states, flying apart under an angle (δ, φ) with respect to the polarization direction $\hat{e}$ of the dissociation laser.

In order to derive the kinetic energy distribution $N(E_{kin})$ from the measured internal energy distribution $N[E_{int}(NO)]$ one has to take into account the initial internal energy of the NO_2 and the internal energy of the oxygen atom, due to the fine structure splitting of the ^3P state. Assuming a population of the ^3P_i states according to their degeneracy factor the average internal oxygen atom energy would be $E_{int}(O)$ = 78 cm^{-1}. The internal energy of NO_2 consists of vibrational and rotational energy. Only 3 % of the NO_2 molecules at room temperature are in excited vibrational states, which gives $\overline{E}_{v''}(NO_2)$ = 24.25 cm^{-1}. The contribution to the internal energy of NO_2 due to the rotation of the molecule can be considered, according to Wilson [1], by

$$N(E_{rot}) = c \cdot E_{rot}^{1/2} \exp\{- E_{rot}/kT\}$$

This gives an average value of $\overline{E}_{rot}(NO_2)$ = 309 cm^{-1} at room temperature. The total average internal energy of NO_2 is thus at room temperature $\overline{E}_{int}(NO_2)$ = 333 cm^{-1}. $\overline{E}_{int}(O)$ and $\overline{E}_{int}(NO_2)$ are both rather small compared with the excess energies used in this work, so that they may be neglected to a first approximation. The kinetic energy distribution can now be obtained by transforming the energy

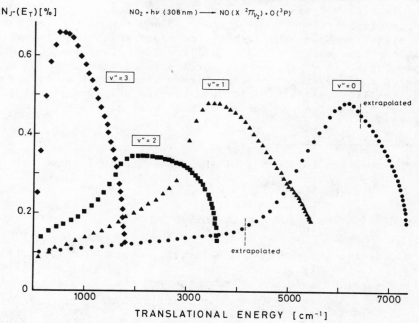

Fig. 10 Kinetic energy distribution of both fragments NO(X $^2\Pi_{1/2}$ and O($^3P_2$) in the center-of-mass system for photodissociation of NO₂ at λ = 308 nm.

scales of the internal energy distributions according to E_{kin} = E_{exc} - E_{int} (NO), where E_{exc} is the excess energy of the dissociating photon over the dissociation energy of NO₂. As an example is shown in Fig. 10 the distribution of the total translational energy of NO($^2\Pi_{1/2}$) + O($^3P_2$) for dissociation at λ = 308 nm, the corresponding distribution to the internal energy distribution shown in the preceding Fig. 9. It should be noted that due to the quantized nature of the internal energy states of NO the kinetic energy distribution appears also to be "quantized".

One can now derive average distributions of the available energy over the internal and translational degrees of freedom for all three dissociation wavelengths. The available energy is given by E_{avl} = E_{exc} + E_{int} (NO₂). The average vibrational, rotational and electronic energy of NO can be determined by

$$\overline{E}_{vib}(NO) = \sum_{v''} N_{v''} E_{vib}(v'')$$

$$\overline{E}_{rot}(NO) = \sum_{v'',J''} N_{v''}(J'') \, E_{rot}(J'')$$

$$\overline{E}_{el}(NO) = \sum_{\Omega'',v'',J''} N_{v'',\Omega''}(J'') \cdot E_{el}(\Omega'')$$

where $\Omega'' = 1/2$ and $3/2$, and $E_{el}(\Omega'' = 1/2) = 0$ and $E_{el}(\Omega'' = 3/2) = 124$ cm^{-1}. The values obtained are summarized in Table III.

DISCUSSION

The energy and angular distribution of the NO$_2$ dissociation has been studied previously at $\lambda = 347$ nm by Busch and Wilson [1]. The differences between their results and the results obtained in this work has been discussed in detail previously for the 337 nm dissociation [7]. They found nearly equally populated vibrational states, whereas the results presented here show that dissociation at $\lambda = 351$ nm and $\lambda = 337$ nm produces preferentially slow fragments, and since such fragments can escape detection relatively easy in the time-of-flight technique, the yield of such fragments may have been underestimated by them. In addition their energy resolution was about 200 cm^{-1}, too small for a state selective observation. This results in a grouping together all fragments within a given translational energy band, which comprises a more or less large number of rotational energy states.

Table III   Distribution of the available energy over the internal and translational degrees of freedom

E_{avl} [cm^{-1}]	$\lambda = 351$ nm 3691		$\lambda = 337$ nm 4866		$\lambda = 308$ nm 7669	
	[cm^{-1}]	[%]	[cm^{-1}]	[%]	[cm^{-1}]	[%]
$\overline{E}_{int}(NO)$	2026	55	3281	67	4112	54
$\overline{E}_{vib}(NO)$	1216	33	2477	51	2745	35
$\overline{E}_{rot}(NO)$	779	21	762	16	1419	19
$\overline{E}_{el}(NO)$	31	1	42	1	48	0.6
$\overline{E}_{int}(O)$	78	2	78	2	78	1
$\overline{E}_{kin}$	1587	43	1507	31	3479	45
$\overline{E}_{kin}(O)$	1035	28	983	20	2268	29
$\overline{E}_{kin}(NO)$	552	15	524	11	1211	16

The rotational population distributions presented in this work
deviate very much from equilibrium distributions. This is indicated
in Figures 5 and 6 by the non-linearity of the dependence of
$\ln [N_{v''}(J'')/(2J'' + 1)]$ on the rotational energy in all vibrational
states at the three dissociation wavelength used in this work.
Taking a closer look to the figures one can readily distinguish
three rotational energy regions. At very low rotational energy
levels a very steep population decrease is observed, followed by
a flat and nearly constant dependence up to about 2000 cm^{-1} rotational energy. Thereafter a third region can be distinguished which
shows a monotonical decrease with nearly constant slope up to maximum rotational energy which can be reached in a particular vibrational state. One therefore may associate a "cold" rotational distribution at low J" values, a "hot" distribution up to quantum
numbers of about 30 1/2 to 35 1/2, and a "moderately hot" rotational
distribution at high rotational quantum numbers. For illustration
the decrease in the first few rotational levels may be approximated
by a straight line, the slope of which would correspond to a "rotational temperature" of about 120 K or below, in all cases. Doing
the same for the other two regions one would obtain for the "hot"
distributions "temperatures" between 4000 K and 6000 K or above,
and between 450 K and 2000 K for the region of high rotational
quantum number. It should be appreciated at this point that the
term "temperature" is inappropriate in describing these rotational
population distributions which certainly are non-equilibrium distributions. They are furthermore in disequilibrium with the other
two degrees of freedom, vibration and translation. The results
found suggest that two or even three basically different decay
channels are involved in the dissociation process. A summation
over the appropriate levels shows that the "cold" channel partic-
ipates in the dissociation to about 3 % to 10 %, the "hot" channel
by about 60 to 80 %, and the "moderate" takes the remaining fraction. A detailed list is given in Table IV. There appears a striking similarity between the onset of the "moderate temperature"
distribution and the internal energy distribution. It was found
that the population in a vibrational level decreases strongly
when the next higher vibrational state became accessible (see Fig.
9). The vibrational spacing of NO is about 1875 cm^{-1}, which is
just the energy range (normally between 1800 and 2000 cm^{-1})
where the change from a "hot" into a "moderate temperature" distribution is observed. It appears thus as if both belong to the
same decay channel, a further population of which is strongly
inhibited by a preferentially population of the next higher vibrational state with the same total energy. Similar arguments for
incorporating the "cold" channel are not found.

The dynamics of the dissociation process may be described
qualitatively by an excitation of the NO_2 molecule from its $\tilde{X}\ ^2A_1$
electronic ground state to the $\tilde{A}\ ^2B_2$ and $\tilde{B}\ ^2B_1$ excited states.

Table IV Average rotational quantum numbers <J''>, average impact parameter <b>, and relative population in the three distinguished regions with "cold", "hot", and "moderate" rotational distributions

λ [nm]		<J''>			<b> [Å]			Relative populations [%]		
		"cold"	"hot"	"moderate"	"cold"	"hot"	"moderate"	"cold"	"hot"	"moderate"
351:										
v'' = 1,	$^2\Pi_{1/2}$	4.4	18.3	-	0.16	0.79	-	7	93	-
,	$^2\Pi_{3/2}$	5.5	17.4	-	0.21	0.79	-	8	92	-
337:										
v'' = 2,	$^2\Pi_{1/2}$	3.5	14.4	-	0.18	0.89	-	9	91	-
,	$^2\Pi_{3/2}$	3.6	12.7	-	0.19	0.84	-	3	97	-
v'' = 1,	$^2\Pi_{1/2}$	5.0	16.5	31.5	0.14	0.46	1.3	10	58	32
,	$^2\Pi_{3/2}$	5.4	16.7	30.9	0.15	0.48	1.4	4	65	31
308:										
v'' = 3,	$^2\Pi_{1/2}$	5.3	18.1	30.1	0.18	0.68	2.3	5	78	17
,	$^2\Pi_{3/2}$	4.9	17.8	(29.6)	0.17	0.69	(2.9)	4	84	(12)
v'' = 2,	$^2\Pi_{1/2}$	11.0	20.8	39.5	0.25	0.50	1.6	9	70	21
,	$^2\Pi_{3/2}$	5.4	19.4	36.8	0.13	0.47	1.4	1	69	30
v'' = 1,	$^2\Pi_{1/2}$	(5.5)	23.4	44.0	(0.10)	0.45	1.2	(8)	65	27
,	$^2\Pi_{3/2}$	6.7	22.8	42.8	0.13	0.44	1.2	10	62	28

Since the excitation process obeys the Franck-Condon law it leads
to highly excited vibrational levels, especially in the bending
mode, because the equilibrium bond length and angle of the ground
state (1.1934 Å and 134.1° [10]) are substantially different from
the excited states, 1.31 Å and 111°[11] for the $\tilde{A}\ ^2B_2$, and 1.23 Å
[12] and 180° [13] for the $\tilde{B}\ ^2B_1$ state. The high rotational excitation of NO can thus result from the decay of a highly excited
bending vibrational mode of NO_2. The total angular momentum is
then conserved by a correspondingly large orbital angular momentum
[14] of both fragments

$$L \equiv \mu vb = \hbar \sqrt{J''(J'' + 1)}$$

with

$$(J'' - J^P) \leq L \leq (J'' + J^P)$$

where J" is the rotational quantum number of NO, J^P the rotational
quantum number of the NO_2 parent, μ the reduced mass, v the relative
velocity of O and NO, and b the impact parameter. Values of the
average impact parameter <b>, obtained by neglecting J^P and averaging over the rotational quantum number,

$$<J''> = \{\sum_{J''v''} N_{v''}(J'')J''\}/\{\sum_{J''v''} N_{v''}(J'')\},$$

have been calculated for different vibrational states at the three
dissociation wavelength. The results are given in Table IV for the
"cold", the "hot" and the "moderate" channel. The values have been
derived for the O atom in its lowest state $^3P_2$ and with neglecting
the rotational excitation of the NO_2 parent molecule. It is found
that <J"> and <b> are very similar in each of the three distinguished regions in the rotational distributions in all vibrational
states at the three dissociation wavelengths. This result also is
supporting a two (or even three) channel decay mode of NO_2.

The vibrational excitation of the NO fragment can be qualitatively understood by considering the change of the equilibrium
N - O bond length from the ground to the excited states of NO_2,
and the equilibrium bond length of 1.15077 Å [15] in the free
radical, according to a model proposed by Simons [16]. The strong
population inversion at λ = 351 nm and λ = 337 nm dissociation
wavelength, and the puzzeling decrease in population for v" = 2
at λ = 308 nm, which was confirmed using Ar as buffer gas for only
rotational relaxation, evidently show that the NO_2 dissociation
can not be described by a statistical decay model [3]. This conclusion is also supported by the rotational population distribution.
A dynamical calculation of the NO_2 decay, as it was performed recently by Shapiro [17] for the H_2O dissociation, would be welcome.

ACKNOWLEDGEMENT

The authors are pleased to thank C. H. Dugan for helpful discussions and a critical reading of the manuscript.

REFERENCES

1. G. E. Busch and K. R. Wilson, Triatomic photofragment spectra. I. Energy partitioning in NO_2 photodissociation, J. Chem. Phys. 56:3626 (1972); Triatomic photofragment spectra. II. Angular distributions from NO_2 photodissociation, J. Chem. Phys. 56:3638 (1972)

2. M. Dzvonik, S. Yang, and R. Bersohn, Photodissociation of molecular beams of aryl halides, J. Chem. Phys. 61:4408 (1974) M. J. Coggiola, P. A. Schulz, Y. T. Lee, and Y. R. Shen, Molecular beam study of multiphoton dissociation of SF_6, Phys. Rev. Lett. 38:17 (1977)

3. M. Quack and J. Troe, Unimolecular Processes IV: Product state distribution after dissociation, Ber. Buns. Ges. Phys. Chem. 79:469 (1975)

4. M. J. Sabety - Dzvonik and R. J. Cody, The internal state distribution of CN free radicals produced in the photodissociation of ICN, J. Chem. Phys. 66:125 (1977) A. P. Baronavski and J. R. McDonald, Electronic, vibrational and rotational energy partitioning of CN radicals from the laser photolysis of ICN at 266 nm, Chem. Phys. Lett. 45:172 (1977)

5. J. Danon, S. V. Filseth, D. Feldmann, H. Zacharias, C. H. Dugan, and K. H. Welge, Laser induced fluorescence of CH_2 $(\tilde{a}\ ^1A_1)$ produced in the photodissociation of ketene at 337 nm. The CH_2 $(\tilde{a}\ ^1A_1 - \tilde{X}\ ^3B_1)$ energy separation, Chem. Phys. 29:345 (1978) W. G. Hawkins and P. L. Houston, 193 nm photodissociation of H_2S: The SH internal energy distribution, J. Chem. Phys. 73:297 (1980)

6. H. Zacharias, R. Schmiedl, R. Böttner, M. Geilhaupt, U. Meier, and K. H. Welge, Spectroscopy of photodissociation products, in: "Laser Spectroscopy IV", H. Walther and K. W. Rothe, eds., Springer Series in Optical Sciences, Vol. 21, Springer, Berlin, Heidelberg, New York (1979)

7. H. Zacharias, M. Geilhaupt, K. Meier, and K. H. Welge, Laser photofragment spectroscopy of the NO_2 dissociation at 337 nm. A nonstatistical decay process, J. Chem. Phys. 74:000 (1981)

8. R. J. M. Bennett, Hönl-London factors for doublet transitions in diatomic molecules, Mon. Nat. R. astr. Soc. 147:35 (1970)

9. K. E. J. Hallin, J. W. C. Johns, D. W. Lepard, A. W. Mantz, D. L. Wall, and K. Narahari Rao, The infrared emission spectrum of $^{14}N^{16}O$ in the overtone region and determination of Dunham coefficients for the ground state, J. Mol. Spec-

trosc. 74:26 (1979)

10. G. Herzberg, "Molecular Spectra and Molecular Structure",
    Vol. III, p. 602, Van Nostrand Reinhold, New York (1966)

11. C. G. Stevens and R. N. Zare, Rotational analysis of the
    5933 Å band of NO_2 J. Mol. Spectrosc. 56:167 (1975)

12. J. L. Hardwick and J. C. D. Brand, The $^2B_1 \leftarrow {}^2A_1$ system of
    nitrogen dioxide, Chem. Phys. Lett. 21:458 (1973)

13. G. P. Gillispie, A. U. Kahn, A. C. Wahl, R. P. Hosteny, and
    M. Krauss, The electronic structure of nitrogen dioxide.
    I. Multiconfiguration self-consistent-field calculation of
    the low-lying electronic states, J. Chem. Phys. 63:3425
    (1975)

14. T. Carrington, Angular momentum distribution and emission
    spectrum of $OH(^2\Sigma^+)$ in the photodissociation of H_2O, J.
    Chem. Phys. 41:2012 (1964)

15. K. P. Huber and G. Herzberg, "Molecular Spectra and Molecular
    Structure", Vol. IV, p. 476, Van Nostrand Reinhold, New
    York (1979)

16. R. C. Mitchell and J. P. Simons, Energy distribution among
    the primary products of photo-dissociation, Disc. Faraday
    Soc. 44:208 (1967)

17. E. Segev and M. Shapiro, Resonances in the H_2O photodissocia-
    tion: A converged three-dimensional quantum mechanical
    study, J. Chem. Phys. 73:2001 (1980)
    M. Shapiro, Dynamics of photodissociation processes, Third
    Minerva Symposium in Chemistry, Spitzingsee, (1980)

18. E. T. Antropov, V. N. Kolesnikov, L. Ya. Ostrovskaya, and
    N. N. Sobolev, Dipole moment of the γ-band system of NO as
    a function of the internuclear distance of transition, Opt.
    Spectrosc. (USSR) 22:109 (1967)

19. H. M. Poland and H. P. Broida, Fluorescence of the γ, ε and
    δ systems of nitric oxide; Polarization and use of cal-
    culated intensities for spectrometer calibration, J. Quant.
    Spectrosc. Radiat. Transfer 11:1863 (1971)

20. R. W. Nicholls, Franck-Condon factors to high vibrational
    quantum numbers IV: NO band systems, J. Res. Nat. Bur.
    Stand. Sect. A 68:535 (1964).

PREPARATION, LASER SPECTROSCOPY AND PREDISSOCIATION

OF ALKALI DIMERS IN SUPERSONIC NOZZLE BEAMS

Friedrich Engelke

Fakultät für Physik der Universität Bielefeld

D-48oo Bielefeld, FRG

INTRODUCTION

Molecular spectroscopy is one of the oldest fields in physics and physical chemistry. With the discovery of molecular beams and lasers it has assumed a new importance. The development of coherent, intense, highly monochromatic light sources from the IR to the UV has promised to put at the disposal of "molecular beamists" new and powerful tools that will enable the performance of highly refined and sophisticated experiments not previously feasible.

The brightest young theoretical brains used to scorn molecular spectroscopy and reaction study as little more than trial-and-error tools of physical chemistry. One of the worst things the physicist could say about some problem was that it was "mere chemistry", meaning that it was a complicated mess and that, even if you solved it, you would not learn anything. Lately, however, the study of molecular spectroscopy and reaction dynamics has acquired a lot of glamour for physical chemists (or chemical physicists), partly because of an infusion of new ideas and techniques from physics.

Among the borrowings are lasers. These intense light sources can be tuned to the precise frequencies of specific types of molecules to make them vibrate, rotate, emit light, change electric charge - in short to make them jump to the researchers' bidding. Chemists employ lasers to probe the structure of atoms and molecules, to synthesize new compounds, to detect impurities, and to identify infinitesimal traces of substances.

Consequently, for a spectroscopist, it is worthwhile to re-examine spectroscopic studies in the light of what molecular beams

125

and lasers can do to improve these studies. Under those felicitous
circumstances where both can be utilized effectively, the progeny
of this combination is a delight in that it provides some of the
most detailed information on spectroscopic knowledge yet available.

In what follows, we describe, after a short general description,
how molecular beam and spectroscopic techniques allow detailed studies
in the field of heteronuclear and homonuclear alkali dimers. The scope
of these new methods is now very wide and includes the use of laser
induced fluorescence (LIF) as a detector for the measurement of
complex spectra, the distribution of internal states of the products
of a supersonic expansion, predissociation dynamics etc. There have
been significant new observations, and the great impact of laser
spectroscopy upon beam experiments in the field of high resolution
molecular spectroscopy is illustrated by four selected examples.

We start with the presentation of a molecular beam-laser setup,
which has already been described in a recent publication.[1] In brief,
a horizontal alkali metal beam crosses a well collimated laser beam,
as shown in Fig. 1. The metal dimers are produced in different types
of metal ovens. Separate temperature controls allow us to maintain
defined pressures of the alkali metal mixture, which escapes through
a o.3 to o.5 mm nozzle and a skimmer or a set of slits into the excitation chamber. This is a separately pumped vacuum chamber which
contains carefully constructed light baffles to avoid scattered laser
light, fluorescence observation flanges and a Langmuir-Taylor detector (LTD). When the metal beam is on, the pressure in the excitation chamber is typically less than 10^{-6} Torr; therefore the setup is
operated as a true molecular beam, which permits high resolution

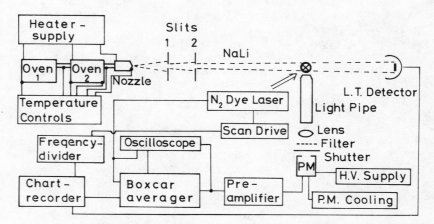

Fig. 1.   Schematic diagram of the molecular beam - laser setup.

spectroscopic studies. The fluorescence is observed by eye to originate from the volume intersected by the alkali beam and the laser.
Either a scanning spectrometer or a fast photon detection system
views the fluorescence through one (or two) light pipe(s) at right
angles to the beams. The excitation light sources we use include
commercial fixed frequency lasers or different (c.w. and pulsed)
dye lasers.

If we prepare a molecule in a few known internal states (v",J"),
detection of the excited states is straightforward and within present
capabilities. Such experiments represent an important advance towards
the determination of excited state properties.

First, we report LIF studies on NaK formed in a supersonic nozzle
beam. By combining this technique with a high resolution grating spectrometer we directly observe the $a^3\Sigma^+$ state, the lowest triplet state
of NaK. In addition, very accurate dissociation energies are obtained.

Next, successful experiments on the heteronuclear alkali dimer
NaLi in a molecular beam were carried out in our laboratory. We have
excited NaLi by means of a nitrogen pumped dye laser, as shown in
Fig. 1. The results are presented in Section B, together with com-
plementary experiments using a new decive: the injection heat pipe
(IHP).

Furthermore, under conditions described above the (molecular)
absorption line widths are sub-Doppler close to the natural line
width (i.e. a few MHz). Therefore, we are able to carry out high
resolution molecular spectroscopy using a single mode dye laser for
excitation. An example is presented in Section C illustrating the
LIF recorded for the Rb_2 $B^1\Pi_u$ - $X^1\Sigma_g^+$ band system.

Last, the collision-free flow of atoms and molecules in a supersonic beam may be used to study intramolecular energy transfer processes. Once again the rubidium dimer serves as an example in Section
D. Here the predissociation dynamics in the $C^1\Pi_u$ and $D^1\Pi_u$ states are
examined. From the knowledge of both the resulting molecular fluorescence and the internal state of the Rb atom populated we determine
the molecular states responsible for the predissociation.

These selected studies serve just as an example of the far extending conclusions that can be derived from a laser induced fluorescence spectrum obtained under collision free conditions in beams.

A. THE NaK $a^3\Sigma^+$ STATE

By combining the LIF technique with a high resolution grating
spectrometer we have directly observed transitions between the $d^3\Pi_i$
and the $a^3\Sigma^+$ state of NaK. The NaK produced in a supersonic nozzle

beam[2] or in a heat pipe oven,[3] is irradiated by the 4765 Å line of
the argon ion laser, and the resulting fluorescence is spectroscopically analyzed. The emission consists of molecular fluorescence from
the NaK D-X system as well as fluorescence from the NaK d-a system.
The latter transition is caused by strong perturbations in the $D^1\Pi$
state. From these spectra we evaluate the molecular constants of the
$a^3\Sigma^+$ state of NaK and estimate the parameters involved in the perturbations of the D state.[3] We combine these results with those reported earlier from LIF studies on NaK formed in a supersonic nozzle
beam[2] and compare them with the results of spin exchange scattering
experiments on the Na-K system[4,5] and with theoretical calculations
using pseudopotential[6,7] or ab initio[7] methods. In addition, very
accurate dissociation energies are obtained.

Al Experimental

Details of the apparatus used for carrying out LIF experiments
in the molecular beam are described in Ref. 2. The high resolution
measurements performed in the heat pipe oven require some description.
A mixture of about 7o% potassium and 3o% sodium was placed in a stainless steel heat pipe oven and heated to a temperature of approximately 85o°K. The calculated concentrations of gaseous species in the
pipe were 1.6% K_2, 1.6% NaK, o.8% Na_2, 66% K and 3o% Na. Since all of
the alkali metals, except Lithium, are completely miscible with each
other, it is possible to optimize the heteronuclear dimer intensity
by varying the composition of the metal charge, and the above composition corresponds to the optimum. In our case the pressure of the
buffer gas was between 4 and 25 Torr, and the vapor zone was about
6oo mm long. The spectra were excited by the different lines of a
commercial 8 W argon ion laser. These lines had a FWHM of typically
o.3 cm^{-1} (gain profile). The laser beam was adjusted so as to pass
through the heat pipe close to the axis. The fluorescence is observed
along an axis only slightly displaced from this one and is focused
onto the slit of the spectrometer with lenses. Since by this method
fluorescence is observed along the entire length of the laser beam,
it leads to a much higher intensity than that obtained by the more
common perpendicular observation. Initially, all fluorescence lines
were scanned using a 3/4 m SPEX monochromator and detected photoelec-
trically with a GaAs (RCA C 31o34A) photomultiplier tube thermoelec-
trically cooled to -25°C. A 12oo lines/mm grating blazed at 5ooo Å
and lo μm slits gave a resolution of about o.1 Å. These were used to
make a "map" of the different fluorescence spectra in the D-X and the
perturbed d-a transitions. The scans were then analyzed to determine
the wavelength positions of the spectral lines to within o.1 Å and to
identify the members of each resonance fluorescence progression. This
preliminary step to determine the wavelength positions of the spectra
is a prerequisite to the high resolution photographic measurements.
The fluorescence spectra were photographed on the 5 m JARRELL-ASH
grating spectrometer of the DFVLR in Stuttgart at a resolution of

about 2oo ooo. For fixed experimental conditions each spectrum was taken for three different exposure times on 1o3A-F plates with reciprocal dispersion of about 1.4 Å/mm together with the comparison spectrum at the top and bottom. All plates were measured on a ZEISS comparator, and the lines measured on different photographic plates were generally consistent to within about o.o2 cm^{-1}. The absolute frequencies of lines, free of distortion, were determined to better than o.o3 cm^{-1}.

A2. Analysis

Three years ago we reported the first results of LIF studies on NaK formed in a supersonic nozzle beam.[2] In addition to strong emission in the green region of the spectrum we observed bound-continuum emission between 625 and 71o nm (see Fig. 2). It originates from perturbations of the $D^1\Pi$ state by a $^3\Pi_i$ manifold, causing radiation to different parts of the $a^3\Sigma^+$ potential curve, the lowest triplet state of NaK.[2,3] Here we concentrate on the investigation of the "bound - bound" part of this emission, i.e. the structure in the first three maxima which belongs to these bound-bound transitions into the Van der Waals minimum of the a state. The high resolution measurements analyzed here make it possible to determine this minimum for the first time with spectroscopic accuracy.

The 4765 Å exciting radiation selectively pumps the NaK molecule from thermally populated ground state vibronic levels to two different vibronic levels in the heavily perturbed D state. The resulting transitions to a state give rise to two fluorescence series of lines, one containing doublets, the other triplets. The first problem in the treatment of our data is that of determining the vibrational and rotational quantum numbers of each of the observed lines. This is done by searching the singlet-singlet transitions D-X in the green-yellow

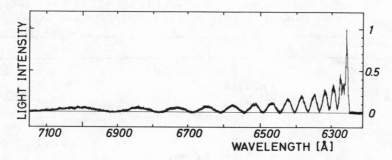

Fig. 2.   Low resolution (5 Å) fluorescence spectrum of NaK following 4765 Å argon ion laser excitation (4oo mW/cm^2).

part of the spectrum excited by the same laser line and somparing
the spacing and distance between doublets for the P-R progression
with the known rotational and vibrational spacings of the $X^1\Sigma^+$ ground
state of NaK. The Q series cannot be identified in the same way as the
P-R progression, since the upper state level is so heavily perturbed
that most of the fluorescence intensity appears in the triplet-triplet
transition. Once having established the upper state quantum numbers fo
the P-R series (v'=12, J'=7), each of the observed fluorescence lines
belonging to this series can be used to determine the energy of a rota
tional level associated with different vibrational levels of the a
state. In this way we established the energies of rotational levels
between v"=0 and v"=14 of the lowest triplet state. It proved difficult to evaluate the molecular constants by the traditional methods
since rotational constants (γ_e and D_v) are significant but difficult
to determine from the limited number of experimentally established
energy levels. A quick survey of the data at this stage showed that
the potential curve of the a state is actually represented, even for
large values of v", quite accurately by a Morse expression. We have
used the properties of the potential curbe, calculated by the RKR
method, to assist us in evaluating the higher molecular constants.
This is done in a cyclic procedure whereby the constants obtained
from the experimental data yield a potential curve which, after
suitable corrections, gives the improved constants listed in Table 1
in the form of Dunham coefficients. Using these constants determined
from the Dunham expression an RKR potential is calculated for J"=14.
Direct solution of the wave equation using the Numerov-Cooley technique give energy levels which differ less than o.o3 cm^{-1} from those
measured for the Q series. After iterative procedures to improve the
potential, we obtain energy levels which agree well with all observed
lines. Of the 64 observed levels, for only 5 does the calculated value
differ from the observed one by more than o.o2 cm^{-1}.

Table 1.   The Dunham coefficients Y_{ik} for a $a^3\Sigma^+$ state of NaK for
           v"$\leq$ 14 calculated in the analysis described in Sec. A2.
           The precision is indicated by the quoted standard error.
           All quantities are in cm^{-1}. The number in parentheses is
           the exponent of lo that multiplies the quantity.

Y_{ik}	Value		Error	
Y_{1o}	o.22816	(+o2)	1.1	(-o2)
Y_{2o}	-o.5916	(+oo)	3.1	(-o3)
Y_{3o}	-o.346	(-o2)	3.2	(-o4)
Y_{4o}	o.47	(-o4)	1.2	(-o5)
Y_{o1}	o.3915	(-o1)	2.8	(-o4)
Y_{11}	-o.1o9	(-o2)	2.5	(-o5)
Y_{21}	-o.33	(-o4)	1.8	(-o6)
Y_{o2}	o.43	(-o6)	2.5	(-o7)

Even though we have not observed levels v">14 it is interesting to compare the potential curve with long range interaction potentials due to van der Waals forces. It should be noted that the Dunham expression diverges at large R values instead of asymptotically approaching a dissociation limit, and hence this approach is not acceptable near the top of the potential well.

Another procedure which is accurate over the whole attractive potential well is the RKR method. Utilization of this approach for levels near the dissociation limit is given in the literature.[8]

A3. Long-Range Forces in NaK

The interaction energy between two atoms at large internuclear distances can be expressed as a series

$$V(R) = D - \sum_n (C_n/R^n) \quad , \tag{1}$$

where R is the internuclear distance, D is the energy at infinite separation, n are integers, and C_n are constants. If an RKR potential of a molecule can be determined for energies near the dissociation limit, D-V(R) can be determined for values of R which are sufficiently large for the long-range potential to be valid. Using this concept LeRoy et al.[8] have examined the constants C_n of the long-range potentials associated with different atomic states. For the interaction between two uncharged S-state atoms, e.g. Na and K, no first-order perturbation terms arise, i.e. there is no contribution from terms $\leq$ 5; second-order perturbation theory gives rise to dispersion terms corresponding to n = 6, 8 and lo which (a) always contribute to the potential and (b) are all positive (attractive) for atoms in their ground electronic states.

Two molecular states arise in the course of interaction between two alkali atoms in their $^2S_{1/2}$ ground states, $^1\Sigma^+$ and $^3\Sigma^+$. However, since the long-range van der Waals interaction is purely electrostatic in origin, it is independent of the configuration of the electron spins. Therefore, it is possible to calculate the long range portion of the potential curves of the $a^3\Sigma^+$ as well as of the $X^1\Sigma^+$ states of NaK from the same constants C_6, C_8, C_{10}, etc. Dalgarno and Davison[9] have calculated C_6, Davison[10] the ratio C_8/C_6, and Fontana[11] the ratio C_{10}/C_6. Using their values, we have calculated the long range portion of the potential curve of the $a^3\Sigma^+$ state. A comparison of the calculated and observed levels of D-V(R) for the $a^3\Sigma^+$ state shows good agreement for R $\leq$ lo Å. The dissociation energy of this state is thereby found to be

$$D_e(\text{NaK}, a^3\Sigma^+) = 2o3.1 \pm o.5 \text{ cm}^{-1}. \tag{2}$$

The analysis also gives the internuclear distance at the well depth,

$$R_m(\text{NaK, } a^3\Sigma^+) = 5.457 \pm 0.001 \text{ Å,} \tag{3}$$

a value consistent with the observed rotational eigenvalues. The D_e value can be straightforwardly used to calculate to nearly the same uncertainty the dissociation energy of the ground state,

$$D_e(\text{NaK, } X^1\Sigma^+) = 5268.1 \pm 0.8 \text{ cm}^{-1}, \tag{4}$$

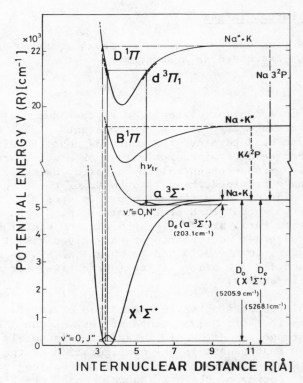

Fig. 3.   Potential curves for different states of NaK. Three independent cyclic determinations of the dissociation energy
of NaK X ground state are given by the straight, dashed
and dash-dotted lines. The energies are expressed in cm^{-1}
and are not all to scale. Based on the excellent agreement,
we consider that the dissociation energies of NaK, $X^1\Sigma^+$
and $a^3\Sigma^+$ states are well established to within the stated
uncertainties of 0.8 and 0.5 cm^{-1}, respectively.

and the dissociation energies for the electronically excited states
of NaK. Eq. (4) agrees almost exactly with a former value determined
in Ref. 12, the uncertainty now being considerably smaller.

The knowledge of the $a^3\Sigma^+$ potential curve for NaK over a wide
range of internuclear distance contributes significantly to calculations of alkali spin exchange cross sections[13] and to the interpretation of alkali scattering experiments.[4,5] The present experiments
do, also, permit a comparison with pseudopotential calculations performed on NaK. Our results show that the recent calculations of
Janoschek and Lee[7] and of Habitz[14] are much better able to predict
the $a^3\Sigma^+$ state interaction potential for NaK than earlier calculations
by Roach.[6]

As mentioned above, the present value for the dissociation energy
of NaK $X^1\Sigma^+$ agrees almost exactly with the value determined in Ref. 12.
Fig. 3 shows that the two different cycles used to determine the two
values have no measurement in common. The various wavelength measurements are even performed in very different parts of the spectrum. Such
excellent agreement between two independent determinations of the same
quantity using different methods is the strongest type of "proof"
that experiment can offer.

B. THE Na^6Li AND Na^7Li $B^1\Pi - X^1\Sigma^+$ BAND SYSTEM

B1. The Injection Heat Pipe (IHP)

A new device based on the heat pipe oven[15,16] has been demonstrated to be extraordinarily useful in LIF studies of mixed alkali
metal molecules. The identification of these heteronuclear molecules
MM' has always been complicated by the existence of stron M_2 and M'_2
band systems in the spectral range of interest.[17] Our new device
allows fluorescence studies under known, uniform and easily adjustable
metal compound concentrations even for immiscible metals which have
quite different vapor pressures at a given temperature.

The basic idea is to take a conventional heat pipe for the metal
with the smaller vapour pressure at a given temperature and then in-
ject continuously a small amount of the other metal vapor which is
produced outside the heat pipe in an external oven held at a different,
usually lower temperature. Part of the added vapor condenses in the
cooler end parts of the heat pipe, the other part contributes to the
normal heat pipe operation. By adjusting the amount of vapor injected
by varying the temperature of the external oven, steady state conditions are obtained and held for many hours. The apparatus is shown
schematically in Fig. 4.

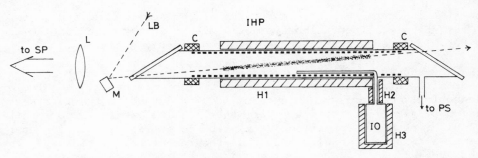

Fig. 4.   Schematic diagram of the Injection Heat Pipe (IHP).
          A detailed description is given in the text and in ref. 18.

The heat pipe consists of a stainless steel tube with Brewster windows
on both ends and a stainless steel mesh inside, which serves as a wick.
The central part is heated with heater H1 and the ends are positioned
by cooling flanges C. The external injection oven IO filled with the
second material is heated to a different temperature by the heater
H3. The small diameter injection tube is always kept at a higher temperature to prevent clogging. A small amount of rare gas of about
o.1 Torr is added through the pump system PS to keep the windows free
from contamination. The excitation laser beam is directed through
the IHP close to the optical axis by the mirror M. The fluorescence
light is collected by lens L and then focused on the entrance slit
of a spectrometer. First we have tested this system with the alkali
metals Na and Li by observing the different dimers Na_2, Li_2 and NaLi
by LIF from an argon ion laser.[18,19] Typical operating conditions are
530°C, 580°C and 410°C for heat pipe, injection tube and external
oven, respectively. Fluorescence of Na_2 is nearly almost suppressed
under this condition. To test the axial distribution of vapor densities
of the three molecular species Li_2, NaLi and Na_2 we vary the distance
lens to heat pipe and focus at each distance the light emerging from
a distinct point of the IHP axis onto the entrance slit of the
spectrometer. Measuring the light intensity of a given molecular line
for all three species, we find a nearly homogeneous density distribution all over the heated zone of the IHP. Raising the rare gas
pressure, from o.1 to about 1o Torr, we are able to detect for the
first time the band head formation by collision induced rotational
transitions, and in one case even a vibrational transition (v'=1 -
v'=2) with the accompanying rotational quantum jumps.

B2. The NaLi $B^1\Pi - X^1\Sigma^+$ Band System Analysis

Six different argon ion laser lines excite 15 different vibronic
levels in Na^6Li and Na^7Li. We have measured the wavelength of about
7oo lines, fluorescence "parent lines" and collision induced satellite
lines, with an accuracy of ± o.1 Å. Detailed analysis is given in a
recent publication.[1]

The most interesting feature of the NaLi fluorescence spectra is the fact that the fluorescence progressions of some excited levels extend much farther into the red spectral region than the fluorescence series of the homonuclear molecules Li_2 and Na_2. Even beyond the Li $(2^2P - 2^2S)$ resonance line at 6707 Å, which is the dissociation limit of the NaLi B state, molecular fluorescence lines are observed. This can be explained if one remembers that the homonuclear molecules have a potential "hump" in the $B^1\Pi_u$ state[20] due to the repulsive R^{-3} resonance interaction term in the long range potential expression. This leads to a rather "narrow" potential curve of the $B^1\Pi_u$ state. Heteronuclear alkali dimers on the other hand do not show this contribution to their excited state long range potentials. Here the leading term is a pure van der Waals R^{-6} attractive potential term. Therefore, their $B^1\Pi$ potential curves are much wider and shallower than the homonuclear ones. Following the Franck-Condon principle we expect radiative transitions from higher vibrational levels of the B state to nearly all vibrational levels, even the highest ones, of the X ground state, what in fact is observed. One striking example is analyzed in Fig. 5. It shows the last portion of a Birge Sponer plot obtained from the fluorescence excited with the 4765 Å laser line. Neglecting the three last (measured) points, the Birge Sponer plot follows the dotted line and gives v"=39 as the last bound level. The observed deviation from this behaviour is due to the influence of the long range van der Waals forces. Following Bernstein and LeRoys theory[8] of long range potentials, a straight line should be obtained in a plot of $(\Delta G_{v"})^{1/2}$ versus v" for long range forces proportional to R^{-6} (see Fig. 6). In addition, we compare our results with the theoretical values we obtain from the C_6 constant given by calculations of Dalgarno et al.[9] (full points in Fig. 6). They agree very well with our measurements. In this manner we find the vibrational level v"=43 as the last bound level of the Na^7/Li ground state potential. Since we have measured the vibrational levels up to 42, a highly precise determination of the dissociation energy of the $X^1\Sigma^+$ state is possible. Knowing the separation of the potential wells and the dissociation products we also obtain to nearly the same accuracy the dissociation energy of the $B^1\Pi$ state. Table 2 gives our results compared with some theoretical[6,21-25] and experimental[26,27] results.

B3. The K^7Li, Rb^7Li and Cs^7Li $B^1\Pi - X^1\Sigma^+$ Band Systems

Using the IHP we are able to observe LIF spectra even from molecules such as CsLi whose constituents have vapor pressures which differ by a factor of about 10^4. With different lines of the krypton ion laser we excite several vibrational-rotational levels of the heteronuclear lithium containing molecules KLi, RbLi, and CsLi. From a first rough analysis of the data we find preliminary values of the vibrational constant ω_e'' of the respective ground states. The concept of mass reduced potential determination[28] as a mean of usefully

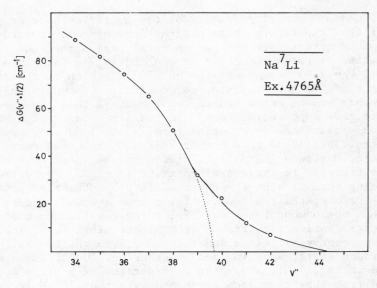

Fig. 5.   Enlarged part of a Birge Sponer plot, vibrational term energy
differences $\Delta G_{v''}$ versus v'', for the $X^1\Sigma^+$ state of Na^{7}Li.

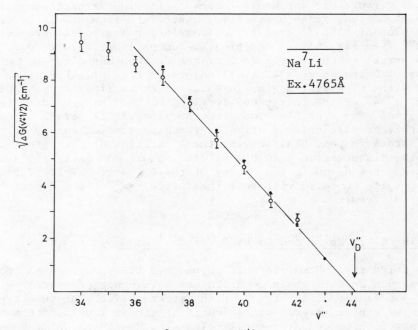

Fig. 6   LeRoy-Bernstein plot[8], $(\Delta G_{v''})^{1/2}$ versus v'', of the term
values from Fig. 5. v_D'' is the non-integer fictitious vibrational quantum number at which the molecule dissociates.

Table 2.   Summary of calculated and measured dissociation energies of the $X^1\Sigma^+$ and $B^1\Pi$ states of NaLi in cm^{-1}

Reference	D_e (NaLi, $X^1\Sigma^+$)			D_e (NaLi, $B^1\Pi$)			Method
6	7o17	$\pm$	13oo	---			pseudopot.
21	6877	$\pm$	32o	o			ab initio
22	7178	$\pm$	16o	---			ab initio
23		6491		---			MTX
24		9598		---			?
25		6936		968			pseudopot.
26		7364		---			mass spectr.
27	634o	$\pm$	1oo	115o	$\pm$	1oo	LIF
this work	685o	$\pm$	2o	1758	$\pm$	25	LIF

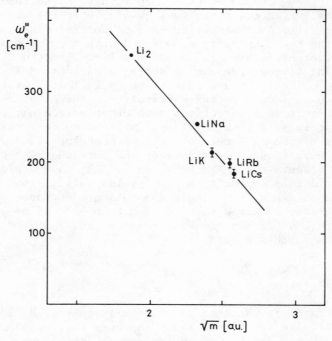

Fig. 7.   Plot of the vibrational constant of the ground states of the lithium containing alkali dimers versus the square root of the reduced mass of the molecule MLi, M: Li,....,Cs.

combining different variants of these dimers could be tested. Fig. 7
shows a plot of ω_e'' versus the square root of the reduced mass of the
molecules as is suggested by this theory. The condition that this
plot should give a straight line is not well fulfilled. It seems to us
that this theory only can give a rough guide to spectroscopic constants within 1o%. In conclusion, we have developed a new device for
spectroscopic studies, the IHP, and employed it to examine first
heteronuclear alkali dimers containing lithium atoms. The emission
spectra for several of these molecules are observed for the first
time.

HIGH RESOLUTION SPECTROSCOPY IN SUPERSONIC NOZZLE BEAMS:

C. THE Rb_2 $B^1\Pi_u$ - $X^1\Sigma_g^+$ BAND SYSTEM.

It is well known that in usual excitation spectroscopy in cells
or heat pipes the ultimate resolution is limited by the linewidth due
to Doppler broadening. This limitation has been overcome by using
Lamb-dip[29,30] and two photon absorption techniques.[30-32] The disadvantage of these methods is that they require rather high power
tunable lasers, which are not always available in the wavelength
region of interest. Another obvious method for reducing the linewidth
and simplifying the complexity associated with the large number of
populated rovibronic states is to use a well-collimated molecular
nozzle beam and to cool the sample during expansion.[33] The disadvantage of using molecular beams is the low optical density, which makes
fluorescence experiments rather difficult. On the other hand, the
spectrum is free from interactions with other molecules, and the
spectroscopist is sure that novel and interesting features are due
to a property of the isolated molecule without the disadvantages of
collisions, their induced interactions and perturbations. In the
following we discuss the B-X band system of Rb_2 as example where the
above-mentioned sub-Doppler technique yields information that could
not have been obtained from simple absorption spectroscopy of rubidum
in a cell. We have observed the spectrum from 673-655 nm at a
resolution of about 25 MHz and have resolved and analyzed eleven
vibronic bands in this region.

C1. Experimental

The apparatus is sketched schematically in Fig. 8. The output
from a commercial c.w. tunable laser (Spectra Physics 58oA, pumped
with all lines of a 165 4 W argon ion laser) intersects, orthogonally,
the well collimated molecular beam. We use a mixture of Rhodamine 1o1
and 6G, which gives a single mode output power of maximal 15 mW in
the spectral region of interest. The Doppler broadening due to finite

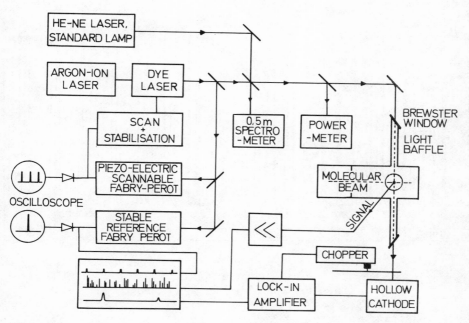

Fig. 8.   Laser Frequency Monitoring and Calibration Measurement
           combined with Crossed Laser - Molecular Beam Setup.

collimation ratio of the beam is less than the linewidth of the dye
laser used. The resulting fluorescence is viewed at right angles to
both the excitation laser beam and the molecular beam with a cooled
high efficiency photomultiplier RCA 8852 (ERMA III response); the
photoelectric signal is detected with a Keithley 417 fast picoammeter;
and the fluorescence spectrum is displayed on a strip chard recorder
simultaneously with frequency marks from a stable Fabry-Perot etalon
and with absolute frequencies from an opto-galvanic calibration
spectrum of a thorium lamp filled with neon. With this improved calibration, the system is capable of high accuracy measurements (25 MHz).
Continuous scans were performed as large as 8ooo GHz.

     The molecular beam, after passing through the excitation zone,
is detected by a Langmuir-Taylor detector. This detector can be moved
perpendicular to the beam to measure the beam profile and the flux.

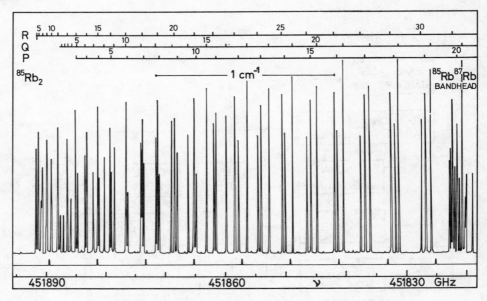

Fig. 9.   LIF spectrum of Rb_2 produced in a supersonic nozzle beam.
The (9,0) band is shown within the B-X system. The rotational lines are assigned on top, and at the bottom the
frequency marks of the external Fabry-Perot interferometer
are displayed.

C2. Excitation Spectrum of Rb_2 $B^1\Pi_u - X^1\Sigma_g^+$

Sub-Doppler LIF spectroscopy is applied to the measurement of
the rotational structure of vibronic bands in the Rb_2 B-X band system.
Fig. 9 is a spectrum obtained for the (9,0) transition. The signal-
to-noise ratio is excellent. All P-, Q-, and R-lines are resolved and
measured for the first time. Similar results, with an accuracy of a
few MHz, are obtained for all of the rotational levels with $J \leq 60$
in the $v' = 4, 5, 6, 7, 8, 10$ and 11 vibronic levels connected by
allowed transitions to the $v''=0$, in the $v'=6,7$ connected to the
$v''=1$, and $v'=9$ connected to the $v''=2$ level of the ground state,
respectively. As can be seen from Fig. 9, considerable narrowing of
the absorption has been achieved by using the nozzle beam. The Doppler
width of bulk vapor of rubidium is about 500 MHz. The linewidth of
the transitions measured here is about 35 MHz.

The first step in our analysis is to identify the correct v'',J''
and v',J' quantum numbers which are to be assigned to each line and
this task is not easy for the dense rubidium molecular spectrum. Unfortunately, in this case the previous absorption and fluorescence
experiments[34-41], which furnish approximate values for vibrational

constants for the ground and excited state, are not very precise. The vibrational numbering we use agrees with the vibrational number adopted by Matuayma.[34] Later Kusch[35] and recently Kotnik-Karuza[41] confirmed this vibrational numbering by isotope studies. In our case of completely resolved rotational lines the <u>rotational</u> numbering is greatly facilitated (see Fig. 9). From the <u>vibrational</u> and rotational analysis, fully described elsewhere,[42] we have fitted the molecular constants of the X and B state. The results in Table 3 for the most important molecular constants are given in the form of conventional Dunham coefficients, and we have used all measured vibronic bands with approximately 1ooo lines excited in Rb_2 to improve them. The potential curves for the X and B states of Rb_2 have been constructed from the spectroscopic constants given in Table 3 using the RKR method. From the potential curves we have evaluated Franck-Condon factors for the observed bands. Using the measured intensities and the calculated Franck-Condon factors we are able to determine the "internal" temperature the rubidium beam reaches during the expansion. This will be discussed in the following section.

C3. Internal distributions of Rubidium Dimers in a Nozzle Beam

Various methods have been used to characterize the quantum states of the nozzle beam molecules and their relaxation processes in great detail.[43] Different distributions have been obtained during and after the expansion. Since the technique of near Doppler-free spectroscopy described above gives directly the excitation spectrum of the molecules in the beam, it is perfectly suited for determining the distribution of nozzle beam particles. The v"=0 progression of the B-X band system is used to obtain the rotational state distribution. The line intensities are taken to be proportional to the height. The experimental data have been fitted to a Boltzmann distribution yielding

$$T_{rot} = 45 \pm 5^{\circ}K . \tag{5}$$

Table 3. The Dunham coefficients Y_{ik} for the Rb_2 $B^1\Pi_u - X^1\Sigma_g^+$ states calculated in the present analysis. The precision is indicated by the quoted stantard error. All quantities are in cm^{-1}. The number in parentheses is the exponent of 1o.

Y_{ik}	$^{85}Rb_2(X^1\Sigma_g^+)$		Error		$^{85}Rb_2(B^1\Pi_u)$		Error	
Y_{1o}	0.57747	(+o2)	0.2	(−o2)	0.47431	(+o2)	0.3	(−o2)
Y_{2o}	−0.15817	(+oo)	0.8	(−o2)	−0.15326	(+oo)	0.4	(−o2)
Y_{o1}	0.2278o	(−o1)	0.15	(−o3)	0.19988	(−o1)	0.1	(−o3)
Y_{11}	−0.474o2	(−o4)	0.1	(−o4)	−0.7oo92	(−o4)	0.1	(−o4)
Y_{o2}	−0.15o6o	(−o7)	0.3	(−o8)	−0.14o34	(−o7)	0.2	(−o8)

These rotational distributions are the same regardless of the band in
the v"=0 progression being fitted or the method used to get rotational
line intensities. At this stage it is not very meaningful to deduce
a vibrational temperature for the rubidium dimers, because the number
of different excited vibrational levels of the ground state is not
large enough within the restricted tuning range of the dye laser.

   With the highest resolution presently available to us, it is
possible to analyse the rotational structure of many bands of heavy
diatomic molecules in the visible. We have shown the complexity of the
spectra associated with different rubibium dimers. Although this section has been limited to a discussion of the B state of these dimers,
it is clear that similar, and perhaps greater difficulties will exist
in analysing the higher-lying states associated with Rydberg series,
which are often perturbed. A detailed analysis of additional bands of
Rb$_2$ in the next section shows the complications resulting from predissociation of the C and D states and from overlapping triplet series.

D. PREDISSOCIATION IN THE Rb$_2$ ($D^1\Pi_u$, $C^1\Pi_u - X^1\Sigma_g^+$) BAND SYSTEMS

   The past ten years have seen a revival of theoretical interest
in predissociation phenomena, coupled with increasing sophistication
in experimental techniques[44] and a continuous search for related
effects in molecular scattering.[45] Given the established characterization of predissociation types[46] and selection rules,[47] and the
accepted value of predissociation measurements in locating the dissociation limit,[48] the object of our recent experimental work has been
to underline further information available from high resolution LIF
measurements in molecular beams. These informations relate primarily
to the forms and the symmetries of the relevant potential energy
curves. The purpose of this section is to present preliminary results
of predissociation of Rb$_2$ D and C states by curve crossings with a
hitherto unknown $^3\Sigma_u^+$ state and another unknown molecular state.

D1. Experimental

   The apparatus for these measurements is essentially the same as
shown in Fig. 1. The LIF experiments are performed in a similar way
to the Rb$_2$ B-X studies described above. During a typical run the oven
chamber is maintained at a temperature of about 83o°K, corresponding
to a rubidium vapor pressure of about 1oo Torr. The light beam from
a pulsed dye laser (Avco C 5ooo, Lambda Phyik FL 1ooo, 2oo Hz repe-
tition rate, 6 nsec pulses, 25 kW peak power) intersects the nozzle
beam at right angles, producing visible fluorescence approximately
14 cm downstream from the nozzle source. In this region free molecular flow exists. The fluorescence is detected at right angles to
both the nozzle beam and the laser beam by two photomultipliers.
The molecular bands are measured with an EMI 6256S (S 11 photocathode).

This photocathode sensitivity is limited to wavelengths shorter than
600 nm. The atomic emission at 780 nm or at 794.8 nm has been measured
at the same time with a cooled RCA 8852 PM tube (ERMA III response)
in conjunction with interference filters. These filters, centered at
780 $\pm$ 2.6 nm and 795 $\pm$ 2.6 nm, limit the response to the D1 or D2
line of the rubidium atom.

D2. The Rb_2 ($C^1\Pi_u - X^1\Sigma_g^+$) Band System

The dye laser pumps the Rb_2 C-X transition in the blue (480 -
460 nm). The resulting fluorescence from the nozzle beam consists of
molecular fluorescence and the Rb $5^2P_{3/2} - 5^2S_{1/2}$ atomic line (780 nm),
the Rb D2 line. The Rb D1 line at 794.8 nm is absent. There is also
no evidence for the A-X and B-X emission in the rubidium beam or the
quasi-continuum feature, all of which had been observed earlier in a
cell or in a heat pipe.[37] Our findings support the conclusion, found
earlier by Feldman and Zare,[38] that these emissions are caused in the
cell by collisional energy transfer and are not primary photo-pro-
cesses. The information derived from the molecular and atomic fluorescence is displayed in Fig. 10.

The different spectra, associated with the dye laser excitation
and selective detection of the fluorescence light, are plotted versus
v'. Note that each v' level populated by the laser gives rise to
fluorescence which determines the magnitude of the predissociative
perturbations. From the relative intensities useful first approximations of the curve crossings can be made.[49] The intensity of the
molecular fluorescence and of the Rb D2 line emission is found to
be proportional to the dye laser intensity, thus ruling out multiphoton processes. Hence it is concluded that predissociation of the
Rb_2 C state is responsible for the appearance of the Rb D2 line.

D3. The Rb_2 ($D^1\Pi_u - D^1\Sigma_g^+$) Band System

Similar excitation spectra are obtained for this band system,
detecting the blue-violet molecular radiation, and atomic (D1 and D2)
line emission. The dominant peaks in the spectra correspond to band
heads originating from higher vibrational v' levels of the v" = 0
progression of the Rb_2 D-X band system. Closer examination of the
spectra reveals the appearance of band heads originating from higher
vibrational levels v" of the ground state too. Since the fraction of
Rb $5^2P_{1/2}$ coming from the lower vibrational levels of the D state is
very small, the D2 spectra of the two states appear similar. However,
examination of the short wavelength tail of the D-X spectrum reveals
the strong appearance of D1 line emission from bands of higher vibrational numbering v'. The D1 line emission from these bands indicates
that higher vibrational states of Rb_2 D are predissociated by a new
and unknown state which gives first Rb $4^2D_{3/2}$ and then by cascading,

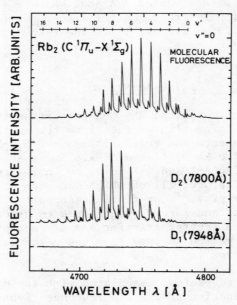

Fig. 1o. Summary of the Rb_2 C–X laser induced molecular and atomic
          fluorescence. The latter is caused by predissociation due
          to a hitherto unknown $c\,^3\Sigma_u^+$ state belonging to the Rb $5\,^2P_{3/2}$
          and Rb $5\,^2S_{1/2}$ configuration. Observed molecular fluorescence
          is given on top together with the vibrational numbering,
          the atomic fluorescence is represented below. These spectra
          shown are not corrected for the varying laser power, which
          is $\leq 15\%$ in this spectral region. The pulsed dye laser
          bandwidth is measured as o.15 $\pm$ o.o3 Å.

Rb $5\,^2P_i$ atoms. It is interesting to speculate about the form of the
observed predissociation processes in relating them to different
potential curves. We suggest that the results can be explained by two
predissociating channels, which in further detail are discussed in
a recent publication.[49] Concluding this, it seems imperative in these
studies to establish what processes are responsible for the emission
observed. Viewed alternatively, this information is fundamental to
determining the structure and properties of high lying states of

alkali dimers, and, at the same time, we learn from these spectroscopic studies the detailed way in which predissociation processes
yield excited atomic states.

CONCLUSION

High resolution diatomic molecular spectroscopy is the definitive method for determining molecular potentials; the present experiments are major steps forward in the spectroscopy of weakly bound and
repulsive molecular states. Bound-free transitions and predissociation
dynamics observed by us in different systems permit the spectroscopy
of repulsive potentials, until today a domain of scattering experiments. Basically, we combine molecular spectroscopy with molecular
beam techniques. Under those felicitous circumstances, as reported
here, the progeny of the "marriage" of lasers and beams is a delight
in that it provides some of the most detailed information on intramolecular forces yet available.

ACKNOWLEDGEMENT

We gratefully acknowledge many helpful discussions with
R.N. Zare, D.E. Pritchard and R.W. Field. Special thanks go to
L. Krauss for support to carry out a portion of this project while
the author was at the Deutsche Forschungs- und Versuchsanstalt für
Luft- und Raumfahrt (DFVLR) in Stuttgart and to D. Beck for his
valuable encouragement throughout the course of this work. Last not
least it is a pleasure to thank all my coworkers, E.J. Breford,
C.D. Caldwell, G. Ennen, H. Hage, K.H. Meiwes and H. Rudolf, devoted
to these studies of laser spectroscopy, and to acknowledge support
of this work by the Deutsche Forschungsgemeinschaft.

REFERENCES

1. E.J. Breford, F. Engelke, G. Ennen and K.H. Meiwes, Disc. Farad.
      Soc. 71 (1981)
2. E.J. Breford and F. Engelke, Chem. Phys. Letters 53:282 (1978).
3. E.J. Breford and F. Engelke, J. Chem. Phys. 71:1994 (1979).
4. D.E. Pritchard and F.Y. Chu, Phys. Rev. A2:1932 (197o).
5. L.T. Cowley, M.A.D. Fluendy and K.P. Lawley, Trans. Farad. Soc.
      65:2o27 (1969).
6. A.C. Roach, M. Mol. Spectrosc. 42:27 (1972).
7. R. Janoschek and H.U. Lee, Chem. Phys. Letters 58:47 (1978).
8. R.J. LeRoy and R.B. Bernstein, J. Chem. Phys. 52:3869 (197o);
      R.J. LeRoy and R.B. Bernstein, J. Mol. Spectrosc. 37:1o9 (1971)
      R.J. LeRoy, Can. J. Phys. 52:246 (1974).
      R.J. LeRoy, "Molecular Spectroscopy 1", Specialist periodical
      report, the Chemical Society, London (1973).

9. A. Dalgarno and W.D. Davison, Mol. Phys. 13:479 (1967).
1o. W.D. Davison, J. Phys. B1:139 (1968).
11. P.R. Fontana, Phys. Rev. 123:1865 (1961).
12. E.J. Breford and F. Engelke, "Laser Induced Processes in Mole-
    cules", Springer Series in Chem. Phys., Vol. 6 ed. K.L. Kompa
    and S.D. Smith, Springer, Berlin, Heidelberg, NY (1978).
13. A. Dalgarno and M.R.H. Rudge, Proc. Roy. Soc. A286:519 (1965).
14. P. Habitz, to be published.
15. C.R. Vidal and F.B. Haller, Rev. Sci. Instr. 42:1779 (1971).
16. M.M. Hessel and P. Jankowski, J. Appl. Phys. 43:2o9 (1972).
17. J.M. Walter and S. Barratt, Proc. Roy. Soc. London A119:257 (1928).
18. F. Engelke, G. Ennen and K.H. Meiwes, Ber. Bunsenges. Phys. Chem.
    subm. for publication.
19. M.M. Hessel, Phys. Rev. Letters 26:215 (1971).
2o. R. Velasco, Ch. Ottinger and R.N. Zare, J. Chem. Phys. 51:5522
    (1969).
21. P.J. Beroncini, G. Das and A.C. Wahl, J. Chme. Phys. 52:5112
    (197o).
22. P. Rosmus, W. Meyer, J. Chem. Phys. 65:492 (1976).
23. D.D. Konowalow and M.E. Rosenkrantz, Chem. Phys. Letters 49:54
    (1977).
24. H.S. Fricker, J. Chem. Phys. 55:5o34 (1971).
25. P. Habitz, W.H.E. Schwarz and R. Ahlrichs, J. Chem. Phys.
    66:5117 (1977).
26. K.F. Zmbov, C.H. Wu and R.H. Ihle, J. Chem. Phys. 67:46o3 (1977).
27. G. Baumgartner, G. Gerber and F. Pfluger, Vhdlg.DPG, 577 (1979).
28. W.C. Stwalley, J. Chem. Phys. 63:3o62 (1975).
29. W.E. Lamb Jr., Phys. Rev. A134:1429 (1964).
3o. W. Demtröder, Phys. Rep. 7C:223 (1973).
31. L.S. Vasilenko, V.P. Chebotaev, and A.V. Shisheav, JETP Letters
    12:113 (197o).
32. B. Cagnac, G. Grynberg and F. Biraben, Phys. Rev. Letters, 32:643
    (1974).
33. R.E. Smalley, B.L. Ramakrishna, D.H. Levy and L. Wharton,
    J. Chem. Phys. 61:4363 (1974).
34. E. Matuyama, Nature 14:567 (1934).
35. P. Kusch, Phys. Rev. 49:218 (1936).
36. Ny Tsi-Ze and Tsien San-Tsing, Phys. Rev. 52:91 (1937).
37. J.M. Brom and H.P. Broida, J. Chem. Phys. 61:982 (1974).
38. D. Feldman and R.N. Zare, Chem. Phys. 15:415 (1976).
39. A.C. Tam and W. Happer, J. Chem. Phys. 64:4337 (1976).
4o. D.L. Drummond and L.A. Schlie, J. Chem. Phys. 65:2116 (1976).
41. D. Kotnik-Kariza and C.R. Vidal, Chem. Phys. 4o:25 (1979).
42. C.D. Caldwell, F. Engelke and H. Hage, Chem. Phys. 54:1 (198o).
43. K. Bergmann, U. Hefter and P. Hering, J. Chem. Phys. 65:488 (1976);
    Chem. Phys. 32:329 (1978).
44. E.O. Degenkolb, J.I. Steinfeld, E. Wassermann and W. Klemperer,
    J. Chem. Phys. 51:615 (1969).
45. R.B. Bernstein, Phys. Rev. Letters 16:385 (1966).

46. R.S. Mulliken, J. Chem. Phys. 33:247 (196o).
47. G. Herzberg, "Spectra of Diatomic Molecules", 2nd edition, Van Nostrand, Princeton (195o).
48. A.G. Gaydon, "Dissociation Energies and Spectra of Diatomic Molecules", 3rd ed. Chapman and Hall, (1968).
49. E.J. Breford and F. Engelke, Chem. Phys. Letters, 75:132 (198o).

QUANTUM-STATE-RESOLVED SCATTERING OF LITHIUM HYDRIDE

Paul J. Dagdigian*

Department of Chemistry
The Johns Hopkins University
Baltimore, Maryland 21218
U.S.A.

ABSTRACT

A technique for the determination of integral rotationally inelastic state-resolved cross sections, which involves electric quadrupole state selection of a molecular beam and tunable laser fluorescence detection of the final states, is reviewed. This method has been applied to the scattering of lithium hydride, and results with a number of targets (helium and several polar molecules) are presented. The experimental results are compared with calculations by M.H. Alexander and D.M. Silver which employed a number of theoretical treatments of varying sophistication. One clear result of this collaborative effort is the conclusion that rotational energy transfer, both for LiH-He and for polar molecule collisions, cannot be described adequately by simple models but requires detailed consideration of the interplay between translational and rotational motion, as well as a realistic description of the intermolecular forces.

1. INTRODUCTION

Rotational energy transfer is the most facile and conceptually the simplest of a large class of gas-phase relaxation processes.[1] Until recently, experimental limitations have hindered detailed study of this fundamental collisional process.

---

*Alfred P. Sloan Research Fellow; Camille and Henry Dreyfus Teacher-Scholar.

One of the earliest techniques employed was microwave double resonance.[2,3] However, the connection between experimental observables and state-to-state cross sections, which are the fundamental quantities of interest, is not straight-forward.[4] Spectroscopic techniques have been more recently employed successfully for the study of state-resolved rotational energy transfer in excited vibrational or electronic states. Here initial state preparation is achieved by infrared[5] or visible[6-12] laser excitation, and the final states are determined by resolved fluorescence. Alternatively, time-resolved infrared-infrared double resonance has been used for studies in excited vibrational states.[13,14] Recently, sub-Doppler optical-optical double resonance has been employed to study simultaneously rotational and translational relaxation in excited electronic states.[15]

Molecular beam techniques have been the most successful method of studying rotational energy transfer in ground vibelectronic states. Early experiments[16,17] were severely hindered by the lack of sufficient translational energy resolution to detect the small changes signaling rotationally inelastic collisions. Because of the relative ease in handling ions as opposed to neutral molecules, ion-molecule collisions provided the first rotational-state-resolved results.[18,19] Advances in techniques for producing intense nearly monochromatic molecular beams have recently allowed determination of differential state-to-state cross sections in neutral particle collisions involving H_2 and its isotopes[20,21] and, more recently, in heavier molecules with smaller rotational spacings.[22] Pronounced structure has also been observed[23] in the differential cross section in $K-CO,N_2$ collisions with incomplete state resolution.

State selection and detection with electric quadrupole fields[24] has also been used to observe rotationally inelastic processes in molecular beam experiments. Small-angle scattering of the polar molecules TlF[25] and CsF[26] with various collision partners have been investigated in this way. Electric quadrupole selection has also been used to measure with high resolution the differential scattering of polarized $j=1$ LiF molecules.[27] Finally, optical pumping state selection[28] has recently been utilized in combination with a rotatable fluorescence detector by Bergman and coworkers to measure differential state-to-state rotationally inelastic cross sections involving Na_2 molecules.[29] State resolved differential cross sections have also been obtained for Na_2 with a stationary detector by measurement of Doppler profiles.[30] This versatile optical pumping technique, unlike selection with inhomogeneous electric fields, is limited to low j levels and is equally applicable to polar and nonpolar molecules with accessible electronic transitions.

In our laboratory, we have developed a technique for the measurement of integral state-to-state rotationally inelastic cross sections, which utilizes electric quadrupole state selection[24] of an incident molecular beam and laser fluorescence detection[31] of scattered

molecules. While the initial rotational state is limited to low j
values, laser fluorescence allows determination of an almost complete
final state distribution. We have applied this technique to energy
transfer involving lithium hydride. This molecule was chosen for
study because not only is it convenient experimentally, but it is
suited for theoretical studies, being the simplest polar molecule.
As a result, it has been possible to collaborate closely with theoretical studies of LiH inelastic collisions, principally by
M.H. Alexander (University of Maryland). A very important advantage
in such a collaboration has been the ability to compare the experimental measurements with calculations based on sophisticated theoretical models, of the type impossible for an experimentalist alone
to carry out.

In this article, we review our technique for obtaining state-
resolved inelastic cross sections.[32] We then discuss the scattering
system ^{7}LiH–He, in which the experimental results are compared with
scattering calculations which employed an accurate ab initio potential
surface somputed by D.M. Silver (Applied Physics Laboratory, The Johns
Hopkins University). The validity of simpler theoretical treatments
are also discussed. We then discuss the scattering of ^{7}LiH with the
polar molecules HCl, DCl, and HCN, systems which are dominated by
the long-range dipolar interaction.

2. EXPERIMENTAL METHOD

Because of unfavourable thermodynamics, it is necessary to
prepare[33,34] lithium hybride within a molecular beam source by passing
H_2 gas continuously over molten Li metal at $\sim 1200°K$ and allowing the
resulting gas mixture to expand into vacuum through a small orifice.

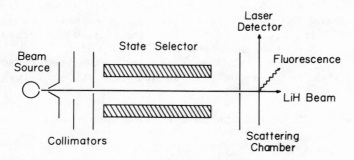

Fig. 1. Plan view of the apparatus

The resulting supersonic[35] beam consists primarily of H_2, with a lesser amount of Li and still less of LiH (approx. o.1%) and Li_2. It should be noted that because of the hydrodynamic expansion from the orifice, the LiH component is rotationally cold ($T_{rot} \simeq 5o^oK$) and has translational energy (approx. o.9 eV) considerably above a thermal value because of acceleration in the principally H_2 gas.

As Figure 1 illustrates, the collimated LiH beam is injected into an electric quadrupole field after several stages of differential pumping. The rotational state dependence of the Stark effect of a polar molecule is utilized to select[24] the j=1 rotational level. Downstream of the quadrupole within the scattering chamber, a tunable dye laser excites fluorescence in the near uv LiH A-X band system in order to determine populations of individual rotational levels.

Figures 2 and 3 present typical laser excitation spectra of the incident beam and of the state-selected beam, without and with target gas in the scattering chamber. The primary data in these experiments are such laser excitation spectra as a function of scattering gas pressure. The increase in the j≠1 populations with HCl pressure is quite apparent in Fig. 3 and is due to rotationally inelastic collisions. As has been shown previously[25,34] the populations n_i and n_f of the initial and final rotational state, respectively, are related to an apparatus-effective state-to-state inelastic cross section in the following way:

$$n_f/n_i - n_f^{(0)}/n_i^{(0)} = n\sigma_{i \to f}^{eff} \ell \tag{1}$$

provided the target gas density n is low enough and the incident beam is sufficiently pure. Here, ℓ is the target gas path length, and the superscript (0) denotes populations with no scattering gas present.

These cross sections $\sigma_{i \to f}^{eff}$ include all collisions in which LiH is inelastically scattered forward in the laboratory frame into the laser excitation zone. Thus, these are lower limits to the integral state-to-state cross sections. However, if the $\sigma_{i \to f}^{eff}$ are found to be independent of path length and hence angular resolution, then we may equate our results to integral cross sections. For the polar molecule targets studied, for which the scattering is expected to be predominantly forward because of the long-range dipolar interaction, this is found to be the case. For LiH-He, in which the potential is mainly repulsive, incomplete cross sections have been measured; nevertheless, we believe our results are at most ∿ 3o% low.

3. ROTATIONALLY INELASTIC LiH-He COLLISIONS

Because of the LiH-He system containing only six electrons, it has been possible for Silver[36] to compute an accurate correlated ab initio potential surface for the interaction of He with a rigid

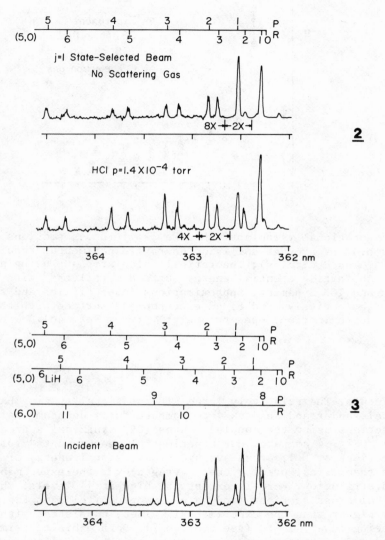

Fig. 2 and 3. Laser excitation spectra of ^{7}LiH in the $A^1\Sigma^+ - X^1\Sigma^+$
band system, with quadrupole state selector turned off
(Fig. 2) and with state selector tuned to refocus j=1
molecules, without and with HCl in the scattering
chamber (Fig. 3). Increases in detector sensitivity
are denoted by "...X". (from Ref. 32).

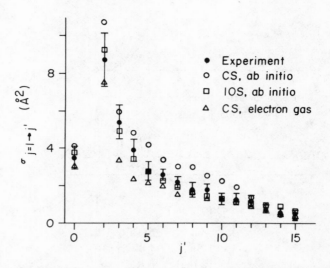

Fig. 4.   Rotationally inelastic LiH-He j=1→j' cross sections at
          o.32 eV average collision energy: Experimental results[40]
          (closed circles); Theoretical calculations[37] using a) the
          ab initio potential energy surface of Silver[36] with the CS
          and IOS dynamical approximations (open circles and squares,
          respectively) and b) an electron-gas surface with the CS
          approximation (open triangles). (from Ref. 4o).

[7]LiH molecule. The relatively large LiH rotational spacing has
allowed Alexander and Jendrek[37] to compute fully quantum inelastic
cross sections using the coupled states (CS) dynamical approcima-
tion.[38,39] Fig. 4 compares our experimental measurements[40] of the
cross sections $\sigma_{j=1 \to j'}$ at an average translational energy of o.32 eV
with the theoretical predictions, averaged over the experimental
energy distribution. For all except the highest j' levels, these
theoretical cross sections (indicated by open circles in Fig.4) are
slightly higher than the experimental values, consistent with the
latter being lower limits (see Section 2). Both exhibit a similar
j' dependence, indicative of the extreme anisotropy of the potential
energy surface: The most probable final state is j'=2 and the cross
sections fall off slowly with increasing j'.

    An important consideration in the calculation of cross sections
is the computational difficulty of an accurate treatment of the
collision dynamics. Hence, it is of interest to check the validity
of simpler, more approximate models. Fig. 4 also presents results
obtained from two simpler quantum mechanical treatments, one

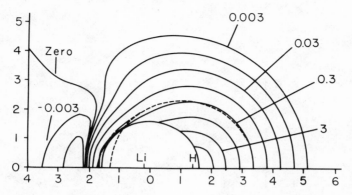

Fig. 5.   Equipotential contours (in eV) of the rigid rotor LiH-He
          potential energy surface of Silver.[36] The origin of the
          coordinate system (in Å) is the LiH center of mass, and the
          z axis is along the LiH bond. Within the small circle of
          radius 1.6 Å the analytic fit of the potential given in
          Ref. 36 is no longer valid. Rigid shell fits to the o.3 eV
          contour: (a) Offset sphere (r = 2.3 Å, d = 1.o3 Å), dashed
          line; and (b) Offset sphere is in (a) for z > -o.53 Å and a
          paraboloid of revolution for -d > z > -o.53 Å [$x^2 + y^2 =
          a(a+d')$, where a = 2.44 Å, d' = 1.7o Å] , dot-dashed line.

employing the simpler infinite-order-sudden (IOS) treatment[39,41,42]
of the dynamics and the other a CS treatment employing a more
approximate potential energy surface. The IOS approximation yields
values that are somewhat lower but still reasonable; however, it
does not predict a maximum in $\sigma_{j=1 \to j'}$ vs. j' which appears in the CS
calculations at a single energy.[37]

     The dependence on the accuracy of the potential energy surface
has been investigated by calculations using the electron-gas model
of Gordon and Kim,[43] wherein the electron densities of the isolated
molecules are used to compute interaction energies. We found that
CS calculations[37] with this surface yield cross sections which are
considerably lower than experiment and the more accurate theoretical
calculation. This is a result of the less repulsive character of
this surface compared with the ab initio one. We believe that the
poorer agreement with experiment of these simpler theoretical models
demonstrates that such models must be used extremely carefully. In
particular, the accuracy of the potential energy surface employed

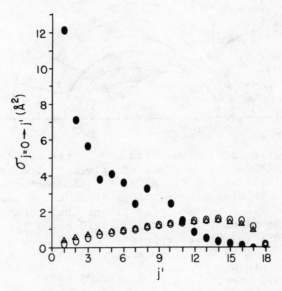

Fig. 6.   Integral cross sections for rotationally inelastic LiH(j=0)
          + He collisions at o.3 eV translational energy: Accurate
          quantum calculations,[37] (closed circles); rigid shell model
          calculations using shells (a) and (b) in Fig. 4 (open circles
          and triangles, respectively).

appears crucial. This in turn suggests difficulties in extrapolating
to new collision systems.

      Because of its simplicity, classical mechanics has been exten-
sively utilized in treatments of scattering dynamics. In particular,
rotationally inelastic atom-molecule collisions have been simulated
[44,46] by the purely classical impulsive scattering of a point particle
by an anisotropic rigid shell, most recently by Beck, Ross, and
Schepper,[47] who have characterized in detail the scattering from
rigid cylindrically symmetric shells. We have investigated[48] whether
a classical rigid-shell model provides a reasonable description of
the dynamics of ^{7}LiH-He collisions, in which only the repulsive part
of the potential plays the dominant role. Fig. 5 displays a contour
plot of the ab initio ^{7}LiH-He surface calculated by Silver.[36] Superimposed are two fits to the contour of o.3 eV energy, which is the

average translational energy in the experiments. In one, we have
employed an offcenter spherical shell;[48] the other is a slightly more
accurate fit and includes a paraboloid of revolution at the Li and
spliced to a sphere. We have computed integral cross sections vs. j'
for the scattering from an initially nonrotating (j=0) ^{7}LiH molecule
by summing over many trajectories[48] or by integrating the doubly
differential cross section $J(j',\theta)$ derived by Beck et al.[47] for
arbitrary rigid shells.

In Fig. 6, we compare the resulting cross sections with CS
quantum results[37] using the ab initio surface. Clearly, the rigid-body
model yields completely inaccurate cross sections, even with a good
fit to the equipotential corresponding to the collision energy. This
disagreement is not due to quantum effects since classical trajectory
calculations[49] employing the ab initio surface yield cross sections
which agree very well with the quantum results. From Fig. 5, it is
obvious that the total cross section is greatly underestimated by
the rigid-shell model. This suggests that a significant fraction of
the inelasticity results from collisions which never fully penetrate
to the repulsive wall, a conclusion corroborated by examination of
the partial opacities.[48] While the rigid-shell model does not provide
a reasonable quantitative description of the dynamics of rotationally
inelastic collisions, at least for LiH-He, in which the interaction
potential is very anisotropic, nevertheless this model has been
extremely helpful in understanding the physical origin of "rotational
rainbows",[23,50,51] which are structures observed in doubly differential
cross sections.[23,29,30,52]

4. ROTATIONALLY INELASTIC POLAR MOLECULE COLLISIONS

We have also investigated ^{7}LiH collisions with polar molecules,
for which the difficulties in devising an accurate theoretical description are quite different than in the previous example. As the
results[34,53] in Fig. 7 demonstrate, the cross sections in these polar
molecule collisions even at hyperthermal energies are very large,
with the implication that the dominant inelasticity occurs at large
impact parameters. Thus, it is reasonable to consider only the long-
range part of the potential, primarily the dipole-dipole electrostatic
interaction. The crux of the theoretical description then is an
accurate bimolecular dynamical model.

The most commonly used treatment for polar molecule sollisions
has involved the time-dependent Born approximation.[54-56] This is a
perturbative solution of the usual time-dependent equations resulting
from classical and quantum descriptions of the relative (orbital) and
internal (rotational) motions, respectively. Unfortunately, the Born
approximation, to first- or second-order, yields ^{7}LiH state-to-state
cross sections (averaged and summed over initial and final target
rotational levels, respectively) which are much larger than the

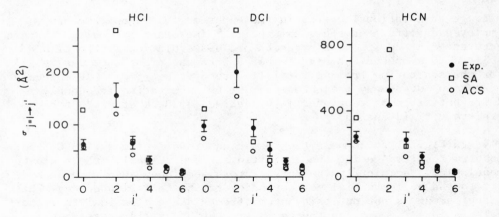

Fig. 7.   Integral cross sections for rotational-state-resolved
          collisions of LiH(j=1) with HCl, DCl, and HCN at o.7 eV
          translational energy. (from Ref. 32)

experimental integral cross sections (by a factor of ~ 3 for HCl)
for the dipole-coupled transitions (j'=j±1). This suggests[57] that
the dipole-dipole coupling is too strong for a perturbative solution
to be valid, even at these high collision energies. Moreover, cross
sections for transitions allowed in second-order are found[53] to be
larger than first-order transitions, in some cases. Thus, we find
the Born approximation to be completely inadequate to describe polar
molecule collisions, at least in state-to-state studies involving
[7]LiH, although this model has often been used to interpret[58] the
results of less highly resolved experiments.

     At the high translational energies in these experiments, it is
reasonable to consider solution of the classical path equations within the sudden[59-62] rather than the Born approximation. Here the
effective collision time τ_{col} is assumed to be much smaller than a
rotational dephasing rime τ_{rot}, and the coupled time-dependent equations for the rotational state amplitudes can be solved analytically
to infinite order. As Fig. 7 shows, the time-dependent sudden
approximation (SA) provides a considerably more accurate description
of the scattering,[34,53] although the cross sections for the directly
coupled transitions (j'=j+1) are significantly overestimated. In addition, the isotopic enhancement with DCl vs. HCl targets is not predicted.

     The pure sudden approximation runs into difficulties for polar
molecule collisions because the dipole-dipole interaction is so long
ranged that τ_{col} becomes comparable to τ_{rot}. Alexander and DePristo[63]
have developed a modification of the SA, described as an "adiabatically corrected" sudden (ACS) approximation, which corrects for

the breakdown of the SA in large impact parameter collisions. This model, which is based on ideas similar to those advanced by others,[64-66] has been used to calculate ^{7}LiH-polar molecule inelastic cross sections[34,53] which are compared in Fig. 7 with the experimental results. The ACS cross sections agree with the latter rather well, and moreover the enhancement with DCl vs. HCl is reproduced as well. This effect turns out to be an energy resonance effect since the DCl rotational spacings better match those of ^{7}LiH than does HCl.[34]

This collaborative study suggests that simple, yet computationally efficient, dynamical models for rotationally inelastic collisions can be devised. The principle deficiencies in the presently employed ACS theory involve the description of large Δj transitions, where the assumption of straight-line trajectories and the neglect of shorter-range components in the interaction potential become inappropriate.

Recently there has also been much interest in the development of simple parameterized models for rotational energy transfer based on the original suggestion of Polanyi and Woodall[67] that such rate constants might vary inversely with the change in rotational energy. We have applied[68] two such energy gap models (exponential[69] and power law[11]) to LiH-polar molecule scattering. While the experimental partially averaged cross sections can be fitted extremely well by a power law dependence on the energy gap of the resolved LiH molecule, there is no way to predict fitted parameters a priori. Moreover, the fits to partially averaged cross sections cannot be used to predict the more fundamental fully resolved quantities (as determined by ACS calculations[63]). Additionally, fits to fully-quantum state-resolved cross sections[70] for HF-HF revealed a fundamental inaccuracy in the statistical factors[71] used which cannot be corrected by the introduction of a factor dependent solely on the energy gap. This indicates that the fine details of polar molecule rotational energy transfer cannot be reduced to a simple dependence on the inelastic energy gap. An alternative approach to the interpolation and extrapolation of R→T/R cross sections is the scaling relations developed by DePristo et al.[72] These are based on a physically reasonable extension of the sudden approximation.

5. CONCLUSION

Recent experimental advances now allow accurate determination of state-to-state integral and differential rotationally inelastic cross sections. An important tool in the detailed study of molecules heavier than H_2 and its isotopic variants has been tunable dye lasers, which have been employed both for state selection and detection. In this paper, we have described our technique, applied to LiH, for obtaining integral cross sections, which involves electric quadrupole rotational state selection and laser fluorescence

detection of scattered molecules. Bergmann and coworkers have developed [28,29] an elegant technique, applied to Na_2, for determining
differential cross sections, which uses lasers for both selection
and detection.

In the study of such collisional energy transfer processes, it
is greatly advantageous to interpret experimental results with sophisticated theoretical models. In our work on lithium hydride, we have
been very fortunate to collaborate with M.H. Alexander. We have thus
far studied in detail the scattering of LiH from helium and several
polar molecules. A clear result of our collaborative effort is the
demonstration that rotational energy transfer cannot be described
adequately by simplistic models but requires detailed consideration
of the interplay between translational and rotational motion, as well
as the intermolecular forces. We have also recently begun study of
vibrationally inelastic processes,[73] for which the difficulties in
formulating an accurate theoretical description are even greater.

ACKNOWLEDGEMENTS

The author is deeply indebted to his collaborators, Millard
Alexander and David Silver, for their contributions, encouragement,
and interest. This research has been supported in part by the
National Science Foundation under grants CHE-7725283 and CHE-78o8729.

REFERENCES

1.   Recent reviews of rotational energy transfer include
         J.P. Toennies, Ann. Rev. Phys. Chem. 27, 225 (1976);M. Faubel
         and J.P. Toennies, Adv. At. Mol. Phys. 13, 229 (1977);
         H.J. Loesch, Adv. Chem. Phys. 42, 421 (198o).
2.   T. Oka, J. Chem. Phys. 45, 752 (1966); Adv. At. Mol. Phys. 9,
         127 (1973).
3.   A.M. Ronn and E.B. Wilson, Jr., J. Chem. Phys. 46, 3262 (1967).
4.   R.G. Gordon, P.E. Larsen, C.H. Thomas, and E.B. Wilson, Jr.,
         J. Chem. Phys. 5o, 1388 (1969).
5.   B.A. Esche, R.E. Kutina, N.C. Lang, J.C. Polanyi, and A.M. Rulis,
         Chem. Phys. 41, 183 (1979); J.A. Barnes, M. Keil, R.E. Kutina,
         and J.C. Polanyi, J. Chem. Phys. 72, 63o6 (198o).
6.   Ch. Ottinger, R. Velasco, and R.N. Zare, J. Chem. Phys. 52,
         1636 (197o).
7.   R.B. Kurzel, J.I. Steinfield, D.A. Hatzenbuhler, and G.E. Leroi,
         J. Chem. Phys. 55, 4822 (1971).
8.   Ch. Ottinger and D. Popper, Chem. Phys. 8, 513 (1971);
         G. Ennen and Ch. Ottinger, ibid. 3. 4o4 (1974).
9.   K. Bergmann and W. Demtröder, Z. Phys. 243, 1 (1971); J. Phys.
         B 5, 1386, 2o98 (1972); K. Bergmann, W. Demtröder, M. Stock,
         and G. Vogl, ibid. 7, 2o36 (1974).

1o. R.K. Lengel and D.R. Crosley, J. Chem. Phys. 67, 2o85 (1977).

11. T.A. Brunner, R.D. Driver, N. Smith, and D.E. Pritchard,
    Phys. Rev. Lett. 41, 856 (1978); J. Chem. Phys. 7o, 4155 (1979).

12. M.D. Rowe and A.J. McCaffery, Chem. Phys. 43, 35 (1979).

13. J.J. Hinchen and R.H. Hobbs, J. Chem. Phys. 65, 2732 (1976).

14. Ph. Brechignac, Opt. Commun. 25, 53 (1978); Ph. Brechignac,
    A. Picard-Bersellini, R. Charneau, and J.M. Launay (to be publ.)

15. R.A. Gottscho, R.W. Field, R. Bacis, and S.J. Silvers,
    J. Chem. Phys. 73, 599 (198o).

16. A.R. Blythe, A.E. Grosser, and R.B. Bernstein, J. Chem. Phys.
    177, 84 (1964).

17. H.G. Bennewitz, K.H. Kramer, W. Paul, and J.P. Toennies,
    Z. Phys. 177, 84 (1964).

18. H. Van den Bergh, M. Faubel, and J.P. Toennies, Faraday Discuss.
    Chem. Soc. 55, 2o3 (1973); K. Rudolph and J.P. Toennies,
    J. Chem. Phys. 65, 4486 (1976).

19. H. Schmidt, V. Hermann, and F.Linder, J. Chem. Phys. 69, 2734
    (1978).

2o. W.R. Gentry and C.F. Giese, Phys. Rev. Lett. 39, 1259 (1977);
    J. Chem. Phys. 67, 5389 (1977).

21. U. Buck, F. Huisken, J. Schleusener, and H. Pauly, Phys. Rev.
    Lett. 38, 68o (1977); U. Buck, F. Huisken, and J. Schleusener,
    J. Chem. Phys. 68, 5654 (1978); U. Buck, F. Huisken,
    J. Schleusener, and J. Schäfer, ibid. 72, 1512 (198o).

22. M. Faubel, K.H. Kohl, and J.P. Toennies, J. Chem. Phys. 73,
    25o6 (198o).

23. W. Schepper, U. Ross, and D. Beck, Z. Phys. A29o, 131 (1979);
    D. Beck, U. Ross, and W. Schepper, Phys. Rev. A19, 2173 (1979).

24. H.G. Beenewitz, W. Pau-1, and Ch. Schlier, Z. Phys. 141, 6 (1955).

25. J.P. Toennies, Z. Phys. 182, 257 (1965).

26. U. Borkenhagen, H. Malthan, and J.P. Toennies, Chem. Phys. Lett.
    41, 222 (1976); J. Chem. Phys. 71, 1722 (1979).

27. L.Y. Tsou, D. Auerbach, and L. Wharton, Phys. Rev. Lett. 38,
    2o (1977); J. Chem. Phys. 7o, 5296 (1979).

28. K. Bergmann, R. Engelhardt, U. Hefter, and J. Witt, J. Phys. E
    12, 5o7 (1979).

29. K. Bergmann, R. Engelhardt, U. Hefter, P. Hering and J. Witt,
    Phys. Rev. Lett. 4o, 1446 (1978); K. Bergmann, R. Engelhardt,
    U. Hefter, and J. Witt, J. Chem. Phys. 71, 2726 (1979);
    K. Bergmann, U. Hefter, and J. Witt, ibid. 72, 4777 (198o).

3o. J.A. Serri, A. Morales, W. Morkowitz, D.E. Pritchard, C.H. Becker,
    and J.L. Kinsey, J. Chem. Phys. 72, 63o4 (198o).

31. R.N. Zare and P.J. Dagdigian, Science 185, 739 (1974);
    J.L. Kinsey, Ann. Rev. Phys. Chem. 28, 349 (1977).

32. P.J. Dagdigian, in XI Intern. Conf. on the Physics of Electronic
    and Atomic Collisions: Invited Papers, edited by N. Oda and
    K. Takayanagi (North-Holland, Amsterdam, 198o), p. 513.

33. B.E. Wilcomb and P.J. Dagdigian, J. Chem. Phys. 67, 3829 (1977).

34. P.J. Dagdigian, B.E. Wilcomb, and M.H. Alexander, J. Chem. Phys.
    71, 167o (1979).

35. J.B. Anderson, R.P. Andres, and J.B. Fenn, Adv. Chem. Phys. 1o, 275 (1966); J.B. Anderson, in Molecular Beam and Low Density Gas Dynamics, edited by P.P. Wegener (Dekker, New York, 1974), Vol. IV.
36. D.M. Silver, J. Chem. Phys. 72, 6445 (198o).
37. E.F. Jendreck and M.H. Alexander, J. Chem. Phys. 72, 6452 (198o).
38. P. McGuire, Chem. Phys. Lett. 23, 575 (1973); P. McGuire and D.J. Kouri, J. Chem. Phys. 6o, 2488 (1974).
39. R.T. Pack, J. Chem. Phys. 6o, 633 (1974).
4o. P.J. Dagdigian and B.E. Wilcomb, J. Chem. Phys. 72, 6462 (198o).
41. D. Secrest, ibid. 67, 1394 (1977).
42. L.W. Hunter, J. Chem. Phys. 62, 2855 (1975).
43. R.G. Gordon and Y.S. Kim, J. Chem. Phys. 56, 3122 (1972).
44. R.A. LaBudde and R.B. Bernstein, J. Chem. Phys. 55, 5499 (1971).
45. W.L. Dimpfl and B.H. Mahan, J. Chem. Phys. 6o, 3238 (1974).
46. C.E. Kolb and J.B. Elgin, J. Chem. Phys. 66, 119 (1977).
47. D. Beck, U. Ross, and W. Schepper, Z. Phys. A 293, 1o7 (1979).
48. M.H. Alexander and P.J. Dagdigian, J. Chem. Phys. 73, 1233 (198o).
49. A. Metropolous and D.M. Silver (to be published).
5o. L.D. Thomas, J. Chem. Phys. 67, 5224 (1977).
51. R. Schinke, J. Chem. Phys. 72, 112o (198o).
52. W. Eastes, U. Ross, and J.P. Toennies, Chem. Phys. 39, 4o7 (1979).
53. P.J. Dagdigian and M.H. Alexander, J. Chem. Phys. 72, 6513 (198o).
54. P.W. Anderson, Phys. Rev. 76, 647 (1949).
55. C.G. Gray, and J. Van Kranendonk, Can. J. Phys. 44, 2411 (1966).
56. H.A. Rabitz and R.G. Gordon, J. Chem. Phys. 53, 1815 (197o).
57. A.E. DePristo and M.H. Alexander, J. Chem. Phys. 66, 1334 (1977).
58. R.D. Sharma and C.A. Brau, J. Chem. Phys. 5o, 924 (1969).
59. R.B. Bernstein and K.H. Kramer, J. Chem. Phys. 4o, 2oo (1964).
6o. R.J. Cross, Jr., J. Chem. Phys. 55, 51o (1971).
61. M.H. Alexander, J. Chem. Phys. 71, 1683 (1979).
62. G.G. Balint-Kurti in Theoretical Chemistry, MTP Intern. Review of Science, Phys. Chem., Series 2, Vol. 1 (Butterworths, London, 1975), p. 285.
63. M.H. Alexander and A.E. DePristo, J. Phys. Chem. 83, 1499 (1979); M.H. Alexander, J. Chem. Phys. 71, 1683 (1979).
64. T.A. Dillon and J.C. Stephenson, Phys. Rev. A6, 146o (1972); J. Chem. Phys. 58, 3849 (1973).
65. D.P. Olsen and M.A. Wartell, J. Chem. Phys. 68, 5294 (1978).
66. R.J. Cross, Jr., J. Chem. Phys. 69, 4495 (1978).
67. J.C. Polanyi and K.B. Woodall, J. Chem. Phys. 56, 1563 (1972).
68. M.H. Alexander, E.F. Jendrek, and P.J. Dagdigian, J. Chem. Phys. 73, 3797 (198o).
69. I. Procaccia and R.D. Levine, Physica A82, 623 (1975); R.D. Levine, R.B. Bernstein, P. Kahana, I. Procaccia, and E.T. Upchurch, J. Chem. Phys. 64, 796 (1976); R.D. Levine, Ann. Rev. Phys. Chem. 29, 59 (1978).
7o. M.H. Alexander, J. Chem. Phys. 73, 5135 (198o).
71. M.H. Alexander, J. Chem. Phys. 71, 5212 (1979).

72. A.E. DePristo, S.D. Augustin, R. Ramaswamy, and H. Rabitz,
     J. Chem. Phys. 71, 850 (1979); A.E. DePristo and H. Rabitz,
     ibid. 72, 4685 (1980).
73. P.J. Dagdigian, Chem. Phys. 52, 279 (1980).

EXCITED STATES OF SMALL MOLECULES -

COLLISIONAL QUENCHING AND PHOTODISSOCIATION

Roberta P. Saxon

Molecular Physics Laboratory
SRI International
Menlo Park, CA 94025

INTRODUCTION

In the early days of quantum chemistry research, major emphasis was placed on method development and calculations were performed on physical systems more to evaluate calculational methods than to gain new insight into these systems. The field has now matured to the point where present-day methods are capable of providing reliable information for systems of experimental interest, including those where large numbers of excited states are involved. Predicting the cross sections and products of photodissociation requires knowledge of large numbers of excited state potential curves or surfaces, particularly repulsive potential curves that are not easily studied spectroscopically. Also, theoretical study of collisions between excited species requires information about excited state potential curves. Calculations of excited state potential curves discussed in this paper support the assertion that current quantum chemistry techniques are now providing physical insight for systems of interest.

This paper only briefly describes the calculational methods used, with major emphasis on the results. First, survey calculations on the many valence states of O_2 are described. Next, the problem of collisional quenching of excited atoms, which arose in the effort to develop a uv laser based on transitions in the Group IV atoms, O, S, and Se, is presented. This problem led to calculations on the Rydberg states of O_2 and on the Rydberg and valence states of S_2 which are subsequently described. The

application of these results to the collisional quenching problem
is given. Finally, the theoretical study of the photodissociation
of the astrophysically interesting CH^+ molecule is discussed. It
is a pleasure to acknowledge collaboration with Dr. Bowen Liu (IBM
Research Laboratory, San Jose, California) on all of these
calculations and with Dr. Kate Kirby (Harvard-Smithsonian Center
for Astrophysics, Cambridge, Massachusetts), who participated in
the CH^+ study.

SURVEY OF VALENCE STATES OF O_2

    The oxygen molecule, which is the second most abundant
element in the earth's atmosphere, plays an important role in
terrestrial chemistry. Detailed knowledge of excited state
potential curves is frequently needed to help understand the
complex oxygen spectrum, collision processes between oxygen atoms,
and photodissociation processes, as well as mechanisms for
chemical reactions involving the oxygen molecule. Consequently,
this molecule has been studied extensively both theoretically and
experimentally.[1]

    Seven low-lying bound valence states and several Rydberg
series have been identified in the O_2 spectrum. Although some
information about repulsive states may be inferred from the
observed spectrum, they are best known, for the most part, from
theoretical calculations. Surveys of the valence states of O_2 at
the minimal basis set, valence CI level have been performed by
Schaefer and Harris[2] and by Beebe, Thulstrup, and Andersen,[3] the
first of these published more than 10 years ago. The results
obtained in these calculations, however, are only of qualitative
accuracy. Although a number of calculations of higher accuracy
were performed[4,5] for a few states of O_2, no systematic study
existed of all the valence states of O_2 at a higher level of
accuracy than the minimal basis set valence CI studies. Therefore, we performed state-of-the-art configuration interaction
calculations on the 62 valence states of 25 molecular symmetries
of O_2 arising from two oxygen atoms in their lowest ^3P, ^1D, and ^1S
states.[6]

    Our calculations were at the level of a first-order CI (FOCI)
wavefunction, with configurations constituted by distributing 12
electrons among the $2\sigma_g$, $2\sigma_u$, $3\sigma_g$, $3\sigma_u$, $1\pi_u$ and $1\pi_g$ valence
orbitals, corresponding to oxygen atom 2s and 2p orbitals and by

distributing 11 electrons in valence orbitals and one in virtual orbitals. The $1\sigma_g$ and $1\sigma_u$ core orbitals were kept doubly occupied. With an extended Slater basis set consisting of 5s, 3p, and 2d functions on each center, the first-order wavefunctions amounted to between 397 and 5444 configurations, with the majority of the symmetries lying between 1000 and 3500 configurations. The same set of molecular orbitals was used for all the FOCI calculations. They were determined by a two-step process consisting of a properly dissociating MCSCF calculation on the ground $X^3\Sigma_g^-$ state of O_2, followed by the determination of natural orbitals from a FOCI calculation on the ground state using the MCSCF orbitals. This procedure eliminated the undesirable asymptotic splitting in energy between different states arising from the same asymptotic limit.

The potential curves resulting from these calculations for singlet and triplet states are shown in Figures 1 and 2. The calculated separated-atom limit energy levels of O_2 are given in

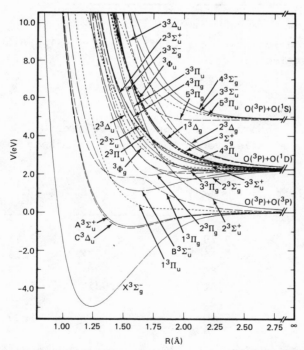

Fig. 1   Calculated potential curves for triplet valence states of O_2. Energies are given relative to $O(^3P) + O(^3P)$.

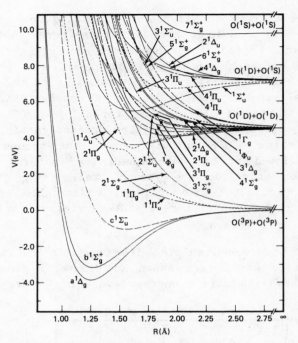

Fig. 2   Calculated potential curves for singlet valence states of
          O_2. Energies are given relative to $O(^3P) + O(^3P)$.

TABLE 1   Separated-Atom Energy Levels of O_2
Given in eV Relative to the $^3P + ^3P$ Asymptote

Separated-Atom Limit	This Work	Experiment	SH[*]	MG[+]
$^1S + ^1S$	9.79	8.380	9.46	8.05
$^1D + ^1S$	7.14	6.157	7.28	6.36
$^3P + ^1S$	4.88	4.190	4.73	4.02
$^1D + ^1D$	4.51	3.935	5.10	4.67
$^3P + ^1D$	2.25	1.967	2.55	2.34
Average ΔE	.79		1.09	.38

[*]Schaefer and Harris, Reference 2, Minimal basis, Valence CI
[+]Moss and Goddard, Reference 5, GVB + CI

Table 1, along with experimental values and the results of previous calculations. Our calculation yielded the correct ordering of the asymptotic states, in contrast to the minimal basis valence CI calculation of Schaefer and Harris[2] and the GVB+CI calculation of Moss and Goddard[5], these latter calculations erroneously place the $^3P + {}^1S$ asymptotic below the $^1D + {}^1D$ asymptote. The seven lowest bound states, in the correct order, were predicted by these calculations. Table 2 shows a comparison of some of the calculated molecular constants with those deduced from experimental data for four of these states. Table 3 shows the average error in these four molecular constants for the seven lowest bound states.

TABLE 2.   Comparison of Calculated and Experimental
Molecular Constants

State		R_e (Å)	D_e (eV)	T_e (eV)	ω_e (cm^{-1})
$X^3\Sigma_g^-$	This Work	1.236	4.957		1498.8
	Exptl.*	1.208	5.213		1580.2
$a^1\Delta_g$	This Work	1.250	3.857	1.098	1403.4
	Exptl.*	1.216	4.231	0.982	1509.3
$c^1\Sigma_u^-$	This Work	1.555	1.062	3.888	759.8
	Exptl.*	1.517	1.114	4.098	794.3
$B^3\Sigma_u^-$	This Work	1.627	1.136	6.079	724.9
	Exptl.*	1.604	1.007	6.173	709.1

*Experimental values from Krupenie, Reference 1.

TABLE 3   Average Error for Molecular Constants
of Seven Lowest Bound States

		This Work	SH*	MG*
R_e	0.033 Å	2.6%	5.7%	1.7%
D_e	0.20 eV	8.8%	33%	16%
T_e	0.20 eV	5.9%	17%	4.4%
ω_e	61 cm^{-1}	5.1%	13%	6.8%

*Defined in Table 1.

These comparisons show that the present calculations yielded realistic potential curves for the bound states of O_2 and, as expected, the results represent a significant improvement over the minimal basis set calculations.[2] This improvement is expected to hold true for all other states of O_2. The comparison also shows that our calculation and the GVB+CI work of Moss and Goddard[5] are comparable in accuracy.

Our calculation yielded ten additional bound states, five of which were also obtained in the minimal basis calculations. The most valuable information obtained in this work are the potential curves for many repulsive states of O_2, as well as some of the high-lying excited states that have not been observed experimentally. The most reasonable way to make use of these potential curves is to assume that each calculated interaction potential curve is correct and to shift the asymptote to agree with the observed separations. The potential curves can then be used to predict the products of photodissociation or the outcome of a collision between excited species. An example of such a collision problem is given in the next section.

GROUP VI LASER PROBLEM

A few years ago, Murray and Rhodes[7] proposed the auroral ($^1S \rightarrow {}^1D$) and transauroral ($^1S \rightarrow {}^3P$) transitions of Group VI elements as candidates for a high energy storage short pulse laser. Lasers with these characteristics have been proposed as reactor drivers for inertial confinement fusion. Attention was initially focused on the 5577 Å auroral line of atomic oxygen. The stimulated emission cross section was estimated to be 9×10^{-20} cm^2, small enough to permit high energy storage; the lower state, ^1D, is rapidly deactivated by most gases, and the upper state was expected to be strongly resistant to collisional deactivation.

To have high storage densities of $O(^1S)$, the cross section for self-quenching must be small. As shown in Fig. 2, the only potential curve arising from the $O(^1S) + O(^1S)$ asymptote is the $7\,{}^1\Sigma_g^+$ state and it clearly does not encounter any other valence state at an energy accessible in thermal collisions. However, Rydberg and ion pair states may need to be considered. Therefore, we performed calculations on the Rydberg and ion pair states of O_2 for all molecular symmetries that are coupled by some operator to the $^1\Sigma_g^+$ symmetry, the initial channel of a $O(^1S) + O(^1S)$

collision. Calculations were performed for the $^3\Pi_g$ and $^3\Sigma_g^-$ symmetries, which are coupled to the $^1\Sigma_g^+$ symmetry by the spin-orbit operator; the $^1\Pi_g$ symmetry, which is coupled by rotational coupling; and the $^1\Sigma_g^+$ symmetry itself. These calculations are described in the next section. Because of experimental problems associated with the production of $O(^1S)$, practical attention was soon directed to the analogous sulfur system. We subsequently performed calculations on the valence, Rydberg, and ion pair states of S_2, as discussed below.

RYDBERG STATES OF O_2

   To address the problem of $O(^1S)$ self-quenching, we performed ab initio calculations on the oxygen molecule in which the Rydberg and valence states were considered together. Because the ion pair states are valence states, they are automatically included in the calculation; it is necessary only to compute a sufficient number of states. Relatively little was known about the Rydberg states of O_2. Several Rydberg series have been identified spectroscopically, mainly in symmetries dipole-connected to the lowest three states. Theoretical studies have been limited to a region near the potential minimum of the Rydberg states. Since we are interested in a collision problem, we need to obtain potential curves that are correct asymptotically, as well as near the equilibrium region.

   Although we have attempted to give a balanced treatment of the valence and Rydberg states, our calculations are not of sufficient accuracy to yield quantitative results when valence-Rydberg mixing is significant. However, it is important to note that there are many cases where significant valence-Rydberg mixing does not occur, or occurs only over a very small range of internuclear separations. In these cases it is reasonable and convenient to adopt the notion of diabatic Rydberg states. This diabatic Rydberg state is exactly what is desired in addressing the $O(^1S)$ self-quenching problem.

   We first performed three calculations on the lowest $^3\Pi_g$ Rydberg state of O_2 with different amounts of electronic correlation energy for the valence and Rydberg states.[8] The goal of these calculations was to establish general computational procedures suitable for determining the potential curves of diabatic Rydberg states.

For all of these calculations, the oxygen atom Slater basis set used previously in the O_2 valence state was augmented by two 3s, two 3p, and one 3d basis function. Calculation I included the same configurations for the valence states as the FOCI survey calculations, with the restriction that the $2\sigma_g$ and $2\sigma_u$ orbitals remained fully occupied. Very little electronic correlation of the Rydberg states was included; that is, no configurations with excitations to the virtual space with respect to Rydberg configurations were included in the configurations list. Calculation II was the same as calculation I, but without the restriction of freezing the 2σ orbitals. Calculation III included the same amount of correlation for the valence states as calculation I (i.e., the 2σ orbitals were frozen), but also included correlation of the Rydberg state. The molecular orbitals for calculations I and II were taken from the orbitals for our valence state calculations, augmented by symmetrized unit vectors corresponding to the additional Rydberg series basis functions. The orbitals for calculation III were determined by a properly dissociating MCSCF calculation on the lowest $^3\Pi_g$ Rydberg state. The dimensions of the CI calculations were 1766 (I), 9222 (II), and 8350 (III).

In summary, the three sets of calculations differ in the amount of 2σ correlation for the valence states, and the amount of correlation for the Rydberg states. A comparison of their results would show how these two types of correlation energy affect the shape of diabatic Rydberg state potential curves.

The CI wavefunctions obtained in our calculations were classified according to their weights on the valence configuration, which consisted solely of core and valence orbitals. Those states with little or no weights on the valence configurations were classified as Rydberg; those with large weights were termed valence.

Results of the three CI calculations are shown in Figs. 3-5, where the calculated energy points have been connected adiabatically by cubic spline interpolation. The dots denote the lowest state of Rydberg character at each internuclear separation. In each case, connecting the dots gives a potential curve for the lowest Rydberg state in the diabatic sense.

The valence state potential curves obtained in these three calculations are very similar to each other. There are five valence states of $^3\Pi_g$ symmetry arising from combinations of

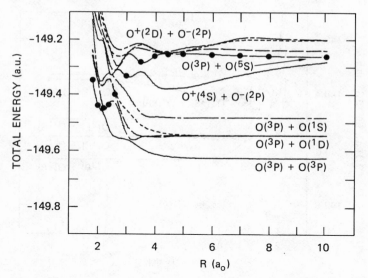

Fig. 3   Adiabatic potential curves for the nine lowest $^3\Pi_g$ states of O_2 from calculation I described in text. denotes the lowest Rydberg state at each internuclear separation. Separated atom limits to which the ion-pair states correlate diabatically are indicated near their minima.

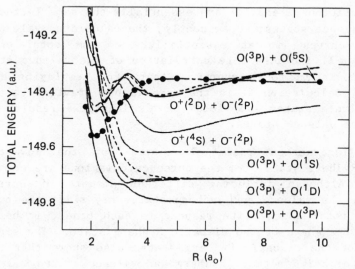

Fig. 4   Adiabatic potential curves for the nine lowest $^3\Pi_g$ states of O_2 from calculation II described in the text.

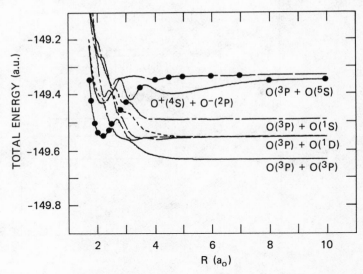

Fig. 5 Adiabatic potential curves for the seven lowest $^3\Pi_g$ states of O_2 from calculation III described in the text.

$O(^3P)$, $O(^1D)$, and $O(^1S)$. The next higher valence states arise from ion pair asymptotes. The ion-pair states may be identified by their attractive behavior at long range.

Calculations I and III are essentially the same in the treatment of valence states; consequently, the calculated valence state potential curves lie at essentially the same total energies. Calculation II includes more correlation of the valence state and thus is lower in total energy. Apart from employing a larger basis set, calculation II is the same as our previous calculation on the valence states and yields nearly identical results for the five lowest valence states.

The interaction potential curves for the lowest $^3\Pi_g$ diabatic Rydberg state from the three calculations are compared in Fig. 6. Although including different amounts of correlation energy for the valence and Rydberg states may shift the potential curves in total energy with respect to each other, it has little effect on the shape of the diabatic Rydberg state. The similarity in shape of the diabatic Rydberg states also shows that there is little interaction between Rydberg and valence states. It may be concluded that in the absence of significant Rydberg valence

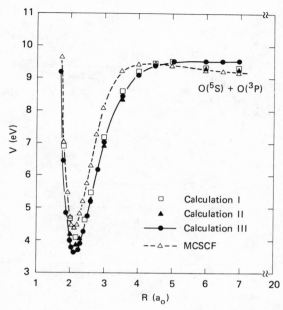

Fig. 6 Rydberg interaction potentials from the three calcula-
        tions described in text. □, calculation I; ▲, calcu-
        lation II; and ●, calculation III (Connected by a solid
        line). The interaction potential from the MCSCF calcula-
        tion that determined the Rydberg orbitals for calculation
        III is denoted by △ (connected by a dotted line). The
        zero of energy is two ground state oxygen atoms.

interaction, when one is interested in describing a diabatic
Rydberg state, a relatively small calculation for the Rydberg
state is sufficient.

    Following this comparative study, configuration interaction
calculations were performed for the Rydberg, valence, and ion pair
states of the $^3\Sigma_g^-$, $^3\Sigma_u^-$, $^3\Pi_g$, $^1\Pi_g$ and $^1\Sigma_g^+$ symmetries
of O_2.[9] These calculations were at the level of calculation II,
which includes little correlation for the Rydberg state and
includes all the configurations for a full first-order CI for the
valence states. Results for the $^3\Sigma_g^-$ symmetry, for example, are
shown in Fig. 7, where the deeply bound $X^3\Sigma_g^-$ ground state,
three excited valence states, and four ion pair states may be
easily identified. Again, the dots denote the lowest Rydberg
state at each internuclear separation.

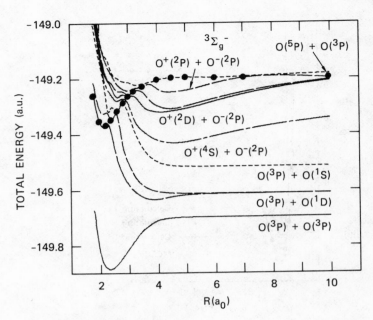

Fig. 7   Adiabatic potential curves for the nine lowest $^3\Sigma_g^-$ states of O_2.

All of the potential curves relevant to the quenching of $O(^1S)$ by $O(^1S)$ are shown in Fig. 8, where they have been translated to their spectroscopic asymptotes with respect to ground state atoms. From these calculations, the only potential curves that cross the $7\ ^1\Sigma_g^+$ state at an energy accessible in thermal energy collisions are the $^3\Pi_g$ and $^3\Sigma_g^-$ ion pair states. We will defer a discussion of the implications for $O(^1S)$ collisional quenching until after examining the situation in the analogous S_2 molecule.

CALCULATIONS ON S_2

Although the oxygen molecule has received a great deal of theoretical attention, there have been very few calculations on the analogous second-row molecule S_2 or on second row diatomics in general. Recent work by Swope et al.[10] focused on the lowest few valence states of each symmetry of S_2, but there had been no calculation on highly excited valence, Rydberg, or ion pair states of S_2. To study the $S(^1S)$ self-quenching problem, we performed calculations on the $^1\Sigma_g^+$, $^3\Pi_g$, $^3\Sigma_g^-$, and $^1\Pi_g$ symmetries of

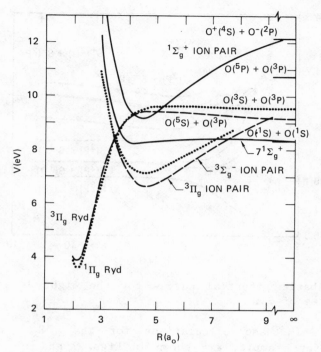

Fig. 8 O_2 potential curves relevant to $O(^1S)$ self-quenching
translated to the spectroscopic asymptotic separation.
Ion pair curves have been translated under the assumption
that they are Coulombic at $20a_o$.

S_2,[11] analogous to those for O_2 described above.   In addition to
determining potential curves for the study of the collision
problem, this work provided the opportunity to compare the O_2 and
S_2 systems under analogous treatment.

The elementary basis set for this calculation consisted of
9s, 8p and 3d type Slater functions, including two 4s and two 4p
Rydberg functions on each atom.   The molecular orbitals were
determined by a properly dissociating MCSCF calculation on the
ground $X^3\Sigma_g^-$ state.   The configuration list included the configurations of a first-order CI for the valence states, with the
orbitals correlating to the sulfur K and L shells kept fully
occupied.   Very little correlation was included for the Rydberg
state.

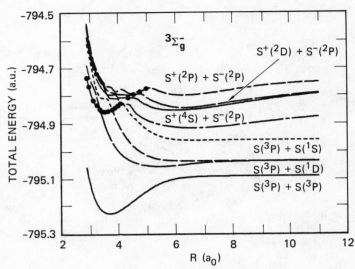

Fig. 9 Adiabatic potential curves for the eight lowest $^3\Sigma_g^-$ states of S_2.

Results of these calculations for the $^3\Sigma_g^-$ and $^3\Pi_g$ symmetries, for example, are shown in Figs. 9 and 10. They are qualitatively similar to the corresponding O_2 potential curves (Figs. 7 and 4). Since oxygen and sulfur have the same valence configurations, the number of states arising from each asymptote is the same for both molecules. The differences arise from the differences in spacing and ordering of the asymptotes. Calculated and experimental values of the spectroscopic parameters were compared for the ground $X^3\Sigma_g^-$ state. In the sulfur system, the first-order CI produces larger errors in the well depth and equilibrium separation than the analogous calculation for the O_2 ground state.

IMPLICATIONS FOR ^1S SELF-QUENCHING AND THE GROUP VI LASER

All of the potential curves relevant to the study of $S(^1S)$ self-quenching are shown in Fig. 11. The situation is completely analogous to that found in O_2. In both cases, we need only consider the crossings by the $^3\Sigma_g^-$ and $^3\Pi_g$ ion pair states that occur at very large internuclear separations. If collisional quenching does occur, it would take place by a multistep process, the first step of which involves transfer to an ion pair curve. Flux could then transfer from an ion pair curve to a Rydberg

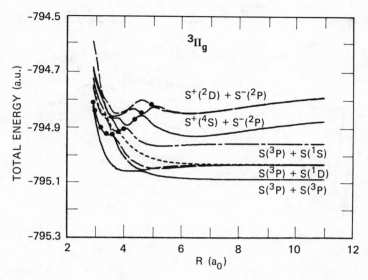

Fig. 10 Adiabatic potential curves for the eight lowest $^3\Pi_g$ states of S_2.

state, which would most probably subsequently autoionize or predissociate, resulting in ^1S deactivation. The only operator that couples states of $^1\Sigma_g^+$ symmetry is the spin-orbit operator. The rate for this complicated multistep process, then, depends initially on the magnitude of the spin-orbit matrix element.

We have made a preliminary evaluation of the one-electron part of the spin-orbit operator; for S_2 at $10a_o$, the distance of the crossing, we found the matrix element between the $^3\Pi_g$ ion pair and the $7^1\Sigma_g^+$ state to be very small (~ 0.4 cm^{-1}). One may reasonably expect the matrix element between the $^3\Sigma_g^-$ ion pair and the incoming channel to be similarly small. The matrix element was the same order of magnitude for the same two states in O_2 at their crossing distance, $\sim 7a_o$.

Although the Landau-Zener theory is not expected to be entirely reliable when the probability of making a transition is extremely small, it may be used to obtain an estimate of the quenching rate from the calculated matrix element. In making this estimate, we assume that the first crossing with the ion pair curve controls the quenching rate; that is, all further steps take

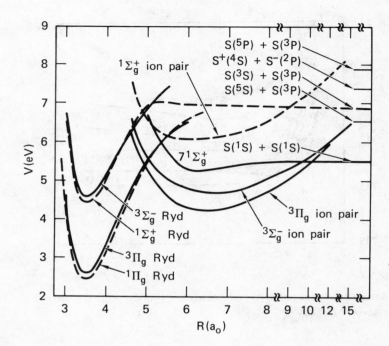

Fig. 11 S_2 potential curves relevant to $S(^1S)$ self-quenching have been translated to the spectroscopic asymptotic separation. Ion pair curves have been translated under the assumption that they are Coulombic at $30a_o$.

place with unit probability. With this assumption, one would estimate that the self-quenching rate for $S(^1S)$ and for $O(^1S)$ is $\ll 10^{-12}$ cm^3 sec^{-1} $molecule^{-1}$. Thus, we concluded that ^1S self-quenching will not significantly affect excited state storage times.

In the end, problems other than the ^1S self-quenching considered here proved to be the major stumbling block in the development of Group VI lasers.

CALCULATIONS ON CH^+: PHOTODISSOCIATION

There has been great astrophysical interest in CH^+ since it was identified in diffuse interstellar clouds. It has also been observed in comets. Furthermore, transitions to excited states of CH^+ not yet discovered in the laboratory have been suggested as possible origins of unidentified absorption lines in interstellar clouds. In modeling the formation and destruction of CH^+ in diffuse interstellar clouds, it is assumed photodissociation occurs only by absorption into the excited $A^1\Pi$ state and is negligible. It would be desirable to know the rates of photodissociation of CH^+ through other excited states.

Therefore, we undertook a study to characterize those excited states of CH^+ that, in dipole transitions from the ground state, can give rise to band spectra or to direct dissociation of the molecule.[12] Since the observed abundance of CH^+ appears to be as much as 30 times larger than predictions based on probable formation and destruction mechanisms, prediction of additional destruction by photodissociaion will not resolve the discrepancy, but the process should be included in the model.

In this work, we were interested in states that lie within 13.6 eV of the ground state because there are no photons in the interstellar medium with greater energy, as a result of ionization of atomic hydrogen. The lowest two Rydberg limits, $C(^1P, 2p3s)$ + H^+ and $C^+(^2P)$ + H (2s, 2p) lie within about 10 eV of the ground state asymptote. States arising from these Rydberg asymptotes may be of astrophysical interest. In addition, lower-lying states may experience some valence-Rydberg mixing at small internuclear separations. Thus it was necessary to design a calculation that gives a balanced treatment of both valence and Rydberg states.

The elementary basis set of Slater type functions included Rydberg exponents. The CI calculations included all single and double excitations with respect to a set of reference configurations in which both valence and Rydberg orbitals were occupied and which correctly dissociated the molecular states. The molecular orbitals were determined by a procedure designed to provide physically realistic Rydberg as well as valence orbitals. After determining valence orbitals by MCSCF calculations on pure valence states, we obtained Rydberg σ orbitals as natural orbitals from a two-electron CI calculation for Σ Rydberg states; an analogous procedure was followed for π orbitals.

The results of these calculations are shown in Fig. 12, where the five lowest $^1\Sigma^+$ and four lowest $^1\Pi$ states are plotted. Of these nine states, only the $X^1\Sigma^+$, $A^1\Pi$, and $2^1\Sigma^+$ states are bound with respect to their asymptotic limits. The small $^1/R^4$ attraction of the $2^1\Sigma^+$ state leads to an energy minimum of about

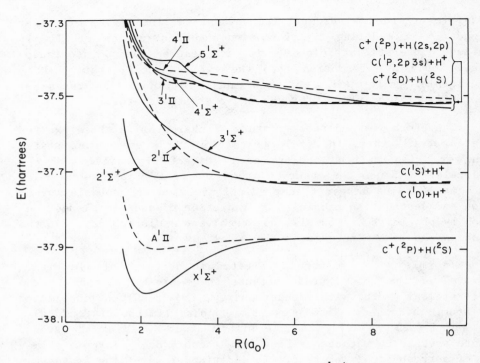

Fig. 12   Potential curves of CH$^+$: five of $^1\Sigma^+$ symmetry and four
          of $^1\Pi$ symmetry.

0.2 eV at R ~ 6.2a_0, which is, however, well outside the Franck-
Condon region of the ground state. The avoided crossing of the
$2^1\Sigma^+$ and $3^1\Sigma^+$ states gives rise to a hump in the $2^1\Sigma^+$ state
that can support quasibound levels. The $5^1\Sigma^+$ state also appears
to have a quasibound vibrational level.

It is clear that with the exception of these quasibound
levels, no new CH$^+$ band spectra are to be found. The excited
states may be important, however, for photodissociation. With the
asymptotes shifted to agree with experiment, in addition to the
previously studied A$^1\Pi$ state, the $2^1\Sigma^+$, $3^1\Sigma^+$, and
$2^1\Pi$ states lie within 13.6 eV of the v = 0 level of the ground
state. Transition moments and photodissociation cross sections
for absorption to these three states have been calculated.[12,13]
The most important state in the photodissociation of CH$^+$ in the
interstellar medium is the $3^1\Sigma^+$ state with a peak cross section
of 3 × 10^{-17} cm^2 at 12.6 eV. Half of the total integrated cross

section for the $X^1\Sigma^+ - 2^1\Pi$ transition, which has a peak value of 1.3×10^{-17} cm^2 at 13.6 eV, also will be effective in photodissociation of CH$^+$ in the interstellar medium. The effect of these large cross sections on the various models for formation and destruction of interstellar CH$^+$ is currently being explored.

CONCLUDING REMARKS

These examples illustrate that processes involving highly excited states are amenable to study by ab initio techniques. At present, impressive advances are being made by a number of workers in the development of MCSCF and configuration interaction techniques. It is probable that in the next few years, the range of systems that can be studied by ab initio methods will be greatly expanded over what is now possible.

ACKNOWLEDGMENT

Portions of this work were supported by the National Science Foundation Aeronomy Program and by the Air Force Office of Scientific Research.

REFERENCES

1. P. Krupenie, The spectrum of molecular oxygen, J. Phys. Chem. Ref. Data 1:423 (1972).
2. H. F. Schaefer III and F. E. Harris, Ab initio calculations on 62 low-lying states of the O_2 molecule, J. Chem. Phys. 48:4946 (1968).
3. N.H.F. Beebe, E. W. Thulstrup and A. Andersen, Configuration interaction calculations of low-lying electronic states of O_2, O_2^+ and O_2^{2+}, J. Chem. Phys. 64:2080 (1976).
4. H. F. Schaefer III and W. H. Miller, Curve crossing of the $B^3\Sigma_u$ and $^3\Pi_u$ states of O_2 and its relation to predissociation in the Schumann-Runge Bands, J. Chem. Phys. 55:4107 (1971).
5. B. J. Moss and W. A. Goddard III, Configuration interaction studies of low-lying states of O_2, J. Chem. Phys. 63:3523 (1976).
6. R. P. Saxon and B. Liu, Ab Initio configuration interaction study of the valence states of O_2, J. Chem. Phys. 67:5432 (1977).

7. J. R. Murray and C. K. Rhodes, The possibility of high-energy-storage lasers using the auroral and transauroral transitions of column-VI elements, J. Appl. Phys. 47:5041 (1976).

8. R. P. Saxon and B. Liu, Ab Initio configuration interaction study of the Rydberg states of O_2. I. A general computational procedure for diabatic Molecular Rydberg states and test calculations on the $^3\Pi_g$ states of O_2, J. Chem. Phys. 73:870 (1980).

9. R. P. Saxon and B. Liu, Ab Initio configuration interaction study of the Rydberg states of O_2. II. Calculations on the $^3\Sigma_g^-$, $^3\Sigma_u^-$, $^3\Pi_g$, $^1\Pi_g$ and $^1\Sigma_g^+$ symmetries, J. Chem. Phys. 73:8760 (1980).

10. W. C. Swope, Y.-P. Lee, and H. F. Schaefer III, Diatomic sulfur: low-lying bound molecular electronic states of S_2, J. Chem. Phys. 70:947 (1979).

11. R. P. Saxon and B. Liu, Ab Initio calculations on the $^3\Sigma_g^-$, $^1\Sigma_g^+$, $^3\Pi_g$ and $^1\Pi_g$ symmetries of S_2: valence, ion pair, and Rydberg states, J. Chem. Phys. 73:5174 (1980).

12. R. P. Saxon, K. Kirby and B. Liu, Excited states of CH^+: Potential curves and transition moments, J. Chem. Phys. 73:1873 (1980).

13. K. Kirby, W. G. Roberge, R. P. Saxon, and B. Liu, Photodissociation cross sections and rates for CH^+ in interstellar clouds, The Astrophysics J. 239:855 (1980.)

MOLECULAR NEGATIVE IONS

Ronald E. Olson[*]

Molecular Physics Laboratory
SRI International
Menlo Park, California 94025

INTRODUCTION

The determination of the collision mechanisms responsible for electron detachment in low-energy collisions of negative ions with atoms is an interesting and active field. Several different collision mechanisms have been proposed to explain the experimental scattering data.

One is the complex potential method given by Lam et al[1] which requires a deep penetration of the negative-ion state into the continuum for applicability. This electron-detachment model is similar to that first used by Mason and Vanderslice[2] with the extension that the transition probabilities are computed in a manner similar to that employed in Penning ionization calculations. Another mechanism, termed the "zero-radius model," is similar to a charge transfer process which only requires that the negative-ion and neutral states merge reasonably close to one another for there to be a significant probability for electron detachment during the collision. Calculations using the latter mechanism have been presented by Gauyacq.[3] Finally, the influence of electron capture channels in the electron removal process has been described by Olson and Liu[4] in their treatment of H⁻ scattering from alkali atoms.

However, inherent in the application of any scattering model

[*]Present address: Department of Physics, University of Missouri-Rolla, Rolla, MO 65401, U.S.A.

185

to describe electron detachment, is the need to accurately know the interaction potentials that the reactants and products follow. This information is the basic input to any scattering calculation. In this report we present some of our recent molecular calculations on negative-ion and neutral systems which illustrate the different scattering mechanisms. Major emphasis will be given to the numerical results and predicted cross section behavior with the calculational details being only briefly described.

H⁻ + RARE GASES

Experimentally, some of the most extensively studied systems are collisional detachment of H⁻ by rare gases. Much of the low energy work has been conducted at William and Mary by the group of Champion and Doverspike and colleagues. One of their interesting observations[5] was an isotope effect in H⁻ and D⁻ scattering from He and Ne. These data displayed reverse isotope effects with the D⁻ + He total electron-detachment cross section lying above that of H⁻ + He when plotted as a function of collision energy, while the H⁻ + Ne data were above those for D⁻ + Ne. Such observations prompted us to calculate the interaction energies for these systems.

Our first calculations for the ground state interaction energies for HeH and HeH⁻ were performed using the self-consistent-field (SCF) method.[6] The results showed the HeH⁻ state penetrated into the HeH continuum for $R \lesssim 2.9$ a_o. The values were somewhat questionable because of the neglect of electron correlation in the SCF method. Thus, we redid the calculations at the configuration interaction (CI) level including all single and double electron excitations.[7] The results are shown in Fig. 1.

The CI calculations confirmed the SCF results that there is a sharp crossing of the HeH⁻ state into the HeH continuum. The CI value for the internuclear separation at the crossing was 2.7 a_o with a threshold energy of 1.34 eV. This energy is consistent with the experimental data,[5] along with the fact that one would expect to observe a distinct angular threshold for electron detachment on the differential cross sections for angles corresponding to turning points less than 2.7 a_o.

The conclusion that the HeH⁻ penetrates deeply into the continuum and an electron detachment model such as the complex potential method[1] is valid at threshold energies for this system is confirmed by the isotope dependence of the total electron

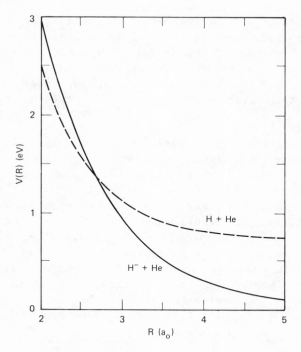

Fig. 1. Potential energy curves for the $X^1\Sigma$ state of HeH$^-$ and the $X^2\Sigma$ state of HeH calculated using the CI method.

detachment cross sections. At a given collision energy, the complex potential model predicts the detachment probability is proportional to the amount of time the collision partners remain in the continuum or inversely proportional to the relative velocity. Thus, as shown by Lam et al.,[1] their theory confirms the larger cross section for D$^-$ + He as compared to H$^-$ + He.

To further substantiate our prediction that the interaction potential followed by the H$^-$ + He collision partners does cross into the continuum, we performed additional calculations on the HeH$^-$ and HeH systems. Additional diffuse hydrogenic orbitals of the type $ns\sigma$, $np\sigma$, $np\pi$ ($3 \leqslant n \leqslant 10$) were added to the H basis set to represent continuum orbitals. A complete two-electron CI was then performed on HeH$^-$ and an SCF calculation on HeH (in this test we did not correlate the inner He(1s^2) shell because doing a balanced calculation on both systems would require quadrupole excitations for HeH$^-$). In every test we found the negative ion curve crossed sharply into the continuum at $\sim 2.7\ a_o$ with a threshold energy of ~ 1.4 eV. At all R values, analysis of the natural orbitals showed H^- (1sns) + He and H^- (1snp) + He continuum states (3 ⩽ n ⩽ 10) lying above the H + He neutral state. However, for 1.5 ⩽ R ⩽ 2.5 a_o a higher-lying eigenvalue was always found with a configuration close to that of H^- ($1s^2$) + He whose energy remained higher than that of the neutral state even under variation in the added hydrogenic orbitals.

The interaction potentials for the NeH and NeH$^-$ systems, however, displayed a much different behavior, Fig. 2. For NeH the negative ion curve tends to merge with the continuum around R ≈ 2.25 a_o rather than cross sharply into it. Even the SCF calculations predict very shallow penetration into the continuum. Hence, one would predict a different cross section behavior for NeH than was observed for HeH. From the potential energy curves, we would expect smaller cross sections for NeH than HeH at threshold energies since the NeH negative ion state does not penetrate deeply into the ocntinuum to allow for strong coupling; such cross section behavior is observed.[5] We would also not expect the complex potential method to be applicable to NeH.

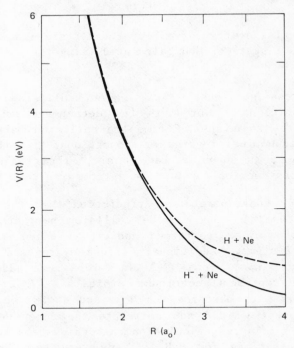

Fig. 2. Potential energy curves for the $X^1\Sigma$ state of NeH$^-$ and the $X^2\Sigma$ state of NeH calculated using the CI method.

However, the method of Gauyacq,[3] which is similar to a charge transfer formalism, should lead to a reasonable comparison with experiment. Inherent in this model would be the prediction that the H^- + Ne total detachment cross section will be larger than that for D^- + Ne at the same collision energy, in agreement with experiment.

An interesting aspect of the potential energy results is that the effective radius for electron detachment is greater for H^- + He than for H^- + Ne. If we use a rough estimate of the total electron detachment cross section of πR_c^2, where R_c is the distance where the negative ion and neutral state merge together, we would predict maximum cross sections of approximately $\pi(2.7)^2 a_o^2 = 6.4 \times 10^{-16}$ cm^2 for H^- + He and $\pi(2.25)^2 a_o^2 = 4.5 \times 10^{-16}$ cm^2 for H^- + Ne. Values close to these have been observed by Risley and Geballe[8] and Williams[9] and thus further substantiate the potential energy calculations.

H^- + ALKALI ATOMS

Collisions of H^- with alkali atoms present the interesting possibility that the electron detachment may be due to electron capture to form negative alkali ions

(1) H^- + Alk → H + Alk$^-$

or simply due to direct detachment

(2) H^- + Alk → H + Alk + e^-.

In order to assess these possibilities, the interaction energies for the low-lying states of NaH and NaH$^-$ were calculated.[4]

The CI results for several of the states of NaH are displayed in Fig. 3. The potential minimum of the $X^1\Sigma$ state was calculated to be located at 3.558 a_o with a well depth of 1.922 eV. The equilibrium position agrees with spectroscopic data to within 0.01 a_o and the calculated vibrational spacings are within 6 cm^{-1} to published RKR results.[10]

The results of the CI calculations on the low-lying $^2\Sigma$ molecular states of NaH$^-$ are presented graphically in Fig. 4. We should caution the reader that the calculations were not stabilized in any way. Thus, the interaction curves presented

which are in the H^O + Na + e^- continuum can only be considered as
qualitative. The general characteristics of the NaH$^-$ curves are
that the ground state is stable relative to autoionization, there
is very strong long-range configuration mixing between the $X^2\Sigma$
and $A^2\Sigma$ states arising from the H^- + Na and H + Na$^-$ charge
transfer conbinations, and that there are a series of strong
avoided crossings in the continuum.

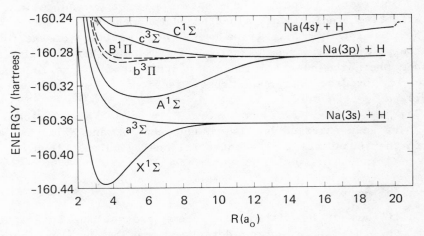

Fig. 3.   CI interaction energies for the low-lying molecular
states of NaH.

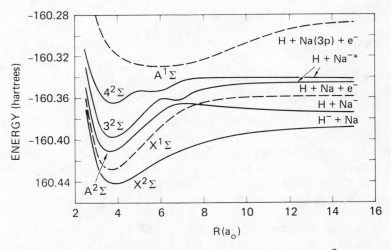

Fig. 4.   CI interaction energies for the low-lying $^2\Sigma$ molecular
states of NaH$^-$. As a reference, the calculated $X^1\Sigma$ and
$A^1\Sigma$ states of NaH are given by dashed lines.

The calculated potential well parameters for the $X^2\Sigma$ state of NaH$^-$ are R_e = 3.851 a_o and D_e = 1.522 eV. We obtain the adiabatic molecular electron affinity by taking the difference in the calculated negative ion and neutral state potential minima, E_{min}(NaH) $-$ E_{min}(NaH$^-$). The theoretical value is 0.358 eV. Since negative ion photodetachment experiments will observe the v' = 0 to v = 0 energy difference, we have corrected for the zero point energies and find 0.374 eV. A more accurate value for the molecular electron affinity may be obtained by adjusting for the error in our calculated electron affinity of H which is 0.015 eV. This procedure leads to a best estimate of the adiabatic electron affinity of 0.373 ± 0.02 eV and a value of 0.389 ± 0.02 eV for the v' = 0 to v = 0 transition.

A major reason for our calculation of several of the low-lying states of NaH$^-$ was to try to determine the low energy electron detachment mechanism, Reaction (1) or (2). There are no experimental measurements at low energies of the total cross sections, yet, these reactions are very important in determining the efficiency and the intensity that can be realized in H$^-$ ion sources. We employed impact parameter coupled equations to describe the scattering so that it was possible to follow the probability evolution on each channel at every impact parameter. Thus, we can determine the probability that the particles will be on the $A^2\Sigma$ potential and cross into the neutral state continuum (see Fig. 4) We then assumed unit probability for electron ejection for the trajectories that cross into the continuum. This assumption is based on expected strong coupling to H + Na + e$^-$ continuum states and to curve crossing interactions in the continuum, which lead to autodetaching H + Na^{-*} states. The results of these calculations are given in Fig. 5.

From the cross section calculations, we can see the electron capture reaction (1) dominates over the electron ejection reaction (2) in low energy electron loss collisions of H$^-$ with alkalis. The cross sections are also very large, reflecting the fact that the collision process is very long-range in nature with impact parameters of 15 a_o contributing to the cross sections.

In the collision energy range investigated, a general trend of the cross sections is for the lighter alkalis to yield the largest cross sections. This trend is determined by the smaller exothermicities, $\Delta V(\infty)$, and dipole polarizabilities of the lighter alkalis which allows reaction to occur at larger impact parameters and with higher probability. The large ratio of

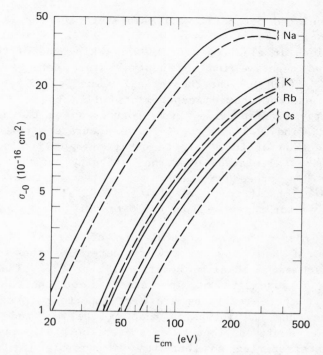

Fig. 5 Calculated σ_{-o} cross sections for the electron loss
reactions $H^- + Alk \rightarrow H^o + \dots$ where $Alk \equiv Na$, K, Rb,
and Cs--solid lines. The components of the electron loss
that   are   due   to   electron   transfer,
$H^- + Alk \rightarrow H^o + Alk^-$, are given by the dashed lines.
The difference between the above cross sections
represents direct detachment to the $H + Na + e^-$ continuum
and production of autodetaching states of $H + Na^{-*}$.

electron   capture,   Reaction (1),   to   electron   ejection,
Reaction (2), in the overall electron loss process can be
understood in terms that the reaction radius for electron capture
is $\sim 15\, a_o$ while that for electron ejection is only $\sim 8\, a_o$.
Since the total cross sections are roughly related to the square
of the reaction radii, it is understandable that the electron
ejection process will be very much less than that of electron
capture.   In turn, differential cross section measurements should
show that electron capture dominates at small scattering angles
and direct electron detachment dominates at large angles.

H^- + ALKALINE EARTH ATOMS

An interesting aspect of the alkaline earths is that they do not possess a bound and stable negative ion state due to their closed shell configuration. Hence, the electron detachment process can only proceed via the direct ionization channel

$$(3) \qquad H^- + AE \rightarrow H + AE + e^-.$$

Crucial to any prediction of the cross section for (3) is the determination whether or not the interaction energy for the negative ion reactants crosses or merges with the neutral state continuum. If this occurs, as it does for the H^- + rare gas system, the cross section will be very large even down to low energies.

Thus, we undertook to calculate CI interaction energies for the CaH ionic and neutral sequence because they should provide a good prototype of the heavier alkaline earth-hydride systems. The general relationship of the ionic states to the neutral state should be analogous for the other systems. Except for spectroscopic data on the neutral states, there is little other information on the heavy alkaline earth-hydride systems.

The CI calculations for the ground states of CaH and CaH^- were relatively small, and only included single and double excitations out of the valence electron shells. The CaH^+ calculations, however, also included the correlation between the valence and first inner shells. The results of the computations are displayed in Fig. 6 and a compilation of the molecular constants is given in Table 1.

From Table 1 we see the calculated equilibrium separation for the $X^2\Sigma$ state of CaH is at an R value which is greater than the spectroscopic value by ~ 0.05 Å. This shift is expected since CaH in the well region is primarily an ion-pair state composed of $Ca^+ - H^-$. From calculations on the Ca atom, we find the ion-pair state will dissociate to a level which is 0.20 eV below the true value. Hence, the coulomb potential of the ion-pair state shifts the R_e for the $X^2\Sigma$ state of CaH to too large of an internuclear separation. The calculated dissociation energy, however, is in good agreement with experiment.[12]

The $X^2\Sigma$ state of CaH^- is found to be bound and stable to electron autodetachment. The calculated adiabatic molecular electron affinity for the $v' = 0$ to $v'' = 0$ transition is

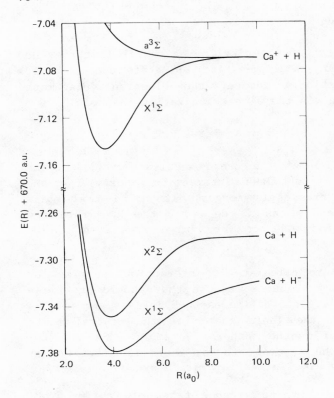

Fig. 6. CI potential energy calc-ulations for the molecular states disso-ciating to the ground states of CaH^+, CaH, and CaH^-.

Table 1 Molecular Constants

Molecule	State	Method	R_e(Å)	D_e(eV)	ω_e(cm^{-1})	B_e(cm^{-1})
CaH^+	$X^1\Sigma$	CI	1.881	2.166	2.073	1504.5	4.85
CaH	$X^2\Sigma$	CI	2.055	1.764	1.686	1268.7	4.06
		Exptl.	2.003	-----	⩽ 1.70	1298.3	4.28
CaH^-	$X^1\Sigma$	CI	2.167	1.900	1.837	1035.1	3.65

0.766 eV. If we shift the negative ion and neutral potentials to their true asymptotic separation, the adiabatic electron affinity increases to 0.905 eV. Hence, it is unequivocal that the CaH^- molecular ion is stable and deeply bound.

Referring to the neutral and negative ion potential curves given in Fig. 6, we predict the electron detachment cross section for

$$(4) \qquad H^- + Ca \rightarrow H + Ca + e^-$$

will be exceedingly small in the molecular collision regime. This prediction is based on the observation that the negative ion and neutral states are well separated at all internuclear distances calculated. Coupling of negative ion state to the continuum will be small since there is no direct crossing into the continuum as observed for H^- + rare gas systems, nor is the electron loss mediated by a charge transfer state as in the H^- + alkali systems. We conclude the electron detachment cross section will decrease rapidly with decreasing energy below 500 eV. At higher energies this cross section will be simply due to impact ionization of the H^-. The other heavy alkaline earth atoms should exhibit the same behavior.

Li^- + ALKALI ATOMS

Theoretical studies on the heteronuclear alkali molecules provide another opportunity to compare calculated and experimental results. The excited states of the neutral molecules are expecially interesting because of their wealth of structure due to configuration interaction between close-lying levels and the influence of two ion-pair states in the $^1\Sigma$ manifold. No information is available about the negative-ion molecular states.

Our CI calculations on the alkali molecules have been very modest in size and are meant to provide a qualitative insight into the behavior of ground and excited molecular states. The CI calculations only include single and double excitations out of the valence electron shell and do not take into account the core-valence interactions which are known to be necessary for an accurate description of the states.[13]

Calculations for the singlet and triplet manifolds of LiNa are presented on Figs 7 and 8. The high-lying $^1\Sigma$ states exhibit a very interesting behavior due to the interaction from the two ion-pair states $Li^- + Na^+$ and $Na^- + Li^+$. The triplet manifold is somewhat more regular except for a series of avoided crossings between the high-lying $^3\Sigma$ states around $R \approx 8\ a_o$.

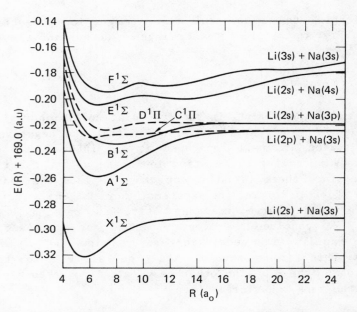

Fig. 7. CI calculations for the low-lying singlet molecular
         states of LiNa.

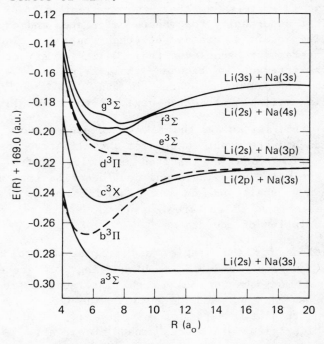

Fig. 8. CI calculations for the low-lying triplet molecular
         states of LiNa.

The singlet and triplet states of LiK display similar shapes to those of the LiNa as seen in Fig. 9. Here, however, the effects of both ion-pair states $Li^- + K^+$ and $K^- + Li^+$ are not as pronounced due to the very much different ionization potentials of Li and K.

A tabulation of the calculated ground state parameters for LiNa and LiK is given in Table 2. We should note that since our

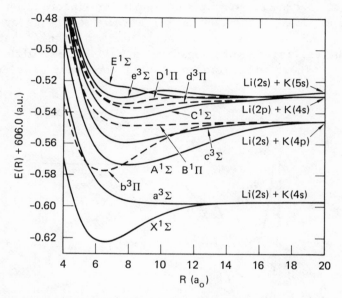

Fig. 9. CI calculations for the low-lying molecular states of LiK.

Table 2   Molecular Constants for LiNa and LiK

Molecule	State	R_e (Å)	D_e (eV)
LiNa	$X^1\Sigma$	2.95	0.83
LiNa$^-$	$X^2\Sigma$	3.35	0.63
LiK	$X^1\Sigma$	3.44	0.70
LiK$^-$	$X^2\Sigma$	3.87	0.58

CI calculation only correlated the valence shell electrons, the equilibrium separations will be overestimated by ~ 0.1 Å and the potential wells will be too shallow by the amount of 0.05 to 0.1 eV. For LiNa, more accurate calculations by Rosmus and Meyer[13] place the $X^1\Sigma$ well parameters at R_e = 2.87 Å and D_e = 0.87 eV.

The calculations on the negative-ion states of LiNa and LiK show a ground state with a very broad potential well which is shifted to larger internuclear separations, but more shallower than the neutral state, Figs. 10 and 11. As in the NaH⁻ calculations, there is strong configuration interaction between the $X^2\Sigma$ and $A^2\Sigma$ states which force the two states apart as R decreases. Such an interaction will lead to significant charge transfer in collisions of negative alkali atoms with neutral alkali atoms. There will also be a small component of electron

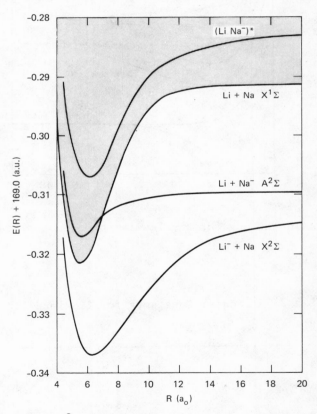

Fig. 10. Low-lying $^2\Sigma$ molecular states of LiNa⁻ relative to the neutral state continuum.

detachment which is due to the penetration of the $A^2\Sigma$ molecular state into the continuum.

The adiabatic molecular electron affinities of LiNa and LiK can be determined from Table 2.   No correction is necessary to allow for proper asymptotic dissociation since the calculated electron affinity of Li is 0.618 eV, while the true value is very close, 0.620 eV.

Li⁻ + ALKALINE EARTH ATOMS

In order to ascertain whether or not the relationship between the negative-ion and neutral molecular states of Li⁻ and alkaline earth atoms is similar to that observed for H⁻, we completed a series of calculations on the ground states of LiCa.   As in our work on the heteronuclear alkali molecules, the CI calculations

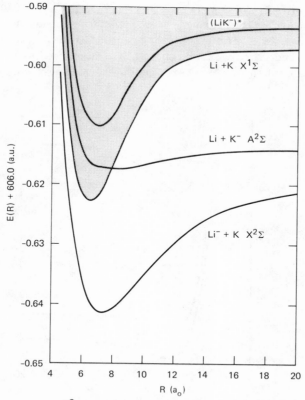

Fig. 11.   Low-lying $^2\Sigma$ molecular states of LiK⁻ relative to the neutral state continuum.

only included correlation of the valence shell electrons.

The results of the CI calculations are presented graphically in Fig. 12. As in the CaH case, the LiCa⁻ remains well removed from the neutral continuum throughout the investigated internuclear separation range. Thus, we conclude that electron detachment in collisions of Li⁻ with alkaline earth atoms will have very small cross sections at energies $E \lesssim 1000$ eV.

The calculated potential well constants are given in Table 3. The equilibrium separations are probably overestimated by ~ 0.1 Å. It is of interest to note that the neutral state well is very shallow, which indicates a weak interaction with the Li⁻ + Ca⁺ ion pair state.

CONCLUSIONS

A series of molecular calculations have been performed on H⁻ and Li⁻ with rare gas, alkali, and alkaline earth atoms. The purpose of the calculations was to determine the electron detachment mechanisms for low energy collisions. Interpretation of the calculated interaction energies indicates strong coupling to the neutral continuum for rare gases, the importance of

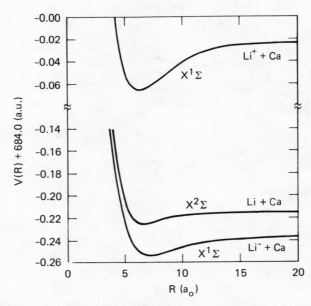

Fig. 12. Calculated interaction energies for the ground and ionic molecular states of LiCa.

Table 3 Molecular Constants for Neutral and Ionic States of LiCa

Molecule	State	R_e (Å)	D_e (eV)
LiCa$^+$	X$^1_3\Sigma$	3.36	1.17
	a$_3\Sigma$	4.03	0.58
	b$^3\Pi$	3.23	0.81
LiCa	X$^2\Sigma$	3.50	0.25
LiCa$^-$	X$^1\Sigma$	3.85	0.38

electron capture for alkali earth atom targets, and the lack of a direct coupling mechanism for alkaline earth atom targets.

ACKNOWLEDGEMENTS

The author would like to thank Professor Howard Taylor for helpful discussions related to the use of the stabilization method and Dr. Bowen Liu for his guidance in the molecular calculations. Partial support of the National Science Foundation is gratefully acknowledged.

REFERENCES

1. S. K. Lam, J. B. Delos, R. L. Champion, and L. D. Doverspike, Electron detachment is low-energy collisions of H$^-$ and D$^-$ with He, Phys. Rev. A 9 : 1828 (1974).

2. E. A. Mason and J. T. Vanderslice, Interaction energy and scattering cross sections of H$^-$ ions in helium, J. Chem. Phys. 28 : 253 (1958).

3. J. P. Gauyacq, A simple approximation for low-energy total detachment cross sections, J. Phys. B 12 : L387 (1979).

4. R. E. Olson and B. Liu, Interaction energies for low-lying electronic states of NaH and NaH$^-$: scattering of H$^-$ by alkali atoms, J. Chem. Phys. 73 : 2817 (1980).

5. R. L. Champion, L. D. Doverspike, and S. K. Lam, Electron detachment from negative ions: the effects of isotopic substitution, Phys. Rev. A **13** : 617 (1976).

6. R. E. Olson and B. Liu, Self-consistent-field potential energies for the ground negative-ion and neutral states of HeH, ArH, and ArCl, Phys. Rev. A **17** : 1568 (1978).

7. R. E. Olson nd B. Lui, Interactions of H and H$^-$ with He and Ne, Phys. Rev. A **22** : 1389 (1980).

8. J. S. Riley and R. Geballe, Absolute H$^-$ detachment cross sectioms, Phys. Rev. A **9** : 2485 (1974).

9. J. F. Williams, Single and double electron loss cross section for 2-5 keV H$^-$ ions incident upon hydrogen and the inert gases, Phys. Rev. **154** : 9 (1967).

10. F. B. Orth, W. C. Stwalley, S. C. Yang, and Y. K. Hsieh, New spectroscopic analyses and potential energy curves for the X $^1\Sigma^+$ and A $^1\Sigma^+$ states of NaH, J. Mol. Spectrosc. **79** : 314 (1980).

11. R. E. Olson and B. Liu, Interaction energies for the ground states of CaH$^+$, CaH and CaH$^-$ (private communication).

12. K. P. Huber and G. Herzberg, <u>Constants of Diatomic Molecules</u>, (Van Nostrand Reinhold Co., New York, 1979).

13. P. Rosmus and W. Meyer, Spectroscopic constants and the dipole moment functions for the $^1\Sigma^+$ ground state of NaLi, J. Chem. Phys. **65** : 492 (1976).

THE HYPOTHETICAL PH_5 MOLECULE AND ITS REACTION TO $PH_3 + H_2$

Werner Kutzelnigg, Jan Wasilewski[+)] and Holger Wallmeier

Lehrstuhl für Theoretische Chemie
Ruhr-Universität Bochum, Universitätsstr. 150
D-4630 Bochum/FRG

INTRODUCTION

The molecule PH_5 is interesting from various points of view.

1. PH_5 is the prototype of phosphoranes and moreover of systems with pentavalent phophorous. Questions concerning binding, e.g. the role of d-AO's vs. multicenter bonds arise in the most clearcut way.

2. PH_5 is even the simplest of all electron rich (hypervalent) molecules.

3. PH_5 is a typical and one of the simplest 'non-rigid' molecules. The isomerization barrier is only ~2 kcal/mol. This is in obvious contrast to the isomerization barrier for PH_3, which is about 38 kcal/mol[1].

4. PH_5 has not yet been observed experimentally. Is it a non-existing compound? Why has it not been observed? The energy of PH_5 lies ~40 kcal/mol above that of $PH_3 + H_2$, it is hence not thermodynamically stable, but most molecules that we live with are thermodynamically unstable. What really matters is the kinetic stability.

The question of the metastability of PH_5 has three aspects.
(a) Is there a local minimum of the potential hypersurface that corresponds to a PH_5 molecule?

[+)]permanent address:Nicholas Copernicus University,Institute of Physics, Quantum Chemistry Group,ul Grudziadzka 5,87-100 Torun/Poland

(b) How high is the barrier for the decomposition of PH_5 to PH_3+H_2 or for any other desintegrations[2]

(c) Are there bimolecular processes for the decomposition of PH_5 with a lower barrier?

Point (a) requires the evaluation of the full force field. As a byproduct the equilibrium geometry is obtained.

Point (b) turns out to be much more complicated than anticipated, mainly due to the number of degrees of freedom that one has to vary. The structure of the saddle point is namely highly asymmetric.

One might have guessed that qualitative considerations of the Woodward-Hoffmann (WH) type would help one to find the least energy path. However it turned out not only that there is a formally WH allowed path with a very high barrier, but also that there is - due to the lack of molecular rigidity - no clearcut distinction between WH allowed and WH forbidden paths.

It is quite puzzling that the location of the saddle point depends very much on the sophistication of the method employed. With a double-zeta quality basis in SCF one obtains a zwitterionic saddle point PH_4^+/H^-; with polarization functions two competitive saddle points, separated by a very small 'super saddle' are found, one zwitterionic and the other concerted. With correlation only the concerted one survives.

The basis dependence should not surprise one too much, it is obvious that a decent description of pentavalent P without d-AO's is impossible. To compute a potential surface without d-AO's as Howell[2] did, is a priori quite hazardous.

But the basis dependence is even more subtle. Even the equilibrium geometry of PH_5 in its trigonal bipyramidal form depends very much on details of the basis[3]. Geometries are usually much less sensitive. The normalvalent molecule PH_3 behaves normal in this respect. An accuracy of a few thousandths of an Å is easily achieved, and the results for the harmonic force constants are excellent, much like in similar molecules[3].

However, in PH_5 the variation of the d-exponent from 0.57 to 0.925 (with p-AO's on the hydrogens present) changes the equilibrium geometry by more than 0.01 Å.[3]

Since no experimental data are available for comparison we needed intrinsic criteria for the basis quality, i.e. we had to use still larger basis sets for comparison.

The first part of this talk will deal with the study of the equilibrium structure and the force field. I can be short since all this is published[3]. In a short 2nd part I shall comment on the barriers for internal conversion, mainly on Berry vs. turnstile processes. There we essentially confirm what has been found by Russegger and Brickmann[4], some time ago in a CNDO study of PF_5, namely that there is no turnstile process.

The 3rd and main part of this talk will be on the potential hypersurface for the reaction $PH_5 \rightarrow PH_3 + H_2$.

In a short 4th part I shall comment on bimolecular and auto-catalytic desintegration processes and on the chances whether PH_5 will ever be observed.

EQUILIBRIUM GEOMETRY AND FORCE FIELD OF PH_5

For PH_3 the CEPA values of r_e and α_e are 1.417 Å and 92.5°, the best experimental values 1.419 Å and 93.8°(in SCF the bond length is 1.409 Å). The harmonic force constants are in reasonable agreement with experiment, except for the non-diagonal force constants which are very uncertain experimentally (values between −0.298 and +0.75 for F_{12} can be found in the literature while we get +0.09 mdyn/Å; for F_{34} they vary between −0.54 and +0.16, we get −0.04).

More illustrative is a comparison of computed and harmonized experimental vibration frequencies

	CEPA[3]	exp
PH_3, A_1	2482	2448 − 2456
	1081	1041 − 1045
PH_3, E	2487	2390 − 2457
	1170	1150 − 1154
PD_3, A_1	1778	1761
	788	756
PD_3, E	1788	1766
	834	822

We expected to get similarly accurate results for $PH_5(D_{3h})$.

However, we found that two (double-zeta plus polarization type) basis sets A and B that differed only in the exponent of the d-AO on P lead to equilibrium geometries that differed by ~0.02 Å for r_{ax} and ~0.01 Å for r_{eq}. A calculation with a triple zeta plus polarization type basis G,with two d-sets on P,confirmed essentially the results with basis A ($\eta = 0.57$), such that the geometrical parameters of PH_5 are predicted as

$$r_{ax} = 1.471 \pm 0.003 \text{ Å}$$
$$r_{eq} = 1.419 \pm 0.003 \text{ Å}$$

This prediction differs significantly from previous ones,especially from that of a recent calculation[5] in which less attention was paid to a careful choice of the basis. Completely unreliable are the results from calculations without d-AO's on P [6].

For the energy difference between PH_5 and $PH_3 + H_2$ a value of 38 kcal/mol is obtained. (The reaction $PH_5 \rightarrow PH_3 + H_2$ is exoergic).

The harmonic vibration frequencies of PH_5 and PD_5 have been predicted[3], but shall not be repeated here.

It is interesting to compare the PH bond lengths (r in Å) and the force constants (f in mdyn/Å) for symmetric PH stretch for different types of PH bonds [3].

	r	f
three-center bond (r_{ax} in PH_5)	1.47	2.48
'equatorial' (r_{eq} in PH_5)	1.42	3.40
'pure' p-bond (PH_3)	1.42	3.55
sp^3 (PH_4^+)	1.39	4.29
sp^2 (PH_3 planar)	1.37	4.61

INVERSION BARRIER OF PH_5

For the Berry[7] inversion barrier, i.e. the energy difference between the lowest C_{4v} structure and the D_{3h} equilibrium structure, we get from a CEPA calculation with basis A a value of 2 kcal/mol (while in ref. 5 a value of 0.9 kcal/mol, that we regard as somewhat less reliable, is predicted). The Berry inversion (pseudorotation) is initiated by the lowest E' vibration, which has a harmonic vibration frequency of 648 cm^{-1} that corresponds to a harmonic zero-point energy of 0.0015 h = 0.9 kcal/mol. The potential is, of course, largely anharmonic and the actual zero-point energy may differ significantly from the harmonic estimate. The lowest E' vibration has probably a very large amplitude in the direction of the Berry inversion, even in the zero-point vibration. The closeness of the barrier height and the harmonic estimate of the zero-point energy is rather irritating. If the zero-point energy were above the pseudorotation barrier the description of the molecule with a D_{3h} equilibrium structure would probably break down and the five hydrogen atoms would become equivalent in the vibronic ground state. It seems to us, however, that our results indicate that this is not the case.

One may wonder whether the turnstile isomerization[8] is a serious alternative to Berry's pseudorotation[7]. The turnstile rotation in the sense of Ugi[8] proceeds formally in three steps:

1. Deformation of the molecule such that three H atoms (eea) and the P atom form a regular trigonal pyramid that has a common axis with the equilateral triangle formed by P and the remaining two H atoms (ea) (The abbreviations e,a stand for equatorial and axial).

2. Torsion of the two subunits by $60°$.

3. Reversion of process 1.

Step 1 requires 8.8 kcal/mol (in CEPA), while step 2 only needs 0.3 kcal/mol[9]. There is hence nearly free rotation in the 'turnstile configuration'.

However, as is seen from fig. 1, the pseudorotation and the turnstile processes are topologically equivalent[9], as has already been pointed out by Russegger and Brickmann[4]. Both processes are of (aa) ↔ (ee) type, i.e. they exchange two axial with two equatorial ligands. The 'turnstile transition state' is stationary only if certain artificial geometry constraints are imposed. If these restrictions are relaxed the energy drops, i.e. it represents not a saddle point but rather a point on the slope of the energy surface. Shih et al.[7] have tried to get a lower barrier by relaxing the geometry of the 'turnstile transition state' and were partially successful. They did not realize that full geometry relaxation – in C_s-symmetry with only one H-atom in the symmetry plane – would have led to the Berry transition state, i.e. that the whole concept of the turnstile process becomes spurious.

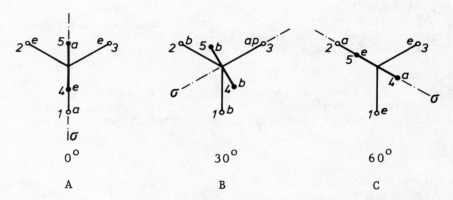

Fig. 1   Illustration of the equivalence of the pseudorotation and
turnstile isomerizations. e and a refer to equatorial and
axial in the D_{3h} structure, b and ap to basal and apical
in the C_{4v} structure. Open circles characterize the pyramidal
and closed circles the trigonal subunits in the turnstile
process. A,B and C are starting, transition and final
structures. The essential symmetry plane for any of these
structures is σ.

POTENTIAL SURFACE FOR THE REACTION $PH_5 \rightarrow PH_3 + H_2$

A priori there are three possibilities for the concerted H_2
abstraction from PH_5

1) (aa) two axial H atoms are removed
2) (ee) two equatorial H atoms are removed
3) (ae) one equatorial and one axial hydrogen are removed

The least motion path is (ae), but this is WH forbidden, not
strictly by symmetry, but by approximate local symmetry[10].

PH$_3$ would be obtained almost in its equilibrium form, but an
(sa) electronic configuration had to cross an (ss) configuration

(s for symmetric, a for antisymmetric with respect to the approximate local twofold axis that bisects the angle between the axial and the equatorial bond). In fact the 'crossing point' lies about 200 kcal/mol above PH_5, so the barrier should be very large.

The (aa) and (ee) processes should be indistinguishable, since an (aa) ↔ (ee) exchange is always possible by a Berry pseudorotation. In (ee) the H-atoms are already closer and this is the appropriate starting point for the concerted abstraction. If the PH_3 fragment remains in its T-shaped form, the configuration is (sa) on both sides, the PH_3 can relax afterwards. (s and a refer now to the symmetry plane that is conserved.)

Other processes may be competitive

4) A non-least motion variant of the (ae) abstraction (like in the reaction $CH_2 + H_2 \rightarrow CH_4$) with symmetry lowering in the course of the reaction.
5) A two-step reaction, either $PH_5 \rightarrow PH_4^+ + H^- \rightarrow PH_3 + H_2$ or $PH_5 \rightarrow PH_4 + H \rightarrow PH_3 + H_2$.
6) Bimolecular or autocatalytic processes.

Overimposed is always the possibility of Berry pseudorotations. In fact (ae) ↔ (ee) is possible by two Berry steps:

$$(123)_e \ (45)_a \quad \leftrightarrow \quad (145)_e \ (23)_a \quad \leftrightarrow \quad (235)_e \ (14)_a$$

the pair (35) was (ea) and becomes (ee).

One may also start the process from the C_{4v} structure, but this does not seem to be a serious alternative.

The system PH_5 has 12 internal degrees of freedom and an investigation of the potential hypersurface as a function of all degrees of freedom is hopeless. If one imposes C_{2v}-symmetry one is left with only 5 degrees of freedom. Two of these, namely the PH distances in the remaining PH_3 fragment (of which two are equal) can be safely kept fixed, such that 3 inner coordinates are necessary for the C_{2v} surface. We can limit ourselves to only two inner coordinates, namely the H-H bond length r of the leaving H_2 molecule and the distance R between the P atom and either atom of the leaving H_2, if we accept that PH_3 is left in T-shaped form and relaxes to its equilibrium geometry only after the H_2 has left.

A contour line diagram of this two-dimensional surface is shown on fig. 2 on the level SCF with polarization functions. One finds the barrier for R=3.0 a_o and r=1.7 a_o. The barrier height E_{bar} and the exoergicity E_{react} of the reaction on three levels of computational sophistication are

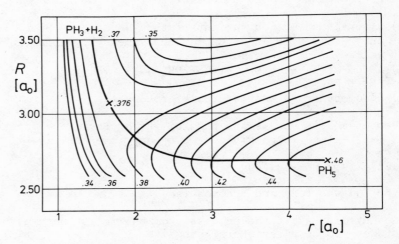

Fig. 2   Two-dimensional cut of the potential hypersurface with C_{2v}
constraint and the remaining PH_3 fragment in T-shaped form.
SCF level with polarization functions.   A label such as .35
for a contour line means -343.35 hartree (atomic units).

	E_{bar}	E_{react}
SCF (without polarization functions)	38 kcal/mol	− 82 kcal/mol
SCF (with polarization functions)	51 -"-	− 43 -"-
CEPA (with polarization functions)	43 -"-	− 39 -"-

If one looks at the minimum energy path ('reaction coordinate') one sees that there is nearly a two-step reaction. First the HPH angle is closed until the HH distance becomes just somewhat larger than that in H_2, then in a small region both R and r change, finally the H_2 is removed.

Two remarks are in order

1. Both the barrier E_{bar} and the reaction energy E_{react} depend sensitively on the level of computational sophistication. Without d-AO's one is not too bad for PH_3+H_2 but one is unable to describe the bonding situation in PH_5 (error $\sim$40 kcal/mol in E_{react}). The error in the barrier is smaller, since at the saddle point d-AO's are about equally important as in PH_5. Correlation is more important for the saddle point because there are 'long' bonds.

2. There is a rather large barrier although the process is WH allowed. This is unusual. However, in the second row the WH rules are less relevant. The PH_3 inversion is WH allowed and has a barrier of 38 kcal/mol (NH_3 only 6 kcal/mol). Changes in the valence state (hybridization) may involve enormous reorganization energies. In fact the process just discussed leaves PH_3 in a T-shaped structure, which is in its lowest energy geometry $\sim$80 kcal/mol above the pyramidal equilibrium geometry of PH_3.

The exact value obtained for the barrier height with the restrictions that C_{2v} symmetry is conserved and that the PH_3 fragment remains T-shaped, would be an upper bound for the barrier height that one would obtain without any symmetry restriction. Relaxation of the symmetry can only lower the barrier. If we assume that our CEPA value of 43 kcal/mol is sufficiently accurate, we can expect that the actual barrier is significantly lower. However, very extensive studies of a geometry relaxation have only led to a relatively small lowering of the barrier.

If one tries to relax the PH_3 fragment too early towards its equilibrium geometry, one destabilizes the cyclic three-center HPH-bond through which the concerted process has to go, so geometry relaxation has a large effect only after one has passed the barrier.

To understand the 'reaction path' better it is recommended to look at the reaction from the other side (fig. 3).

If we choose ß = 0° and $\gamma = 90^{\circ}$ we have the WH forbidden (ae) process and this does in fact have a large barrier. If we let PH_3 and H_2 approach, they show a typical closed shell repulsion. One can overcome this repulsion only if one makes a charge transfer interaction possible, namely from the lone pair of PH_3 to the unoccupied antibonding MO of H_2. In this way one gets a bonding interaction between P and H^1 (possibly H^2) while the bond between H^1 and H^2 is weakened.

This interaction is optimum (over a wide range of R) for ß = 73° and $\gamma = 81^{\circ}$.

The H_2 is then not far from parallel to the lone pair of PH_3, but not exactly so, due to steric hindrance.

The first step of the reaction is thus a nucleophilic one (PH_3 is a nucleophile, it donates electrons). The reaction must be completed by a charge transfer in the opposite direction, PH_3 must accept electrons from H_2.

Such a two-phase electron transfer is characteristic for non-least motion processes, the simplest examples of which are the

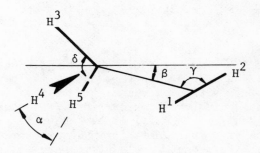

Fig. 3   The approach of H_2 towards PH_3.
         The horizontal line is the 3 fold axis of PH_3 in its
         equilibrium geometry and is appropriately defined in
         deformed structures

addition of CH_2 to H_2 [11,12] or C_2H_4 [13]. In these processes the first step is electrophilic (CH_2 accepts electrons) and the reaction is completed by a nucleophilic one.

The second reaction step for the formation of PH_5 is now possible in two different ways.

1) 'Concerted'. The angle α opens from $\sim 93^\circ$ to 180° such that PH_3 becomes T-shaped, the lone pair of PH_3 becomes of π-type, but T-shaped PH_3 has also an unoccupied δ-AO pointing towards H_2. The WH allowed process $H_2 + \pi \rightarrow 2\delta(PH)$ can take place. For H^1, H^2, H^3 to become equatorial and to form an equilateral triangle, β has to change from 73.3° to 57.2°, γ from 81 to 90°, δ from 93 to 90°. These are small changes.

It is reasonable to assume that C_s-symmetry is conserved; in C_s symmetry PH_5 has 8 internal degrees of freedom of which only two, namely the PH distances in the PH_3 fragment (of which two are equal) can be safely regarded as constant.

There are three main reaction coordinates

$$R = \sqrt{(r_1^2 + r_2^2)/2} \quad , \qquad r = r(H^1 H^2), \alpha \quad \text{with } r_1 = r(PH^1), r_2 = r(PH^2)$$

while the other three relevant coordinates (β, γ and δ) can be optimized for each triple of the first three.

The search for a saddle point in this 6-dimensional potential hypersurface is rather laborious. A saddle point (which is very flat) was found for the coordinates

$$R = 2.8 \, a_o, \quad r = 2.1 \, a_o, \quad \alpha = 150^\circ, \quad \beta = 70^\circ, \quad \gamma = 80^\circ, \quad \delta = 90^\circ$$

on the level SCF with polarization functions. The barrier height E_{bar} and its lowering ΔE_{bar} with respect to the former calculation in C_{2v} geometry with a T-shaped PH_3 fragment are

	E_{bar}	ΔE_{bar}
SCF (without polarization functions)	36 kcal/mol	-2 kcal/mol
SCF (with polarization functions)	47 -"-	-4 -"-
CEPA (with polarization functions)	39 -"-	-4 -"-

Again one notes the dependence on the sophistication of the
calculation, but also that the energy lowering as a result of the
relaxation from C_{2v} to C_s symmetry is rather small. The geometry
of the 'transition state' is rather close to the equilibrium
geometry of PH_5 which is understandable since it is also energetically
closer to PH_5 than to PH_3+H_2.

Again we first close the H^1PH^2 angle and only when H_2 is nearly
formed it can separate and the PH_3 fragment can continue to relax.

2) There is, however, a second possibility to add H_2 to PH_3.
Rather than to want that H^1,H^2 and H^3 become equatorial (ee process)
we may want H^4, H^5,H^2 to become equatorial and H^1,H^3 axial (ae
process). Now α need only open from 93.5° to 120°, but β has to vary
much, namely from 73.6 to 12.2°. The H_2 unit must, so to say turn
around the PH_3.

There is no unoccupied orbital of PH_3 and no electrophilic step
can follow the nucleophilic one. This is actually the WH forbidden
process, which becomes allowed by symmetry distortion. If we vary
γ as well we can achieve that H^1 gets much closer to P than H^2,
such that an ion pair $PH_4^+H^-$ is formed, and that the H^- turns around
the PH_4^+ and finally forms PH_5.

It is surprising that

(a) the symmetry distorted WH forbidden way becomes identical
    with the heterolytic process with an ion-pair-transition
    state,
(b) that this process is competitive with the concerted one.

The homolytic process going via PH_4+H requires much higher
energy since PH_4 is not stable with respect to PH_3+H, and can hence
be discarded.

There are a few more puzzling things.

(c) There is no clearcut distinction between the two processes,
    they occur in the same rather broad reaction valley.
(d) In SCF with polarization functions one finds two saddle
    points that are separated by a 'super barrier' that is
    extremely flat, although the geometries of the two saddle
    points differ enormously (fig. 4).
(e) In SCF without polarization functions one is unable to
    account for the 'concerted' saddle point and only gets
    the zwitterionic one – which has been found previously
    by Howell[2].

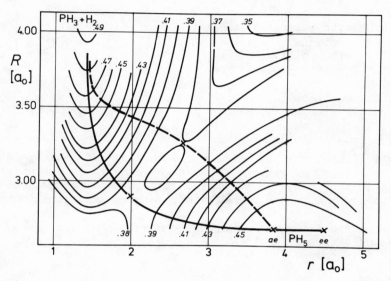

Fig. 4   The energy as function of the two 'reaction coordinates'
minimized with respect to the other internal coordinates
on SCF level with polarization functions. Energies as in
fig. 2. The thick full line represents the least energy
path for the concerted reaction, the broken line that via
an ion pair.

(f) Including electron correlation only the 'concerted' saddle
     point survives.

If we evaluate the energy of the two configurations, that
represent the two saddle points in the SCF (with polarization)
calculation, with respect to the equilibrium geometry of PH_5 also
on the two other levels of sophistication we get the following table
for the energy differences.

	concerted	zwitterionic
SCF (without polarization functions)	36 kcal/mol	32 kcal/mol
SCF (with polarization functions)	47 -"-	46 -"-
CEPA (with polarization functions)	39 -"-	45 -"-

and the geometries

concerted: $R = 2.8\ a_o$, $r = 2.1\ a_o$, $\alpha = 150^{\circ}$, $\beta = 70^{\circ}$, $\gamma = 80^{\circ}$, $\delta = 90^{\circ}$

zwitterionic: $R = 3.2\ a_o$, $r = 2.7\ a_o$, $\alpha = 120^{\circ}$, $\beta = 40^{\circ}$, $\gamma = 60^{\circ}$, $\delta = 100^{\circ}$.

Trajectory calculations would be very difficult because of
the possibility of motions with large geometry change and small
energetic barriers, but we expect that part of the trajectories
will follow a more concerted, others a more zwitterionic path.

A more detailed study of this potential hypersurface will be
published elsewhere[9].

FINAL REMARKS

If there were only the concerted decomposition possible, PH_5
should be kinetically quite stable. The unexpected ease of the
formation of the zwitterionic transition state is an indication that
in the presence of Lewis acids (like BF_3) that stabilize H^- or also
in the presence of H^+ (that abstracts H_2^- to yield PH_4^+) the hetero-
lysis of PH_5 may get such a small barrier that PH_5 is easily
desintegrated.

This also indicates how difficult it should be to prepare PH_5.
The way from $PH_3 + H_2$ is hopeless in view of the barrier of ~80 kcal/
mol, that via $PH_4^+ + H^-$ would also preferentially lead to $PH_3 + H_2$
rather than PH_5.

For other phosphoranes, say PF_5, the heterolytic process is even more likely than in PH_5, because the zwitterion PF_4^+/F^- is likely to be less unstable relative to PF_5 as is PH_4^+/H^- relative to PH_5 and because the bond in F_2 is so much weaker than in H_2.

It is - in view of the complications that arise here - by no means trivial to make valid predictions concerning the reaction $PX_5 \rightarrow PX_3 + X_2$ for arbitrary X from a study of $PH_5 \rightarrow PH_3 + H_2$ only and to claim that one is able to do so is not fair.

ACKNOWLEDGEMENT

These studies were started when Dr. J. Wasilewski had got a fellowship from the Alexander von Humboldt Foundation to which we are grateful. The computations were mainly done on the INTERDATA 8/32 'minicomputer', for which we are indebted to Deutsche Forschungs-gemeinschaft. We finally thank Dr. H. Kollmar and Dr. V. Staemmler for discussions and many valuable suggestions.

REFERENCES

1. R. Ahlrichs, F. Keil, H. Lischka, W. Kutzelnigg, V. Staemmler, J. Chem. Phys. 63:455 (1975)
2. J.M. Howell, J. Am. Chem. Soc. 99:7447 (1977)
3. W. Kutzelnigg, H. Wallmeier, J. Wasilewski, Theoret. Chim. Acta 51:261 (1979)
4. P. Russegger, J. Brickmann, Chem. Phys. Letters 30:276 (1975)
5. S.K. Shih, S.D. Peyerimhoff, R.J. Buenker, J. Chem. Soc. Faraday II 75:379 (1979)
6. J.M. Howell, J.F. Olsen, J. Am. Chem. Soc. 98:7119 (1976)
7. R.S. Berry, J. Chem. Phys. 32:933 (1960)
8. I. Ugi, D. Marquarding, H. Klusacek, G. Gokel, P. Gillespie, Angew. Chem. 82:741 (1970)
9. W. Kutzelnigg, J. Wasilewski, to be published
10. R. Hoffmann, J.M. Howell, E.L. Muetterties, J. Am. Chem. Soc. 94:3047 (1972)
11. H. Kollmar, Tetrahedron 28:5893 (1972)
12. H. Kollmar, V. Staemmler, Theoret. Chim. Acta 51:207 (1979)
13. B. Zurawski, W. Kutzelnigg, J. Am. Chem. Soc. 100:2654 (1978)

AB-INITIO STUDIES OF THE MOLECULAR SYMMETRY BREAKING PROBLEM:

LOW LYING STATES OF CO_2 AND NO_2

Lüder Engelbrecht* and Bowen Liu

IBM Research Laboratory,

San Jose, California 95193

ABSTRACT

The potential energy surfaces near the equilibrium conformation of the two lowest bound triplet states of CO_2 and of the NO_2 ground state have been calculated based on a multiconfiguration-SCF (MCSCF) and subsequent first order CI approach. The elementary basis sets were of double-zeta and double-zeta-plus-polarization quality. This work presents the first accurate characterization of the CO_2 triplet states and is expected to facilitate their experimental investigation. For NO_2 this work presents the extension of ab-initio calculations to the full three-dimensional problem not too far from the equilibrium conformation.

The MCSCF wavefunction is chosen to allow for a correct description of the energy dependence on the variation with unequal C-O (or N-O) bond lengths to avoid the symmetry breaking problem encountered in RHF-calculations. This problem is analyzed.

The spectroscopic properties derived from the calculated NO_2 surface show that the method used is capable of yielding quantitatively correct results.

---

* On leave from
  Fakultät für Chemie, Universität Bielefeld
  D 4800 Bielefeld, FRG

I. INTRODUCTION

a) The CO_2 Molecule

Although the carbon dioxide molecule is a very common simple molecule, little is known about its excited states. Information about CO_2 both from experimental investigation and ab-initio calculations is available only for the ground state to excited valence and Rydberg states at linear nuclear conformations. England, Rosenberg, Fortune, and Wahl[1] presented a linear-bent correlation diagram for the valence states of CO_2 establishing the relative ordering of valence states at bent C_{2v}-conformations. They showed that the two lowest states above the linear ground state are the two triplet states $^3B_2$ and $^3A_2$. The existence of a bent equilibrium geometry of these states is already predicted by qualitative arguments based on the Walsh-Mulliken rules.

It is the purpose of this work to characterize the two lowest triplet states by accurate ab-initio calculations. They are of particular interest for the dissociation reaction $CO_2 \rightarrow CO(^1\Sigma^+) + O(^3P)$ where the products are in their ground states which goes over the triplet surfaces of CO_2. This work may serve as a basis for considering this process in a future ab-initio calculation.

While the lower $^3B_2$ state was expected to be stable with respect to the dissociation into CO + O this was not a priori clear for the upper $^3A_2$ state. Using the experimental value of 5.45 eV for the dissociation energy[2] for the ground state the calculation of England et.al.[2] place the $^3A_2$ state minimum at nearly the same energy as that of the dissociation products and the $^3B_2$ state minimum slightly below the dissociation limit where both states are in their C_{2v}-geometries. Furtheron we may infer from the global properties of these surfaces that at large values of the mean bond distance R we have to find a double minimum curve plotting the potential energy against the asymmetric coordinate ΔR. These relative minima correspond to one short CO-distance reflecting the presence of the dissociation channel. (Coordinates are defined in Fig. 1).

This raises the question as to how the potential surfaces behave quantitatively with respect to deformation from C_{2v}-symmetric conformations. For instance one has to investigate whether these states have absolute minima at geometries with equal or unequal bond lengths assuming their stability with respect to dissociation.

The correct description of the potential surface behavior with respect to the asymmetric stretch coordinate presents a great problem in ab-initio calculations known as the symmetry breaking problem. We give a detailed analysis of this problem as found in RHF-calculations in Section II. In the following Section III we present one

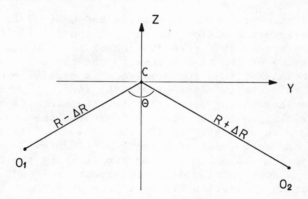

Fig.   1.   Definition of internal coordinates. The orientation of
            the Cartesian coordinate system (x,y,z) is chosen to be
            consistent with the conventional definitions for the C_{2v}
            point group symmetry (compare ref. 2).

solution to that problem on the basis of the multiconfiguration-SCF
(MCSCF) approach. The construction of the MCSCF wavefunction is
based on the qualitative interpretation of the RHF-results.

   Since the symmetry breaking problem is of fundamental importance
for a whole class of molecular systems with similar electronic
structure we will discuss it in detail.

b) The NO_2 Molecule

   Through vibrational[3] and rotational[4] spectroscopy within the
lowest rovibrational states of the $^2A_1$ ground state a firm characterization of the NO_2 molecule in its ground equilibrium conformation
has been establieshed; Electronic spectra of transitions to excited
states are highly complex and irregular and, thus have defied detailed
analysis up to now. There have been some attempts to facilitate the
interpretation of experimental investigations by ab-initio calculations. Most of the calculations were restricted to C_{2v} conformations.
The most accurate calculations of this type were done in an OVC-CI
study by Gillespie et.al.[5] on seven low lying valence states.
Previous calculations[6-13] at C_{2v} conformations were less accurate.
In all these calculations emphasis was put on the dependence of the
energy on the bending angle Θ while the dependence on the antisymmetric internal coordinate was neglected. This restriction,
however, imposes severe limitations on their usefulness since the
complexity of vibronic NO_2 states in the energy range corresponding

to the observed absorption in the visible and near UV is mainly due
to the complicated structure of the three-dimensional potential
surfaces. Within the adiabatic separation of nuclear and electronic
motion the crossing of the $^2A_1$ state with the $^2B_2$ state at small
angles of about $1o5^o$ and C_{2v} conformations is part of a conical
intersection within the two-dimensional $^2A_1$ surface connected with
strong nondiabatic coupling terms.[14] After transformation to the
diabatic representation of the electronic states one would find
diabatic surfaces which might justify the neglect of explicit considerations of the antisymmetric stretch dependence. However, there
is a very strong non-diagonal coupling representing the off-diagonal
Hamilton matrix element of the diabatic electronic functions which
is to the lowest approximation linear in the antisymmetric stretch
coordinate. These effects have been investigated by Jackels and
Davidson[14] for the first time, by the method described in II.d.
They showed that the actual vibronic states can by no means be
attributed to one or the other electronic state alone. In a second
paper Jackels and Davidson[15] give a survey over the valence states
of NO_2 including $4o$ doublet and 12 quartet states at C_{2v} and in
several cases also at C_s conformations. These calculations, too,
were not of quantitative nature. For references to the various
experimental investigations up to that time we refer to their
paper.[15]

The goal of this work is to investigate the threedimensional
potential surface: The antisymmetric internal coordinates have never
been considered before in ab-initio calculations. We show in Sec.
II.c. that standard procedures based on RHF orbitals are severely
spoiled by the existence of several solutions in C_s constrained
calculations. Since experimental values for the harmonic antisymmetric stretch are available the reliability of our calculation
method can be tested (see Ch.III).

II. THE SYMMETRY BREAKING PROBLEM IN RHF-CALCULATIONS

The term symmetry breaking here refers to two different effects.
On the one hand it means the breaking of the nuclear frame symmetry,
i.e. the nuclear conformation of the potential energy minimum is not
invariant under the point group operations of the high point group
symmetry, here C_{2v}. This lack of invariance leads to a set of equivalent stable conformations, here two, on the global energy surface.
(In principle there is no need to call this effect symmetry breaking
since no general theorem would require the minimum to be located at
conformations of the highest accessible symmetry. In general one can
only state that the potential energy surface has a zero derivative
with respect to asymmetric coordinates at the high symmetry conformations corresponding to a local minimum or maximum. Thus this
effect should be considered more as a deviation from the intuitively
expected symmetry of the equilibrium conformation.) On the other

hand symmetry breaking describes the more technical problem of a Hartree-Fock instability leading to a symmetry breaking of the wavefunction. In that case the solution of the electronic Schrödinger equation does not transform according to any irreducible represen- tation of the point group symmetry. Although in principle it is required to do so since the Hamiltonian is invariant with respect to the spatial symmetry operations of the point group, this may occur because only the projection of the full Hamiltonian onto a specific choice of one- and n-particle space is used in ab-initio calculations. When the wavefunction breaks symmetry the breaking of the nuclear frame symmetry is an immediate consequence.

In the following we discuss the symmetry breaking effects within the spin restricted Hartree-Fock (RHF) approach which is the most common method for the determination of orbitals for a subsequent CI calculation and is often used as an approximation to the full solution directly. We have performed RHF-calculations in the vicinity of the equilibrium geometries of both the CO_2 triplet states and the NO_2 ground state and have also investigated the effect of correlation within a CI calculation including all single and double excitations from the RHF configurations for the CO_2 states.

a) $^3A_2$ State of CO_2

The $^3A_2$ state of CO_2, labeled by the C_{2v} point group species which correlates to $^3A''$ in the C_s point group, is the second lowest triplet state. Its C_{2v} minimum is found at a bending angle near $130°$. The state is bound with respect to dissociation into CO + O. The $^3A_2$ state correlates diabatically to the $^3\Delta_u$ state of linear CO_2. (See however England and Ermler[16] who have determined the $^3\Pi_g$ as the lowest $D_{\infty h}$ state correlating with $^3A_2$.)

The RHF-configuration in C_{2v}-symmetry is given by the excitation $1a_2 \to 6a_1$ in C_{2v}, $3a'' \to 1oa'$ in Cs, or $1\pi_g \to 2\pi_u$ in $D_{\infty h}$ from the closed shell ground state configuration.

We start our discussion of the results of RHF-calculations with the $^3A_2$ state because its electronic configuration of the out-of-plane π-space is very much akin to that of the allyl radical ground state in which the symmetry breaking ("doublet instability") of the RHF wavefunction has been investigated previously by Paldus and Veillard.[17] Both the CO_2 $^3A_2$ state and the allyl $^2A_2$ state are characterized by the distribution of three π-electrons in a doubly occupied bonding orbital (for CO_2: $1b_1 = 1a''$) and a singly occupied lone pair orbital (for CO_2: $1a_2 = 2a''$).

Before we try to understand the reasons for the symmetry breaking effect in more detail we give an outline of the main features of our calculations on $CO_2(^3A_2)$:

a) The C_{2v}-constrained solution is obtained either by performing an explicitely C_{2v}-constrained calculation or by starting the SCF iteration in the lower symmetry calculations with symmetric trial functions. In Fig. 2 we have plotted the energy of the symmetry constrained solution as the large dot on the energy axis.

b) Calculations in C_s symmetry at C_{2v} symmetric conformations produce a wavefunction with localized orbitals with an energy lower than that of the C_{2v}-constrained solution when an unsymmetric trial function is used. This symmetry broken solution can be continued smoothly into the C_s-symmetry conformations ($\Delta R \neq 0$) giving for instance curve 1 in Fig. 2. The energy goes through the point $\Delta R = 0$ with a non-zero slope. At $R = 0$, the mirror image solution formed by reflection of the electronic coordinates through the symmetry plane perpendicular to the molecular plane is also a solution of the RHF at the same energy. This gives rise to curve 2 in Fig. 2. By reason of symmetry both curves are related by

$$E_1 (\Delta R) = E_2 (-\Delta R)$$

for fixed R and Θ. The well deopth of the absolute minimum with respect to the C_{2v} symmetric conformation increases with R.

The locally stable C_{2v} constrained wavefunction could not be continued into C_s conformations. The stability region is just the point $\Delta R = 0$.

d) The convergence of the symmetry broken solutions in the SCF iteration is generally very bad. Even within modest convergence criteria the number of iterations was larger by an order of magnitude compared to the C_{2v} constrained SCF.

e) The upper branch of the two crossing symmetry broken solutions is stable only up to a rather small ΔR-value where it breaks down to the lower solution. We find a value of about o.o5 a_o for this ΔR-limit in our calculations which is smaller than the minimum value of about o.2 a_o.

CI-calculations consisting of all single and double excitations out of the HF configuration, leaving the 1s-electrons uncorrelated were also performed using both orbital sets. The results using the orbitals of the symmetry broken C_s type structure show that CI changes the picture not significantly, again a curve crossing at $\Delta R = 0$ obtains. Equivalent CI calculations using the C_{2v} constrained orbitals at $\Delta R = 0$ yield an energy lower than the crossing point, in contrast to the SCF picture. This observation suggests that the symmetry breaking problem can be treated as a correlation effect using CI where the orbitals are allowed to adjust optimally.

The interpretation of these results at the symmetric nuclear conformation ($\Delta R=0$) may be based on simple valence bond structure pictures. For this discussion we neglect completely the in-plane $\sigma(a')$-orbital space. We assume the doubly occupied localized σ-orbitals of the symmetry broken solution to be related by a rotation of the doubly occupied symmetric σ-orbitals and the open shell σ-orbital to be identical. This assumption is supported by the RHF-orbital.

The symmetry constrained solution may be represented, among other less favourable possibilities, by the resonance structure

$$|\bar{O} \rightleftharpoons \dot{C} - \underset{\bullet}{\bar{O}}| \quad - \quad |\underset{\bullet}{\bar{O}} - \dot{C} \rightleftharpoons \bar{O}| \qquad\qquad (a)-(a')$$

while the symmetry broken solution is given by just one structure, for instance

$$|\bar{O} \rightleftharpoons \dot{C} - \underset{\bullet}{\bar{O}}| \qquad\qquad (b)$$

In these structures 16 valence electrons are depicted. The localized π-orbitals are indicated by the heavy lines and large dots. The remainder are σ-electrons.

Clearly the resonance structure (a) - (a') has a lower energy than any single structure (a) or (a') formed with the same orbitals due to the resonance energy gain. Since (b) looks identical to (a) one would conclude that the symmetry constrained solution is actually lower than the symmetry broken solution. However, one cannot compare (a) and (b) directly because they are constructed with different orbital sets. In order to write the single determinant symmetry constrained solution in terms of two determinants with localized orbitals, these orbitals have to be related by a transformation between the occupied C_{2v} constrained orbitals. The localized π-orbitals in the structures (a) and (a') are therefore restricted by the C_{2v} symmetry constraints. Releasing the symmetry constraints the two occupied π-orbitals may relax, which brings the hypothetical structure (a) to the real symmetry broken solution (b). Since the symmetry broken solution has a lower energy than the symmetry constrained one we conclude that the relaxation energy $\Delta E(a \rightarrow b)$ is larger than the resonance energy $\Delta E(a - a')$.

Since the relaxation of the π-orbitals in the C_{2v} constrained calculation can only be achieved by mixing in π-antibond character of the symmetry constrained orbitals we conclude that the symmetry breaking is a pure correlation effect. In order to include the relaxation energy present in the symmetry broken solution in a symmetry constrained calculation one has to increase its n-particle basis to allow the occupation of the π-antibonding orbital. Due to the lag of the resonance energy in the symmetry broken solution the ΔR-dependence of the energy represented by a single configuration is also erroneous.

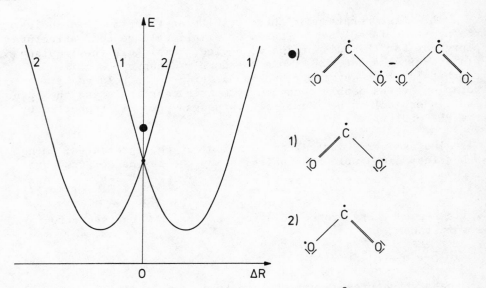

Fig. 2.   The potential energy dependence of the $^3A_2$ state on ΔR in
RHF calculations is shown. The heavy dot represents the C_{2v}
symmetry constrained solution while curves 1 and 2
correspond to the symmetry broken solutions.

b) The $^3B_2$ State of CO_2

The other investigated triplet state is the lowest CO_2 triplet state.
Its C_{2v} minimum is found at a bending angle near 120°. The electronic
wavefunction is of $^3B_2$ symmetry in the C_{2v} point group and $^3A'$ in the
C_s subgroup. It correlates to the $^3\Sigma_u^+$ state in $D_{\infty h}$ symmetry. The
HF-configuration is given by the excitation $4b_2 \rightarrow 6a_1$ with respect to
the closed shell ground state configuration in C_{2v} symmetry.

The RHF calculations show a behaviour different from that of
the $^3A_2$ state. In contrast to the RHF symmetry breaking in the $^3A_2$
state only one unique symmetry correct function is found as stable
solution at the C_{2v} conformations. However, the SCF convergence at
conformations with $\Delta R = 0$ is as bad as for the $^3A_2$ state and even
worse for small ΔR-values. The slope of the energy as function of
ΔR for R and Θ near the equilibrium values is given by the solid
line in Fig. 3.

The HF potential curve of the $^3B_2$ state has a double minimum
corresponding to a distorted equilibrium geometry. Thus, in the
case of the $^3B_2$ state we observe "symmetry breaking" of the nuclear
frame but not of the RHF wavefunction. However, similar to the
situation in the $^3A_2$ state we observe that the RHF orbitals become
strongly localized going from $\Delta R = 0$ to distorted conformations.

At the ΔR-values corresponding to minimum the orbital localization
is as strong as in the $^3A_2$ state.

Again a CI calculation including all single and double excitations with respect to the HF configuration does not change the
qualitative slope of E (ΔR) as shown by the solid line in Fig. 4.

The interpretation of these results in terms of valence bond
structures is presented in the following. In this case the highest
occupied σ-orbitals are important. The symmetry constrained SCF
solution may be represented by the resonance structure

$$|\bar{o} \rightleftharpoons \dot{c} - \underset{\cdot}{\underline{o}}| \quad - \quad |\underset{\cdot}{\underline{o}} - \dot{c} \rightleftharpoons \bar{o}| \qquad\qquad (c) - (c')$$

where again the π-orbitals are indicated by the heavy lines. This
resonance structure is valid when the localized orbitals are generated
by a transformation between the occupied symmetric orbitals. This
transformation leaves the σ-space unchanged because it is a closed
shell. Releasing these constraints on the localized orbitals the
single structures (c) and (c') each gain relaxation energy. However,
in the $^3B_2$ state this relaxation energy is not large enough to overcome the resonance energy. The relaxation of the localized orbitals
may be considered as the splitting of the resonance structure (c),

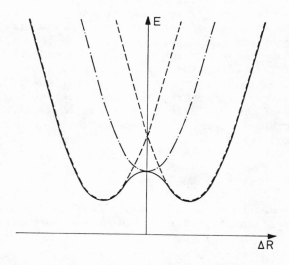

Fig. 3.   Schematic representation of hypothetical RHF solutions for
          the $^3B_2$ state leading to the solid curve, which represents
          the actually calculated solution. See the text for dis-
          cussion.

(c') into the sum of two resonance structures given by

$$|O \rightleftharpoons \overset{\cdot}{C} - \overset{\cdot}{\underline{O}}| \quad - \quad |\overset{\cdot}{\underline{O}} - \overset{\cdot}{C} \rightleftharpoons \overline{O}| \qquad\qquad \text{(d) - (d')}$$

$$+ \; |\overset{\ominus}{\underline{O}} - C \rightleftharpoons \overset{\cdot}{\underline{O}}|^{\oplus} \quad - \quad |\overset{\cdot\oplus}{\underline{O}} \rightleftharpoons C - \overset{}{\underline{O}}|^{\ominus} \qquad \text{(e) - (e')}$$

The sum (d) - (d') + (e) - (e') with relaxed localized orbitals is within a reasonable approximation equivalent to (c) - (c') with unrelaxed localized orbitals. If the energy loss due to the inclusion of the high energy structures (e), (e') is larger than the resonance energy due to both structures the symmetry broken solution would be lower. From the fact that the symmetry constrained solution is the only stable solution at C_{2v} geometries one may conclude that the energy loss due to the inclusion of the high energy ionic structures, which is equivalent to the gain of relaxation energy of (d) with respect to (e), is smaller than the resonance energies. This result is different from that of the electronically very similar situation in NO_2 ($^2A'$) discussed in II.c.

The change of C_{2v} symmetric orbitals to strongly localized orbitals on a rather small ΔR interval may be explained by a mechanism presented in Fig. 3.

There we plotted the energy of the hypothetical continuation of the symmetry constrained solution neglecting the relaxation effects associated with the symmetry lowering in the C_s symmetry as the dash-point curve. The energies of the hypothetical symmetry broken solutions are given by the two dashed curves. From this figure one expects the solid line as final solution which corresponds to an avoided curve crossing. This situation may well explain the extreme convergence problems for small ΔR-values: In conclusion one finds that the nuclear frame "symmetry breaking" in the $^3B_2$ state may be (and really is) an artifact of the RHF calculations due to an unbalanced treatment of relaxation and resonance energies which cannot be remedied easily by a large scale CI calculation.

c) Ground State Surface of NO_2

The single determinant C_{2v} constrained wavefunction for the NO_2 ground state is dominantly represented by the resonance structures

$$|\overline{O} \rightleftharpoons \overset{\cdot\oplus}{N} - \overset{}{\underline{O}}|^{\ominus} \quad - \quad |\overset{\ominus}{\underline{O}} - \overset{\cdot\oplus}{N} \rightleftharpoons \overline{O}| \qquad\qquad \text{(a) - (a')}$$

with a mixture of

$$|\bar{O} \rightleftharpoons \bar{N} - \dot{\underline{O}}| \quad - \quad |\dot{\underline{O}} - \bar{N} \rightleftharpoons \bar{O}| \qquad (b) - (b')$$

$$+ |\dot{O}^\oplus \rightleftharpoons \bar{N} - \underline{\bar{O}}|^\ominus \quad - \quad |\underline{\bar{O}}^\ominus - \bar{N} \rightleftharpoons \dot{O}|^\oplus \qquad (c) - (c')$$

where all the latter structures possess the same weight and express the mixing of oxygen character into the lone pair orbital.

The C_{2v} constrained solution for the $^2B_2$ state is expressed by the sum of (b) - (b') + (c) - (c'). Calculations with C_s symmetry constraints, the symmetry broken solution corresponding to (b) or equivalently (b') give lower energies at angles at least up to 130° than those obtained for the C_{2v} conformations.

With unequal bond length the structure (b) or (b') corresponding to the double bond on the short distance is the energetically lowest solution.

At larger angles towards linear NO_2 one may expect a crossing of the symmetry broken solution with the symmetry constrained solution, which would lead to a further complication.

The favouring of the localized structure is obvious for the $^2B_2$ throughout all nuclear conformations. The inclusion of high energy structures (c-c') is connected with an energy loss higher than the energy gain due to incorporating the resonance energy (which actually is a loss, too, at angles larger than about 110°). For the $^2A_1$ ground state the situation is not that clear, but obviously also here the resonance energy and the energy gain due to the dominance of the structures (a) and (a') is not enough to compensate for the unfavourable ionic structures (c) and (c').

A measure for the relaxation energy gain due to localization is given by the splitting of the covalent structure (b) and the ionic structure (c).

The Hartree Fock equations now have more than one solution at most geometries. The symmetry constrained solution of the $^2A_1$ state is locally stable at all angles and bond distances R while its continuation into the C_s symmetric conformations is stable only within a restricted region of internal coordinates. This solution will collaps into the lower symmetry broken type solution going to smaller angles and larger asymmetry. The wavefunction corresponding to this surface is expected to yield a reasonable description for the C_{2v} conformation but is going to the wrong asymptote at conformations with unequal bond lengths.

Thus we find the situation that the RHF model is incapable of describing both symmetric and asymmetric conformations. Both the C_s and the C_{2v} variants have serious deficiencies.

d) Symmetrization of Symmetry Broken Solutions

One straightforward possibility to go beyond the erroneous one-determinant description is to solve the secular equation of the 2x2 eigenvalue problem in the space spanned by the two symmetry broken RHF solutions. At C_{2v} conformations the two symmetrized solutions are formed from the positive and negative combination of the two degenerate symmetry broken solutions. The calculation of the off-diagonal elements of the 2x2 CI matrix and the overlap matrix involve non-orthogonal orbital sets.

This method was proposed by Jackels and Davidson[14] in their investigation of the $^2A'$ surface of NO_2 near the crossing point of the $^2A_1$ and the $^2B_2$ states.

The results of such a procedure are qualitatively described in Fig. 4 for a small and a large resonance effect.

This method is expected to give reasonable results for the energy versus ΔR-dependence in all cases where only one resonance structure is sufficient to describe the electronic structure. However this assumption is not fullfilled in some cases. A further improvement of the results of the 2x2 CI also seems to be necessary in view of the results presented in the next Sections where correlation effects are found to be important. However, a large scale CI calculation with non-orthogonal orbitals seems to be prohibitive. A way out might be to generate natural orbitals from the 2x2 CI and to use them in the construction of a later CI; but this has never been tested.

The most striking disadvantage, besides the technical problem of using non orthogonal orbitals in a CI calculation, of the suggestion by Jackals and Davidson is that it cannot be applied to states like the $^3B_2$ of CO_2, where the RHF wavefunction does not break symmetry. As discussed in II.b. the RHF description is inadequate for this state also, for basically the same reasons which apply for the $^3A_2$ state. For the NO_2 molecule the procedure can be used only for angles below 110°. This leads us to the conclusion that the only way to treat these electronic systems consistently is to include relaxation and resonance effects in one step which is possible with an MCSCF calculation.

III. ELECTRONIC WAVEFUNCTION CALCULATIONS BASED ON AN MCSCF METHOD

To overcome the problems presented in the previous Section II an MCSCF + CI calculation method is the ultimate choice in our estimation. The MCSCF is designed to avoid the symmetry breaking of molecular orbitals and to ensure a basically balanced treatment at high and low symmetry conformations. Section III.a. is devoted to the derivation of the MCSCF wavefunction. The MCSCF orbitals are used in a subsequent CI calculation which is discussed in III.b. This CI is based on excitations from a multireference wavefunction selected from the MCSCF calculation. All calculations were performed with the Alchemy system of programs developed by P.S. Bagus, B. Liu, A.D. McLean, and M. Yoshimine.

a) MCSCF Method

In the preceeding Section II we interpreted the RHF results in terms of a competition between energy gain due to orbital relaxation of localized orbitals and energy gain due to resonance stabilization of delocalized symmetry orbitals. The RHF wavefunction is not able to balance these two tendencies as a function of the deviation from symmetric conformations. Therefore, strongly asymmetric conformations are preferred in all cases.

Our aim is to derive an MCSCF wavefunction flexible enough to incorporate both effects in a balanced treatment. The resulting MCSCF orbitals would then be correctly symmetrized at C_{2v} conformations and go smoothly to localized orbitals at strongly asymmetric conformations. The derivation of such an MCSCF wavefunction may be based on the simple valence bond structures presented in Section II. For example one may proceed in the following systematic manner:

i) Assign one or more differently weighted resonance structures to the electronic structure of the investigated state at the C_{2v} symmetric conformations.

ii) Choose simple localized orbitals (atomic type lone pairs and localized bonds) to represent each particular structure by a single determinant wavefunction in these orbitals.

iii) Choose a set of reasonable symmetric orbitals (linear combinations of atomic lone pairs and delocalized bonding and anti-bonding orbitals) or determine them as natural orbitals from the first order density matrix generated by the sum of all structures.

iv) Express the localized orbitals of the two non-orthogonal orbital sets by the symmetric orbitals and write out the sum of all determinants defined by the localized orbitals.

The result of this procedure will be a multiconfiguration (MC) wavefunction where the different configuration state functions (CSF) are built from symmetric orbitals. By construction this MC wavefunction includes the dominant resonance effects. The relaxation of the localized orbitals is translated into an MC expansion.

The only remaining problem to be solved is the proper definition of the core and valence orbital space. The doubly occupied core orbitals of the localized and the symmetric descriptions have to be related to each other by orbital transformations. The minimum number of valence orbitals to be included is easily determined within the approach given above. For the $^3A_2$ state of CO_2 one needs the symmetric orbitals $1b_1$, $1a_2$, $2b_1$ to represent the localized π-bonds and lone pairs. For the $^3B_2$ state of CO_2 and the lowest states of NO_2 at least six orbitals, i.e. $5a_1$, $4b_2$, $6a_1$, $1b_1$, $1a_2$, and $2b_2$ are required to represent the three localized in plane lone pairs and the two π-bond and lone pair orbitals. Our calculations show that MCSCF wavefunctions using the valence orbitals listed above do have all the desired properties.

We neglected the possibility that the representation of localized σ-bonding orbitals might require the inclusion of symmetric σ-antibonding orbitals. However, this further relaxation of the core orbitals is allowed in the subsequent CI discussed in III.b. The only problem with restricting the σ-bonding orbitals to the core space for the MCSCF calculation is that dissociation of CO_2 into $CO + O$ and NO_2 into $NO + O$ is not properly described. We assume that in the equilibrium region of the molecular states the configurations needed for bond breaking are negligible compared to those needed for the correct description of the resonance effects discussed above. In our actual calculations we used the same set of six valence orbitals listed above for all states. It is noteworthy that these six orbitals correlate to the $1\pi_u$, $1\pi_g$, and $2\pi_u$ orbitals in $D_{\infty h}$ symmetry. The n-particle basis sets for the MCSCF wavefunctions contain all spin and spatial symmetry adapted CSF's which can be constructed from distributing eight valence electrons within these six valence orbitals. Thus, we generated more CSF's than needed for the $^3A_2$ state. The additional CSF's allow for a further relaxation of the σ-lone pair orbitals with respect to the π-configurations. The MC wavefunction derived described in points (i) to (iv) provide a good description of the MCSCF wavefunction which was actually calculated. All dominant CSF's except those of the π-bond to π-antibond double excitation are properly obtained. This demonstrates that the bond correlation is not important for the symmetry breaking problem. The crucial point is the representability of localized orbitals by symmetry orbitals. For the $^3B_2$ state of CO_2 qualitative agreement with the MCSCF wavefunction was achieved when a ratio of 3/2 for the covalent structures (d-d') to the ionic structures (e-e') was assumed. This demonstrates that more than one single resonance structure is important for a proper description.

Table 1.   List of reference configurations in the FOCI calculations

electronic state	$5a_1$ $8a'$	$4b_2$ $9a'$	$6a_1$ $10a'$	$1b_1$ $1a''$	$1a_2$ $2a''$	$2b_1$ $3a''$	C_{2v} C_s
$^3B_2$	2	1	1	2	2	o	
	2	1	1	2	o	2	
	2	1	1	o	2	2	
	1	2	1	2	1	1	
	1	1	2	1	2	1	
$^3A_2$	2	2	1	2	1	o	
	2	2	1	o	1	2	
	2	2	1	1	1	1	
	2	1	1	1	2	o	
	1	2	2	1	1	1	
NO_2	2	2	1	2	2	o	
	2	2	1	2	o	2	
	2	2	1	o	2	2	
	2	1	2	2	1	1	
	2	1	2	1	1	2	

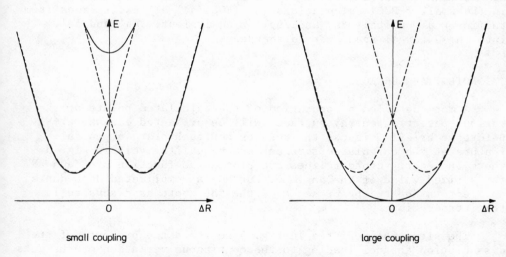

small coupling   large coupling

Fig. 4.   Eigenvalues (solid lines) of the 2x2 CI as a function of
ΔR at values of R and Θ near the calculated equilibrium
values. The dashed lines refer to the curves for the
symmetry broken solutions.

b) The n-particle Basis Sets for the CI

The electronic structure calculated within the MCSCF procedure
was refined by a subsequent large scale CI calculation. The CI
calculation was designed to include all CSF's necessary for describing
valence polarization and semi-internal correlation effects.[18] This
first order CI[19] is expected to give a good description of the shape
of the potential surfaces.

To construct the n-particle basis sets (CI configuration list)
for our calculations, the one-particle space spanned by the MCSCF
orbitals was partitioned into two subspaces: The internal space and
the external (virtual) space. This partitioning is uniquely defined
by the MCSCF orbital space. The internal space was further partitioned into a core space and a valence space, distinct from the core
and valence space of the MCSCF calculation. The core orbitals $1a_1$,
$1b_2$, $2a_1$ or $1-3$ a', which correlate to the atomic 1s orbitals, are
doubly occupied in each CSF. In the MCSCF calculation seven σ-orbi-
tals are kept doubly occupied. A set of electronic reference configurations was chosen from the MCSCF wavefunction list for each
electronic state. They are presented in Table 1. The configurations
with four open shells give rise to three different CSF's due to the
possible spin couplings.

In the FOCI calculation (b) (FOCI(b)) all CSF's were generated
by single and double excitations from the reference CSF's into the
internal valence and the external space under the restriction that
at most one external orbital may be singly occupied in each CSF.
In the smaller FOCI calculations (a) (FOCI(a)) all excitations from
the internal orbitals $3a_1$ and $2b_2$, or equivalently 4a' and 5a',
into the external space were discarded.

IV. POTENTIAL SURFACES

A more detailed presentation of the calculated points of the
various potential energy surfaces will be presented elsewhere.[20]
Before we briefly discuss the results for each state a general
feature of the calculated surfaces shown qualitatively in Fig. 5
where the energy contour lines are plotted as function of R and ΔR
for a fixed value of the angle Θ. For NO_2 a complicated dependence
on Θ enters due to the presence of the $^2B_2$ state as discussed in
the introduction.

The situation depicted in Fig. 5 reflects the presence of the
dissociation channel leading to the continuous variation of the cuts
parallel to the ΔR-axis from quadratic curves at small R-values to
the double minimum potential curve at large R-values.

Table 2.   Calculated equilibrium values for the $^3B_2$ state

calculation	E_{min} (h)	R (A)	θ ($^\circ$)	ΔR
MCSCF/DZ	-187.495832	1.266	120.5	0.0
FOCI(a)/DZ	-187.583037	1.281	119.4	0.0
FOCI(b)/DZ	-187.607254	1.275	119.4	0.0
MCSCF/DZ+P	-187.585968	1.237	118.4	0.0
FOCI(a)/DZ+P	-187.690800	1.244	117.7	0.0

Table 3.   Calculated equilibrium values for the $^3A_2$ state

calculation	E_{min} (h)	R (A)	θ ($^\circ$)	ΔR (A)
MCSCF/DZ	-187.488229	1.285	127.3	0.05
FOCI(a)/DZ	-187.560010	1.290	127.2	0.0
FOCI(b)/DZ	-187.579304	1.286	128.3	0.0
MCSCF/DZ+P	-187.571707	1.259	126.4	0.05
FOCI(a)/DZ+P	-187.658759	1.255	126.8	0.0

Table 4.   Calculated energy differences between the $^3A_2$ and the $^3B_2$ states of CO_2

calculation	$E_{min}(^3A_2) - E_{min}(^3B_2)$ (eV)	$\Delta E_{vertical}$ (eV)
MCSCF/DZ	0.21	0.29
FOCI(a)/DZ	0.63	0.74
FOCI(b)/DZ	0.76	0.86
MCSCF/DZ+P	0.38	0.53
FOCI(a)/DZ+P	0.87	1.03

Table 5.   Calculated values for the nimimum of the $^2A_1$ potential
           surface and harmonic frequencies

		R_{min} A	Θ_{min} deg	$\nu 1$ cm^{-1}	$\nu 2$ cm^{-1}	$\nu 3$ cm^{-1}
DZ	MCSCF	1.23	132.3	1291	716	1591
	FOCI	1.24	133.9	1267	697	1534
DZ+P	MCSCF	1.19	132.7	1484	795	1930
	FOCI	1.20	133.2	1450	776	1801
Ref. 3		1.20	134	1351	758	–
exp.		1.193$_4$[a]	134.07[a]	1325[b]	750[b]	1634[b]

[a] Ref. 4

[b] Ref. 3

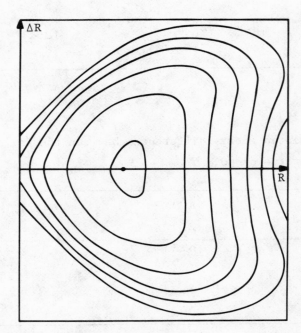

Fig. 5.   Schematic potential energy contour diagram as a function
          of R and ΔR for fixed Θ.

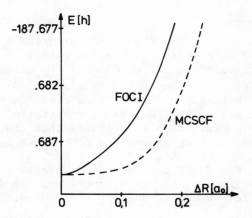

Fig. 6.   A cut through the potential surface at fixed values of R
          and Θ with R = 2.36 a_o and $\Theta = 120^{\circ}$ for the MCSCF (dashed
          lines) and FOCI(a) calculations on the $^3B_2$ state of CO_2
          using the DZ+P-basis. The parameters R" and Θ" are near
          the calculated equilibrium values of the FOCI(a) calcula-
          tion. The MCSCF curve is shifted up by o.1o5o9 h.

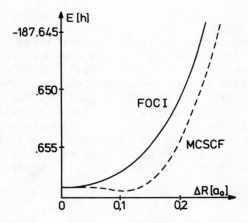

Fig. 7.   A cut through the potential surface at fixed values of R
          and Θ with R = 2.36 a_o and $\Theta = 128^{\circ}$ for the MCSCF (dashed
          line) and the FOCI(a) calculations on the $^3A_2$ state of CO_2
          using the DZ+P-basis. The parameters R and Θ are near the
          calculated equilibrium values of the FOCI(a) calculation.
          The MCSCF curve is shifted up by o.o874o h.

a) $^3B_2$ State of CO_2

In Fig. 6 a cut through the three-dimentional surface for fixed
values of R and Θ near their equilibrium values is depicted for the
MCSCF and FOCI calculation of the CO_2 and $^3B_2$ state in the DZ+P basis.
The results are similar for the DZ calculation. A symmetric equilibrium conformation is favoured now in contrast to the RHF and the
RHF + SDCI calculation. The FOCI has a substantial effect on the
shape of the curve. This result indicates that the MCSCF active
orbital space was not large enough to compensate for the relaxation
gain completely. The FOCI has to correct for this deficiency by
further relaxing the MCSCF core orbitals. The influence of the
polarization functions is of minor importance. In Table 2 the calculated equilibrium values of the $^3B_2$ are listed.

b) $^3A_2$ State of CO_2

In Fig. 7 a cut corresponding to that of Fig. 6 is presented
for the $^3A_2$ state of CO_2. As before the FOCI calculation yields a
symmetric minimum whereas the MCSCF calculation favours slightly
distorted conformations. The further remarks of the previous section
apply here as well. In Table 3 the calculated equilibrium values of
the $^3A_2$ states are listed whereas Table 4 presents information about
the separation of the $^3B_2$ and the A_2 states.

c) Ground State ($^2A_1$) of NO_2

Near the symmetric equilibrium conformation the dependance of
the potential energy on ΔR for fixed R and Θ is approximately quadratic over a wide range of deformations. The calculated equilibrium
values together with the derived spectroscopic constants in the
harmonic approximation are listed in Table 5. For the asymmetric
stretch frequency ν_3 the difference between calculated and experimental value is sufficiently small and, what is important, in the
same order as those for the two symmetric stretch vibrational modes.
This result demonstrates the advantages of the presented MCSCF + FOCI
method for all applications where the full surface has to be considered.

V. RECAPITULATION

Our calculations show that systems like the triplet states of
CO_2 and the states of NO_2 require a careful treatment of the energy
dependence on the asymmetric stretch coordinate. The balanced treatment of orbital relaxation on the C_{2v} and C_s symmetry was possible
by a specially designed MCSCF calculation refined by a subsequent
FOCI calculation. The results of this method differ qualitatively

from the results of RHF calculations followed by a large single
and double CI.

We believe that the MCSCF wavefunctions used in this work are
adequate to describe the ΔR-dependence of the electronic structure
qualitatively correct irrespective of the substantial influence of
the FOCI calculations. We have shown that the flexibility of the
MCSCF wavefunctions is needed to represent all important resonance
structures which may be derived for these electronic states. Without that flexibility opposing effects cannot be balanced leading
to ill defined molecular orbitals. Furthermore the MCSCF calculation
yields a set of important electronic reference configurations on
which the FOCI calculations can be based.

It seems interesting that no near degeneracies are involved
which would explain the necessity to use an MCSCF determination
of orbitals.

There is a large class of molecular systems with somparable
symmetry related problems in the determination of the electronic
structure. In this work we presented tools to deal with these
problems in a systematic and we believe correct manner.

REFERENCES

1.   W.B. England, B.J. Rosenberg. P.J. Fortune and A.C. Wahl,
         J. Chem. Phys. 65, 684 (1976).
2.   G. Herzberg, Molecular Spectra and Molecular Structure,
         III. Electronic Spectra and Structure of Polyatomic Molecules,
         Van Nostrand, N.Y. (1966).
3.   W.J. Lafferty and R.L. Sams, J. Molec. Spec. 66, 478 (1977).
4.   G.R. Bird, J.C. Baird, A.W. Jache, J.A. Hogeson, R.F. Curl Jr.,
         A.C. Kunkle, J.W. Bransford, J. Raskup-Anderson, and
         J. Rosenthal, J. Chem. Phys. 4o, 3378 (1964).
5.   G.D. Gillispie, A.U. Khan, A.C. Wahl, R.P. Hosteny, and
         M. Krauss, J. Chem. Phys. 63, 3425 (1975).
6.   C.F. Jackels and E.R. Davidson, J. Chem. Phys. 63, 4672 (1975).
7.   H.W. Fink, J. Chem. Phys. 49, 5o54 (1968).
8.   H.W. Fink, J. Chem. Phys. 54, 2911 (1971).
9.   L. Burnelle, A.M. May, and R.A. Gangi, J. Chem. Phys. 49, 561
         (1968).
1o.   R.A. Gangi and L. Burnelle, J. Chem. Phys. 55, 843, 851,
         (1971).
11.   L. Burnelle and K.P. Pressler, J. Chem. Phys. 51, 2758 (1969).
12.   J.E. Del Bene, J. Chem. Phys. 54, 3487 (1971).
13.   P.J. Hay, J. Chem. Phys. 58, 47o6 (1973).
14.   C.F. Jackels and E.R. Davidson, J. Chem. Phys. 64, 29o8 (1976).
15.   C.F. Jackels and E.R. Davidson, J. Chem. Phys. 65, 2493 (1976).

16.   W.B. England and W.C. Ermler, J. Chem. Phys. $\underline{7o}$, 1711 (1979).
17.   J. Paldus and A. Veillard, Mol. Phys. $\underline{35}$, 445 (1978).
18.   H.J. Silverstone and O. Sinanoglu, J. Chem. Phys. $\underline{44}$, 1899 (1966).
19.   H.F. Schaefer III, R.A. Klemm, and F.E. Harris, Phys. Rev. $\underline{181}$,
       137 (1969).
2o.   L. Engelbrecht and B. Liu, manuscript prepared for publication.

DETERMINATION OF PROPERTIES OF CLOSE-LYING EXCITED STATES OF OLEFINS

Robert J. Buenker
Lehrstuhl für Theoretische Chemie
Universität-Gesamthochschule Wuppertal
D-5600 Wuppertal 1, West Germany

Vlasta Bonačić-Koutecký
Institut für Physikalische Chemie
Freie Universität Berlin
D-1000 Berlin 33, West Germany

ABSTRACT

Ab initio calculations dealing with the sudden polarization effect in olefinic systems demonstrate that the properties of close-lying electronic states ($\Delta E \leq 5\text{-}10$ kcal/mole) obtained in limited CI treatments are very dependent on the choice of one-electron functions employed thereby. In the case of the V and Z states of ethylene in simultaneously twisted and pyramidalized nuclear arrangements simple perturbation theory arguments in the framework of a 2x2 diabatic state model are presented which explain the general qualitative appearance of such calculated results and at the same time allow for a straightforward assessment of their reliability in a given application. In particular the suddenness of the changes in the polarity of these states with geometrical variations is shown to be directly related to the existence of a Jahn-Teller degeneracy slightly removed from the mutually perpendicular CH_2 conformation ($\theta = 82°$) of ethylene. It is concluded that while orbital sets which give a balanced description of both electronic states rather than an optimum representation for only one of them are much more likely to lead to reliable dipole moment results for these species, the opposite situation holds in the case of potential surface calculations. Finally the role of improved CI treatments in minimizing discrepancies arising from the use of different one-electron basis sets is stressed and the consequences of the breakdown of the Born-Oppenheimer Approximation in regions of a Jahn-Teller degeneracy is also discussed.

241

I. INTRODUCTION

The term sudden polarization refers to a situation in which large variations in the dipole moments of particular molecular excited states occur as a result of a relatively small change in geometrical conformation (1-8). This phenomenon has been studied almost exclusively in the context of olefinic structures whose ground states develop biradical character after significant torsion about a double bond, and for which at the same time pairs of singlet excited states with essentially complementary charge distributions exist, but closer examination of the causes for this behavior emphasizes that similar sets of circumstances may arise for virtually any class of molecules. In many instances the rapid onset of polarization effects can be traced directly to the fact that two closely lying electronic states exhibit significantly different potential surfaces along one or more internal coordinates of the molecule, for example, something which is not at all restricted to the class of olefinic systems and is seen to be closely related to the well-known Jahn-Teller effect in the theory of electronic structure.

Recognition of this state of affairs has made the study of the lowest excited singlet states of 90° twisted ethylene with a single pyramidalized CH_2 group introducing a non-equivalency in its two biradical centers particularly attractive from a theoretical point of view (1,6,7). As a relatively small molecular system ethylene serves as a convenient test object for detailed investigation by means of ab initio calculations (6-8). At the same time geometrical relaxation effects such as simulateous torsion around the double bond and pyramidalization of a single CH_2 group in this system are closely related to structural changes in larger olefins such as propylene, butadiene and diallyl which are more likely to be of actual experimental interest in this connection. Recent calculations for propylene (9) have found a great similarity between the dipole moment surfaces for its two lowest excited singlet states and those for the analogous ethylene species in highly pyramidalized geometries, for example. In addition the rapid changes in polarity noted earlier (1,10) upon relative rotation of s-cis-s-trans-diallyl about its central C-C bond are simulated to a large degree by ethylene in slightly pyramidalized nuclear arrangements for values of the torsion angle in the 80°-85° range.

In general experience has shown that the sudden polarization effect and indeed many other phenomena related to the Jahn-Teller theorem can be most easily understood in terms of a 2x2 CI for appropriate diabatic states, whereby not only the energy difference between these species but also the magnitude of the H_{12} interaction matrix element between them needs to be considered before a comprehensive analysis can be realized. At the same time the fact that

the most interesting cases, in which the change in properties occurs
over a relatively small geometrical region, arise when ΔE is quite
small raises the basic question as to whether the conventional
Born-Oppenheimer approximation inherent in such CI treatments is
valid in such an investigation. The latter point can be of crucial
significance since vibronic coupling of two electronic states of
opposite polarity can clearly lead to individual vibrational states
with nearly vanishing dipole moment, or to a situation in which
this property varies rapidly from one value of the vibrational
quantum number to another. To consider these various theoretical
aspects in more detail, as well as to develop computational methods
which are capable of describing such effects on a reliable basis,
there is merit in examining the results of ab initio CI calculat-
tions of the ethylene model system in a series of nuclear arrangements in which the effects of varying both the torsion angle for
its methylene groups as well as the degree of pyramidalization
at one of its carbon centers are taken into account.

II. POTENTIAL SURFACES FOR SIMULTANEOUS TORSION AND PYRAMIDALIZATION OF ETHYLENE

Previous CI studies (11,6,7) of the ethylene molecule in conformations with perpendicular CH_2 groups have indicated that the $Z\,^1A_1$
state is 2-3 kcal/mole more stable than the $V\,^1B_2$ species in such
geometries, even though the opposite energy ordering is found at
the SCF level of treatment. This result implies that the V and
Z species undergo a potential curve crossing as torsional motion
takes the system out of the perpendicular orientation, and CI calculations (8) employing a double-zeta AO basis have found this
to occur for a torsional angle of $\Theta = 82°$. The treatment employed
thereby is of the MRD-CI type (12) with three reference configurations and three-root selection; more details may be found in the
original reference (8). The location of the crossing point is basically the same whether separately optimized orbital sets are employed for each of the singlet states in the CI calculations or
if SCF MO's of the lower-lying triplet (T) are used for the same
purpose, despite the fact that the energies of the singlet species
both lie some 8 kcal/mole higher in the latter treatment.

The crossing between the V and Z states is an example of the
Jahn-Teller effect in polyatomic molecules, whereby the requisite
degeneracy is only accidental in nature. Computationally this situation brings with it the general problem of how to describe these
two states when they assume the same symmetry as a result of asymmetric vibrational motion. Under these circumstances in order to
remain within the context of a conventional CI treatment it is
necessary to employ the same set of one-electron functions to describe both electronic states, thereby making it increasingly difficult to obtain an optimal description for both species.

Since the use of triplet-state SCF MO's leads to the same crossing
point between the V and Z states in the relatively high (D_2) symmetry as does the employment of separately optimized orbitals (NO's)
for these two singlet excited states, however, the former species
would appear to offer a good starting point for considering non-
symmetric geometries such as result when the C_2H_4 molecule simultaneously undergoes torsion around the CC bond and pyramidalization
at one of the carbon atoms, and results of these calculations are
shown in Fig. 1. What is apparent from this diagram is that the
spacing between the two potential curves (S_1 and S_2) for a given
pyramidalization angle ϕ increases rather quickly with the magnitude of this quantity. In the region of the crossing point (Θ =82°,
ϕ =0°) it is seen that the lower S_1 state becomes more stable with
increasing pyramidalization, whereas the S_2 species become rapidly
less stable as a result of the same geometrical change. Away from
Θ =82° the situation changes in this respect, however, with both
states favoring the non-pyramidalized structure for Θ =75° and
also for the limiting case of perpendicular CH_2 groups (Θ =90°),
as can be seen from Fig. 2. The S_1 barrier to pyramidalization
is considerably reduced at Θ =90° when this motion is accompanied
by an appropriate reduction in the magnitude of } HCH which reflects
a change in hybridization from sp^2 to sp^3 (see the dashed curves
in Fig. 2).

Examination of the charge distributions of these two states
at relatively large ϕ values shows that they become polar in character in such geometries, with the S_1 species possessing a negative
charge at the carbon terminus at which pyramidalization occurs.
The highest occupied orbital in the latter case is essentially
a pure carbon 2p AO directed perpendicularly to both the CC axis
and the adjacent CH_2 plane and is referred to as the A localized
MO (the corresponding electronic configuration is abbreviated as
A^2). For the same nuclear arrangements the S_2 state possesses a
complementary charge distribution, doubly occupying the B orbital,
which is an equivalent 2p species localized at the other carbon
atom; its electronic configuration is denoted as B^2. The degree
of mixing between the A^2 and B^2 configurations increases significantly as ϕ is decreased toward 0° and D_2 or D_{2d} molecular symmetry, and at Θ =90° in the limit of an unpyramidalized structure
the coefficients of these species become equal in both states
(the $A^2 + B^2$ and $A^2 - B^2$ combinations having been denoted by Mulliken as the Z and V states respectively (13)). At the same geometry
the open-shell AB configuration corresponds to both the ground
(N) and triplet (T) states of ethylene, both of which are significantly more stable than either of the above singlet excited states
(calculated average energy difference is 3.4 eV (11)).

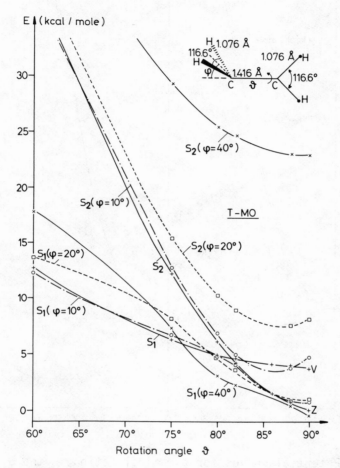

Figure 1: Torsion potential curves for various pyramidalization angles ϕ for the S_1 and S_2 states of ethylene obtained in the CI treatment (3M3R, T=5μH) employing T SCF MO's.

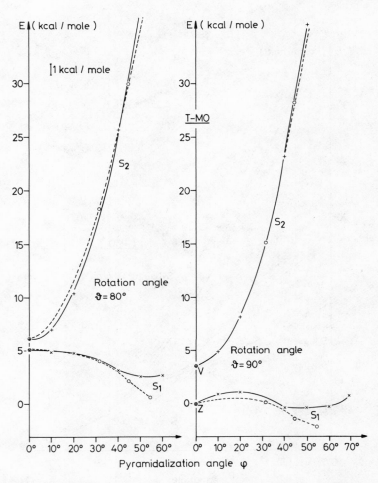

Figure 2: Potential curves for pyramidalization in the S_1 and S_2
states of ethylene for two different twisting angles
as obtained in the CI treatment (3M3R, T=5μH) employing
T SCF MO's. The solid lines correspond to constant HCH
while the dashed curves are for a gradual variation in
this quantity from sp^2 to sp^3 hybridization (see text).

The main effect brought about by pyramidalization at one carbon is to stabilize the A orbital which is localized at this atom, while having the opposite effect on the B MO, thereby leading to the unequal mixing of the A^2 and B^2 configurations in the singlet excited states mentioned above. Up to this point these states have been represented exclusively in terms of the T SCF orbitals, but since these species do not produce the lowest possible energies for the S_1 and S_2 states in unpyramidalized geometries (see Fig. 1 of Ref. 8), it must be expected that similar behavior will result from their use for non-zero values of ϕ as well. The corresponding potential curves for these states obtained with separately optimized natural orbitals (12) are shown in Fig. 3, whereby each state's data have been obtained in a separate CI calculation. Comparison of these results with those obtained through the exclusive use of the triplet-state (T) MO's (Fig. 1) shows a number of clear distinctions, particularly in the amount by which the S_1 state has its energy lowered as a result of pyramidalization. In the S_1 NO basis there is no longer a barrier to pyramidalization for the perpendicular CH_2 orientation for this state (Fig. 4), for example, with or without optimizing the value of ⊰ HCH as this motion proceeds.

Instead a relatively deep potential minimum is found near $\theta = 60°$, representing a relaxation energy of 7.2 kcal/mole relative to the corresponding non-pyramidalized structure; a similar result has been found by Malrieu and Trinquier (15) in a treatment employing singlet-state SCF MO's. As a consequence the adiabatic electronic energy difference (T_e) between the ethylene ground state and the S_1 species is calculated to be 5.84 eV in the MRD-CI treatment, which upon taking account of zero-point energy effects places the corresponding T_o value between 5.6 and 5.7 eV. The latter finding in turn is in good agreement with McDiarmid's experimental determination of this quantity in the 5.5-5.6 eV range (16), thereby indicating that the potential surfaces calculated with optimized natural orbitals accurately reflect the true physical situation. The analogous treatment of the S_2 state in its own NO basis also produces significantly different results from what is obtained with the T MO's, with this species now also showing a (slight) potential minimum as a result of the pyramidalization motion for $\theta = 90°$ (Fig. 4).

III. DIPOLE MOMENT SURFACES

The foregoing discussion of the wavefunctions and potential surfaces of the ethylene singlet excited states has been cast for the most part in a localized framework, with particular attention drawn to the relative stabilities of the A and B carbon 2p orbitals. It has been shown in previous work (8) that the dipole moments of such systems are to a good approximation proportional to the

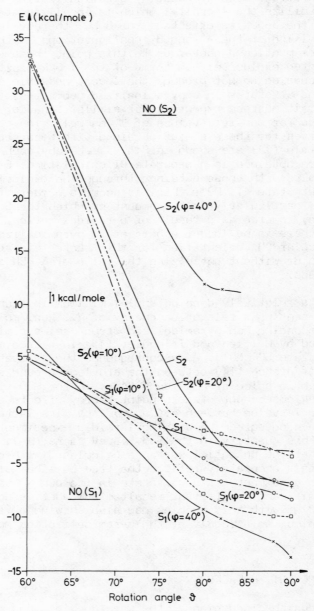

Figure 3: Torsion potential curves for various pyramidalization
angles ϕ for the S_1 and S_2 states of ethylene obtained in
separate CI treatments (each 3M3R, T=5μH) using their
respective natural orbital basis sets.

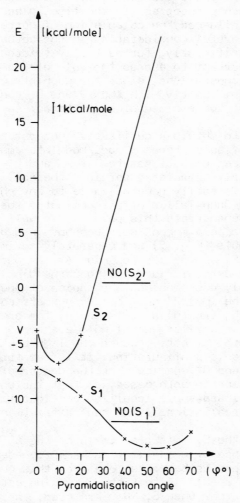

Figure 4: Variation of the total energy with pyramidalization angle for the S_1 and S_2 states of 90° twisted ethylene for the same treatment as in Fig. 3.

difference of the squares of the CI coefficients of the A^2 and B^2 configurations in this basis, so there is considerable advantage in retaining a localized description in discussing such one-electron properties. At the same time it should not be concluded on this basis that only the above two configurations need be considered in determining such properties. In the first place secondary configurations which influence the calculation of correlation effects need to receive careful consideration in the actual CI treatment, but even more significantly, for partially twisted geometries it is also quite important to assess the role of the key AB configuration in these computations, since for such structures this species begins to mix significantly with the A^2 and B^2 configurations in forming the lowest two excited singlet states.

Nonetheless to obtain a qualitative understanding of the dipole moment behavior of such states a good model to employ is that of a 2x2 mixing between diabatic states of A^2 and B^2 character respectively, particularly correlated versions thereof. Furthermore this type of approach is easily generalizable to any physical situation in which the Jahn-Teller effect dominates the theoretical description. To demonstrate this point it is well to consider the corresponding CI dipole moment surfaces obtained by employing respectively T SCF MO's (Fig. 5) and separately optimized natural orbitals for the S_1 and S_2 states (Fig. 6). As before with the potential surface data, it is found that the dipole moment results are rather strongly dependent on the choice of one-electron basis, particularly in the $\Theta \leq 70°$ and $\phi \leq 20°$ geometrical regions. For virtually all such structures it is found, for example, that the S_1 dipole moment is smaller when T MO's are employed instead of its own natural orbital set. Because the T MO basis is optimized for the AB electronic configuration, it is far from optimal for either of the A^2 and B^2 species. Yet the degree of inadequacy would appear to be similar in both cases, so that the energy difference between the key closed-shell localized configurations probably comes out fairly realisticallly in such a treatment.

By contrast the S_1 NO's tend to favor the A^2 configuration fairly strongly once pyramidalization has proceeded even moderately far, since this species benefits more from this geometrical change than does B^2. The result is that the ΔE value between A^2 and B^2 is considerably larger when S_1 NO's are used, thereby significantly reducing the amount of CI mixing between these two configurations in forming the S_1 and S_2 states in this basis relative to the T MO case. In terms of first-order perturbation theory the ratio of the corresponding CI coefficients (c_1 for A^2 and c_2 for B^2) is given by the formula:

$$\frac{c_2}{c_1} \sim \frac{H_{12}}{\Delta E}, \tag{1}$$

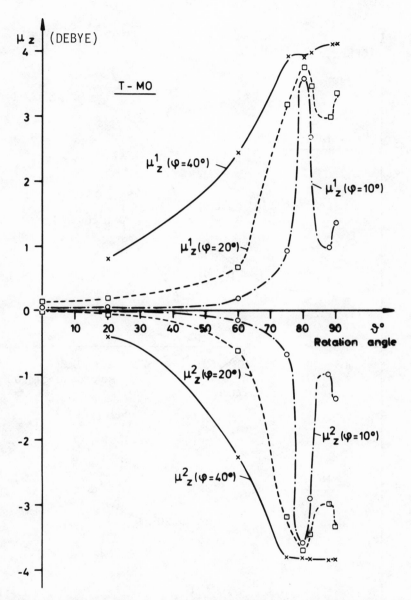

Figure 5: Variation for twist angle Θ and the pyramidalization
angle ϕ of the μ_z component of the dipole moment of the
S_1 and S_2 states of ethylene for the same treatment as
for Fig. 1.

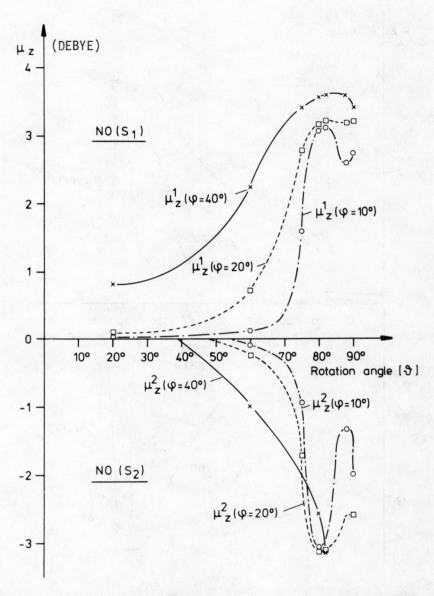

Figure 6: Variation for twist angle Θ and pyramidalization angle ϕ of the μ_z component of the dipole moment of the S_1 and S_2 states of ethylene for the same treatment as in Fig. 3.

which for constant H_{12} and E_1 E_2 leads to the above result. Since the μ_z component (z axis collinear with C-C bond) is given as roughly (1):

$$\mu_z = (c_1^2 - c_2^2) \cdot Q \qquad (2)$$

where Q is a fixed point charge, it is thus easily understandable in this model why the use of its own NO's leads to a larger dipole moment for the S_1 state than in the T MO basis. Furthermore similar arguments can be employed to explain why the use of S_2 NO's lead to a relatively small amount of polarization for these states. From the above observations there is thus good reason to believe that the T MO's lead to a Δ E value which is much more realistic than is obtained through the use of a single set of natural orbitals. As a consequence it seems quite likely that the dipole moment surface indicated in Fig. 5 is more reliable than that in Fig. 6 based on individually optimized singlet NO's, even though the triplet species employed to compute them allow for decidedly less than minimal energies for the S_1 or S_2 states themselves.

In addition to clarifying the question of which one-electron basis is more likely to give reliable property results, the 2x2 diabatic state model is also of value in helping to explain the general appearance of the calculated dipole moment surfaces. The rate in which polarity sets in is seen to depend on a second quantity other than Δ E, however, namely the H_{12} interaction matrix element between the A^2 and B^2 configurations, or more precisely stated, between correlated versions of these two parts of the final wavefunctions. The distinction is useful because in the SCF or single-configuration treatment the H_{12} matrix element is simply equal to the exchange integral K_{AB} between the two localized orbitals. Because such quantities are necessarily positive definite it follows that at this level of approximation the $Z=A^2+B^2$ state must be less stable than the $V=A^2-B^2$ species.

As noted at the beginning of the previous section, however, the experience with correlated treatments employing CI techniques is that the opposite energy ordering is correct, with Z lying 0.08 eV below the V state in previous work (11). Formally one can think of this result as also having been obtained from a 2x2 mixing procedure, but it is first necessary to do a separate CI calculation for each of the A^2 and B^2 species in mutually orthogonal complement spaces, and then to use the resulting multi-configurational functions as basis for the final 2x2 CI. Since the V-Z energy splitting is still given by $2H_{12}$ in such a treatment, it therefore follows that the interaction matrix element is negative in this case, since otherwise it would not be possible to explain why Z turns out to be the more stable of the two states. Dividing the above calculated CI V-Z splitting by two gives a value for the desired H_{12} matrix element of only about -1 kcal/mole at $\Theta=90°$. Since for unpyrami-

structures the ΔE value separating the A^2 and B^2 configuration
is necessarily zero, whereby it rapidly climbs to values much greater
than 1 kcal/mole as the angle ϕ is increased, one need only assume
that H_{12} remains fairly constant as such structural changes occur
in order to explain in this simple model why such a rapid increase
in the polarity of the system is found (Fig. 5).

Away from $\Theta = 90°$ it is no longer rigorously possible to equate
the V-Z energy difference to $2H_{12}$ in the sense defined above, because the Z state begins to take on a gradually increasing contribution from the open-shell AB configuration as torsion proceeds,
but at least in the near neighborhood of the perpendicular confor-$_*$
mation such a relationship still would seem to be fairly realistic*

---

* Strictly speaking some attention should be paid to the fact
   that the A and B localized orbitals are not rigorously orthogonal for $\Theta < 90°$ in formulating the desired 2x2 problem

---

Proceeding on this basis therefore one quickly comes to the conclusion that in the region of the Jahn-Teller crossing at $\Theta = 82°$
the pertinent H_{12} quantity takes on values of very nearly vanishing
magnitude, which by the above 2x2 CI arguments leads one to expect
the largest possible sudden polarization effect for pyramidalization
at this value of the torsion angle. Examination of Figs. 5 and
6 shows that this type of behavior is actually computed and furthermor
that in this case the two different sets of one-electron functions
lead to virtually identical dipole moment results even for ϕ values
as low as 10°. The latter result is also quite plausible in light
of the simple 2x2 diabatic state model since for such a small H_{12}
value it is reasonable to expect that ΔE is already large enough
in either basis to produce a minimum of CI mixing between the A^2
and B^2 species, so that essentially equal and at the same time
quite large $|\mu_z|$ dipole moment values for the S_1 and S_2 states result
thereby.

At still smaller values of Θ the AB configuration takes on
an ever greater role in the description of the V and Z states and
exercises a moderating effect on the usual tendency toward decreased
A^2-B^2 mixing as pyramidalization occurs. In addition the H_{12} interaction matrix element rapidly increases from its apparently minimal
absolute value in the Jahn-Teller crossing region, thereby further
promoting mixing between the A^2 and B^2 localized species at a given
value of ϕ, with the end result that throughout this stage of
the torsion process the degree of polarity in the S_1 and S_2 states
is greatly diminished.

Finally although a slight minimum in the $|\mu_z^1|$ and $|\mu_z^2|$ curves near $\Theta = 88°$ does not immediately fit in with the above qualitative arguments since one expects $|H_{12}|$ to monotonically decrease from $\Theta = 90°$ to the crossing region, this result may simply be explained by the fact that the ΔE value between the A^2 and B^2 parts of these correlated wavefunctions increases with ϕ at a somewhat faster rate for one value of the torsion angle than for another. Altogether the 2x2 diabatic state mixing model appears to explain the rather complex character of the calculated dipole moment results rather well, and it also seems to be easily generalizable for larger olefinic systems, as discussed below.

In twisted propylene, for example, the key carbon 2p AO's analogous to the A and B species in ethylene are not equivalent, so a dipole moment is present even without pyramidalization at one of the hybrid carbon sp^2 centers. The 2p orbital on the central carbon (C_1) experiences a more negative environment and thus is less stable than that of the carbon (C_2) farthest away from the methyl σ electron-donating group. As a result the closed-shell configuration with excess electronic charge at the C_2 (terminal) carbon is favored in the S_1 state of propylene, while S_2 has a complementary charge distribution as usual. Pyramidalizing at C_2 increases the energy difference between the 2p AO's, thereby leading to an increase in polarity in both S_1 and S_2, while the analogous geometrical change at C_1 has the opposite effect. An ab initio CI study on this system recently reported by Bonačić-Koutecký et al. (9) bears out these qualitative conclusions in all cases and also shows that in this system the degree of suddenness in the polarization characteristics is generally smaller than in ethylene because of the strong substitution effect.

In the s-cis-s-trans-diallyl system a strong variation in diole moment occurs strictly as a rsult of twisting around the central bond (10), but again this behavior is easily explained in the 2x2 diabatic state model. In this case, however, the ΔE value remains fairly constant over the geometrical region in which the degree of polarization is rapidly changing, and instead it is the H_{12} matrix element which is responsible for the sudden variation in the mixing of the pertinent localized configurations. Thus from this example it is clear that the existence of a Jahn-Teller effect is not really essential for the occurence of the sudden polarization phenomenon, but given the small magnitude of H_{12} matrix elements in general, it must nonetheless be emphasized that the ΔE values between such pairs of states must take on relatively small values ($\Delta E \leq 5-10$ kcal/mole) before a sizeable (sudden) effect can be expected. This condition of close-lying states of the same symmetry carries with it the methodological problem mentioned in the Introduction, i.e., that the Born-Oppenheimer Approximation is likely to fail in such situations, and thus considerations of this nature also should be taken into account in arriving at a comprehensive theoretical basis for the sudden polarization effect.

IV. DESCRIPTION OF VIBRONIC INTERACTIONS AND CONCLUDING REMARKS

The Born-Oppenheimer Approximation achieves its well-known
separation of nuclear and electronic motion primarily through the
assumption of vanishing derivatives of the electronic (adiabatic)
wavefunctions with respect to all nuclear coordinates. One of the
most practical means of going beyond this level of treatment involves
the use of conventional CI wavefunctions and the calculation of
matrix elements over $\partial / \partial Q$ and higher nuclear derivative operators
(17,18); programs for computing first-derivative quantities have
been reported earlier (19,20) and analogous codes have also been
developed (19) for the corresponding second-derivative results

$$< \psi^\alpha \mid \frac{\partial^2}{\partial Q^2} \mid \psi^\beta >$$

Before such data can be used effectively, however, it is ne-
cessary to obtain the requisite CI wavefunctions ψ^α and ψ^β and
their corresponding potential surfaces to good accuracy, particularly the relative spacings of the latter. In the previous section
and in earlier work (8) arguments have been presented which suggest
rather strongly that the triplet-state MO's, because of their relatively balanced description of key ionic configurations (or correlated diabatic states), allow for a good description of the dipole
moment surfaces of the excited singlet states of twisted and pyramidalized ethylene. It is by no means clear, however, that similar
reliability can be expected for the potential energy surfaces in
this basis, since this property depends on much more than just
obtaining the proper degree of mixing between the A^2 and B^2 configurations.

Indeed, the 2x2 diabatic state model discussed above indicates
that the individually optimized NO basis sets for each state are
superior for the calculation of their potential surfaces, whereby
the T MO's themselves are probably best looked upon as averaged
natural orbitals for the S_1 and S_2 states of twisted ethylene.
Because the natural orbitals for the singlet states tend to produce
highly localized charge distributions for these states, it is found
that the S_1 and S_2 CI wavefunctions which result from use of these
species in a conventional CI calculation are of essentially pure
A^2 or B^2 character for even slightly pyramidalized geometries in
the neighborhood of 90° twisted ethylene. If a 2x2 (non-orthogonal)
CI is carried out to obtain the proper mixing ratio between these
two wavefunctions in this structural region, it is true that large
changes in the computed dipole moment values will result relative
to the diagonal values in this basis. Because the H_{12} interaction
matrix element in this computation is very small, however, the
corresponding potential energy surfaces should not differ greatly
from those obtained in the conventional CI treatments with such
natural orbitals (Fig. 3).

As the level of the MRD-CI treatment gradually improves toward the full CI limit, the results obtained using triplet MO's and specially optimized natural orbitals respectively must eventually converge toward one another. To test the hypothesis that natural orbitals give more realistic potential surfaces than do T SCF MO's then, it was decided to increase the number of reference configurations relative to the three main species used in the original calculations (8). For this purpose a structural region was chosen in which the T MO's and S_1 (or S_2) NO's give distinctly different results, namely for the S_1 state at $\Theta = 90°$; in this region a barrier to pyramidalization is observed in the former one-electron basis (Fig. 2), but a steadily decreasing energy surface is found when its own natural orbitals are employed (Fig. 4). The clear result of this experimentation is that by the time seven reference configurations are employed the energy surfaces obtained with T MO's are no longer greatly different from those resulting from the use of natural orbitals. In particular the barrier to pyramidalization observed in the original T MO calculations at $\Theta = 90°$ (Fig. 2) is found to disappear entirely when the level of the CI treatment is significantly improved.

Once the potential surfaces of Fig. 4 are accepted as being the most realistic, the importance of vibronic coupling between the S_1 and S_2 states can be assessed by means of another type of CI treatment[22], in which the two sets of vibrational wavefunctions obtained from the above potential data serve as basis set and the corresponding Hamiltonian operator contains the non-adiabatic $\partial/\partial Q$ and $\partial^2/\partial Q^2$ terms explicitly. Since significant mixing between vibrational levels of different electronic states is only likely to occur when such species lie in the same energy region, it follows that the lowest S_1 vibrational states near its potential minimum for pyramidalization at $\phi = 60°$ (Fig. 4) are very probably left unchanged in such a calculation, which is to say that the Born-Oppenheimer Approximation functions quite satisfactorily in this instance and the dipole moments of such levels are obtained to good accuracy by simply averaging (geometrically) over the S_1 dipole surface in this region.

At higher energies for which the lowest S_2 vibrational levels occur the situation clearly becomes more complicated, however, because an interaction with nearly isoenergetic S_1 species is now likely. Since the electronic dipole moments of the S_1 and S_2 states are of opposite sign and of fairly large magnitude at all but the smallest ϕ values (whereby again the T MO results of Fig. 5 should most accurately reflect the situation at the full CI level of treatment), it seems clear that vibronic mixing between such species will often result in levels with dipole moments which differ greatly from the diagonal values of the Born-Oppenheimer vibrational states which make the largest contributions thereto. In general

one therefore expects that in this energy region the polarity of
the vibronic species drastically changes from one level to another,
distinguishing this situation quite clearly from what is found
at energy values lying well below the S_2 potential minimum.

Calculations to demonstrate this effect quantitatively are
currently underway (23), but it is important to add that the above
conclusions are premised upon a two-dimensional theoretical treatment in which pyramidalization at only one of the carbon centers
of ethylene is undertaken. While this approach has been quite useful in demonstrating both the nature of the sudden polarization
effect and the methodological problems which arise in describing
it, care must be taken in making experimental predictions solely
on this basis, since pyramidalization at the other carbon atom
tends to produce charge polarization in the opposite direction.
It is unlikely at the S_1 potential minimum that pyramidalization
at the other carbon would lower the energy further, since there
is a definite charge deficiency at this center in this electronic
state. This point does not, however, rule out the possibility that
at other regions in the excited singlet-state potential surfaces
simultaneous pyramidalization may have a significant effect on
the resulting vibrational state manifold.

From the above discussion two conclusions appear to be worthy
of special emphasis. First of all, dipole moments of vibrational
levels falling well outside the region in which a Jahn-Teller intersection of two or more states occurs are nonetheless well represented in the Born-Oppenheimer Approximation. It is therefore only
for energy locations in which the spacings between the potential
curves of these electronic species are relatively small that significant vibronic mixing, and subsequently large variations in the
polarity of successive levels, should be expected.

Secondly, in attempting to describe such phenomena with limited
CI calculations it is possible that a given one-electron basis
may lead to realistic results for the dipole moments of the various
electronic states, while proving to be far less satisfactory for
the description of the corresponding energy data. Thus while triplet-state MO's, or alternatively some set of averaged natural (or
MC-SCF) orbitals, appear to give more reliable results for the
dipole moments of the S_1 and S_2 states of twisted ethylene, evidence strongly indicates that separately optimized natural orbitals
for each of these species are more effective in the computation
of their potential surfaces.

Closer examination has shown that when the dimensions of the
CI secular equations are increased, the discrepancies between potential data obtained in the two basis sets are reduced, whereby it
is clear that they must vanish entirely at the full CI level of
treatment. This observation suggests in turn that it is dangerous

to rely on the results of relatively small CI calculations for any one-electron basis, without verifying that such data are stable with respect to greatly increasing the dimensions of the CI secular equations. Failure to take this precaution may lead to results which differ considerably from those of a full CI in a given AO basis, even though the orbitals employed in the limited CI may have been optimized with respect to an energy criterion at that level of treatment, with the consequence that the corresponding quantitative predictions deduced from such calculations might be unreliable. The present results for the singlet excited states of ethylene indicate that such a state of affairs is most likely to occur when two or more electronic species of the same symmetry lie very close together, but even when this is not the case the danger in basing the theoretical treatment on relatively small CI solutions should not be underestimated.

ACKNOWLEDGMENT

The authors wish to thank Professors S. Peyerimhoff (Bonn and J. Koutecký (Berlin) for numerous helpful discussions in carrying out this study. One of us (V. B.-K.) would also like to express her gratitude for the partial financial support of the Deutsche Forschungsgemeinschaft in the form of research grant BO 627-1 for this study.

REFERENCES

1. V. Bonačić-Koutecký, P. Bruckmann, P. Hiberty, J. Koutecký, C. Leforestier and L. Salem, Angew. Chem., Inst. Ed. Engl. 14, 575 (1975)

2. a) L. Salem and P. Bruckmann, Nature (London) 258, 526 (1975); b) L. Salem, Science 191, 822 (1976)

3. P. Bruckmann and L. Salem, J. Am. Chem. Soc. 98, 5037 (1976)

4. M.C. Bruni, J.P. Daudey, J. Langlet, J.P. Malrieu and F. Momicchioli, J. Am. Chem. Soc. 99, 3587 (1977)

5. C.-M. Meerman von Bethan, H.J.C. Jacobs and J.J.C. Mulder, Nouveau J. Chim. 2, 123 (1978)

6. V. Bonačic-Koutecký, R.J. Buenker and S.D. Peyerimhoff, J. Am. Chem. Soc. 101, 5917 (1979)

7. B.R. Brooks and H.F. Schaefer III, J. Am. Chem. Soc. 101, 307 (1979)

8. R.J. Buenker, V. Bonačic-Koutecký and L. Pogliani, J. Chem. Phys. 73, 1836 (1980) and references therein

9. V. Bonačic-Koutecký, L. Pogliani and M. Persico, preprint

10. V. Bonačic-Koutecký, J. Am. Chem. Soc. 100, 396 (1978)

11. R.J. Buenker and S.D. Peyerimhoff, Chem. Phys. 9, 75 (1975)

12. a) R.J. Buenker and S.D. Peyerimhoff, Theoret. Chim. Acta 35, 33 (1974);
    b) R.J. Buenker and S.D. Peyerimhoff, Theoret. Chim. Acta 39, 217 (1975);
    c) R.J. Buenker, S.D. Peyerimhoff and W. Butscher, Mol. Phys. 35, 77 (1978)

13. R.S. Mulliken, Phys. Rev. 41, 751 (1932)

14. C.F. Bender and E.R. Davidson, J. Phys. Chem. 70, 2675 (1966)

15. J.P. Malrieu and G. Trinquier, Theoret. Chim. Acta 54, 59 (1979)

16. R. McDiarmid, J. Chem. Phys. 69, 2043 (1978)

17. M. Desouter-Lecompte, J.C. Leclerc and J.C. Lorquet, Chem. Phys. 9, 147 (1975)

18. M. Desouter-Lecompte and J.C. Lorquet, J. Chem. Phys. 66, 4006 (1977)

19. a) C. Galloy and J.C. Lorquet, J. Chem. Phys. 67, 4672 (1977)
    b) C. Galloy, Ph. D. Thesis, Université de Liège, Belgium (1977)

20. G. Hirsch, P.J. Bruna, R.J. Buenker and S.D. Peyerimhoff, Chem. Phys. 45, 335 (1980)

21. R.J. Buenker, to be published

22. R.J. Buenker, Gazz. Chim. Ital. 108, 245 (1978)

23. M. Persico and V. Bonačic-Koutecký, private communication

CEPA CALCULATIONS FOR ROTATION BARRIERS ABOUT CC DOUBLE BONDS:

ETHYLENE, ALLENE, AND METHYLENE-CYCLOPROPANE

Volker Staemmler and Ralph Jaquet

Lehrstuhl für Theoretische Chemie, Ruhr-Universität
Universitätsstrasse 150, D-4630 Bochum, FRG

INTRODUCTION

The calculation of rotation barriers about CC double bonds
has been a challenge to numerical quantum chemistry for a long time.
While inversion barriers and rotation barriers about CC single
bonds can be obtained by conventional quantum chemical methods to a
rather high degree of accuracy ($\pm$ 0.5 kcal/mol) all calculations for
rotation barriers about CC double bonds that have been performed
prior to 1977 yielded either completely wrong (errors of $\pm$ 30 kcal/
mol) or only occasionally reasonable results. This holds for ab
initio calculations as well as for semiempirical ones and is illustrated by two review articles[1,2] (which appeared in 1974 and 1977)
on ab initio calculations of barrier heights. The situation improved
largely in 1977 when three independent calculations[3-5] on the rotation barrier of allene were successful in obtaining results close
to each other and to the experimental value.

The reason for the inadequacy of the more conventional ab initio
methods - mainly restricted Hartree-Fock (RHF) calculations - for
rotation barriers about double bonds is quite simple and well understood: At their respective equilibrium geometries molecules like
ethylene, allene etc. have closed-shell structure, i.e. all molecular
orbitals are doubly occupied, in particular the CC bonding π-orbital.
When the molecule is twisted this π-bond has to be broken, and in
the rotated geometry (perpendicular ethylene, planar allene) one is
left with an electronic structure that contains two singly occupied
orbitals, i.e. open shells.

Any treatment aiming at a qualitatively and quantitatively

261

correct description of the rotation process has to take this change
in the electronic structure into account. Performing just conventional closed-shell RHF calculations both for the equilibrium and
for the rotated geometry - as has been done in almost all calculations
quoted in Ref. 1 and 2 - will necessarily lead to much too high
rotation barriers. But even if one uses the appropriate RHF methods
(i.e. closed-shell RHF at equilibrium, open-shell RHF at rotated
geometries) it is not guaranteed that the rotation barriers are more
reliable since electron correlation may then change quite significantly. In this connection it has to be stressed that Dewar's semi-
empirical calculations both on MINDO/2[6] and MINDO/3[7] level though
yielding good[6] or even excellent[7] results for the barrier heights
do not describe correctly the electronic structure and geometry
in the rotated biradical configuration.

In this paper we present the results of large-scale ab initio
calculations including electron correlation effects of the rotation
barriers in ethylene, allene and methylene-cyclopropane (MCP).
After the description of the method of calculation and of the
optimization of the geometry in section II and the discussion of the
overall results in section III we analyse the correlation contributions to the barrier in section IV. Section V deals with the problem of the out-of-plane deformation in MCP and the cyclopropyl radical. Finally, in section VI we discuss the question whether the singl
or the triplet state is the lowest one in the rotated geometries.

METHOD OF CALCULATION

All the calculations have been performed using orbital basis
sets consisting of Gaussian lobe functions. As standard basis set
we used a Huzinaga 7s,3p/3s set contracted to double zeta (DZ)
quality. We have performed further calculations with this DZ basis
set augmented by polarization functions (d on C with $\eta_d = 0.8$, p
on H with $\eta_p = 0.65$) in order to investigate the basis set dependence
of the results.

The SCF calculations are just standard closed-shell and open-
shell Roothaan-type RHF calculations; the technique used for open-
shell singlet states has been described earlier[5]. Electron correlation was taken care of by the CEPA-PNO method which has been
originally proposed by Meyer[9-11]. We used the program recently
written by ourselves[12].

The electronic configurations of the π-electrons in the equilibrium (eq.) and rotated (rot.) geometries of the three molecules
can be written as

	eq.	rot.
ethylene	π^2_{cc}	$\pi_c \pi'_c$
allene	$\pi^2_{cc} \pi'^2_{cc}$	$\pi^2_{ccc} \pi^{(n)}_{ccc} \sigma_c$
MCP	π^2_{cc}	$\pi_c \pi'_c$

with the meaning of π_{cc}, π_{ccc} etc. now to be explained.

In ethylene and MCP we have a bonding two-center π-orbital doubly occupied at eq. while at rot. there are two singly occupied π-orbitals (π_c, π'_c) which are orthogonal to each other and localized on either C-atom each. In allene, however, at eq. the terminal CH_2 groups are perpendicular to each other and two perpendicular two-center π-orbitals are doubly occupied each. When allene is rotated, it becomes planar, one π-orbital can extend over all three C-atoms to form a bonding three-center orbital (π_{ccc}). The remaining two electrons occupy the non-bonding three-center π-orbital $\pi^{(n)}_{ccc}$ localized on the terminal C-atoms and the non-bonding σ_c-orbital at the central C-atom. It has been verified on 3x3 CI level that this configuration has the lowest energy for rotated allene (cf. also Ref. 4).

Since in all three molecules considered here, the π-bond is broken or partially broken upon rotation there will be substantial geometry relaxation in the rotated structures. This applies mainly to the CC bond length, the CCC bond angle in planar allene[4], and the deviation from planarity at the tertiary C-atom in MCP. We have optimized these parameters on RHF level, with CH bond-lengths and HCH angles as well as the geometry of the cyclopropane ring fixed close to their experimental values (cf. Ref. 7). The results are (distances in Å):

		eq.	rot.	rot.bent
ethylene	R_{CC}	1.31	1.47	
allene	R_{CC}	1.31	1.35	1.35
	$\angle_{CCC}$	180°	180°	140°
MCP	R_{CC}	1.31	1.44	1.44
	$\angle_{ring,CC}$	180°	180°	140°

In all cases, we have checked that the planarity at the terminal CH_2 group is preserved, but the force constant for the out-of-plane vibration in the rotated geometry is generally rather small[13].

It has to be noted that correlation does not alter the optimum

geometry of rotated ethylene and allene. But in MCP the force constant for the out-of-plane vibration at the tertiary C-atom is so
small that improvement of the basis and inclusion of correlation
have a marked effect (compare section V). Furthermore, reoptimization
of CH bond lengths etc. in the rotated structures would influence the
barrier heights by about 0.5 kcal/mol and has to be considered only
if one wants to go beyond this accuracy.

OVERALL RESULTS FOR BARRIER HEIGHTS

Table 1 contains the results of our calculations for the barrier
heights for different basis sets and different levels of sophistication: RHF, correlation of the π-electrons only, correlation of
the σ and π-electrons belonging to the CC bond(s), and total valence
shell correlation. For MCP, the valence shell was confined to the
methylene CH bonds and the CC bonds connected to the central C-atom;
extending the valence shell over the whole cyclopropane ring was
too time-consuming.

At first glance there is a tremendous difference in the quality
of the RHF results between ethylene and MCP on one hand and allene
on the other. As we have observed earlier[5] in ethylene (and similarly
in MCP) RHF underestimates the barrier height by almost 20 kcal/mol
while for allene the RHF result is almost correct. As soon as the
correlation of the π-electrons is included the results are comparable
for all three molecules (errors of $\pm$ 5 kcal/mol). Inclusion of the
CC σ-electrons and the remaining valence electrons is necessary if
one is aiming at an accuracy of $\pm$ 1 kcal/mol which is comparable to
the experimental accuracy and if one wants to have an intrinsic
criterion for the quality of the results.

Surprisingly, improvement of the basis beyond DZ quality does
not change the results substantially. One may wonder whether the
"gain" of 2 kcal/mol in the barrier height for ethylene on going from
DZ via DZ+d to DZ+dp is worth the increase in computer time in the
ratio 1:3:8. The reason for this behaviour will be explained later.

From the figures in Table 1 as well as from a series of further
calculations on CI and CEPA level (the results of which are not
presented here) we can deduce an intrinsic estimate of the errors
of our results which is given at the bottom of Table 1. For MCP
the estimated error is larger since the size of the molecule en-
forced more severe restrictions to the calculation. For all three
molecules the ab initio results coincide with the experimental ones
within the respective error bars.

Table 1: SCF and CEPA results for rotation barriers in ethylene, allene, and methylene-cyclopropane (MCP) (in kcal/mol)

	Ethylene	Allene		MCP	
		linear	bent	planar	bent
Basis (size[a])	DZ(28)	DZ(38)		DZ(52)	
RHF	47.2	49.5	42.8	49.5	46.5
CEPA π[b]	67.3	48.5	40.4	66.3	63.1
CC	67.0	44.2	39.5	65.0	63.6
val.	62.8			58.5	59.2
Basis (size)	DZ+d(38)	DZ+d(53)		DZ+d*(62)[c]	
RHF	49.0	51.2	43.4	52.6	47.0
CEPA π	69.5	50.7	42.3	69.9	64.2
CC	69.9	44.9	40.2	68.5	65.1
val.	64.6	47.0	43.6	(59.8)	(59.3)[d]
Basis (size)	DZ+dp(50)				
RHF	48.6				
CEPA π	69.1				
CC	69.5				
val.	64.6				
Estimate	64+2	44+2		59+4	
Experimental	65.0[14]	46+1[15]		56+1[16]	

a) size = total number of basis functions
b) π : correlation for π-electrons only;
   CC : correlation for σ and π-electrons of the olefinic CC bond in ethylene and MCP and the two allenic CC bonds in allene, respectively
   val.: total valence shell correlation
c) d-functions only on the C-atoms adjacent to the double bond
d) estimated from basis set extrapolations

ANALYSIS OF THE CORRELATION CONTRIBUTION TO THE BARRIER HEIGHT

There are several physical effects which are responsible for the large correlation contribution to the barrier and for the big difference between ethylene and allene:

1. Correlation in the π-system (Table 2):
In ethylene and MCP the π-system consists of just two electrons. At eq. they occupy the same bonding π-orbital, hence contribute about 20 kcal/mol to the total correlation energy, very similar to H_2 at its equilibrium distance. At rot. the two electrons occupy two orthogonal π-orbitals localized at different C-atoms. Therefore, their motion is practically not correlated, the contribution to the total correlation energy is almost zero. Increase of the basis size has only a small effect: At eq. the main contribution comes from a

$\pi^2 \rightarrow \pi^{*2}$ excitation which is already well described in the DZ-basis.

The situation is completely different in allene: At eq. the correlation energy of the two perpendicular π-bonds is indeed about twice as large as the corresponding contribution in ethylene. In the rotated geometry we have the external correlation of the three-center π-bond which is again in the order of 20 kcal/mol (including rather small interorbital contributions coupling π_{ccc} with $\pi_{ccc}^{(n)}$ and σ). But the two singly occupied orbitals very effectively polarize the π_{ccc} orbital[17]; the energy contribution of the 'dynamic spin polarization'[1] compensates fully for the loss in external correlation energy such that the π-correlation energies at eq. and rot. are nearly identical.

We further find a small, but significant difference of $\sim$5 mh $\approx$ 3 kcal/mol in the π_{cc}^2 pair correlation energy between ethylene and MCP at their respective equilibrium geometries. The reason for this is that the $\pi^2 \rightarrow \pi^{*2}$ excitation contributes less to the correlation energy in MCP because both the matrix element $(\pi\pi^*|\pi^*\pi)$ is smaller and the respective energy denominator larger than in ethylene. This 'truncation of the correlation space' is caused by an exclusion principle repulsion due to the ring electrons, similar to the one in He_2 discussed first by Kestner and Sinanoğlu[18]. Table 1 shows that this effect is the main reason for the smaller barrier in MCP as compared to the one in ethylene.

2. Total correlation of the CC-bond (Table 3): Since the CC bond lengths change considerably during rotation and since the correlation energy of a σ-bond increases rapidly during lengthening of the bond one would expect a larger value for the correlation energy of the CC-σ-bond in rot. compared to eq.. As it is shown in Table 3 this increase in the σ^2-pair correlation energy of $\sim$2.5 mh in ethylene and MCP is compensated by a decrease of the $\sigma\pi$-interorbital pair energies such that the total σ-contribution remains nearly constant.

Again, in allene the situation is more complicated: First, the σ-bond lengths and consequently the σ-intraorbital correlation energies do not change much. Secondly, the form of the π-orbitals varies so strongly during rotation that it is almost impossible to compare equivalent contributions. Therefore, instead of presenting a large number of pair correlation energies we only quote the two most relevant figures: Inclusion of the σ-bonds lowers the barrier by $\sim$2 kcal/mol (DZ+d, rot. bent, compare Table 1); the dynamic spin polarization of the σ-bonds amounts to 2x4.06 = 8.12 mh, i.e. only half the value per σ-bond as in ethylene and MCP.

3. Complete valence shell correlation:
In the final step we have to include the remaining part of the whole valence shell correlation. As is seen from Table 1, this lowers the barrier height in ethylene and MCP by 5 kcal/mol, but increases it by 3 kcal/mol in allene. A more detailed analysis shows that there are two main contributions: Correlation within the system of CH bonds

Table 2: CEPA π-electron correlation energies (all values negative, in 10^{-3} a.u. = mh; 1 mh = 0.6275 kcal/mol)

Molecule	Basis	eq.	rot.	rot.bent
Ethylene	DZ	32.38	0.16	
	DZ+d	33.26	0.55	
	DZ+dp	33.48	0.73	
MCP	DZ	27.06	0.28	0.61
	DZ+d	28.47	0.87	1.12
Allene ext[a]	DZ	57.71	25.37	26.26
SP			33.88	35.14
ext[a]	DZ+d	66.23	36.50	36.40
SP			30.42	31.51

a) The first entry is mainly the external correlation of π_{CCC}^2, the second the spin polarization of π_{CCC} by the two singly occupied orbitals.

Table 3: CEPA pair correlation energies for CC σ and π pairs (all values negative, in 10^{-3} a.u.; basis DZ+d)

	Ethylene		MCP	
Pair	eq.	rot.	eq.	rot.bent
σ^2	20.48	22.85	20.08	22.77
π^2 [a]	31.37		27.09	
$\pi\pi'$ [a]		0.48		0.87
$\sigma\pi$	31.67	9.27	30.78	9.59
$\sigma\pi'$		9.27		9.78
SP σ [b]		8.33		8.20

a) Slightly smaller than in Table 2 since they originate from a calculation containing a larger number of pairs

b) Dynamic spin polarization of CC σ-bond

and interaction of CH with CC σ- and π-pairs. They affect the barrier height by (all values in kcal/mol)

	ethylene	MCP	allene
CH	+2 to +3	0 to +2	−2
CH vs. CC	−4 to −6	−4 to −6	+5

depending slightly on geometry and basis set.

It is quite difficult to rationalize these small quantities in terms of physically meaningful effects such as 'truncation of correlation space' etc. We only want to stress that the interaction of CH and CC pairs is markedly more important than the change in the CH-pairs themselves, as one would have expected. But even the CH pairs cannot be disregarded completely if an accuracy of 1-2 kcal/mol is to be achieved.

OUT-OF-PLANE DEFORMATION IN MCP

In MCP, planarity at the central C-atom in the equilibrium geometry is required for the olefinic double bond. When MCP is rotated and the double bond is broken a rehybridization from sp^2 to sp^3 at the central C-atom is possible and the ring strain can be reduced by a deviation from planarity. Wiberg[19] estimated a value of ~13 kcal/mol as release of ring strain energy per trigonal C-atom. But this value has been obtained by comparing heats of formation of saturated and olefinic hydrocarbons, and it may be questioned whether this value can also be used for radical or biradical species.

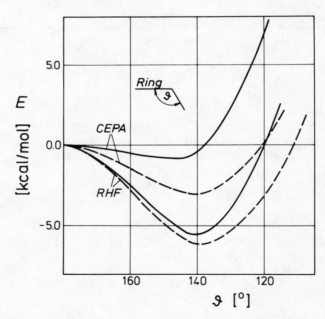

Fig. 1: Out-of-plane deformation for rotated MCP (full curves) and the cyclopropyl radical (dashed curves)

In Figure 1 we present potential energy curves for the out-of-plane deformation for rotated MCP as well as for the cyclopropyl radical. (Basis: DZ+dp for cyclopropyl, DZ+d for MCP;CEPA calculation including all valence electrons at the radical C-atom in cyclopropyl; similar valence-CEPA, but with some additional approximations, for MCP).

One first observes that on RHF level both species have a pronounced minimum with $\vartheta \sim 140°$ connected with a gain in energy of ~ 6 kcal/mol. This is just half of Wiberg's value and may be rationalized by the fact that in both radicals the lone pair orbital at the trigonal C-atom is only singly occupied while in Wiberg's investigation a doubly occupied CH bonding orbital was considered instead.

Correlation, however, has a marked effect on the shape of these potential curves: In both radicals the minimum becomes shallower and shifted to smaller out-of-plane angles, i.e. to larger values of ϑ. While the cyclopropyl radical still exhibits a decent minimum with a depth of ~ 3 kcal/mol, MCP shows a very flat minimum with a depth of only ~ 0.5 kcal/mol. The accuracy of our calculation for MCP does not suffice for ultimately deciding whether or not this minimum will remain if the treatment is improved. From calculations with the DZ basis set - yielding slightly shallower minima both on RHF and CEPA level - we are led to expect that a further increase of the basis size will deepen the minimum in MCP, but the inversion barrier will remain rather small (probably ~ 1 kcal/mol). For the cyclopropyl radical DZ and DZ+dp results were practically identical.

The explanation for the influence of electron correlation on the out-of-plane deformation is the same as the one given previously for CH_3[13]: During bending the singly occupied orbital at the central C-atom changes from a pure p-AO to an sp^3 hybrid; its efficiency in polarizing the adjacent CC and CH σ-bonds becomes smaller since it is pushed off the axes of the σ-bonds and its differential overlap with the corresponding (localized) σ-orbitals decreases. A more detailed analysis of the behaviour of different correlation contributions reveals that all but the spin-polarization terms remain constant while the latter decrease considerably with decreasing ϑ, as it is seen from Table 4.

Since there are two unpaired electrons in the biradical MCP and since the spin polarization of the ethylene CC-σ-bond is larger than that of the ring CC or CH bonds, the absolute contribution and the reduction with decreasing ϑ is larger in MCP than in cyclopropyl.

Thus we have a nearly complete cancellation of the gain in strain energy and the loss in spin polarization in MCP which explains that the potential curve is very flat between $180°$ and $140°$.

Table 4: Spin polarization of the CC and CH σ-bonds in cyclopropyl
and MCP (in mh)

	Basis	$\vartheta = 180°$	$160°$	$140°$	$120°$
Cyclopropyl	DZ	7.19	5.43	3.28	2.44
"	DZ+dp	9.14	7.35	5.21	4.23
rot. MCP	DZ	23.93	21.52	18.08	16.06

SINGLET TRIPLET SPLITTING

In a large scale CI calculation Buenker and Peyerimhoff[20] found
that the singlet diradical state in rotated ethylene is by 2 kcal/
mol lower in energy than the corresponding triplet, contrary to the
prediction of Hund's rule. In a previous preliminary study[17] we have
explained this observation as a consequence of the dynamic spin
polarization of the σ-core by the two singly occupied π-electrons.
In this paper we present a more quantitative confirmation of this
explanation.

The singlet-triplet energy difference of two diradical states
with the orbitals x and y singly occupied is given in RHF approxi-
mation by the exchange integral $2(xy|yx) = 2(x|K_y|x)$. Since this is
positive, the singlet is always above the triplet (Hund's rule).
Correlation can modify the splitting in several ways:

a) Generally, the singly occupied orbitals of an open-shell
singlet state (at least one of them) are more diffuse than those of
the triplet state with the same electronic configuration[21]. Therefore
pair correlation energies ε_{ij} that couple doubly and singly occupied
orbitals should be smaller (in absolute value) for the singlet than
for the triplet state.

b) For the same reason the truncation of the correlation space
should be less severe in the singlet than in the triplet such that
ε_{ij} with i and j doubly occupied are larger (in absolute value) for
the singlet state.

c) Internal excitations ij → xy will favour the singlet since
the exchange integral $(i|K^+_{xy}|j)$ for the singlet is generally larger
than the integral $(i|K^-_{xy}|j)$ for the triplet.

d) The direct correlation between the two singly-occupied or-
bitals involves exchange integrals $(a|K^+_{xy}|b)$ and $(a|K^-_{xy}|b)$ (+ for
singlet, a and b being virtual orbitals). Again, the K^+-integral
should be larger because its density contains less nodes than the
one for the K^--integral. But this may be compensated by the larger

diffuseness of the singlet orbitals. Therefore, it is difficult to predict whether the singlet or the triplet state is favoured.

All these effects, in particular b) - d) are very small in the molecules considered here since the exchange integral $(xy|yx)$ is small and the orbitals x and y for the singlet and the triplet state are very similar.

e) Finally, we have the dynamic spin polarization of the doubly occupied orbitals which will favour the singlet as we have outlined in Ref. 17. Table 5 shows that indeed this is the dominating effect which leads to a singlet "ground state" of rot. ethylene and allene. In particular the spin polarization of the allene π_{CCC} bond does not only lower the rotation barrier of allene (section III) but contributes markedly to the singlet-triplet splitting.

Our final value of 0.92 kcal/mol for ethylene compares favorably with other CI calculations[20] and should be accurate to within about 0.5 kcal/mol. The accuracy of the result for allene is somewhat lower ($\pm$ 1 kcal/mol) since the CH bonds were not included in this analysis and a smaller basis was used. For MCP we obtain a slightly smaller value (0.5 $\pm$ 2 kcal/mol) than for ethylene.

Table 5: Contributions to singlet-triplet splitting in ethylene and allene (in kcal/mol)

Molecule	Ethylene	Allene
Geometry	rot.	rot.bent
Basis	DZ+dp	DZ+d
Correlation	CEPA,valence shell	CEPA, CC σ and π bonds
RHF (Hund's rule)	1.48*	2.54*
Correlation**		
Type a)	+0.57	+1.23
Type b)	-0.01	-0.53
Type c)	0+	-0.02
Type d)	+0.09	0.00
Type e)	-1.41 (CHσ)	
	-1.64 (CCσ)	-1.27 (CCσ)
		-4.71 (CCCπ)
Total	-0.92	-2.79

*) positive numbers mean that the triplet state is lower
**) type of correlation contributions see text
+) zero from symmetry reasons

It should be noted that the value of the singlet-triplet splitting is rather strongly influenced (one finds changes of 0.5 to 1.5 kcal/mol) by the choice of the basis and details of the calculations (i.e. CI or CEPA, single excitations included or not etc.). This is not surprising since the different contributions to the correlation part are very sensitive to details of the calculation.

CONCLUSIONS

1) Qualitatively correct values (errors of $\pm$ 5 kcal/mol) for rotation barriers about CC double bonds can be obtained rather cheaply by RHF calculations followed by inclusion of electron correlation within the π-system.

2) Chemical accuracy (errors of $\pm$ 2 kcal/mol) can only be reached if correlation within the whole valence shell, at least that of the orbitals belonging to and adjacent to the double bond, is included.

3) Dynamic spin polarization effects are responsible for the low rotation barrier in allene as well as for the high stability of singlet biradicals in the rotated molecules.

ACKNOWLEDGEMENT

We are grateful to Dr. H. Kollmar and to Prof. Dr. W. Kutzelnigg for numerous discussions and to Prof. Dr. W. Roth for suggesting the calculations and supporting us with his experimental results prior to publication. All the calculations were performed on the INTERDATA 8/32 minicomputer at the Lehrstuhl für Theoretische Chemie Bochum sponsored by the Deutsche Forschungsgemeinschaft.

REFERENCES

1. A. Veillard, in: Internal Rotation in Molecules (W.J. Orville-Thomas, ed.), p. 385, John Wiley, New York 1974
2. P.W. Payne, L.C. Allen, in: Applications of Electronic Structure Theory (H.F. Schaefer, III, ed.), p. 29, Plenum Press, New York 1977
3. C.E. Dykstra, J. Am. Chem. Soc. 99:2060 (1977)
4. R. Seeger, R. Krishnan, J.A. Pople, P.v.R. Schleyer, J. Am. Chem. Soc. 99:7103 (1977)
5. V. Staemmler, Theoret. Chim. Acta 45:89 (1977)
6. M.J.S. Dewar, M.C. Kohn, J. Am. Chem. Soc. 94:2699 (1972)

7. R.C. Bingham, M.J.S. Dewar, D.H. Lo, J. Am. Chem. Soc. 97:1294
    (1975)   •
8. S. Huzinaga, J. Chem. Phys. 42:1293 (1965) and technical reports
    from the University of Alberta, Canada
9. W. Meyer, Int. J. Quantum Chem. Symp. 5:341 (1971)
10. W. Meyer, J. Chem. Phys. 58:1017 (1973)
11. W. Meyer, Theoret. Chim. Acta 35:277 (1974)
12. V. Staemmler, R. Jaquet, Theoret. Chim. Acta, in press
13. F. Driessler, R. Ahlrichs, V. Staemmler, W. Kutzelnigg, Theoret.
    Chim. Acta 30:315 (1973)
14. J.E. Douglas, B.S. Rabinovitch, F.S. Looney, J. Chem. Phys.
    23:315 (1965)
15. W.R. Roth, G. Ruf, P.W. Ford, Chem. Ber. 107:48 (1974)
16. H. Meier, W.R. Roth, private communication (1980)
17. H. Kollmar, V. Staemmler, Theoret. Chim. Acta 48;223 (1978)
18. N.R. Kestner, O. Sinanoğlu, J. Chem. Phys. 45:194 (1966)
19. K.B. Wiberg, R.A. Fenoglio, J. Am. Chem. Soc. 90:3395 (1968)
20. R.J. Buenker, S.D. Peyerimhoff, Chem. Phys. 9:75 (1975)
21. J. Katriel, R. Pauncz, Adv. Quantum Chem. 10:143 (1977)

ELECTRON CORRELATION AND THE MECHANISM OF ATOMIC AUTOIONIZATION

R. Stephen Berry

Department of Chemistry and The James Franck Institute
The University of Chicago, Chicago, Illinois 60637

BACKGROUND

This review deals with our present understanding of the
nature of the quantum states of two-electron atoms. The exposition, a description of new insights into the dynamics of
electron correlation and origins of correlation-induced autoionization, is in large part a brief summary of a discussion
given recently by Paul Rehmus and myself (Rehmus and Berry,
1981) and of the earlier studies of stationary states of two-
electron atoms that led to the analysis of autoionization (Rehmus and Berry, 1979); Rehmus et al., 1978a,b). The essence of
the approach is the construction of the reduced two-particle
density $\rho(r_1, r_2, \theta_{12})$, forming the conditional probability
density

$$\rho(r_2, \theta_{12} \mid r_1 = \alpha) = \frac{\rho(r_1 = \alpha, r_2, \theta_2)}{\left[\int dr_2 \int d\theta_{12} \, \rho(r_1, r_2, \theta_2) \right]} \ . \tag{1}$$

This distribution is the probability of finding particle 2 at
distance r_2 from the origin (essentially the nucleus, in a two-
electron atom) and at angle θ_{12} from the vector $\vec{r_1}$ when the
distance r_1 of particle 1 from the origin has the value α .

The conditional probability distribution (1) or more
correctly, its numerator, had been introduced in one figure
given by Munschy and Pluvinage (1963) to show how the ground
state of the helium atom is polarized when one electron is at
its most probable distance from the nucleus. This representation of correlation appears to have been put aside until our

275

own studies and the essentially simultaneous use of this distribution by Shim and Dahl (1978) to interpret the physical basis of Hund's first rule. Perhaps part of the reluctance to use the distribution $\rho(r_2, \theta_{12} | r_1 = \alpha)$ or $\rho(r_1 = \alpha, r_2, \theta_{12})$ stems from the apparently tedious computations the task would involve, particularly if one wished to use wave functions of high quality. Another might have been the moderately widespread use of distributions $\rho(r_1, r_2)$ in which the angular correlation information is largely removed by integration over all the angles of the system, not just the three Euler angles (Dickens and Linnett, 1957; Macek, 1967, 1968; Fano and Lin, 1974; Fano, 1976, Lin, 1974, 1976). An extreme alternative to the use of $\rho(r_1, r_2)$ is the use of the angular distribution $\rho(\theta_{12})$, in which all radial information is removed by integration over r_1 and r_2. We ourselves began with the intent of studying $\rho(\theta_{12})$ and, along the way, recognized the power of stopping short with $\rho(r_1, r_2, \theta_{12})$ and $\rho(r_2, \theta_{12} | r_1)$. Banyard and Ellis (1972, 1975) and Tatum (1976) had previously studied $\rho(\theta_{12})$ for two-electron atoms.

The conditional probability was first used by us (Rehmus et al, 1978a) to examine the spatial correlation introduced into the doubly-excited states of helium. We were particularly interested in the spatial correlations of the functions obtained by diagonalizing the operator corresponding in classical mechanics to the length (squared) of the vector connecting the semimajor axes of the Kepler ellipses of two electrons moving initially independently of each other in classical orbits. (Wulfman, 1973, 1976; Wulfman and Kumei, 1973; Herrick and Sinanoglu, 1975; Sinanoglu and Herrick, 1975, 1976). The striking success of these functions for yielding accurate energies of the quasibound states stimulated us to look at the spatial distribution of the two electrons, to see how the exceedingly simple configuration mixing of the Herrick-Sinanoglu-Wulfman approach introduced correlation in such an effective way. In retrospect, we can now see that the spatial correlation attainable by mixing only the functions with specified principal quantum numbers n_i, the "ground rules" of the method (the DESB method), is often far from that of accurate functions. Nevertheless, the spatial distributions of electrons for the DESB wave functions are far closer to the accurate ones than pure configurations, based as they are on the notion of nearly good orbital quantum numbers ℓ_1 and ℓ_2.

BOUND STATES

The next step in our exploration of the conditional probability distribution was the examination of a number of ground-state wave functions for two-electron systems H^-, He, Li^+ and Be^{++} (Rehmus et al., 1978b). These included configurational functions and several Hylleraas-Kinoshita functions, from the very simple up to functions essentially indistinguishable to the eye from the best available functions and giving energies within less than 10^{-7} au of the best available value (Frankowski and Pekeris, 1966). It was possible to do these computations rather easily by using a program written for the Hewlett-Packard 9825 desktop computer by W. England and C. C. J. Roothaan for S-states of two-electron atoms. The results give considerable insight into the nature of polarization, of Coulomb holes and cusps at $\theta_{12}=0$, at least in atomic ground states.

It was natural to extend the computation of Hylleraas-Kinoshita wave functions to excited bound S-states and to doubly excited Feshbach resonances (Rehmus and Berry, 1980). Here we learned how much more polarization occurs in excited states than in ground states. One can see the difference between, for example, the 1s2s ^3S and 1s2s ^1S functions. In Figure 1 are projections of 3-dimensional Cartesian plots of the conditional probability density of the first ^3S state of He; the distance r_1 of electron 1 from the nucleus is indicated with each plot. The two axes in the (projected) horizontal plane correspond to the distance of electron 2 from the nucleus and to θ_{12}, the angle between $\vec{r}_1$ and $\vec{r}_2$. The factor r_2^2 from the Jacobian is included in the plotted function to avoid the pile-up of density at the nucleus.

Note that there is a small but noticeable amount of polarization when r_1 is about 1.5 bohr or more. Also note that when r_1 is small, so that electron 1 is in the region where a 1s electron would be found, the conditional probability density for electron 2 near the nucleus is essentially zero (top two plots). It is not obvious in these diagrams, but the spatial antisymmetry of this ^3S function requires that its wave function $\Psi(\vec{r}_1, \vec{r}_2)$ be zero whenever $r_1 = r_2$, so there is a nodal line in each plot for the value of r_2 satisfying this equality.

The conditional probability distribution of the 1s2s ^1S function, Figure 2, by contrast, has no nodal line where $r_1 = r_2$. Moreover this distribution has nonzero values for small r_2 even when r_1 is very small. Electron 2 is not forced out of the "1s" region by the Pauli principle when electron 1 occupies that region. Note that this distinction between the 1s2s single and triplet is achieved with very accurate Hylleraas-Kinoshita wave functions that in no way have orbital character. Nevertheless orbital-like information emerges. One further distinction between the singlet and triplet should be pointed out. The singlet, with a computed total energy of -2.145586 hartree units, is much more spread out in space than the triplet, whose computed total energy is -2.175221 hartree units (1 hartree = 27.21 eu). This is seen most clearly in the plots with $r_1 = 2.0$ bohr. This is in accord with the interpretation that Hund's rule arises from the greater shielding

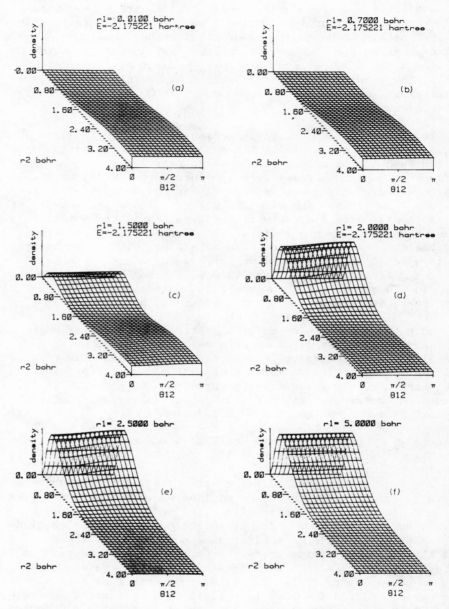

Fig. 1.   The conditional probability density $\rho(r_2, \theta_{12}/r_1)$ for the
1s2s ^{3}S state of He, for several values of r_1, as indicated.
These plots are based on a 26-term Hylleraas-Kinoshita wavefunction with total binding energy of 2.175221 hartree
(Rehmus and Berry, 1979), compared with the value of 2.17522
obtained by Frankowski (1967).

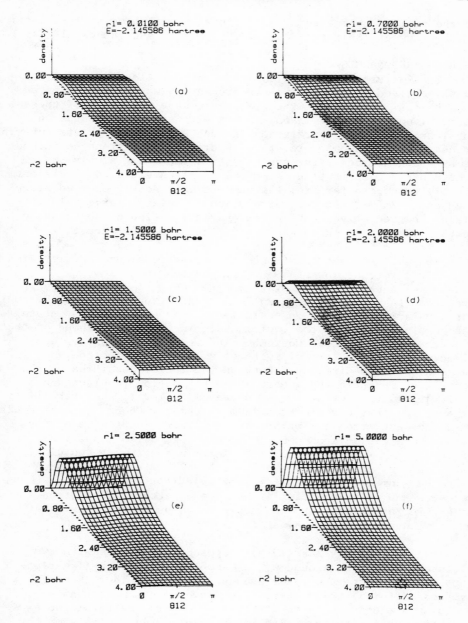

Fig. 2.   The conditional probability density ρ (r_2, θ_{12}/r_1) for the 1s2s ^{1}S state of He, for several values of r_1. These plots are constructed from a 26-term Hylleraas-Kinoshita function giving a total binding energy of 2.145586 hartree (Rehmus and Berry, 1979), compared with 2.145974 hartree obtained by Frankowski (1967).

of the nucleus charge and more diffuse distribution of the electrons
in the singlet (Shim and Dahl, 1978; Katriel and Pauncz, 1977).

FESHBACH RESONANCES

The failure of the orbital or planetary-electron model for
doubly-excited states of two-electron atoms was well recognized
in the first analyses of the electronic wave functions of these
states (Cooper et al., 1963).   The DESB approach brought out the
physics of electron correlation in such states through its inherent
foundations in the picture of two hydrogenic electrons perturbed,
as by a degenerate first-order perturbation, not simply by a
second-order contribution, from the electron-electron repulsions.
The electrons are far from the nucleus in a doubly-excited state
such as a "$2s^2$" ^1S state; correlation can play a large role in
fixing the geometry of the electron distribution without having
an enormous effect on the energy of the quasibound state.

Indeed, when one constructs the conditional probability distributions for the lowest $^1S^e$ (Figure 3) and next-lowest $^1S^e$
(Figure 4) doubly-excited states of He corresponding in a very
crude way to the $(2s)^2$ and $(2p)^2$ $^1S^e$ states, one sees the extremely
large polarizations and, especially for the higher-energy state,
the cusp at $\theta_{12}=0$ and the Coulomb hole in the neighborhood of that
point. [N.B. We use "Coulomb hole" in the sense of depleted
electron density in the vicinity of a localized electron, not in
the sense of the difference between the radial densities of the
correlated function and a Hartree-Fock function (Coulson and Neilson,
1961).]   In the lowest $^1S^e$ doubly-excited level, the probability
is over 85% that one electron is more than 1.1 bohr from the nucleus.
When that condition is satisfied, the other electron is nearly
pinned to a region around $\theta_{12} = \pi$, that is, on the opposite side
of the atom.   This is especially clear in the lowest two plots of
Figure 3. By contrast, the next or "$(2p)^2$" state has no such
e-He^{++}-e geometry, as Figure 4 shows electron 2 may well be found
near $\theta_{12} = 0$ in this state; it may even scatter off electron 1 in
an s-wave, as is shown by the non-zero value of the density at
$r_1 = r_2$ and $\theta_{12} = 0$, i.e. at $r_{12} = 0$.

Kellman and Herrick (1978) proposed that states with two
electrons localized around $\theta_{12} = \pi$ should behave much like rigid
rotors, similar to the rotor states of nuclei. They then made a
synthesis from a) the picture of the two-electron atom as two
hydrogenic electrons, each with 0(4) symmetry, perturbing one
another with the perturbation breaking the 0(4) x 0(4) symmetry
and b) the extension of the geometric picture of some atomic states
as rotor states.   The synthesis (Herrick and Kellman, 1980; Herrick
et al., 1980; Kellman and Herrick, 1980)proposes a description of
two-electron atoms as having a broken 0(4) x 0(4) symmetry related
to that given by the earlier DESB picture, in which the near-con-
stants of motion are a rotor's angular momentum and bending and

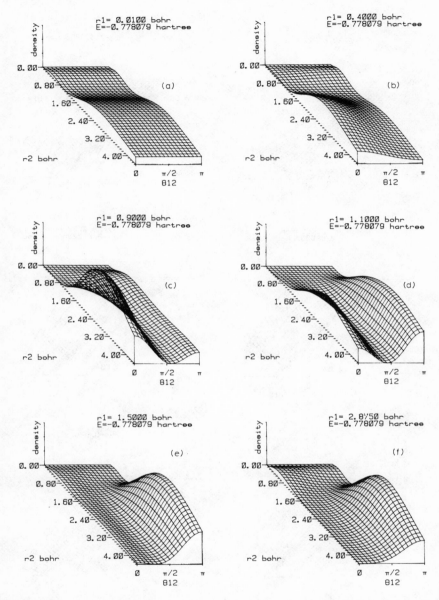

Fig. 3.   The conditional probability distribution for the "$(2s)^2\ {}^1S^e$
Feshbach resonance of He (Rehmus and Berry, 1979). Most of
the statistical weight for r_1 lies in the region $r_1 > 1.1$
bohr, so the lowest two plots describe the most probable
distribution for this state.

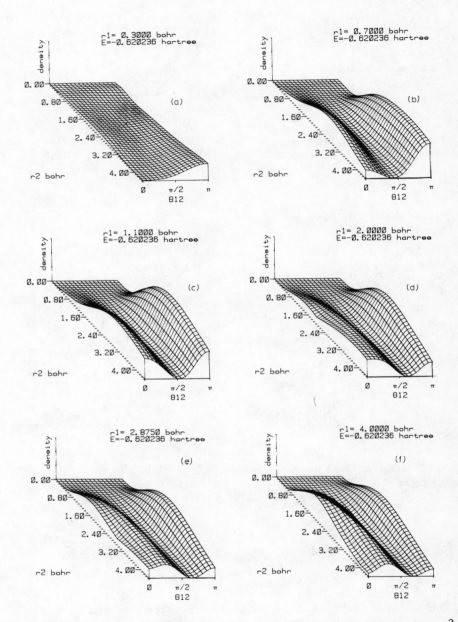

Fig. 4.   The conditional probability distribution for the "$(2p)^2$" $^1S^e$ Feshbach resonance of He (Rehmus and Berry, 1979).

stretching vibrational modes. Naturally there is coupling of the
angular momentum of a doubly-degenerate bending mode with the
rotor's angular momentum. This picture, in very sharp contrast
indeed with the solar-system picture of the atom, has doubly-excited
helium looking much like a linear A-B-A triatomic molecule! The
results of this phenomenological, group-theoretical analysis and
identification with the dynamical picture of an A-B-A model gives
a quite new and apparently quite valid analysis of the quantum
states of highly excited atoms.

According to the analysis of Kellman and Herrick, the "$(2s)^2{}_\text{n}{}^1S^e$"
state of He is the lowest state of a manifold. The "$2s2p{}^3P^o$" corresponds to the state with J=1 and no bending or stretching quanta. The
"$2s2p{}^3P^o$" state corresponds to one of the two states with $J_1 = 0$ but
with one quantum in the bending mode ν_2. We have succeeded in
constructing fairly accurate Hylleraas-Kinoshita wavefunctions for
these states (Yuh et al., 1981), and indeed the lowest $^3P^o$ state of
the manifold is a state with electrons pinned to opposite sides of
the nucleus. The lowest $^1P^o$ state has a distribution with a minimum
at $\theta_{12} = \pi$, but maxima on either side nearly, precisely as one
expects for a state with one quantum of bending motion. One of the
most significant differences between familiar linear triatomic molecules (cf. Kellman, 1980, for an interesting generic approach) and
the doubly-excited states of helium is that the force constants for
bending and for the antisymmetric stretching mode are nearly the
same in helium, in contrast to the molecular situation in which the
bending force constant is invariably small compared with stretching
constants. The result is that bending and antisymmetric stretching
modes are strongly coupled, presumably by Coriolis interactions, in
helium but are only weakly coupled in a molecule such as CO_2. The
classical motion, presumably, and the quantum mechanical distribution
$\rho(r_2, \theta_{12}|r_1)$ reflect this difference (Yuh et al., 1981).

The Mechanism of Atomic Autoionization

Doubly-excited states of helium and most other Feshbach resonances of atoms autoionize by the passage of energy from one electron
to another, allowing the receptor to escape from the attractive
potential of the atom. (Atomic autoionization by hyperfine interaction is possible, in principle, in which energy passes from a
nuclear moment to an electron.) But identifying the degree of freedom in which energy is stored prior to the autoionizing event is only
a step toward identifying the mechanism of atomic autoionization. It
is possible to examine the transition amplitudes for this process or
any other in a manner much like the one just described, and extract
new insights into what we might call "mechanisms." The transition
amplitude

$$\langle a | \mathcal{R} | b \rangle = \int \psi_a^* \mathcal{R} \psi_b \, d\tau \tag{2}$$

involving a 1-electron or 2-electron operator $\mathcal{R}$ can be reduced to an integral over only that one or those two electrons, by integrating over the coordinates of any other electrons. Let us call the remaining integrand

$$R(\vec{r}_1) \quad or \quad R(\vec{r}_1, \vec{r}_2) \qquad , \text{ so that}$$

$$\langle a | \mathcal{R} | b \rangle = \int R(\vec{r}_1) \, d\vec{r}_1 \tag{3a}$$

or

$$\langle a | \mathcal{R} | b \rangle = \int R(\vec{r}_1, \vec{r}_2) \, d\vec{r}_1 \, d\vec{r}_2 \; . \tag{3b}$$

In the case of helium's doubly-excited states, of course there is no prior integration required. It is (3b) that is of interest to us; this function may be reduced from a six-fold integral to a simple triple integral by integrating over Euler angles, as with $\rho(\vec{r}_1, \vec{r}_2)$.

The width or inverse lifetime of a transient state with wavefunction $\psi_b(\vec{r}_1, \vec{r}_2)$ that decays to a continuum state with wavefunction ψ_E orthogonal to is (Fano, 1961)

$$\Gamma_{bE} = 2\pi \left| \int_0^\infty dr_1 \int_0^\infty dr_2 \int_0^\pi d\theta_{12} \sin\theta_{12} \left[r_1^2 r_2^2 \psi_b^*(r_1, r_2) H \psi_E(r_1 r_2) \right] \right|^2 . \tag{4}$$

We have already integrated out the physically irrelevant Euler angles of the two-electron system in Eq. (4).

Now let us define the differential transition amplitude (Rehmus and Berry, 1981);

$$D(r_1, r_2, \theta_{12}) = r_1^2 r_2^2 \psi_b^*(r_1, r_2, \theta_{12}) H \psi_E(r_1, r_2, \theta_{12}). \tag{5}$$

This is the function in square brackets in (4). The differential transition amplitude is the key to interpreting how electron-electron interactions couple ψ_b with ψ_E to cause autoionization. We fix r_1 at a succession of values and examine D as a function of r_2 and θ_{12} for each, just as we examined $\rho(r_2, \theta_{12} | r_1)$ for a succession of values of r_1. That D may be positive or negative is irrelevant, or even sometimes useful. It is helpful for purposes of interpretation to use real wavefunctions ψ_b and ψ_E, whenever this is possible. The important point is that we can see from the structure of $D(r_1, r_2, \theta_{12})$ where, in the space of r_1, r_2 and θ_{12}, the transition amplitude $\langle b | H | E \rangle$ accumulates its magnitude.

We may look at the spatial dependence of D in the space of

r_2, Θ_{12} or we may break the Hamiltonian H into its terms and examine them individually. The former is certainly important to show the overall picture of what makes $\Gamma_{b\epsilon}$ large or small. If, for example, $\Gamma_{b\epsilon}$ is small, D can tell us whether this is because there are simply no large contributions to the width or because there are large but nearly equal contributions of opposite sign. The information obtained by examining the contributions from particular operators in H tells us about the dynamics responsible for the exchange of energy that transforms the system from state b to state E. We shall illustrate by comparing the contribution of the electron-electron repulsion operator with that of the total Hamiltonian, to estimate the relative importance of electron-electron repulsion to the autoionization of helium.

This analysis has thus far been applied to the two lowest $^1S^e$ resonances of He, the quasibound states we called "$(2s)^2$" and "$2p)^2$" in the previous section. Their quasibound state wavefunctions were taken as the Hylleraas-Kinoshita expansions described earlier. The continuum functions were phase-shifted Coulomb functions with corrective terms to handle the behavior at small distances (Rehmus and Berry, 1981). The widths of the two states at their resonant energies of -0.7780 hartree and -0.7371 are 0.12 eV and 0.0011 eV, respectively. Burke and Taylor (1966) give widths of 0.124 and 0.0073 eV for the two states.

Figure 5 shows the differential transition amplitude for autoionization of the "$(2s)^2$" state, for four values of r_1. The spike along the edge where $\Theta_{12} = 0$ occurs at $r_{12} = 0$, but strictly, the transition amplitude becomes infinite there, so we have cut off the surface at $\Theta_{12} = \pi/40$. The spikes at $r_{12} = 0$ are due as one expects to the electron-electron repulsion term e^2/r_{12} of the Hamiltonian. These spikes are dramatic in the plots but contribute much less than the rather isotropic parts of the surface that make up most of this transition amplitude. The smallness of the contribution of the spike becomes apparent when we recall that the surface shown must be multiplied by $\sin \Theta_{12}$ in its integration.

Note the nodal line in the plot for $r_1 = 0.1$ bohr and for $r_1 = 1.5$ bohr. This node arises from the innermost node of the scattering function.

This function can be examined further by integrating over r_1 to construct an average distribution $d(r_2, \Theta_{12})$. Figure 6 is a plot of $d(r_2, \Theta_{12})$ for the "$(2s)^2$" state. The peaks of Figure 5 leave their traces very clearly in this average, but we must be careful to attribute the peaks of Figure 6 to averages that generate maxima where the probability of finding r_1 is greatest. The broad angular

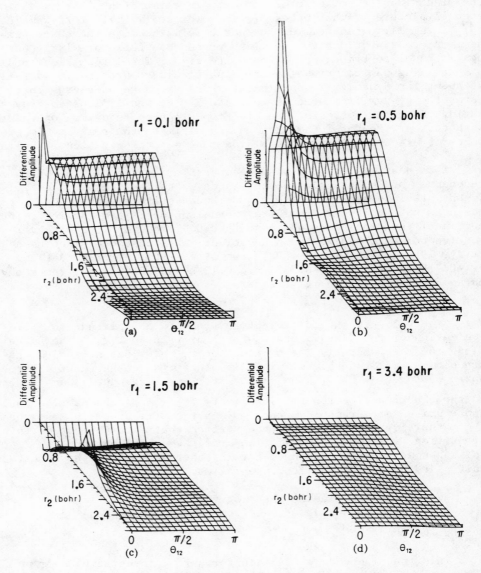

Fig. 5.   The differential transition amplitude for autoionization
of the "(2s)^2" ^{1}S^e Feshbach resonance of He, for four
values of r_1.

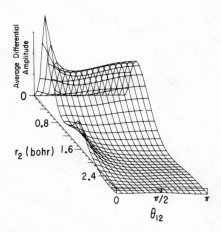

Fig. 6. The differential transition amplitude integrated over the coordinate r_1, for the autoionization of the "$(2s)^2$" $^1S^e$ Feshbach resonance of He.

contribution of Figure 6 is more polarized toward $\theta_{12} = \pi$ than are the plots of Figure 5.

Next we can sort out dynamics by constructing the contribution to $d(r_2, \theta_{12})$ from the electron-electron repulsion above. This contribution is shown in Figure 7. Clearly electron-electron repulsion contributes the peaks at $\theta_{12} \sim 0$, but very little of the large, widely-distributed part of the volume under the surface of Figure 6. Rather, the difference between Figure 5 and Figure 6 is due to electron-nuclear attractions. That is to say that the greater part--about 85%--of Γ for the "$(2s)^2$" state comes from one electron being attracted to the nucleus and driving out the other. Only

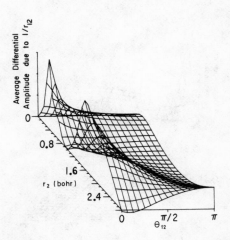

Fig. 7.   The contribution of the electron-electron repulsion term
          (e^2/r_{12}) to the partially integrated amplitude of Fig. 6.

about 15% of the transition rate comes from the direct electron-
electron collisions that had been proposed in the earliest days of
interpretation of the lifetimes of the helium resonances (Cooper
et al., 1963).

The "mechanism" responsible for the 85% of the transition
amplitude is strikingly like the S_N-2 mechanism--"Substitution,
Nucleophilic, 2nd-order"--of organic chemistry. In both cases, a
nucleophile attacks a nucleus or electron acceptor and, in so doing,
drives out another nucleophile.   The two processes differ in that
the organic reaction is inhibited by the space-filling groups that
surround the carbon or other atom being attacked, so that the substitution can only occur from the side opposite the leaving group,
leading to the famous Walden inversion.   In helium, with no "steric

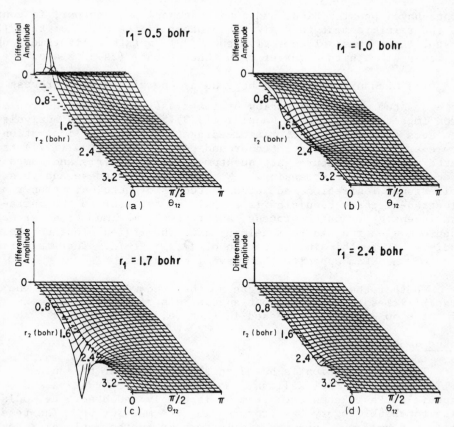

Fig. 8. The differential transition amplitude for autoionization of the "$(2p)^2$" $^1S^e$ Feshbach resonance of He, for four values of r_1. Note how much lower the total amplitudes are compared with those of Fig. 5, and how the peaked contributions have sign opposite the more isotropic, flat parts of the amplitude.

hindrance," attack can come from any direction, so attack from the region $\Theta_{12} = \pi/2$ must dominate due to its large statistical weight.

The differential transition amplitude of much longer-lived "$(2p)^2$"state is illustrated in Figure 8. This function is far flatter and much smaller than the corresponding function for the "$(2s)^2$" state. Moreover, the electron-electron repulsion contribution and the "S_N-2" contribution to the width of the "$(2p)^2$"

state have opposite signs and tend to cancel one another, in contrast to the reinforcement they give each other in the width of the $"(2s)^2"$ state. Both these effects give the $"(2p)^2"$ state a lifetime roughly an order of magnitude larger than that of the $"(2s)^2"$ state.

The mechanism described here as a generalized S_N-2 process is a quantum mechanical analog of the classical mechanism of electron impact ionization by Wannier (1953). In Wannier's analysis, one sees that the trajectories leading to collisional ionization at low energies are concentrated around $\theta_{12} = \pi$. Rau (1971) translated Wannier's picture into quantum-mechanical terms and named the process "dynamical screening." We conclude from these analyses of collisional ionization and our own studies of autoionization that electron-electron repulsion is a small contributor to the intra-atomic energy transfer process, and that the maximum transfer occurs when one electron can be accelerated to the region close to the nucleus, preferably in the region of $\theta_{12} \sim \pi/2$. That is, from the side of the atom opposite the leaving electron.

Whether these processes can be observed will depend on the possibilities for carrying out angular distribution measurements of the slow electrons released in fast (e,3e) reactions.

Acknowledgments

This presentation is based on the research of the author's collaborators Michael Kellman, Huoy-Jen Yuh, Gregory Ezra and especially Paul Rehmus, to whom he is deeply indebted. He would like to thank Gregory Ezra for reading the manuscript. The research described here was supported in large part by the National Science Foundation.

REFERENCES

Banyard, K.E., and Ellis, J.D., 1972, Molec. Phys. 24:1291.
Banyard, K.E., and Ellis, J.D., 1975, J. Phys. B 14:2311.
Burke, P.G., and Taylor, A.J., 1966, Proc. Phys. Soc. 88:549.
Cooper, J.W., Fano, U., and Prats, F., 1963, Phys. Rev. Lett. 13:518.
Coulson, C.A., and Neilson, A.H., 1961, Proc. Phys. Soc. 78:831.
Dickens, P.G., and Linnett, J.W., 1957, Quart. Revs. 11:291.
Fano, U., 1961, Phys. Rev., 124:1866.
Fano, U., 1976, Physics Today 29(9):32.
Fano, U., and Lin, C.D., 1974, Invited Talk for the Fourth International Conference on Atomic Physics, Heidelberg, July, 1979.
Frankowski, K., 1967, Phys. Rev. 160:1.
Frankowski, K., and Fekeris, C.L., 1966, Phys. Rev. 146:46.
Herrick, D.R. and Kellman, M.E., 1980, Phys. Rev. A 21:418.
Herrick, D.R., Kellman, M.E., and Poliak, R.D., 1980, Phys. Rev. A 22:1517.

Herrick, D.R., and Sinanoglu, O., 1975, Phys. Rev. A 11:97.
Katriel, J., and Pauncz, R., 1977, Adv. Quantum Chem. 10:143.
Kellman, M.E., 1980, Chem. Phys. 48:89.
Kellman, M.E., and Herrick, D.R., 1978, J. Phys. B 11:L755.
Kellman, M.E., and Herrick, D.R., 1980, Phys. Rev. 22:1536.
Lin, C.D., 1974, Phys. Rev. A 10:1986.
Lin, C.D., 1976, Phys. Rev. A 14:30.
Macek, J.H., 1967, Phys. Rev. 160:170.
Macek, J.H., 1968, J. Phys. B 1:831.
Munschy, G., and Pluvinage, P., 1963, Revs.Mod . Phys. 35:494.
Rau, A.R.P., 1971, Phys. Rev. A 4:207.
Rehmus, P., and Berry, R.S., 1979, Chem. Phys. 38:257.
Rehmus, P., and Berry, R.S., 1981, Phys. Rev. A 23 (in press).
Rehmus, P., Kellman, M.E., and Berry, R.S., 1978a, Chem. Phys. 31:239.
Rehmus, P., Roothaan, C.C.J., and Berry, R.S., 1978b, Chem. Phys.
    Lett. 58:321.
Shim, I., and Dahl, J.P., 1978, Theor. Chim. Acta 48:165.
Sinanoglu, O., and Herrick, D.R., 1975, J. Chem. Phys. 62:886.
Sinanoglu, O., and Herrick, D.R., 1976, J. Chem. Phys. 65:850.
Tatum, J.P., 1976, Int'l. J. Quant. Chem. X:967.
Wulfman, C.E., 1973, Chem. Phys. Lett. 23:370.
Wulfman, C.E., 1976, Chem. Phys. Lett. 40:139.
Wulfman, C.E., and Kumei, S., 1973, Chem. Phys. Lett. 23:367.
Yuh, H.-J., Ezra, G.S., Rehmus, P. and Berry, R.S., 1981, (to be
    be published).

THE SYMMETRY PROPERTIES OF NONRIGID MOLECULES

Gregory S. Ezra[+]

Department of Chemistry
The University of Chicago
Chicago, Illinois 60637

1. INTRODUCTION

The problem of the symmetry properties of molecules has a long history[1-4], yet the subject continues to attract considerable attention.[5-44]

The apparent geometrical symmetries of ordinary 'rigid' molecules are a consequence of extreme localization of the system wavefunction at the bottom of deep wells in a particular potential energy hypersurface, leading to the possibility of a well-defined equilibrium structure[45] (a restriction to the Born-Oppenheimer approximation is implicit in usual discussions of molecular shape;[69] cf., however, Ref. 70). We present here a discussion of the symmetry properties of nonrigid molecules (NRMs) from a geometrical point of view.[39] The NRMs considered are assumed to be rigid enough to be characterized by notional reference structures explicitly incorporating one or more large-amplitude collective motions, corresponding to soft directions on the energy hypersurface. It is shown that this approach can be regarded as a straightforward generalization of the usual theory of rigid molecule symmetry.[27]

The concept of a symmetry operation in quantum mechanics ultimately refers to some Hamiltonian.[17] The fundamental symmetries of the complete molecular Hamiltonian are therefore briefly reviewed in Section 2. It is pointed out that molecules usually exhibit spontaneously-broken symmetry, characterized by a clustering of molecular energy levels.[8,35] This clustering of levels is the key to

---

+NATO Postdoctoral Fellow

the concept of feasibility.[7] In Section 3, we consider the symmetry
properties of rigid molecules. The static molecular model is introduced, and the use of Eckart constraints to define the molecule fixed
frame is discussed.[27] These ideas prepare the way for the generalization to NRMs given in Section 4, involving use of the semi-rigid
molecular model and the Eckart-Sayvetz constraints.[39,46] The symmetry group of the molecular model is then defined, and identified
as the NRM symmetry group. Section 5 deals with the structure of
the NRM symmetry group. Invariant subgroups are identified, and the
formulation of NRM groups as semi-direct products considered. Several
examples of semi-direct product structure are given. Finally in
Section 6 isodynamic operations are defined for NRMs. These are
closely related to the familiar Wigner vibrational symmetry operations
for rigid molecules, and are shown to generate the invariance group
of the Eckart-Sayvetz frame.

2. FUNDAMENTAL SYMMETRIES OF THE MOLECULAR HAMILTONIAN

In the absence of external fields, the nonrelativistic spin-free
molecular Hamiltonian expressed in terms of lab-fixed coordinates is

$$H_{mol} = - 1/2 \sum_\alpha m_\alpha^{-1} (\partial/\partial R_i)(\partial/\partial R_i) - 1/2 \sum_\varepsilon m^{-1} (\partial/\partial R_i^\varepsilon)(\partial/\partial R_i^\varepsilon)$$
$$+ V_{NN} + V_{NE} + V_{EE} \qquad\qquad 2.1$$

where R^α is the position vector of nucleus α in the lab-frame $\{l_i\}$,
R^ε is the position vector of electron ε, and V_{NN}, V_{NE} and V_{EE} are
Coulomb potential terms. Following Longuet-Higgins[7] (cf. Ref. 48),
the fundamental symmetries of the Hamiltonian 2.1 are (unitary symmetries associated with the nuclear spin spaces are not considered[47]):
   a) Time translations. The generator is $\hat{H}_{mol}$ itself, and the
associated constant of the motion is E_{mol}

$$H_{mol} \tau_{mol} = E_{mol} \tau_{mol} \qquad\qquad 2.2$$

   b) Time-reversal[48,65]
   c) Space translations. The homogeneity of space allows us to
separate off the collective motion of the molecular center of mass
from internal motions.
   d) Rotations. The complete Hamiltonian 2.1 is invariant under
$SO^\ell(3)$, the group of all proper lab-fixed rotations, with associated
angular momentum quantum numbers j-m.
   e) Inversion. Neglecting the weak interactions, the inversion
operation $\mathcal{J}$

$$\mathcal{J}: \quad R^\alpha \to -R^\alpha; \quad R^\varepsilon \to -R^\varepsilon \qquad\qquad 2.3$$

gives rise to the parity quantum number π.

f) Permutations of the space and spin variables of electrons or identical nuclei. If there are N_ε electrons and N_a nuclei of type a, the corresponding symmetry group is the direct product of permutation groups

$$S_{N_\varepsilon} \times {}_a^\pi S_{N_a} \qquad\qquad 2.4$$

Symmetries a) to f) remain valid within the Breit-Pauli approximation,[48] where relativistic and magnetic corrections to electronic motion are taken to second order in the fine-structure constant.

In what follows, we consider a particular subgroup of the complete symmetry group of the molecular Hamiltonian generated by symmetries a) to f), namely the <u>complete</u> <u>nuclear</u> <u>permutation-inversion</u> (CNPI) group.[25]

$$CNPI \equiv {}_a^\pi S_{Na} \times \mathcal{J} \qquad\qquad 2.5$$

In principle, molecular states and energy levels should be classified in the CNPI group, which rapidly increases in size as the number of identical nuclei in the molecule increases. However, it is an empirical fact that molecules usually exhibit spontaneously-broken CNPI symmetry, manifest as a <u>clustering</u> of molecular energy levels (cf. Ref. 35; such clustering is also referred to as structural degeneracy.)[37] The systematics of the clustering reveal that each cluster consists of a <u>set</u> of irreducible representations (IRs) of the CNPI group <u>induced</u> by an IR Γ_H of a subgroup $\mathbb{H}$ [8,31]:

$$\mathbb{H} \subset {}_a^\pi S_{NA} \times \mathcal{J} \qquad\qquad 2.6a$$

$$\Gamma_H \uparrow {}^\pi S_{Na} \times \mathcal{J} \qquad\qquad 2.6b$$

The subgroup $\mathbb{H}$ provides an optimal labelling of the <u>clusters</u> of levels, rather than individual states.[8]

The resolution obtained in a given experiment determines what is recognized as a cluster, so that the appropriate group $\mathbb{H}$ depends on the timescale of observation as well as the intramolecular dynamics.[50] Following Berry[45], we may refer to $\mathbb{H}$ simply as the 'heuristic group' (otherwise known as the molecular symmetry group (MSG),[25] Q-group,[10] group of feasible perrotations,[17] isometric group,[20] symmetry group of the molecular model[27,39]). These ideas lead directly to the important concept of <u>feasibility</u>.[7] An element of the CNPI group is said to be <u>feasible</u> if it gives rise to observable splittings of energy levels; that is, if it is a generator of the symmetry group required to describe a given experiment.

For <u>rigid</u> molecules, $\mathbb{H}$ is necessarily isomorphic with (or a homomorphic image of) a point symmetry group, which is the covering group of the assumed equilibrium structure (see next section).

For nonrigid molecules, there is by definition no single reference or equilibrium structure, and the group IH is in general not isomorphic with a point group.

3. RIGID MOLECULES

We now examine the relation between the fundamental symmetries of the molecular Hamiltonian and the usual notions of point group symmetry for rigid polyatomic molecules[5-7,27,41], bringing out the close connection between descriptions of molecular symmetry and dynamics. These considerations can then be generalized to certain classes of NRM in the next section.

The transformation to molecular coordinates

$$\underset{\sim}{R}{}^{\alpha} = \underset{\sim}{R} + \underset{\sim}{c}{}^{\alpha} + \underset{\sim}{d}{}^{\alpha} \qquad\qquad 3.1a$$

$$R_i^{\alpha} = R_i + C_{ij}(\theta)(a_j^{\alpha} + d_j^{\alpha}(q_{\lambda})) \qquad\qquad 3.1b$$

shown in Figure 1 is of central importance. $\underset{\sim}{R}$ is the position vector of the molecular center of mass, while $\underset{\sim}{c}{}^{\alpha}$ is the equilibrium or reference position of nucleus α and $\underset{\sim}{d}{}^{\alpha}$ is the displacement vector. The direction-cosine matrix $C_{ij} \equiv \underset{\sim}{\ell}_i \cdot \underset{\sim}{f}_j$ defines the orientation of the molecule-fixed frame $\{\underset{\sim}{f}_i\}$, and is a function of 3 coordinates $\{\theta_t\}$ such as the Euler angles. The components $a_j^{\alpha} \equiv \underset{\sim}{f}_j \cdot \underset{\sim}{c}{}^{\alpha}$ are 3N constants defining the static molecular model,[27] and the molecule-fixed components $d_j^{\alpha} \equiv \underset{\sim}{f}_j \cdot \underset{\sim}{d}{}^{\alpha}$ of the displacement vectors are functions of 3N-6 vibrational coordinates $\{q_{\lambda}\}$.

The physical picture underlying the transformation 3.1 is quite familiar: the nuclei are assumed to assume small-amplitude oscillations about well-defined equilibrium positions in a freely rotating equilibrium structure. The traditional route to a Hamiltonian expressed in molecule-fixed coordinates is as follows[51-53]

i) First, write the velocity of nucleus α as

$$\dot{R}_i^{\alpha} = \dot{R}_i + \sum_{t=1}^{3} \dot{\theta}_t \, \partial C_{ij}/\partial\theta_t (a_j^{\alpha} + d_j^{\alpha}(q_{\lambda})) + \sum_{=1}^{3N-6} C_{ij} \partial d_j/\partial q_{\lambda} \, \dot{q}_{\lambda} \qquad 3.2$$

ii) Express the classical nuclear kinetic energy

$$T_N = 1/2 \sum_{\alpha} m_{\alpha} \, \dot{R}_i^{\alpha} \dot{R}_i^{\alpha}$$

in terms of molecular coordinates and velocities

iii) Obtain a Hamiltonian form for the total classical energy $T_N + V(q)$ in terms of conjugate momenta and quasi-momenta.[54]

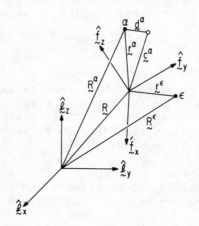

Fig. 1.   The transformation from lab-fixed to molecular coordinates.

iv) Invoke a quantization procedure to obtain the corresponding quantum-mechanical operator.   This leads to the Wilson-Howard-Watson form of the molecular Hamiltonian:[55]

$$\hat{H}_{mol} = 1/2\,(\hat{\Pi}_i - \hat{\pi}_i)\mu_{ij}\,(\hat{\Pi}_j - \hat{\pi}_j) + 1/2\Sigma\,\hat{P}_\lambda^2 + V(q_\lambda) + U(q_\lambda)\quad 3.4$$

v) Finally, approximate the very complicated operator $\hat{H}_{mol}$ by a zeroth-order rigid rotor/harmonic oscillator Hamiltonian

$$\hat{H}_{mol} \rightarrow \Sigma_i \hat{\Pi}_i^2/2I_i^o + 1/2\Sigma_\lambda\,(\hat{P}_\lambda^2 + \omega_\lambda^2 q_\lambda^2) + \hat{H}'\qquad 3.5$$

and treat the remainder $\hat{H}'$ as a perturbation.[51,71]

Note that the static molecular model $\{a_i^\alpha\}$ normally corresponds to a local minimum in a given potential surface, so that we work within the Born-Oppenheimer approximation for nondegenerate electronic states. Motions of the electrons are not considered explicitly, and $\hat{H}_{mol}$ is therefore an effective nuclear Hamiltonian with effective potential $V(q)$.

For rigid molecules, the molecule-fixed frame $\underset{\sim}{f}_i$ is usually embedded into an arbitrary configuration of nuclei using the Eckart conditions, which are[27]

$$\sum_\alpha m_\alpha \underset{\sim}{c}^\alpha \wedge \underset{\sim}{d}^\alpha = \underset{\sim}{o} \qquad\qquad 3.6$$

and imply that the internal or vibrational angular momentum, and hence the Coriolis coupling, vanishes to first order in the vibrational displacements. Attempts to eliminate vibration/rotation coupling completely, rather than just to first order, lead to nonholonomic(nonintegrable) constraints.[56,57]

The Eckart conditions are normally ascribed a dynamical significance as above. However, it is possible to compute the Eckart frame $\{\underset{\sim}{f}_i\}$ itself directly from the instantaneous nuclear configuration $\{\underset{\sim}{R}^\alpha\}$ and the assumed molecular model. Defining three Eckart vectors $\underset{\sim}{F}_i$

$$\underset{\sim}{F}_i \equiv \sum m_\alpha \underset{\sim}{R}^\alpha a_i^\alpha \qquad\qquad 3.7a$$

and the Gram matrix Γ

$$\Gamma_{ij} \equiv \underset{\sim}{F}_i \cdot \underset{\sim}{F}_j \qquad\qquad 3.7b$$

the Eckart frame is obtained by a process of symmetric orthonormalization[27,58]

$$\underset{\sim}{f}_i = \underset{\sim}{F}_j \, \Gamma_{ji}^{-1/2} \qquad\qquad 3.8$$

Furthermore, taking the quantity

$$\zeta \equiv \sum m_\alpha \underset{\sim}{d}^\alpha \cdot \underset{\sim}{d}^\alpha \qquad\qquad 3.9$$

to be a measure of how much a given nuclear configuration deviates from the assumed reference structure, it can be shown that minimization of ζ is entirely equivalent to the Eckart constraints[21,58]

$$\left.\begin{array}{l} \delta\zeta = 0 \\ \delta^2\zeta > 0 \end{array}\right\} \equiv \text{Eckart conditions.}$$

This least-squares criterion for orientation of the Eckart frame is useful, since a) it emphasizes the static aspects of the Eckart conditions, b) generalizes easily to NRMs (Eckart-Sayvetz constraints

cf. next section), and c) clearly displays the <u>invariance</u> property of
ζ under the transformation

$$d_i^\alpha \rightarrow \sum_\beta R_{ij} S_{\alpha\beta} d_j^\beta \qquad 3.10$$

where R is an arbitrary 3x3 orthogonal matrix (rotation or rotation-
inversion) and S is an NxN matrix representing an arbitrary permutation of identical nuclei. This result is important when we come to
consider the invariance group of the Eckart frame.

The molecular point group $\mathfrak{G}$ is defined as the set of all rotations or rotation-inversions g that result in permutations of static
model vectors corresponding to identical nuclei.

$$\mathfrak{G} = \{g | g : a_i^\alpha \rightarrow R(g)_{ij} a_j^\alpha = \sum_\beta a_i^\beta S(g)_{\beta\alpha}\} \qquad 3.11$$

This is simply a formal definition of the intuitive concept of the
point group.[27] There are two representations of $\mathfrak{G}$ of particular
interest:
   i) The so-called defining representation by 3x3 orthogonal
matrices $\{R(g) | g \varepsilon \mathfrak{G}\}$, which is a faithful representation, and
   ii) The representation by NxN or $N_a x N_a$ permutation matrices
$\{S(g) | g \varepsilon \mathfrak{G}\}$, which is not necessarily faithful.

Static model representations for various point groups have
been given by Louck and Galbraith.[27]

For every point group operation g, we define an associated permutation-rotation or <u>perrotation</u>[17] operation L_g :

$$\forall \quad g \varepsilon \mathfrak{G}, \quad L_g : x_i^\alpha \rightarrow L_g \cdot x_i^\alpha \equiv \sum_\beta R(g)_{ij} S(g)_{\alpha\beta} x_j^\beta \qquad 3.12$$

Note that for all static model vectors,

$$L_g : a_i^\alpha \rightarrow a_i^\alpha , \qquad 3.13$$

so that the point group $\mathfrak{G}$ can be characterized by a set of transformations $L_\mathfrak{G} \equiv \{Lg\}$ leaving the static model invariant. The perrotations form a faithful representation of $\mathfrak{G}$.

The induced action of point group operations upon molecular
coordinates is

$$g : (R_i, C, d_i^\alpha, r_j^\varepsilon) \rightarrow (R_i, C\tilde{R}(g), L_g \cdot d_i^\alpha, R(g)_{ij} r_j^\varepsilon) \qquad 3.14$$

The transformation of rotational coordinates

$$C \rightarrow C\tilde{R}(g) \qquad 3.15$$

is a 'molecule-fixed' rotation; that is, an element of the group $O^f(3)$.[27] Perrotations L_g acting on the vibrational displacements $\{d_i^\alpha\}$ are just the familiar Wigner vibrational symmetry operations[1], which leave the potential energy and the vibrational kinetic energy unchanged. The transformation of electronic positions r_i^ε corresponds to the usual action of point group operations upon electronic coordinates.[59]

Inserting the above transformations into the relations 3.16 and

$$R_i^\varepsilon - R_i = C_{ij}\, r_j^\varepsilon \,, \qquad\qquad 3.16$$

it is found that point group operations $g\varepsilon\mathbb{G}$ induce permutations g of identical nuclei:

$$g : \underset{\sim}{R}^\alpha \to \sum_\beta S(g)_{\alpha\beta} \underset{\sim}{R}^\beta \qquad\qquad 3.17$$

where the permutation is determined by $S(g)$, the matrix appearing in our analysis of the static model symmetries. A study of the induced action of these _feasible_ permutations of identical nuclei upon molecular coordinates, and thence upon the rotational, vibrational and electronic parts of the molecular wavefunction, is therefore the basis of an analysis of rigid molecule symmetry[5,6]. For non-planar rigid molecules, the permutations form a group isomorphic with the point group $\mathbb{G}$. In order to avoid considering the behavior or rotational wavefunctions under improper molecule-fixed rotations, the proper transformations of rotational variables

$$g : C \to (\det R(g))\,C\underset{\sim}{R}(g) \,, \qquad\qquad 3.16'$$

can be used, corresponding to introduction of the PI group.[5-7]

Finally, it is important to note that the group $L_G \equiv \{L_g\,|\,g\varepsilon\mathbb{G}\}$ of perrotations of vibrational displacements is the _invariance_ group of the Eckart frame $\{\underset{\sim}{f}_i\}$[27,41]. This means: Take an arbitrary configuration of nuclei $\{\underset{\sim}{R}^\alpha\}$, and calculate the Eckart frame orientation using the orthogonalization procedure described above. Then consider the new configuration $\{\underset{\sim}{\bar{R}}^\alpha\}$, where

$$\bar{R}_i^\alpha - R_i^\alpha = C_{ij}(a_j^\alpha + L_g \cdot d_j^\alpha). \qquad\qquad 3.18$$

Calculating the orientation of the Eckart frame anew, we find

$$\underset{\sim}{f}_i' = \underset{\sim}{f}_i \qquad\qquad i = x,y,z \qquad\qquad 3.19$$

We can therefore consider the set of transformations L_G of vibrational variables entirely independently of any transformations of rotational coordinates C:[27]

$$L_g : (C, d_j^\alpha) \to (c; L_g \cdot d_j^\alpha) \qquad\qquad 3.20$$

The generic term _isodynamic_ is used to describe operations of this
type. The invariance property of the Eckart frame under isodynamic
operations follows easily by noting that (cf. equation 3.10)

$$L_g : \zeta \to \zeta \qquad g\varepsilon\mathbb{G} \qquad 3.21$$

The concept of isodynamic operations has a very natural generalization to the case of NRMs, as we show in Section 6.

4. NONRIGID MOLECULES

The standard PI treatment of the symmetry properties of NRMs
concentrates upon the permutational aspect of the problem, and is
therefore of quite general applicability.[7] We describe here an
approach to the symmetry properties of a large class of NRMs which
is geometrical in spirit, following as closely as possible the account
of rigid molecule symmetry given above. Our analysis clarifies the
relations between previous accounts of NRM symmetry. (cf. Refs. 17,
22, 39).

The underlying physical picture is that described by Sayvetz[46]:
we imagine the nuclei to execute rapid, small-amplitude (vibrational)
motions about an equilibrium configuration that is itself performing
a slow, large-amplitude (internal or contortional[37]) motion of some
sort, in addition to undergoing overall rotation. A timescale separation of internal and vibrational motions is therefore assumed in
first approximation.

This picture leads to the introduction of the _semi-rigid molecular model_ (SRMM),[20,21,39] which is a reference structure depending
upon one or more large-amplitude _curvilinear_ coordinates γ_t,[60] collectively denoted $\gamma \equiv (\gamma_1, \ldots \gamma_t, \ldots \gamma_T)$,

$$\mathcal{A} \equiv \{\underset{\sim}{a}^\alpha(\gamma)\} \qquad\qquad 4.1$$

The SRMM underlies the description of the symmetry and dynamics of
NRMs in precisely the same way that the static model does for rigid
molecules.

The curvilinear coordinates γ may represent internal rotations, inver-
sions or pseudorotations, or may even correspond to reaction coordinates.[61] As an example of internal rotation models, we show in
Figure 2 the single-parameter SRMM for CH_3-NO_2-type molecules;[62]
the curvilinear coordinate γ is here the torsional angle. Figure
3 shows an XY_3-inverter model, e.g., NH_3,[63] while Figure 4 shows a
1-parameter ZXY_4 pseudorotation SRMM, where Z is the pivot ligand.[10,64]

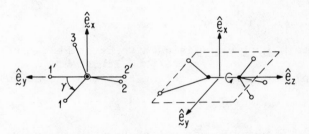

Fig. 2.   The 1-parameter CH_3-NO_2 internal rotor SRMM.   The curvilinear coordinate is the torsional angle γ.

Sørensen has recently given a detailed discussion of NRM dynamics in terms of the SRMM.[60]   The transformation to molecular coordinates given previously for rigid molecules (Equation 3.1) generalizes to

$$R_i^\alpha = R_i + C_{ij}(\theta)(a^\alpha(\gamma)_j + d_j^\alpha(\gamma, q_\lambda)) \qquad 4.2$$

where there are now 3N-6-T vibrational coordinates q_λ, so that the
velocity of nucleus α is

$$\dot{R}_i^\alpha = \dot{R}_i + \sum_{t=1}^{3} \dot{\theta}_t \, \partial C_{ij}/\partial \theta_t (a_j^\alpha + a_j^\alpha) + \sum_\lambda C_{ij} \, \partial d_j^\alpha/\partial q_\lambda \, \dot{q}_\lambda$$

$$+ \sum_{t=1}^{T} C_{ij}(\partial a^\alpha(\gamma)_j/\partial \gamma_t + \partial d_j^\alpha/\partial \gamma_t)\dot{\gamma}_t \qquad 4.3$$

Note the appearance of new terms involving $\dot{\gamma}_t$.   Following the route
outlined in Section 3, we obtain a classical Hamiltonian having a
form entirely analogous to that for

$$H_{mol} = \sum_{\nu\nu'} (\Pi_\nu - \pi_\nu)\mu_{\nu\nu'} (\Pi_\nu' - \pi_\nu') + \sum_{\lambda=1}^{3N-6-T} P_\lambda^2 + V(\gamma; q_\lambda) \qquad 4.4$$

where the index ν takes the values x,y,z,1,T, corresponding to both
rotational and internal large-amplitude motions, so that $\mu_{\nu\nu'}$ is now
a generalized inverse effective inertia tensor and so on.   Quantization to give the corresponding operation $\hat{H}_{mol}$ does not give such
a compact expression as for rigid molecules, since $\mu_{\nu\nu'}$ will in
general depend upon the curvilinear coordinates γ.[60]

Use of T curvilinear coordinates requires the introduction of

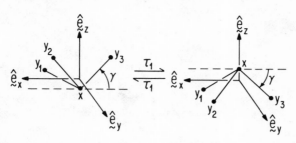

Fig. 3.   The XY_3-inverter model.   The curvilinear coordinate is the
inversion angle γ.

T additional constraints on the displacements d_i^α.   These are taken
to be the Eckart-Sayvetz conditions:[46]

$$\sum_\alpha m_\alpha \partial a^\alpha(\gamma)_i / \partial \gamma_t \, d_i^\alpha = 0 \qquad t = 1, \ldots \tau, \tag{4.5}$$

which ensure that small-amplitude vibrations are uncoupled from large-
amplitude internal motions in first approximation.   The molecule-
fixed frame defined by the Eckart and Eckart-Sayvetz conditions is
appropriately designated the Eckart-Sayvetz frame, and, like the
Eckart frame in rigid molecules, can be characterized in a least-
squares fashion.   Thus, minimization of the quantity

$$\zeta \equiv \sum_\alpha m_\alpha \underset{\sim}{d}^\alpha \cdot \underset{\sim}{d}^\alpha$$

with respect to variation of both the rotational and internal coordinates is completely equivalent to the Eckart-Sayvetz conditions[21,58]

We now define the symmetry group of the semi-rigid molecular
model.[39]   Consider the transformation of SRMM vectors

$$h \equiv (\rho, \tau) : a^\alpha(\gamma)_i \rightarrow R(\rho)_{ij} \, a^\alpha(\tau^{-1}(\gamma))_j \tag{4.6}$$

where $h \equiv (\rho, \tau)$ is an ordered pair of operations, ρ being a proper
or improper rotation of SRMM vectors and τ being a transformation
of internal parameters

$$\tau : \gamma \rightarrow \gamma' \equiv \tau(\gamma) \epsilon \; \Gamma \quad \text{(Parameter domain)} \tag{4.7a}$$

e.g., an inhomogeneous linear transformation of the γ_t:[20]

$$\tau : \gamma_t \rightarrow \gamma'_t = \sum_{\bar{t}} \eta(\tau)_{t\bar{t}} \, \gamma_{\bar{t}} + \chi_t \tag{4.7b}$$

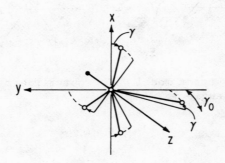

Fig. 4.   The 1-parameter ZXY_4 pseudorotation model.

A transformation h is a <u>symmetry operation of the SRMM</u> if it results in a permutation of SRMM vectors corresponding to identical nuclei (recall the definition of the symmetry group of the static model for rigid molecules):

$$h = (\rho,\tau) : a^{\alpha}(\gamma)_i \rightarrow R(\rho)_{ij}\, a^{\alpha}(\tau^{-1}(\gamma))_j = \sum_{\beta} a^{\beta}(\gamma)_i S(h)_{\beta\alpha} \qquad 4.8$$

The set of all symmetry operations of the SRMM forms a group, denoted $\mathbb{H}$, which we call the <u>symmetry group of the SRMM</u>

$$\mathbb{H} \equiv \{h = (\rho,\tau)\,|\,h:a^{\alpha}(\gamma)_i \rightarrow R(\rho)_{ij}\, a^{\alpha}(\tau^{-1}(\gamma))_j$$

$$= \sum_{\beta} a^{\beta}(\gamma)_i\, S(h)_{\beta\alpha}\,\} \qquad\qquad 4.9$$

To illustrate the above definitions, the symmetry operations of the relatively simple CH_3-NO_2-type SRMM shown in Figure 1 are listed in Table 1. The set of symmetry operations form a group $\mathbb{H}$ of order 12.[7]

   In order to investigate the properties of the symmetry group of the SRMM, it is of course necessary to define a multiplication rule in $\mathbb{H}$. The simplest and most useful is clearly:

Table 1. Symmetry operations of the 1-parameter CH_3-NO_2 internal rotor SRMM shown in Figure 1.

Notation: $(\rho,\gamma') \equiv (\rho,\tau:\gamma \rightarrow \gamma')$ $\omega \equiv 2\pi/3$

$(\hat{E},\gamma)$	$(\hat{C}_{2z},\gamma+\pi)$	$(\hat{\sigma}_{yz},-\gamma)$	$(\hat{\sigma}_{xz},\pi-\gamma)$
$(\hat{E},\gamma+\omega)$	$(\hat{C}_{2z},\gamma+\pi+\omega)$	$(\hat{\sigma}_{yz},-\gamma-\omega)$	$(\hat{\sigma}_{xz},\pi-\gamma-\omega)$
$(\hat{E},\gamma+2\omega)$	$(\hat{C}_{2z},\gamma+\pi+2\omega)$	$(\hat{\sigma}_{yz},-\gamma-2\omega)$	$(\hat{\sigma}_{xz},\pi-\gamma-2\omega)$

$\forall$ $h_1,h_2 \in \mathbb{H}$,

$$h_1 \cdot h_2 = (\rho_1,\tau_1) \cdot (\rho_2,\tau_2)$$

$$= (\rho_1\rho_2,\tau_1\tau_2) \equiv h_{12} \qquad\qquad 4.10$$

where $\rho_1\rho_2$ and $\tau_1\tau_2$ denote the usual composition of operators. A necessary condition for the validity of this rule is that the specification of the rotation ρ associated with h should not depend upon the parameters γ. Given a multiplication rule such as 4.10, the symmetry operations of the SRMM can be manipulated directly. Note that from 4.10 the permutation matrices $\{S(h)\}$ form a representation of $\mathbb{H}$:

$$\forall h_1,h_2 \in \mathbb{H}, \quad S(h_1h_2) = S(h_1)S(h_2) \qquad\qquad 4.11$$

For any $h = (\rho,\tau) \in \mathbb{H}$, consider the transformation of internal parameters

$$\tau : a^\alpha(\gamma)_i \rightarrow a^\alpha(\tau^{-1}(\gamma))_i \qquad\qquad 4.12$$

This operation maps the distance set

$$\Delta \equiv \{ |\underset{\sim}{a}^\alpha(\gamma) - \underset{\sim}{a}^{\alpha'}(\gamma)| \ \ |\alpha < \alpha'\} \qquad\qquad 4.13$$

onto itself; in fact, the transformations τ leave invariant the vertex-valued molecular graph,[20] and are called internal isometric transformations by Günthard et al.[20] The internal isometric group is defined as the set

$$\mathscr{F} \equiv \{\tau | (\rho,\tau) \in \mathbb{H} \ \} \qquad\qquad 4.14$$

For every element of $\mathbb{H}$, we define the associated permutation-rotation L_h:

$$L_h : x^\alpha_i \rightarrow \sum_\beta R(\rho)_{ij} S(h)_{\alpha\beta} x^\beta_j \qquad\qquad 4.15$$

Since the operations L_h leave the Eckart-Sayvetz frame invariant, we have

$$L_h \cdot a^\alpha(\gamma)_i = \sum_\beta R(\rho)_{ij} S(h)_{\alpha\beta}\, a^\beta(\gamma)_j = a^\alpha(\tau(\gamma))_i \qquad 4.16$$

and the induced action of symmetry operations of the SRMM upon molecular coordinates is as follows:

$\forall$ hϵ $\mathbb{H}$,

$$h:(R_i;C;\gamma;d^\alpha_i;\ r^\epsilon_i) \rightarrow (R_i;C\tilde{R}(\rho);\tau(\gamma);L_h \cdot d^\alpha_i;R(\rho)_{ij} r^\epsilon_j) \qquad 4.17$$

Classification of the molecular wavefunction $\psi(\xi)$ ($\{\xi\}$ represents any of the Born-Oppenheimer coordinates) can then be accomplished using the prescription[65]

$$h:\ \psi(\xi) \rightarrow \psi(h^{-1}(\xi)) \qquad 4.18$$

Inserting the coordinate transformations 4.17 into the relations 3.16 and 4.2 shows that

$$\forall \text{ h}\epsilon \text{ } \mathbb{H} \text{ , } h:\underset{\sim}{R}^\alpha \rightarrow \underset{\sim}{R}^{\overline{\alpha}} \equiv \sum_\beta S(h)_{\alpha\beta}\underset{\sim}{R}^\beta \qquad 4.19a$$

$$h:\underset{\sim}{R}^\epsilon \rightarrow \underset{\sim}{R}^\epsilon \qquad 4.19b$$

Elements of the symmetry group of the molecular model $\mathbb{H}$ therefore induce permutations of nuclear position vectors corresponding to identical nuclei, where the permutation is defined by the matrix S(h). These permutations are of course symmetry operations of the complete molecular Hamiltonian.

The permutations 4.19a are the <u>feasible</u> permutations of nuclei considered by Longuet-Higgins.[7] In the approach we have taken, the feasibility or otherwise of a given permutation of identical nuclei is built into the specification of the molecular model. Note also that when ρ is improper it is usual in the PI theory to consider the <u>permutation-inversion</u>[7]

$$h:\underset{\sim}{R}^\alpha \rightarrow -\underset{\sim}{R}^{\overline{\alpha}} \qquad 4.19a'$$

$$h:\underset{\sim}{R}^\epsilon \rightarrow -\underset{\sim}{R}^\epsilon \qquad 4.19b'$$

corresponding to the proper transformation of rotational variables

$$h:C \rightarrow (\det R(\rho))C\tilde{\underset{\sim}{R}}(\rho) \qquad 4.20$$

5. THE STRUCTURE OF THE GROUP $\mathbb{H}$

In this section we consider the structure of the NRM symmetry group $\mathbb{H}$. Even for relatively simple SRMMs, the associated symmetry group can be quite large, e.g., for boron trimethyl $(B(CH_3)_3)$ $\mathbb{H}$ is of order 324[7,31], while for $C(CH_4)_4$ $\mathbb{H}$ is of order 1944[66], so that we wish to examine the structure of $\mathbb{H}$ to determine whether there are any simplifying features that can be exploited when, for example, finding its IRs and character table. Questions in which we shall be interested are: Can we find invariant subgroups of $\mathbb{H}$? and, Is it possible to write $\mathbb{H}$ as a semi-direct product[11-14,20,22,31,33,39,44]?

Given the validity of the multiplication rule 4.10, it is possible to identify two general invariant subgroups (normal divisors) of $\mathbb{H}$:

i) The <u>point group</u> $\mathfrak{C}^P$ is defined by the highest common covering symmetry over the whole of the parameter range :

$$\mathfrak{C}^P \equiv \{(\rho,\tau)\varepsilon\ \mathbb{H}\ |\ \tau=\tau_o,\ \text{arbitrary }\gamma\} \qquad 5.1$$

it follows from 4.10 that

$$\mathfrak{C}^P \lhd \mathbb{H} \qquad 5.2$$

i.e., $\mathfrak{C}^P$ is an invariant subgroup of $\mathbb{H}$. As an example of a non-trivial point group $\mathfrak{C}^P$, consider the XY_3-inverter model (Figure 2) constrained to have C_{3v} covering symmetry from all values of γ (except for the isolated point $\gamma = 0$, where the covering symmetry is D_{3h}). The point group $\mathfrak{C}^P$ is C_{3v} in this case.

ii) The <u>intrinsic group</u> $\mathfrak{C}^I$ is the set of all operations h acting only upon the internal parameters, i.e., $\rho = \rho_0$, the identity

$$\mathfrak{C}^I \equiv \{(\rho,\tau)\varepsilon\ \mathbb{H}\ |\ \rho = \rho_0\} \qquad 5.3$$

it follows from 4.10 that

$$\mathfrak{C}^I \lhd \mathbb{H} \qquad 5.4$$

The CH_3-NO_2 SRMM shown in Figure 1 provides an example of a non-trivial intrinsic group, which is the set of 3 elements (cf. Table 1)

$$\mathfrak{C}^I = \{(\hat{E},\gamma),(\hat{E},\gamma+\omega),(\hat{E},\gamma+2\omega)\} \overset{iso}{=} C_3 \qquad 5.5$$

isomorphic with the cyclic group of order 3. It is clear that the operations of the intrinsic group correspond to rotations of the $-CH_3$ group about the C-N axis.

Since elements of $\mathbb{C}^P$ and $\mathbb{C}^I$ commute with each other, $\mathbb{H}$ has the invariant subgroup

$$\mathbb{C}^I \times \mathbb{C}^P \ \vartriangleleft \ \mathbb{H} \qquad\qquad 5.6$$

i.e., the direct product of $\mathbb{C}^I$ and $\mathbb{C}^P$. We can then define a <u>factor group</u>

$$\mathcal{K} \equiv \mathbb{H} \ / (\mathbb{C}^I \times \mathbb{C}^P), \qquad\qquad 5.7$$

Examples are:

$$XY_3\text{-inverter} \qquad \mathcal{K} \ \overset{\text{iso}}{\cong} \ \mathcal{N}_2 \ , \qquad\qquad 5.8a$$

the abstract group of order 2, and

$$CH_3\text{-}NO_2 \qquad \mathcal{K} \ \overset{\text{iso}}{\cong} \ C_{2v} \qquad\qquad 5.8b$$

Given the invariance of the subgroup $\mathbb{C}^I \times \mathbb{C}^P$, it is natural to ask whether it is possible to write $\mathbb{H}$ as a semi-direct product. That is, can we find a set of coset representatives of the invariant subgroup $\mathbb{C}^I \times \mathbb{C}^P$ that closes to form a group $\mathbb{K}$ isomorphic with the factor group $\mathcal{K}$? In general this is not possible (Example: C_4-C_2 ring-puckering mode[67]). However, when it is possible to find such a group $\mathbb{K}$ we can write $\mathbb{H}$ in semi-direct product form

$$\mathbb{H} \ = \ (\mathbb{C}^I \times \mathbb{C}^P) \otimes \mathbb{K} \qquad\qquad 5.9$$

which can be rearranged as follows:[31]

$$\mathbb{H} \ = \ \mathbb{C}^I \otimes (\mathbb{C}^P \otimes \mathbb{K}) \qquad\qquad 5.10a$$

$$= \ \mathbb{C}^P \otimes (\mathbb{C}^I \otimes \mathbb{K}) \qquad\qquad 5.10b$$

to display either $\mathbb{C}^I$ (5.10a) or $\mathbb{C}^P$ (5.10b) as the invariant subgroup. The form 5.10a corresponds to the analysis given by Woodman for NRMs with internal rotors[12], as well as the most recent version of Altmann's theory[31], whereas 5.10b corresponds to the semi-direct product structure proposed by Günthard et al. for the complete iso-metric group $\mathcal{H}$ [33], with

$$\mathcal{F} \ \overset{\text{iso}}{\cong} \ (\mathbb{C}^I \otimes \mathbb{K}) \qquad\qquad 5.11$$

in Table 2 we show some examples of NRM symmetry groups having semi-direct product structure. It is clear that the recognition of semi-direct product structure can be helpful in the classification of large NRM symmetry groups.

It is important to note that systematic procedures for deriving

Table 2.   The semi-direct product structure of some NRM symmetry groups

		$|H|$
XY_3-inverter	$C_{3v}^P \otimes C_h \equiv C_{3v}^P \times C_h$	12
ZXY_4	$C_{2v}^P \otimes C_{s'}$	8
CH_3-NO_2	$C_3^I \otimes C_{2v}$	12
XY_3-Z-XY_3	$(C_3^I \times C_3^I) \otimes C_{2v}$	36
CH_3-SF_5	$(C_3^I \times C_4) \otimes C_S$	24
XY_4-XY_4	$(D_4^I \times D_4^P) \otimes C_{s'}$	128

the irreducible representations, the class structure and the character table for a semi-direct product in terms of the (known) properties of its invariant subgroup and the associated factor group have been described by Altmann and others.[31]   It is of course much easier to handle a large group when it has been broken into a product of smaller ones, and we have used these techniques to construct character tables for a variety of NRMs, including the XY_4-XY_4 and XY_6-XY_6 coaxial rotors.[67]   In Table 3 we give the character table for the relatively simple semi-direct product $C_3^I \otimes C_{2v}$, which is the symmetry group for CH_3-NO_2-type NRMs and is isomorphic with the point group D_{3h}[7] (the multiplication rule 4.10 is valid here).   The elements of a semi-direct product are written in the form $[t|g]$, where g is an element of the invariant subgroup and t is an element of the factor group.   The IRs are characterized by an <u>orbit</u>, which is specified by an IR of the invariant subgroup, together with an IR of the associated <u>little co-group</u> (a proper or improper subgroup of the factor group).[31]

When applicable, semi-direct product nomenclature for IRs of NRM symmetry groups provides a great deal of useful information concerning the correlation with various subgroups or point group symmetries occurring over the parameter range.   In fact, rigid/nonrigid correlation diagrams[45,68] can be constructed virtually by inspection.   Figure 5 shows a simple example, involving the correlation of the IRs of $C_3^I \otimes C_{2v}$ with those of the subgroups C_3 and C_{2v}, which correspond to the symmetries of the 'top' and 'frame' parts of the molecule, respectively.

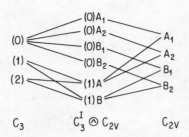

$$C_3 \qquad C_3^I \otimes C_{2v} \qquad C_{2v}$$

Fig. 5.   Correlation of the IRs of $C_3^I \otimes C_{2v}$ with those of its
subgroups C_3 and C_{2v}.

Table 3.   The character table for the semi-direct product
$C_3^I \otimes C_{2v}$

$E \equiv (\hat{E}, \gamma) \qquad 1 \equiv (\hat{E}, \gamma+\omega) \qquad C_{2z} \equiv (\hat{C}_{2z}, \gamma+\pi) \qquad \sigma \equiv (\hat{\sigma}, -\gamma)$

$\hat{\sigma}' = \hat{C}_{2z} \cdot \hat{\sigma}$

	1 $[\hat{E}\vert 0]$	2 $[\hat{E}\vert 1]$	3 $[\hat{\sigma}\vert 0]$	1 $[\hat{C}_{2z}\vert 0]$	2 $[\hat{C}_{2z}\vert 1]$	3 $[\hat{\sigma}'\vert 0]$
$(0)A_1$	1	1	1	1	1	1
$(0)A_2$	1	1	-1	1	1	-1
$(0)B_1$	1	1	1	-1	-1	-1
$(0)B_2$	1	1	-1	-1	-1	1
$(1)A$	2	-1	0	2	-1	0
$(1)B$	2	-1	0	-2	1	0

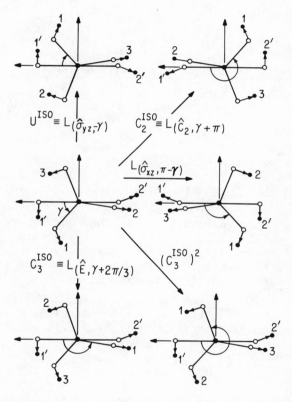

Fig. 6.   Isodynamic operations for the CH_3-NO_2 molecule

6. ISODYNAMIC OPERATIONS

Consider the set L_H consisting of all permutation-rotations L_h

$$L_H \equiv \{L_h | h \varepsilon \mathbb{H} \} \tag{6.1}$$

where

$$L_h : d_i^\alpha \to \Sigma R(\rho)_{ij} S(h)_{\alpha\beta} d_j^\beta \tag{6.2a}$$

and

$$L_h : a^\alpha(\gamma)_i \to \Sigma R(\rho)_{ij} S(h)_{\alpha\beta} a^\beta(\gamma)_j = a^\alpha(\tau(\gamma))_i \tag{6.2b}$$

L_H is the <u>invariance group of the</u> Eckart-Sayvetz <u>frame</u>. To see this, it is only necessary to note that the quadratic form ζ is invariant under all $L_h \varepsilon L_H$, so that we can write

$$L_h : (C; \gamma; d_i^\alpha) \to (C; \tau(\gamma); L_h \cdot d_i^\alpha) \tag{6.3}$$

We can therefore consider the set of perrotations L_H independently of any transformations of rotational variables; the group L_H is said to be the <u>isodynamic group</u> associated with the NRM symmetry group $\mathbb{H}$ [39].

The isodynamic operations L_h, which act upon the vibrational coordinates and the internal parameters, are a natural generalization to NRMs of the Wigner vibrational symmetry operations for rigid molecules (cf. Section 3). In Figure 6 we show the action of isodynamic operations associated with $C_3^I \otimes C_{2v}$ upon an arbitrary distorted configuration of the CH_3-NO_2 molecule. It is very interesting to note that straightforward implementation of transformations 6.2 leads naturally to the 'switches' originally introduced by Altmann.[11]

ACKNOWLEDGEMENTS

Most of the work described here was done in Oxford, and it is a pleasure to thank Dr. P.W. Atkins for his constant guidance and advice. The financial support of the UK Science Research Council and the Governing Body of Christ Church, Oxford, is gratefully acknowledged. I should also like to thank Professor R. S. Berry and Dr. B.T. Sutcliffe for their encouragement.

REFERENCES

1. E.P. Wigner, <u>Nachr. Ges. Wiss. Gottingen</u>, 133 (1930).
2. E.B. Wilson, <u>J. Chem. Phys.</u> 2:432 (1934).
3. E.B. Wilson, <u>J. Chem. Phys.</u> 3:276 (1935).

4. E.B. Wilson, J. Chem. Phys. 3:818 (1935).
5. J.T. Hougen, J. Chem. Phys. 37:1433 (1962).
6. J.T. Hougen, J. Chem. Phys. 39:358 (1963).
7. H.C. Longuet-Higgins, Molec. Phys. 6:445 (1963).
8. J.K.G. Watson, Can. J. Phys. 43:1996 (1965).
9. J.T. Hougen, Pure Appl. Chem. 11:481 (1965).
10. B.J. Dalton, Mol. Phys. 11:265 (1966).
11. S.L. Altmann, Proc. Roy. Soc. A 298:184 (1967).
12. C.M. Woodman, Mol. Phys. 19:753 (1970).
13. J.K.G. Watson, Mol. Phys. 21:577 (1971).
14. S.L. Altmann, Mol. Phys. 21:587 (1971).
15. B.J. Dalton, J. Chem. Phys. 54:4745 (1971).
16. J.T. Hougen, J. Chem. Phys. 55:1122 (1971).
17. J.M.F. Gilles, and J. Philippot, Int. J. Quant. Chem. 6:525 (1972).
18. J. Moret-Bailly, J. Mol. Spec. 50:485 (1974).
19. J.T. Hougen, J. Mol. Spec. 50:483 (1974).
20. A. Bauder, R. Meyer, and Hs.H. Günthard, Mol. Phys. 28:1305 (1974).
21. G.A. Natanson, and M.N. Adamov, Vestn. Leningr. Univ., No. 10:24 (1974).
22. J. Serre, Adv. Quant. Chem. 8:1 (1974).
23. C.D. Cantrell, and H.W. Galbraith, J. Mol. Spec. 58:158 (1975).
24. J.T. Hougen, "MTP International Review of Science," Vol. 3 (Physical Chemistry Series 2), p. 75, (1975).
25. P.R. Bunker, "Vibrational Spectroscopy and Structure," Vol. 3, Edited by J.R. Durig, Marcel Dekker (1975).
26. J-C. Hilico, H. Berger, and M. Loete, Can. J. Phys. 54:1702 (1976).
27. J.D. Louck, and H.W. Galbraith, Rev. Mod. Phys. 23:90 (1976).
28. G.A. Natanson, Opt. and Spec. 41:18 (1976).
29. H. Frei, R. Meyer, A. Bauder, and Hs.H. Gunthard, Mol. Phys. 32:443 (1976). (Erratum: Mol. Phys. 34:1198 (1977)).
30. H. Berger, J. de Phys. 38:1371 (1977).
31. S.L. Altmann, "Induced Representations in Crystals and Molecules", Academic (1977).
32. M. Quack, Mol. Phys. 34:477 (1977).
33. H. Frei, P. Gröner, A. Bauder, and Hs. H. Günthard, Mol. Phys. 36:1469 (1978).
34. H. Frei, and Hs.H. Günthard, Lecture Notes in Physics 79:92 (1978).
35. W.G. Harter, C.W. Patterson, and F.J. da Paixao, Rev. Mod. Phys. 50:37 (1978).
36. J.M.F. Gilles, and J. Philippot, Int. J. Quant. Chem. 14:299 (1978).
37. P.R. Bunker, "Molecular Symmetry and Spectroscopy," Academic (1979).
38. G.A. Natanson, Opt. and Spec. 47:139 (1979).
39. G.S. Ezra, Mol. Phys. 38:863 (1979).
40. Hs. H. Günthard, A. Bauder, and H. Frei, Topics in Current Chemistry 81:7 (1979).

41. J.D. Louck, Lecture Notes in Chemistry 12:57 (1979).
42. A. Dress, Lecture Notes in Chemistry 12:77 (1979).
43. P.R. Bunker, Lecture Notes in Chemistry 12:38 (1979).
44. K. Balasubramanian, J. Chem. Phys. 72:665 (1980).
45. R.S. Berry, Lectures Presented at the NATO ASI 'Quantum Dynamics of Molecules,' Cambridge, England (1979).
46. A. Sayvetz, J. Chem. Phys. 7:383 (1939).
47. R.L. Flurry, and T.H. Siddall, Mol. Phys. 36:1308 (1978).
48. A. Messiah, "Quantum Mechanics," Vol. 2, Wiley (1958).
49. R.T. Pack, and J.O. Hirschfelder, J. Chem. Phys. 49:4009 (1968).
50. R.S. Berry, Rev. Mod. Phys. 32:447 (1960).
51. H.H. Nielsen, Rev. Mod. Phys. 23:90 (1951).
52. E.B. Wilson, J.C. Decius, and P.C. Cross, "Molecular Vibrations," McGraw-Hill (1955).
53. Yu. S. Makushkin, and O.N. Ulenikov, J. Mol. Spec. 68:1 (1977).
54. J.K.G. Watson, Mol. Phys. 19:465 (1970).
55. J.K.G. Watson, Mol. Phys. 15:479 (1968).
56. R. Meyer, and Hs. H. Günthard, J. Chem. Phys. 49:1510 (1968).
57. D.J. Rowe, and P. Gulshani, Can. J. Phys. 54:970 (1976).
58. F. Jorgensen, Int. J. Quant. Chem. 14:55 (1978).
59. R. McWeeny, "Symmetry", Pergamon (1963).
60. G.O. Sørensen, Topics in Current Chemistry 82:99 (1979).
61. W.H. Miller, N.C. Handy, and J.E. Adams, J. Chem. Phys. 72:99 (1980).
62. J.W. Fleming, and C.N. Banwell, J. Mol. Spec. 31:378 (1969).
63. D. Papoušek, and V. Špirko, Topics in Current Chemistry 68:59 (1976).
64. P. Russeger, and J. Brickmann, J. Chem. Phys. 62:1086 (1975).
65. E.P. Wigner, "Group Theory," Academic (1959).
66. A.J. Stone, J. Chem. Phys. 41:1568 (1964).
67. G.S. Ezra, Mol. Phys. (1981), to be published, and unpublished.
68. G. Turrel, J. Mol. Struct. 5:245 (1970).
69. R.G. Woolley, and B.T. Sutcliffe, Chem. Phys. Lett. 45:393 (1977)
70. E.B. Wilson, Int. J. Quant. Chem. S13:5 (1979).
71. L. Engelbrecht, and J. Hinze, Adv. Chem. Phys. XLIV:1 (1980).

MOLECULAR STRUCTURE DERIVED FROM FIRST-PRINCIPLES

QUANTUM MECHANICS: TWO EXAMPLES

Peter Pfeifer

Laboratory of Physical Chemistry
ETH Zürich
8092 Zürich, Switzerland

ABSTRACT

It is shown how the classical concept of molecular structure finds a natural setting in quantum theory in terms of superselection rules, and how the latter can be understood in a variety of cases as result of the universal coupling of molecules to the radiation field. Particularly favorable conditions for this mechanism to be operative arise in large assemblies of closely packed identical molecules (cooperative effect). Specific examples include the symmetry breaking in chiral molecules and the localization of the electron in H_2^+ for widely separated nuclei. Finally, an experiment is proposed to check some of the predictions of this theory.

1.   STATEMENT OF THE PROBLEM

The problem of how to retrieve the chemist's notion of molecular structure from a fullfledged quantum description of molecular matter has many facets (for some early and recent discussions, respectively, see Refs. 1-4 and 5,6). To appreciate some of these,
- consider the usual translation/rotation/etc.-invariant Coulomb Hamiltonian H (with center-of-mass kinetic energy subtracted, but still in space-fixed coordinates) for J nuclei and K electrons and assume that H has at least one isolated eigenvalue;
- for any given set of configurations of the J nuclei, introduce a decomposition into disjoint "structure classes" by assigning two configurations $(\underline{r}_1,\ldots,\underline{r}_J)$ and $(\underline{r}_1',\ldots,\underline{r}_J')$ (with entries $\in \mathbb{R}^3$) to the same class if there is a translation $\underline{a} \in \mathbb{R}^3$ and a rotation

315

$R \in SO(3)$ such that $(Rr_1 + a, \ldots, Rr_3 + a)$ and $(r'_1, \ldots, r'_3)$ differ at most by a permutation of identical nuclei;
- and imagine us to have accomplished each of the following computational steps:

a) Quantum structures[7] from exact stationary states: Determine the J-particle nuclear probability density ϱ_ψ (averaged over electrons and nuclear spins) associated with the (J+K)-particle eigenstate Ψ (e.g. with "constant center-of-mass part") of H. ϱ_ψ is translation/rotation/etc.-invariant. Decompose the set $\mathfrak{M}_\psi$ of all absolute maxima of ϱ_ψ into disjoint structure classes $\mathfrak{M}_\psi(1), \ldots, \mathfrak{M}_\psi(n_\psi)$.

b) Quantum structures from lowest adiabatic energy surface: Set all nuclear masses equal to ∞ in H, regard H as K-electron operator depending parametrically on the nuclear positions, and compute the lowest eigenvalue E of H as function of all nuclear configurations. E is translation/rotation/etc.-invariant. Decompose the set $\mathfrak{M}$ of all absolute minima of E into disjoint structure classes $\mathfrak{M}(1), \ldots, \mathfrak{M}(n)$.

c) Quantum structures from lowest adiabatic nuclear states: Let H' be the Hamiltonian describing, again with center-of-mass kinetic energy subtracted, the motion of the nuclei in the adiabatic potential E. Compute the spin-averaged nuclear probability density ϱ'_φ from the J-particle eigenstate φ (e.g. with "constant center-of-mass part") of H'. ϱ'_φ is translation/rotation/etc.-invariant. Decompose the set $\mathfrak{M}'_\varphi$ of all absolute maxima of ϱ'_φ into disjoint structure classes $\mathfrak{M}'_\varphi(1), \ldots, \mathfrak{M}'_\varphi(n'_\varphi)$.

(Steps b and c may also be modified by replacing E by E plus the centrifugal correction term given by the Born-Huang adiabatic decoupling scheme[8].) Some of the issues may now be phrased as follows:

1) What is the range of the number n of structure classes of $\mathfrak{M}$, and is it equal to n_ψ and n'_φ if, e.g., Ψ and φ are ground states of H and H', respectively? (Presumably, one has $n = n_\psi = n'_\varphi = 1$ for $J \leqslant 3$. But $n = n'_\varphi = 1$ for NH_3 and C_2H_2 with identical isotopes; $n = n'_\varphi = 2$ for NHDT; $n = 2$ and $n'_\varphi = 1$ for C_2H_2 with unequal isotopes. See also question 5.)

2) Are $\mathfrak{M}_\psi(1), \ldots, \mathfrak{M}_\psi(n_\psi)$ and $\mathfrak{M}'_\varphi(1), \ldots, \mathfrak{M}'_\varphi(n'_\varphi)$ *) qualitatively the same if Ψ and φ are ground states of H and H', respectively? (See e.g. Ref. 9 for results towards a partial yes.)

3) Are $\mathfrak{M}(1), \ldots, \mathfrak{M}(n)$ and $\mathfrak{M}'_\varphi(1), \ldots, \mathfrak{M}'_\varphi(n'_\varphi)$ *) essentially the same if φ is ground state of H'? (Probably no for cases as presented in B.T. Sutcliffe's lecture.)

---

*) if they agree in number.

4) Is there, e.g. for the ground state Ψ of H, a direct-integral representation over $\{$translations$\} \times \{$ Euler angles$\} \times \{$ permutations $\}$ so that the resulting integrand is free of delocalization due to translation/rotation/permutation invariance of Ψ ? (The translational part is standard. See e.g. Ref. 10 for separation of overall rotations as indicated. For an actual rotational degeneracy, cf. Ref. 11.)

5) Let Ψ be ground state of H. Does $\mathfrak{m}_\Psi$, for $n_\Psi = 1$, correspond to the empirical structure (if any) of the molecule? What if $n_\Psi > 1$, or if $n_\Psi = 1$ with g_Ψ being close to the maximal value in significant regions different from $\mathfrak{m}_\Psi$ - i.e. if nuclei are delocalized other than by the mechanism of question 4, so that the notion of a leading structure is no longer meaningful? (Examples: As $\mathfrak{m}_\Psi$ is always inversion-invariant, Ψ can never represent a state of a chiral molecule; rather, one expects $n_\Psi = 2$ for this case. The second type of delocalization is typical of "freely rotating" methyl groups etc.)

6) Do $\mathfrak{m}_\Psi(1), \ldots, \mathfrak{m}_\Psi(n_\Psi)$ change drastically as Ψ varies over low-lying eigenstates of H? If yes, how can we account for the fact that transitions between states with widely differing structure classes do not seem to occur? (Example: While it is reasonable to expect the 2 structural isomers for $^{12}C^{13}$CHD to appear as $\mathfrak{m}_{\Psi_1}$ and $\mathfrak{m}_{\Psi_2}$ (where Ψ_1 is ground state and Ψ_2 some excited state of H) because $\mathfrak{m}'_{\varphi_1}$ and $\mathfrak{m}'_{\varphi_2}$ yield such structures (where φ_1 is ground state and φ_2 some excited state of H'), the functions φ_1 and φ_2 definitely cannot provide the clue as to why Ψ_1 and Ψ_2 are so stable - for *the usual answer to this question in terms of potential barriers and slow tunneling rates, etc. simply begs the question - what are the potential barriers?*[12].)

Here we concentrate on questions 5 and 6. More precisely, we consider the following two paradoxes:

I.   Why is it that, say, alanine comes only as inversion-noninvariant d- or l-enantiomer rather than as inversion-invariant eigenstate of the corresponding 61-particle Hamiltonian H, i.e. that superpositions of and transitions between the two do not occur?

II. Why is it that, for fixed nuclei sufficiently far apart, H_2^+ exists only as $H+H^+$ or H^++H (with protons distinguishable by position) rather than as $1\sigma_g$ ground state? (For experimental studies, see Ref. 13.)

The usual reasoning as for an explanation is that the quantum ground states in question are almost degenerate so that very small environmental perturbations (perhaps even weak-interaction induced parity violations[14]) suffice to destroy the symmetry of our systems, and that the accordingly perturbed Hamiltonians will have correctly

Alanine (incl. nuclear motion) H_2^+ (with fixed nuclei)

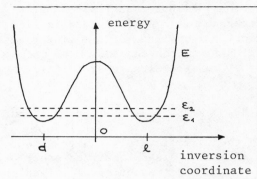
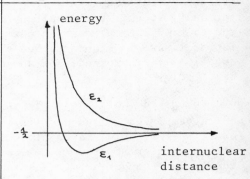

Adiabatic estimate of the two   Two lowest (electronic)
lowest energy levels   energy levels

asymmetric eigenstates which, if subjected to "secondary" influences
of the surroundings, will be quasi-stationary. But the two-level
Hamiltonian

$$\begin{pmatrix} \varepsilon_1 & 0 \\ 0 & \varepsilon_2 \end{pmatrix} + \begin{pmatrix} v_1(t) & v(t) \\ v(t)^* & v_2(t) \end{pmatrix} \qquad (t \in \mathbb{R})$$

customarily invoked for this argument (where the time-dependent matrix elements $v_1(t), v_2(t), v(t)$ are to mimic collisions, external
fields, etc.), shows that the desired states $\binom{1}{\pm 1}/\sqrt{2}$ are quasi-
stationary only if $|\varepsilon_1 - \varepsilon_2 + v_1(t) - v_2(t)| \ll v(t) \cong v(t)^*$ for "almost" all
$t \in \mathbb{R}$. I.e. the paradox remains unsolved because there is no reason
why this condition (which is much stronger than $|\varepsilon_1 - \varepsilon_2| \ll |v(t)|$)
should universally hold for the great variety of perturbations to be
considered in order to make the argument cogent. Thus it is really
the exclusiveness and stability of the chiral states of alanine,
say, which form the heart of the problem.

2.   SUPERSELECTION RULES

     The state of affairs just described corresponds precisely to
what is known as superselection rule (see e.g. Ref. 15 or 16): If
$\mathcal{H}_1, \mathcal{H}_2, \ldots$ are mutually orthogonal subspaces of the Hilbert space
$\mathcal{H}_1 \oplus \mathcal{H}_2 \oplus \cdots$ containing all state vectors of the system in question,
and if only vectors in $\mathcal{H}_1, \mathcal{H}_2, \ldots$ correspond to physical states,
then states lying in different such subspaces (sectors) are said to
be separated by a superselection rule. This is equivalent to admit-
ting as observables only operators of the form

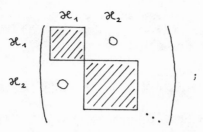

and implies that there are no observables connecting different sectors, but that there exist (classical) observables such as

$$\left(\begin{matrix} 1 & O \\ O & O \end{matrix} \right) \quad , \quad \left(\begin{matrix} O & O \\ O & 1 \end{matrix} \right) \quad , \cdots$$

that commute with all other observables. It follows that a classical observable takes a definite value in every physical state.

Merely to illustrate these concepts, we recall the example Bose/Fermi statistics for two identical particles: Here the total Hilbert space is $h \otimes h$ where h is the one-particle space. The particle indistinguishability requires the observables to belong to the (closed) span of $\{a \otimes b + b \otimes a \, | \, a, b$ are (bounded) operators on $h\}$.[17] It follows that the classical observables are generated by the observable (!) C where $C(\varphi \otimes \Psi) = \Psi \otimes \varphi$ for all $\varphi, \Psi \in h$. The ensuing two sectors $\mathcal{H}_{\pm} = (1 \pm C)(h \otimes h)$ yield the superselection rule that superpositions of symmetric and antisymmetric functions are not physical states (but rather correspond to classical mixtures in the sense that $\langle \Phi_+ + \Phi_- | A(\Phi_+ + \Phi_-) \rangle = \langle \Phi_+ | A\Phi_+ \rangle + \langle \Phi_- | A\Phi_- \rangle$ for all $\Phi_{\pm} \in \mathcal{H}_{\pm}$ and all observables A). Since $\langle \Phi_{\pm} | C\Phi_{\pm} \rangle = \pm \| \Phi_{\pm} \|^2$ for all $\Phi_{\pm} \in \mathcal{H}_{\pm}$, the classical observable C may be visualized as measuring the particle statistics.

Typically for most superselection rules, the underlying physics (here the indistinguishability of the two particles) is most easily specified in terms of observables. Once this is done, the problem of finding the accordingly possible physical states is just a mathematical one.

Taking this and § 1 as a guide, we will now study both a general molecule (not necessarily alanine) and the fixed-nuclei version of H_2^+ under the influence of the radiation field as generated by the corresponding dynamical charges (minimal-environment hypthesis), and analyze the so enlarged systems for superselection rules due to the

quasi-locality of field observables, i.e. the fact that a field operator can be an observable only if it acts essentially on finitely
many (out of the infinitude of) field modes.

3.   EFFECT OF THE RADIATION FIELD ON A SINGLE MOLECULE

I. For a general molecule, suppose that the Hamiltonian

$$H = \sum_{\nu=1}^{N} \frac{1}{2m_\nu} |p_\nu|^2 + \sum_{\substack{\nu,\mu=1\\ \nu<\mu}}^{N} z_\nu z_\mu |q_\nu - q_\mu|^{-1} - \left(2\sum_{\nu=1}^{N} m_\nu\right)^{-1} \left|\sum_{\nu=1}^{N} p_\nu\right|^2$$

(in a.u. and with charge z_ν, mass m_ν, position operator q_ν, momentum
operator p_ν for the ν-th particle ($\nu=1,\ldots,N$) - all of which are assumed as spinless and distinguishable for the sake of a simple exposition) has at least two isolated eigenvalues $\varepsilon_1 < \varepsilon_2$ at the bottom
of the spectrum, and corresponding normalized eigenfunctions $\psi_1, \psi_2 \in$
$L^2(\mathbb{R}^{3N})$ which we take real-valued and of product form, internal function times spherically symmetric center-of-mass function, where the
latter is the same for ψ_1 and ψ_2. It follows that ψ_1 is spherically
symmetric and in particular even; while ψ_2 is expected to be odd. So
ψ_1 and ψ_2 are not chiral (a state $\psi \in L^2(\mathbb{R}^{3N})$ is called chiral if ψ
is not eigenfunction of an improper-rotation operator).

II. For H_2^+ with clamped nuclei, we have the ($N=1$)-particle
Hamiltonian

$$H = \frac{1}{2m_1} |p_1|^2 + z_1 \left(|q_1 + \tfrac{1}{2}r|^{-1} + |q_1 - \tfrac{1}{2}r|^{-1} \right) \qquad (m_1 = -z_1 = 1, \; \underline{r} \in \mathbb{R}^3 \text{ fixed})$$

with lowest eigenvalues $\varepsilon_1 < \varepsilon_2$ and corresponding normalized real-
valued eigenfunctions $\psi_1, \psi_2 \in L^2(\mathbb{R}^3)$ which are even and odd, respectively. It follows that ψ_1 and ψ_2 are not moving to either side (we
call $\psi \in L^2(\mathbb{R}^3)$ left/right-moving if $\langle \psi | (p_1 \cdot \underline{r}) \psi \rangle \lessgtr 0$).

If we now introduce the radiation field (quantized in a cube of
length L), use the standard dipole-velocity interaction, and restrict
the molecular state space to span$\{\psi_1, \psi_2\}$, then we obtain the following model Hamiltonian (for details, see Ref. 18)

$$\begin{aligned}
H^{(L)} = \quad &\tfrac{1}{2}(\varepsilon_2 - \varepsilon_1) S_3 \otimes 1 && \text{(free molecule)} \\
+ &1 \otimes \sum_{n=1}^{\infty} \omega_n^{(L)} b_n^* b_n && \text{(free field)} \\
+ &S_2 \otimes \sum_{n=1}^{\infty} \lambda_n^{(L)} (b_n + b_n^*) && \text{(interaction).}
\end{aligned} \qquad \Biggr\} \quad (1)$$

Here $S_2 = \begin{pmatrix} 0 & -i \\ i & 0 \end{pmatrix}$, $S_3 = \begin{pmatrix} 1 & 0 \\ 0 & -1 \end{pmatrix}$; $\begin{pmatrix} 1 \\ 0 \end{pmatrix}$, $\begin{pmatrix} 0 \\ 1 \end{pmatrix}$ correspond to ψ_1, ψ_2, respectively; and $n=1,2,\ldots$ refers to the n-th field mode with frequency
$\omega_n^{(L)} > 0$, annihilation operator $b_n = 1 \otimes \ldots \otimes 1 \otimes b \otimes 1 \otimes \ldots$ (b in the n-th
position acts in the one-oscillator Hilbert space $\mathcal{H}$ and satisfies

$bb* = b*b+1$), and coupling constant $\lambda_n^{(L)} \in \mathbb{R}$. Thus $H^{(L)}$ formally lives in the infinite-tensor-product Hilbert space $\mathbb{C}^2 \otimes \mathcal{H}^\infty$. As various details of $\omega_n^{(L)}$ and $\lambda_n^{(L)}$ will cancel in the final result, it suffices to know that these are universal constants resp. universal functions of $\Psi_1, \Psi_2, z_1/m_1, \ldots, z_N/m_N$ – but that one cannot set $L=\infty$ at this stage because $\omega_n^{(L)}, \lambda_n^{(L)} \to 0$ as $L \to \infty$ for each $n=1,2,\ldots$. Finally, we will need also the (normalized) coherent oscillator state $|\alpha\rangle \in \mathcal{H}$ defined by $b|\alpha\rangle = \alpha|\alpha\rangle$ for "amplitude" $\alpha \in \mathbb{C}$ (so $|0\rangle$ is the zero-photon state).

A rigorous analysis of the finite-mode truncations of $H^{(L)}$ gives strong evidence[18] that the "Hartree ground state(s)", as given in the following result, and the exact ground state(s) of $H^{(L)}$ coincide asymptotically as we remove the infrared cutoff by letting $L \to \infty$:

Theorem 1. Put

$$\gamma^{(L)} = \min\left\{1, (\varepsilon_2 - \varepsilon_1)\left(4 \sum_{n=1}^\infty \lambda_n^{(L)2}/\omega_n^{(L)}\right)^{-1}\right\}. \tag{2}$$

Then the energy $\langle \chi \otimes \Phi | H^{(L)}(\chi \otimes \Phi)\rangle$ (with normalized $\chi \in \mathbb{C}^2$ and $\Phi \in \mathcal{H}^\infty$) is minimized by taking

$$\chi = \frac{1}{\sqrt{2}}\begin{pmatrix}\sqrt{1+\gamma^{(L)}} \\ \pm i\sqrt{1-\gamma^{(L)}}\end{pmatrix} \tag{3a}$$

$$\Phi = \bigotimes_{n=1}^\infty |\mp\sqrt{1-\gamma^{(L)2}}\,\lambda_n^{(L)}/\omega_n^{(L)}\rangle. \tag{3b}$$

Both for $\lambda_1^{(L)} = \lambda_2^{(L)} = \ldots = 0$ and for $\varepsilon_1 = \varepsilon_2$, $\chi \otimes \Phi$ is (are) the exact ground state(s) of $H^{(L)}$. Moreover,

$$\gamma = \lim_{L\to\infty} \gamma^{(L)} = \min\{1, (\varepsilon_2 - \varepsilon_1)/\Lambda\} \tag{4a}$$

$$\Lambda = (\pi c)^{-2} \int_{\mathbb{R}^3} |\langle \Psi_1| \sum_{\nu=1}^N \frac{z_\nu}{m_\nu} e^{i\underline{k}\cdot\underline{q}_\nu} \underline{p}_\nu \Psi_2\rangle \times \underline{k}|^2 |\underline{k}|^{-4} d(\underline{k}) \tag{4b}$$

where $c = 137.03\ldots$ a.u. is the velocity of light.

Thus if $\varepsilon_2 - \varepsilon_1$ is so small that $\gamma < 1$, then the combined system molecule plus field does indeed yield two degenerate, symmetry-broken ground states $(\sqrt{1+\gamma}\,\Psi_1 \pm i\sqrt{1-\gamma}\,\Psi_2)/\sqrt{2}$ of the molecule which are chiral in case I, and left/right-moving in case II (for $\gamma=1$ we simply recover the free-molecule ground state Ψ_1). Since each of them is tied to a separate field state (cf. (3)) it is clear that the two will be separated by a superselection rule whenever the corresponding field states (3b/$L \to \infty$) are so – which is almost always the case, as the next result shows:[18]

Theorem 2. Let $\mathcal{A}$ be the algebra of quasi-local field observables, defined as (weak closure of)

$\mathrm{span}\{a_1 \otimes \ldots \otimes a_n \otimes 1 \otimes 1 \otimes \ldots | a_1, \ldots, a_n$ are (bounded) operators on $\mathcal{H}$; $n=1,2,\ldots\}$

(note that $\mathcal{A}$ is much smaller than the set of all bounded operators on $\mathcal{H}^{\infty}$). If $\gamma < 1$ and

$$\langle \Psi_1 | \sum_{\nu=1}^{N} z_\nu q_\nu \Psi_2 \rangle \neq 0 , \tag{5}$$

then for $L \to \infty$ the two coherent field states (3b) converge (in the sense of expectation values on $\mathcal{A}$) to coherent field states $\Phi_{\pm} \in \mathcal{H}^{\infty}$ which are separated by a superselection rule (with respect to $\mathcal{A}$).

Equivalently (recall § 2): Any superposition of $\Phi_{\pm}$ corresponds to a classical mixture; $\mathcal{A}$ contains no operators connecting $\Phi_{\pm}$; there is a classical observable $C \in \mathcal{A}$ such that $C\Phi_{\pm} = \pm \Phi_{\pm}$ (this corresponds to an indirect but dispersion-free measurement of the handedness of the molecule in case I, and of the direction of electronic motion in case II). The physical origin of this superselection rule is that the molecule is "dressed" with a cloud of soft photons of infinite number but very small total energy, in such a way that the two field states belonging to the two ground states of the molecule are macroscopically different.

4.   COOPERATIVE EFFECT

It has been suggested[19] that the effect just described might be enhanced by the presence of further molecules (of the same kind). Support for this conjecture comes from comparing the results of § 3 with the fact[20] that for spontaneous generation of a large number of photons, i.e. of large-amplitude coherent states, at fixed frequency as in a laser (onset of superradiance), one needs a large number of identical molecular two-level systems:

   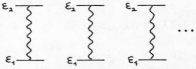

The total number of photons (at progressively lower frequencies) in the ground state of the 1-molecule/∞-mode system (eq. (1), $\gamma^{(L)} < 1$) is proportional to $\log(L)$ as $L \to \infty$.	The total number of photons (at given nonzero frequency) in the ground state of the M-molecule/1-mode system (Ref. 20, phase-transition case) is proportional to M as $M \to \infty$.

To see that the infinitely many (soft) photons required for the superselection rule are indeed more easily "available" when there are several molecules, we consider the M-molecule analogue of the Hamiltonian (1),

$$H_M^{(L)} = \tfrac{1}{2}(\varepsilon_1 - \varepsilon_2) \sum_{m=1}^{M} S_{3,m} \otimes 1$$
$$+ 1 \otimes \sum_{n=1}^{\infty} \omega_n^{(L)} b_n^* b_n$$
$$+ \sum_{m=1}^{M} \sum_{n=1}^{\infty} S_{2,m} \otimes (\lambda_{n,m}^{(L)*} b_n + \lambda_{n,m}^{(L)} b_n^*) \qquad (6)$$

where $S_{j,m} = 1 \otimes \ldots \otimes 1 \otimes S_j \otimes 1 \otimes \ldots \otimes 1$ with S_j in the m-th position acts in $\mathbb{C}^{2M}$, and $\lambda_{n,m}^{(L)} \in \mathbb{C}$ is the product of $\lambda_n^{(L)}$ and an (L,n,m)-dependent phase factor reflecting the spatial position of the m-th molecule (j=2,3; n=1,2,...; m=1,...,M). Note that (6) neglects all direct interactions among the molecules. The following generalization of Theorems 1 and 2 exhibits the cooperative effect in its most pronounced form:

Theorem 3. Let $\lambda_{n,m}^{(L)} = \lambda_n^{(L)}$ (n=1,2,...; m=1,...,M), i.e. let all molecules be at the same location. Multiply each of $\varepsilon_2 - \varepsilon_1$, $\lambda_n^{(L)}$, Ψ_1 by M wherever these expressions occur explicitly in eqs. (2), (3b), and (4). (For future reference, denote the accordingly modified quantity γ by γ_M.) Then Theorems 1 and 2 hold with $H^{(L)}$, $\mathbb{C}^2$, and eq. (2a) replaced by $H_M^{(L)}$, $\mathbb{C}^{2M}$, and

$$\chi = \bigotimes_{m=1}^{M} \tfrac{1}{\sqrt{2}} \begin{pmatrix} \sqrt{1+\gamma^{(L)}} \\ \pm i\sqrt{1-\gamma^{(L)}} \end{pmatrix} ,$$

respectively.

That is, under the hypotheses of Theorem 3 we always have superselection-separated ground states for sufficiently large M (note $\gamma_M \to 0$ as $M \to \infty$). Of course, these hypotheses are somewhat artificial; in fact, they lead to a (negative) ground-state energy of $H_M^{(L)}$ that is proportional to M^2 rather than M, as $M \to \infty$. More realistic constants $\lambda_{n,m}^{(L)}$ - e.g. such that the molecules are separated by some minimal distance - give a ground-state energy proportional to M for $M \to \infty$,[21] but then the Hartree ground states of $H_M^{(L)}$ fail to be explicitly computable. Nevertheless, some information about the general situation is available as follows:

$H_M^{(L)}$ possesses exactly 1 or 2 Hartree ground states depending on whether $\varepsilon_2 - \varepsilon_1$ does or does not exceed the largest eigenvalue of the (M×M)-matrix

$$\left\{ \sum_{n=1}^{\infty} \tfrac{4}{\omega_n^{(L)}} \operatorname{Re}(\lambda_{n,m}^{(L)} \lambda_{n,m'}^{(L)}) \right\}_{m,m'=1}^{M}$$

where, in the first case, this state is given by

$$\left(\bigotimes_{m=1}^{M} \begin{pmatrix} 1 \\ 0 \end{pmatrix} \right) \otimes \left(\bigotimes_{n=1}^{\infty} |0\rangle \right) .$$

It follows that these two cases arise (for $L \to \infty$) if $\gamma_M = 1$ and $\gamma_1 < 1$, respectively.

5.   NUMERICAL ESTIMATES AND EXPERIMENTAL CONSEQUENCES

I. To compare the single-molecule theory of § 3 with experiment
(first for the chiral-molecules case) we need estimates of γ, i.e.
of $\varepsilon_2 - \varepsilon_1$ and $\wedge$ (recall eq. (4)). Since $\wedge$ is a universal functional
of the electrical current density of state $(\Psi_1 + i\Psi_2)/\sqrt{2}$, and since
this density is not expected to change radically for different molecules, the values[18]

$$\wedge = 2.60 \cdot 10^{-6} \text{ a.u. for 1s/2p-eigenstates of the H-atom}$$
$$\wedge = 2.08 \cdot 10^{-7} \text{ a.u. for 2s/2p-eigenstates of the H-atom}$$

suggest that $\wedge = 10^{-8} \ldots 10^{-6}$ a.u. as for a general order-of-magnitude
estimate. If this is so, then the energy differences

$$\varepsilon_2 - \varepsilon_1 \gtrsim 10^{-15} \qquad \text{a.u. for molecules with asymmetric C-atom}$$
$$\varepsilon_2 - \varepsilon_1 = 1.80 \cdot 10^{-6} \quad \text{a.u. for } NH_3$$
$$\varepsilon_2 - \varepsilon_1 = 1.20 \cdot 10^{-7} \quad \text{a.u. for } ND_3 \Big\} \quad \text{(experimental values)}$$
$$\varepsilon_2 - \varepsilon_1 = 2.32 \cdot 10^{-8} \quad \text{a.u. for } NT_3$$
$$\varepsilon_2 - \varepsilon_1 \gtrsim 10^{-2} \qquad \text{a.u. for light atoms (up to Kr)}$$

imply rather satisfactory agreement with observed optical (in)activ-
ity of these examples. Even the fact that 2s/2p-states of the H-atom
do not combine to superselection-separated chiral states fits nicely
into the theory (in spite of the nonrelativistic degeneracy) as the
dipole condition (5) fails for these states.

A direct experimental check of the predicted non-observability
of Ψ_1 and Ψ_2 when $\gamma < 1$ and (5) are satisfied, has been proposed by
Quack:[22] Consider a molecule with lowest two energy surfaces E < E' *)
such that the corresponding (even/odd-parity) pairs $\{\Psi_1, \Psi_2\}$, $\{\Psi_1', \Psi_2'\}$
of lowest rotation-vibration states with energies $\{\varepsilon_1, \varepsilon_2\}$, $\{\varepsilon_1', \varepsilon_2'\}$
satisfy $\varepsilon_2 - \varepsilon_1 < \wedge$ and $\varepsilon_1' - \varepsilon_1' > \wedge$. (The second condition is optional –
it suffices that $\varepsilon_1', \varepsilon_2'$ can be unambiguously resolved and assigned;
in contrast, the splitting $\varepsilon_2 - \varepsilon_1$ is not measurable by hypothesis.)
These prerequisites are likely to be met by some amines.

Now in an attempt to prepare the forbidden ground state Ψ_1, a
molecular beam is excited from E to E' and passed through a state
selector so as to produce an ensemble of molecules in the Ψ_2' state.
By the usual dipole selection rules, only transitions $\Psi_2' \to \Psi_1$ are
allowed upon suitably stimulated emission. In order to see whether
indeed Ψ_1 was obtained one re-excites the decay product back to E'
and invokes the same selection rules again: If yes, then only Ψ_2'
will be detected – if no, both Ψ_2' and Ψ_1' will be found (in a ratio
of $\wedge + \varepsilon_2 - \varepsilon_1$ to $\wedge + \varepsilon_1 - \varepsilon_2$):

_____
*) Unlike in § 1, primes ' refer to the upper surface here!

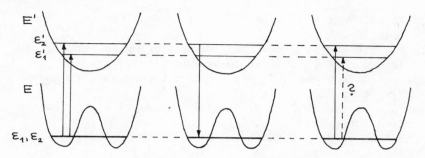

Successive electronic transitions for Quack's experiment

II. For the H_2^+-problem, $\varepsilon_2 - \varepsilon_1 = 4|r| \exp(-|r|-1)(1+O(|r|^{-1}))$ as $|r| \to \infty$ (see e.g. Ref. 23). The result of evaluating $\wedge$ with the usual LCAO-approximation of Ψ_1 and Ψ_2 is bounded above by $O(\exp(-2|r|))$ as $|r| \to \infty$. So if this fall-off persists for the exact $\wedge$ (which to prove would require a refinement of Theorem 1.12 in Ref. 23), then the present theory does not lead to spontaneous electron localization at some finite $|r|$. But for a final judgement of the proposed bond-breaking mechanism one should note that simplifications like the neglect of the so-called A^2-term in the molecule-field interaction, the restriction to a (rather particular) 2-dimensional molecular state space, and the neglect of relativistic terms in the free-molecule Hamiltonian may well cease to be valid as $|r| \to \infty$. See Ref. 24 for a preliminary analysis of some of these points.

ACKNOWLEDGEMENTS

The author would like to thank J. Hinze, W. Kutzelnigg, and H. Primas for valuable discussions; and M. Quack for permission to include here his idea for an experimental check.

REFERENCES

1. F. Hund, Zur Deutung der Molekelspektren, III, Z. Phys. 43: 805 (1927).
2. L. Rosenfeld, Quantenmechanische Theorie der natürlichen optischen Aktivität von Flüssigkeiten und Gasen, Z. Phys. 52: 161 (1929).
3. M. Born and P. Jordan, "Elementare Quantenmechanik," Springer, Berlin (1930): § 47.
4. H. Hellmann, "Einführung in die Quantenchemie," Deuticke, Leipzig (1937): p. 300.
5. "The theory of molecular structure and bonding," R. Pauncz and E.A. Halevi, eds., Israel J. Chem. 19: No. 1-4 (1980).

6. "Quantum dynamics of molecules," R.G. Woolley, ed., Plenum
    Press, New York (1980).
7. P. Claverie and S. Diner, The concept of molecular structure in
    quantum theory: interpretation problems, in: Ref. 5.
8. C.J. Ballhausen and A.E. Hansen, Electronic spectra, Annual Rev.
    Phys. Chem. 23: 15 (1972).
9. J.M. Combes and R. Seiler, Spectral properties of atomic and
    molecular systems, in: Ref. 6.
10. L. Lathouwers and P. Van Leuven, Generator coordinate theory of
    diatomic molecules, in: Ref. 6.
    L. Lathouwers, On body-fixed frames and angular momentum pro-
    jection operators, J. Phys. A 13: 2287 (1980).
11. E.B. Davies, Symmetry breaking for a nonlinear Schrödinger
    equation, Commun. math. Phys. 64: 191 (1979).
    E.B. Davies, Metastable states of molecules, Commun. math. Phys.
    75: 263 (1980).
12. R.G. Woolley, Quantum mechanical aspects of the molecular struc-
    ture hypothesis, in: Ref. 5.
13. E. Gerjuoy, Is the principle of superposition really necessary?,
    in: "Contemporary research in the foundations and philosophy
    of quantum theory," C.A. Hooker, ed., Reidel Publ. Co.,
    Dordrecht (1973).
14. R.A. Harris, Manifestations of parity violations in atomic and
    molecular systems, in: Ref. 6.
15. R.F. Streater and A.S. Wightman, "PCT, spin and statistics, and
    all that," Benjamin, New York (1964): § 1.1.
16. H. Primas, Foundations of theoretical chemistry, in: Ref. 6.
17. J.M. Jauch, "Foundations of quantum mechanics," Addison-Wesley,
    Reading (1968): § 15.3.
18. P. Pfeifer, "Chiral molecules − a superselection rule induced
    by the radiation field," Thesis no. 6551, ETH, Zürich (1980).
19. J. Hinze: At this conference.
20. E.H. Lieb, Exactly soluble models, Physica 73: 226 (1974).
21. K. Hepp and E.H. Lieb, Equilibrium statistical mechanics of mat-
    ter interacting with the quantized radiation field, Phys.
    Rev. A 8: 2517 (1973).
22. M. Quack: At this conference.
23. E.M. Harrell, Double wells, Commun. math. Phys. 75: 239 (1980).
24. P. Pfeifer, A nonlinear Schrödinger equation yielding the 'shape
    of molecules' by spontaneous symmetry breaking, in: "Clas-
    sical, semiclassical and quantum mechanical problems in math-
    ematics, chemistry and physics," K. Gustafson and W.P. Rein-
    hardt, eds., Plenum Press, New York (1980).

KINEMATIC AND DYNAMIC PARTITIONINGS OF THE ENERGY;

COORDINATE AND OTHER TRANSFORMATIONS

Hanno Essén

Theoretical Chemistry Department
University of Oxford, Oxford, England*

ABSTRACT

The basic partitioning of the energy of a molecule into electronic, vibrational and rotational is discussed from the general
point of view of coordinate transformations. Our message is that
when experiment indicates an approximate separability there is also
a natural coordinate system in which it becomes clear how and why
it comes about.

1. SOME DEFINITIONS

Before it is meaningful to speak about energy scrambling the
energy must somehow be partitioned i.e. written as a sum of terms
each with a certain kinematic (geometric) meaning. In Cartesian coordinates a system of N free particles has the partitioned energy

$$H = \sum_{\alpha=1}^{N} T_\alpha \equiv \sum_{\alpha=1}^{N} \vec{P}_\alpha^2 / (2m_\alpha). \tag{1}$$

Introduction of interaction e.g. in the form of a potential energy
contribution $V(\vec{r}_\alpha)$ in general, however, ruins the simple partitioning in (1). The bulk of this workshop is concerned with the vibrational part of the molecular motion where there is a zeroth order
partitioning of the energy into harmonic oscillator normal modes
and scrambling is mainly considered to be due to anharmonic coupling between these (assuming the separation of a vibrational part
as exact). Considerable scrambling between normal (or local) mode

*Work financed by a Royal Society fellowship. Manuscript prepared
at: Institute of Theoretical Physics; University of Stockholm,
Sweden. Present address: Department of Chemistry, McMaster University, Hamilton, Ontario, L8S 4M1, Canada.

oscillators may, however, also result from the fact that the vibrational degrees of freedom can only be approximately separated from a) rotational and b) electronic degrees of freedom. My contribution to the workshop is mainly concerned with these, the two basic separations of molecular physics.

Let us first make some definitions that are intended to bring out the different natures for the two separations. In the general case the potential in the Hamiltonian

$$H = T(\vec{P}_\alpha) + V(\vec{r}_\alpha) \tag{2}$$

for a free molecule does not depend on the position or the orientation of the molecule in space. To take explicit account of this fact is then just a question of rewriting the kinetic energy part of the total energy in terms of suitable coordinates. We suggest that the resulting partitioning is called a kinematic partitioning. Thus, since V does not depend on center of mass coordinates $\vec{R}_o$, introduction of e.g. Jacobi coordinates (see [1], [2]), gives $(\gamma = 1, \ldots, N - 1)$

$$H = T_{CM}(\vec{P}_o) + T(\vec{P}_\gamma) + V(\vec{R}_\gamma) \equiv T_{CM} + H_{int}. \tag{3}$$

The fact that V does not depend on the orientation, i.e. does not depend on some suitably chosen angles η_s, may be explicitly taken account of by an appropriate coordinate transformation which we shall return to in section 3b. A straight forward coordinate transformation will now, however, not lead to a partitioning but to a result of the form

$$H_{int} = T_{rot} + T_{rot-vib} + (T_{vib} + V) \tag{4}$$

Formally one may get rid of the offdiagonal term, $T_{rot-vib}$, by making a certain non-holonomic transformation of the generalized momenta p_a (or momentum operators):

$$p_a \to P_a = \sum_b A_a^b(q)p_b \tag{5}$$

to new such: P_a. These new "momenta" will then, in general, not have the usual Poisson-brackets (or commutators) but instead they obey

$$[P_a, P_b] = \sum_c C_{ab}^c P_c \tag{6}$$

(The reader will easily recognize angular momenta as a special case of such non-holonomic momenta). In this way one can formally always obtain a partitioning:

$$H_{int} = T'_{rot} + (T_{vib} + V) \equiv T'_{rot} + H_{vib} \tag{7}$$

of the energy into rotational and vibrational energy (or, more general: deformational energy in the case of floppy systems). We return to this in sec. 3a. The most interesting (or at least common) case for molecules is, however, that V is such, that the deviations from a certain geometry are small. One can then choose vibrational coordinates in such a way (using the so-called Eckart conditions) that $T_{rot-vib}$ is a small term and that T_{rot} corresponds essentially to rigid rotation of the molecule in the equilibrium geometry. The extent to which $H_{vib} = T_{vib} + V$ can be considered in isolation in solving the equations of motion thus depends on the specific form of V and therefore on the dynamics of the system. This brings us to the second type of energy partitioning which we call dynamic partitioning.

We shall say that we have been able to find a dynamic partitioning of the Hamiltonian (2) if we can write

$$H = \sum_n (T_n + V_n) + \Delta V \qquad (8)$$

where ΔV, under suitable circumstances, can be considered as a perturbation. Note that different states of the system may be consistent with different dynamic partitionings. One example of (8) is obtained if V in H_{vib} has a minimum. One then expands around this minimum and diagonalizes the quadratic part (normal coordinates). ΔV will then correspond to the anharmonic coupling. If V, however, is the full Coulomb potential of an atom or molecule there is no minimum. Nevertheless certain coordinate transformations can indicate adiabatic approximations and corresponding partitioning as we now proceed to discuss.

2. ADIABATIC SEPARATIONS

An adiabatic separation of a dynamical problem is possible (in a subset of configuration space) if (in that subset) the degrees of freedom can be divided into two types: one slow S and one fast F in such a way that the typical frequencies obey

$$\omega_a \ll \omega_i \; ; \; q^a \in S \, , \, q^i \in F.$$

Quantum mechanically this means the energy level spacings resulting from the two sets are of different orders of magnitude. When all forces are of the same order of magnitude it is natural to connect slowness with large mass and vice versa. This is the conventional philosophy behind the Born-Oppenheimer approximation in molecules. One should however keep in mind some weaknesses of this point of view. Firstly, the fact that the nuclei are heavier than the electrons in no way explains why they should stay near equilibrium positions and thus the Born-Oppenheimer approach does not lead to an explanation of molecular structure (see e.g. Woolley [3], Woolley and Sutcliffe [4]). Secondly, the force on most nuclei of the periodic system is in fact one to two orders of magnitude larger than on one of the electrons (because of their charge) while, when there are many electrons, collective electronic degrees of freedom may correspond to quite large masses and weak forces. Thus, if the mass ratio was the only reason for the adiabatic separation in molecules it would not be nearly as successful as it in fact is. The deeper reasons for the success have been discussed by this author earlier [5] and we sketch some of the arguments in sec. 2a.

The insight that the mass ratio is not everything gives hope that adiabatic approximations may have much wider applicability to low energy states of Coulomb interacting systems than normally thought. Thus in sec. 2b we mention some recent applications to atoms.

2a. The Born-Oppemheimer Approximation. The Coulomb potential energy of a (neutral) molecule is

$$V(\vec{r}_i) = \sum_{i<j} \frac{Q_i \, Q_j}{|\vec{r}_i - \vec{r}_j|} \cdot$$

Let us transform it to new coordinates $\vec{R}_o$, $\vec{r}^C$, $\vec{r}^I$ defined by

$$\vec{r}_i = \vec{R}_o + \vec{r}^C_{\gamma(i)} + \vec{r}^I_i \qquad i = 1, \ldots, n + N \qquad (11)$$

$$0 = \sum_{\gamma(i)=\alpha} m_i \vec{r}^I_i \qquad \alpha = 1, \ldots, N \qquad (12)$$

$$0 = \sum_{\alpha=1} M_\alpha \vec{r}^C_\alpha \qquad M_\alpha = \sum_{\gamma(i)=\alpha} m_i \qquad (13)$$

$$\sum_{\gamma(i)=\alpha} Q_i = 0 \cdot \qquad (14)$$

Here $N + n$ is the total number of particles (electrons and nuclei), N is the number of atoms or neutral (according to (14)) subsystems. $\vec{R}_o$ and $\vec{r}^C_\alpha$ are collective coordinates corresponding to the center of mass of the molecule and centers of mass of the neutral subsystems (atoms) respectively. ($\gamma(i)$ is a function of the particle index which takes the value α when i belongs to the αth subsystem). $\vec{r}^I_\alpha$ are coordinates of individual motion of the particles.

The internal Hamiltonian of a molecule, when written in these coordinates takes the (exact) form

$$H_{int} = T^C + \sum_{\alpha=1}^{N} (T^I_\alpha + V^I_\alpha) + \sum_{\alpha<\beta}^{N} V^{CI}_{\alpha\beta} \qquad (15)$$

with self-explaining notation. Here $T^I_\alpha + V^I_\alpha$ is the Hamiltonian of the αth atom while $V^{CI}_{\alpha\beta}$ stands for all Coulomb interactions between particles in atom α with particles in atom β. The last term in (15) is thus not a globally small term but, if the system occupies positions in configuration space corresponding to an arrangement into neutral atoms, it must of course be quite small (compared to V^I_α). Thus a very physical derivation of the adiabatic approximation in molecules can be based on (15). To first order one then makes the approximation

$$V^{CI} \equiv \sum_{\alpha<\beta} V^{CI}_{\alpha\beta} \approx U^C + U^I \tag{16}$$

and treats the remainder as a perturbation. With suitable choices of U^C and U^I this leads to the standard results [5].This was our first example of a dynamic partitioning of the energy.

2b. Antipodial Electron Pairs. In the previous section we obtained a dynamic energy partitioning by explicitly, in terms of a coordinate transformation, taking account of the intuitively obvious fact that the particles, in a low energy state, of a molecule (in general) are arranged, approximately, into neutral atoms. We now try to use a similar trick to get some insight into atomic electronic structure. Assume that we have an even number of electrons. One way to minimize the electron-electron repulsion is then to arrange the electrons pairwise in antipodal (opposite sides of the nucleus) positions. In the atomic Hamiltonian

$$\hat{H} = -\frac{1}{2} \sum_{i=1}^{n} \vec{\nabla}_i^2 - \sum_{i=1}^{n} \frac{Z}{r_i} + \sum_{i<j}^{n} \frac{1}{r_{ij}} \tag{17}$$

we thus assume that $n = 2N$ (i.e. N independent pairs) and make the coordinate transformation

$$\vec{r}_1 = \vec{R}_1 + \vec{r}_1'$$
$$\vec{r}_{1+N} = \vec{R}_1 - \vec{r}_1' \qquad 1 = 1, \ldots, N \tag{18}$$

This gives

$$\hat{H} = -\frac{1}{2} \sum_{1=1}^{N} \left(\frac{\vec{\nabla}_{R_1}^2}{2} + \frac{\vec{\nabla}_{r_1'}^2}{1/2} \right) - \sum_{1=1}^{N} \frac{Z}{|\vec{R}_1 \pm \vec{r}_1'|}$$

$$+ \frac{1}{2} \sum_{1=1}^{N} \frac{1}{r_1'} + \sum_{1<m}^{N} \frac{1}{|\vec{R}_1 - \vec{R}_m \pm (\vec{r}_1' \pm \vec{r}_m')|} \tag{19}$$

(sums taken over all sign combinations indicated). The coordinates $\vec{R}_1$ correspond to centers of mass of the electron pairs and thus to bosonic quasi-particles that are four times as heavy as the fermionic $\vec{r}_1'$ particles corresponding to the relative motion within the pair (see Essén [6]). This approach has been used by the author to explain the special stability of the (filled) inert gas shell $s^2 p^6$. A very simple application of (19) to two-electron atoms is obtained if we put $\vec{R}_1 = 0$. The two electron-atom then becomes an effective hydrogen like system with discrete energy levels

$$E_n = -(2Z - \frac{1}{2})^2 / 4n^2 . \tag{20}$$

For He and He$^-$ this gives the ground state energies -3.0625 and -0.5625 a.u.. The experimental values are -2.90 and -0.528 respectively. It thus seems as if (18) sometimes can lead to an adiabatic separation; $\vec{R}_1$ corresponding to the slow an $\vec{r}_1'$ to the fast degrees of freedom. This was our second example of a dynamic partitioning.

3. SEPARATION OF ROTATION FROM INTERNAL MOTION

This separation consists, as explained in section 1, of two parts: a) the dynamic and b) the kinematic. Here we assume the former to be valid i.e. we assume it known that the molecule's motions are small displacements from a known reference (equilibrium) configuration. The breakdown of the zeroth order vibrating rigid rotor model is then perturbative and our task is to analyse the structure of these perturbations. First, in 3a, we discuss how the exact Hamiltonian, transformed to molecular coordinates, can be written in a partitioned form without off-diagonal terms. In 3b we discuss in more detail how the choice of molecular coordinates is done. In particular we point out that while normal coordinates are meant to maximally simplify the vibrational problem there is another set of vibrational coordinates determined by extensions of the Eckart conditions which maximally simplify the vibration-rotation coupling terms.

3a. How to handle off-diagonal terms in the kinetic energy. Assume that a coordinate transformation has brought the kinetic energy T to the form

$$2T = g^{ab}P_aP_b + 2g^{ai}P_aP_i + g^{ij}P_iP_j \tag{21}$$

where the indices a, b run over one set of variables and i, j over another set. We define a set of new quantities $\mathcal{A}_a^i$ through

$$g^{ai} = g^{ia} \equiv - g^{ab}\mathcal{A}_b^i \tag{22}$$

and also

$$\mathcal{P}_a \equiv P_a - \mathcal{A}_a^i P_i \tag{23}$$

and

$$g'^{ij} \equiv g^{ij} - g^{ab}\mathcal{A}_a^i\mathcal{A}_b^j . \tag{24}$$

In terms of these (21) becomes

$$2T = g^{ab}\mathcal{P}_a\mathcal{P}_b + g'^{ij}P_iP_j . \tag{25}$$

The $\mathcal{P}_a$ given by (23) are of the non-holonomic type displayed in (5) and (6). In (25) we have thus formally partitioned the energy in

(21). One should note that if the Lagrangian form of kinetic energy

$$2T = g_{ab}\dot{q}^a\dot{q}^b + 2g_{ai}\dot{q}_a\dot{q}_i + g_{ij}\dot{q}^i\dot{q}^j \tag{26}$$

(which corresponds to (21) if the matrices

$$\begin{pmatrix} g_{ab} & g_{aj} \\ g_{ib} & g_{ij} \end{pmatrix}, \quad \begin{pmatrix} g^{ab} & g^{aj} \\ g^{ib} & g^{ij} \end{pmatrix} \tag{27}$$

are inverses to each other), has

$$g_{ij} = \delta_{ij} \tag{28}$$

then one finds that

$$g'^{ij} = \delta^{ij} \tag{29}$$

where g'^{ij} is given by (24) (see [1] formulae (6.3) and (6.4)).

After transformation to molecular coordinates, some rotational (corresponding to q^a in (26)) and some orthogonal vibrational (corresponding to q^i in (26) with (28) fullfilled), the Hamiltonian will be of the type (21) as mentioned in section 1. Already Wilson and Howard [7], however, rewrote it using a transformation of the type (22) - (24) and thus expressed it in the wellknown form

$$H = \frac{1}{2}(\Pi_\lambda - \pi_\lambda)\mu_{\lambda\mu}(\Pi_\nu - \pi_\nu) + \frac{1}{2}\sum_{i=1}^{3N-6} P_i + V(Q_i) . \tag{30}$$

This form is not suitable for linear molecules since in them only two rotational degrees of freedom can be expected to separate from the vibrations. (It is not true, as is sometimes said, that there are only two rotational degrees of freedom. What happens is that the rotations around the axis of the molecule do not separate from the bending vibrations). Various different forms of Hamiltonians for linear molecules have been suggested in literature. In a way very analogous to (30), the present author [8] has derived the following expression

$$H = \frac{1}{2} \mu[(P_\theta - p_\theta)^2 + \frac{(P_\varphi - p_\varphi)^2}{\sin^2_\theta}] + \frac{1}{2}\sum_{i=1}^{3N-5} P_i^2 + V(Q_i) . \tag{31}$$

It has the property to reduce to the usual expression in terms of spherical polar coordinates (θ,φ,r) for a di-atomic molecule when N = 2. Introduction of momenta of the type (23) with the help of the concept of quasi-coordinates have been discussed by Watson [9] where also an alternative treatment of linear molecules can be found.

3b. Extensions of the Eckart conditions. The transformation to molecular coordinates can be written explicitly

$$\vec{r}_\alpha = \vec{R}_o + \mathbb{R}(\eta_1,\ldots,\eta_s) \, [\vec{a}_\alpha + \sum_{i=1}^{3N-(3+s)} \vec{b}_\alpha^i Q_i] \tag{32}$$

where $\vec{r}_\alpha$ are the Cartesian position vectors of the particles [8]. $\mathbb{R}$ is a 3×3 rotation matrix and s is 2 or 3 depending on whether the molecule is linear or general. The vectors $\vec{a}_\alpha$ correspond to a reference equilibrium configuration. In order to obtain a Hamiltonian (30) with reasonable properties the vectors $\vec{a}_\alpha$ and $\vec{b}_\alpha^i$ must fulfill a number of conditions and the central ones are the Eckart conditions

$$\sum_\alpha m_\alpha \vec{a}_\alpha \times \vec{b}_\alpha^i = 0 \tag{33}$$

The tensor $\mu_{\lambda\nu}$ in (30), which is the inverse of the so-called effective inertia tensor, will then have a structure that

$$\mu^{-1} = \Pi_{eff} = \Pi^o + \sum_i \mathbb{a}_i Q_i + \sum_{i,j} \mathbb{b}_{ij} Q_i Q_j \tag{34}$$

where (Amat and Henry [10])

$$\mathbb{b}_{ij} = \mathbb{a}_i (\Pi^o)^{-1} \mathbb{a}_j \tag{35}$$

Watson [11] has derived the following commutation rules for the vibrational angular momenta

$$\pi_\lambda \pi_\mu - \pi_\mu \pi_\lambda = ih\epsilon_{\lambda\mu\nu}\pi_\nu - ih \sum_{i,j} (\mathbb{b}_{ij})_{\lambda\mu}(Q_i P_j - P_i Q_j) \tag{36}$$

$$\sum_\lambda [\pi_\lambda, (\mu)_{\lambda\mu}] = 0 \tag{37}$$

The vectors $\vec{b}_\alpha^i$ are not uniquely determined by the conditions imposed by orthogonality (28) and the Eckart condition (33). When one is mainly interested in the vibrational problem one usually thinks of them as chosen so as to diagonalize the quadratic part of the potential energy Taylor expansion (normal coordinates). When the main interest is the coupling to rotation, however, one should instead concentrate on simplifying (34) – (37). It turns out (Essén [8], [12]) that one can add conditions on the b_α^i in terms of the $\vec{a}_\alpha$ i.e. pose more conditions of the type (33), so as to achieve a certain maximum simplicity of the vibration-rotation algebra. Thus it turns out that these coordinates fall into two distinct groups:

$$Q^a = \{Q_1,\ldots,Q_6\} \quad , \quad Q^b = \{Q_7,\ldots,Q_{3N-6}\} \tag{38}$$

which are such that

$$\Pi_{eff} = \Pi_{eff}(Q^a) \tag{39}$$

i.e. μ depends only on the first six coordinates. Further the linear part of Π_{eff} is such that Q_1, Q_2, Q_3 appear diagonally while there is an off-diagonal dependence on Q_4, Q_5, Q_6. Further it turns out that the vibrational angular momenta can be written

$$\pi_\lambda = \pi_\lambda^a(Q^a, P^a) + \pi_\lambda^p(Q^p, P^p) \qquad (40)$$

and here the π_λ^p obey ordinary angular momentum commutation rules

$$\pi_\lambda^p \pi_\mu^p - \pi_\mu^p \pi_\lambda^p = i\hbar\varepsilon_{\lambda\mu\nu}\pi_\nu^p . \qquad (41)$$

The commutators for π_λ^a have been tabulated in [8]. For a spherical top molecule they simplify to:

$$\pi_\lambda^a \pi_\mu^a - \pi_\mu^a \pi_\lambda^a = \frac{1}{2} i\hbar\varepsilon_{\lambda\mu\nu}\pi_\nu^a . \qquad (42)$$

Meyer and Redding [13] have investigated which normal modes in various molecules induce changes in the so-called Eckart vectors (see Louck and Galbraith [14]). In terms of the coordinates introduced here this problem becomes very simple: Q_1, Q_2, Q_3 induce diagonal changes while Q_4, Q_5, Q_6 off-diagonal changes, in the matrix of Eckart vector components. This ends our discussion of the kinematic partitioning of vibration and rotation.

4. CONCLUSIONS AND A SMALL GUIDE TO PARTS OF THE LITERATURE

In this lecture I have stressed the central role of the coordinate transformations in bringing out the physics behind the two central separations of molecular physics: the adiabatic separation of nuclear and electronic motion and the separation of rotation and vibration in reasonably rigid molecules. In doing this I have essentially reviewed a large part of my own work. Many other relevant aspects of these problems have, however, not been mentioned at all up to now, so I will take the opportunity here to point to some work by others of which I happen to be aware.

The traditional Born-Oppenheimer expansion following the $\varkappa = (m_{el}/m_{nuc})^{1/4}$ approach has been exhaustively reviewed by Kiselev [15]. Even if the electronic to nuclear mass ratio is not of central importance for the actual separation a similar mass ratio will still exist for the individual and collective motion quasiparticles resulting from the treatment in section 2a, and it should thus be a valuable parameter for book-keeping in quantitative expansions even from our point of view. A variational derivation of the separation has been given by Longuet-Higgins [16] and Ballhausen and Hansen [17] have tried to clarify the relationship between different approaches. Some of the remaining problems have been reviewed by Sutcliffe [18]. A rather different way of looking at the problem is the generator coordinates method (see e.g. Lathouwers et al. [19]) which has been used in nuclear physics to some extent.

For an account of adiabatic approximations in two-electron atoms see Fano [20]; a recent paper is by Herrick and Kellman [21].

A recent review of the theory of separation of rotational motion from internal can be found in Buck et al. [22]. For discussions of the role and nature of the Eckart conditions, the Eckart vectors and the Eckart frame see [14] and also Louck [23] and Jørgensen [24]. Finally I would like to recommend the classic book Molecular Vibrations by Wilson et al. [25] which recently has appeared in a Dover edition.

REFERENCES

1 Essén, H., Am. J. Phys. 46 (1978) 983
2 Essén, H., Topics in Molecular Mechanics (Thesis), Report 79-08, University of Stockholm, Institute of Physics (1979).
3 Woolley, R.G., Adv. Phys. 25 (1976) 27
4 Woolley, R.G. and Sutcliffe, B.T., Chem. Phys. Lett. 45 (1977) 393.
5 Essén, H., Int. J. of Quant. Chem. 12 (1977) 721.
6 Essén, H., to appear, Int. J. of Quant. Chem.
7 Wilson, F.B. Jr. and Howard, J.B., J. Chem. Phys. 4 (1936) 260
8 Essén, H., Chem. Phys. 44 (1979) 373.
9 Watson, J.K.G., Molec. Phys. 19 (1970) 465.
10 Amat, G. and Henry, L., Cah. Phys. 12 (1958) 273.
11 Watson, J.K.G., Molec. Phys 15 (1968) 479
12 Essén, H., Am. J. Phys. 49 (1981) in press.
13 Meyer III, F.O. and Redding, R.W., J. Mol. Spectry. 70 (1978) 410.
14 Louck, J.D. and Galbraith, H.W., Rev. Mod. Phys. 48 (1976) 69
15 Kiselev, A.A., Canad. J. of Phys. 56 (1978) 615.
16 Longuet-Higgins, H.C. in "Advances in Spectroscopy Vol. 2." (Interscience, London, 1961) 429.
17 Ballhausen, C.J. and Hansen, A.E., in "Annual Reviews of Physical Chemistry" 1972, Editor Eyring. (Annual Reviews Inc., Palo Alto, Cal., USA).
18 Sutcliffe, B.T., in "Computational Techniques in Quantum Chemistry and Molecular Physics (Reidel, Dodrecht, 1975) 1.
19 Lathouwers, L., van Leuven, P., and Bouten, M., Chem. Phys. Lett. 52 (1977) 439.
20 Fano, U., Physics Today 29 (Sept., No 9) (1976) 32.
21 Herrick, D.R. and Kellman, M.E., Phys. Rev. A21 (1980) 418.
22 Buck, B., Biedenharn, L.C. and Cusson, R.Y., Nucl. Phys. A317 (1979) 205.
23 Louck, J.D., J. Mol. Spectry. 61 (1976) 107.
24 Jørgensen, F., Int. J. of Quant. Chem. 14 (1978) 55.
25 Wilson, E.B. Jr., Decius, J.C. and Cross, P.C., Molecular Vibrations (Dover Publ. Inc., New York, 1980).

THE CALCULATION OF NUCLEAR MOTION WAVE FUNCTIONS IN RATHER FLOPPY TRIATOMIC MOLECULES WITH CH_2^+ AS AN EXAMPLE

B.T. Sutcliffe

Department of Chemistry
University of York
York YO1 5DD. England

1.1. INTRODUCTION

The problem that interested us (and by "us" is meant the author and his co-workers, Dr. D. Martin and Dr. R. Bartholomae), was if it was possible, starting from the full Hamiltonian for a triatomic molecule, to develop eigen functions of this full problem which, in some sensible way, represented a molecule that could be said to have a structure.

The molecule CH_2^+ was chosen to experiment on because it is an interesting molecule (at least to astrophysicist) and because simple trial calculations indicated that it was reasonably floppy, at least in bending. That there is little experimental data on the molecule, is a rather mixed blessing, but results were obtained that could be compared with experiment, if and when performed.

It is perhaps as well to say at once that we were not completely successful in our task. We were not in fact able to complete a <u>full</u> calculation, that is one in which no electron-nucleus separation was effected. We were however able to force a solution on the CH_2^+ problem after making the electron nucleus separation.

1.2. THE FORMULATION OF THE PROBLEM

The full Hamiltonian for the molecular problem was considered to be the ordinary Schrödinger Hamiltonian without any relativistic corrections. This Hamiltonian is usually written down in a space-fixed cartesian co-ordinate system, where the classical position vectors of the particles are represented as

$$\bar{x} = \hat{\underline{e}} \, \underline{x}_i \tag{1.2.1.}$$

where $\hat{\underline{e}}$ is a row matrix of cartesian unit vectors $\hat{e}_\alpha$ (α=x,y and z) and $\underline{x}_i$ is a column matrix of components $x_{\alpha i}$.

By introducing the centre of mass co-ordinates

$$\underline{X} = M^{-1} \sum_{i=1}^{N} m_i \, \underline{x}_i \; , \; M = \sum_{i=1}^{N} m_i \; , \tag{1.2.1.}$$

and N-1 new co-ordinates t_i such that.

$$\left(\frac{\underline{X}}{\underline{t}} \right) = \left(\frac{\hat{\underline{m}}}{\underline{V}} \right) x \; , \tag{1.2.2.}$$

it is always possible to re-write the Hamiltonian as

$$\hat{H} = \hat{K}_T + \hat{K}_I + V(\underline{t}_i) \; , \tag{1.2.3.}$$

with

$$\hat{K}_T = - \frac{\hbar^2}{2M} \nabla^2 (\underline{X}) \; , \tag{1.2.4a}$$

$$\hat{K}_I = - \frac{\hbar^2}{2} \sum_{i,j=1}^{N-1} \tilde{G}_{ij} \, \vec{\nabla}(\underline{t}_i) \cdot \vec{\nabla}(\underline{t}_j) \tag{1.2.4b}$$

Thus it is always possible (in the absence of fields) to separate off the centre-of-mass motion and to concentrate attention entirely on the internal motions of the molecule. That this should be possible is essential to any progress in solving the full problem, because the space fixed Hamiltonian, being translationally invariant, has a completely continuous spectrum. The internal Hamiltonian

$$\hat{H}_I = \hat{K}_I + V(\underline{t}_i) \; , \tag{1.2.5.}$$

has an infinite number of bound state solutions[1,2] at least if $V(\underline{t}_i)$ corresponds to a positive ion or a neutral system. (It may have no bound states at all, if it represents a negative ion.)

Of course this separation is not achieved for any arbitrary 3N - 3 by 3N matrix $\underline{V}$, but in the forgoing it has been assumed that $\underline{V}$ is a constant matrix and in that case it is easy to show that any $\underline{V}$ matrix that satisfies the condition

$$\sum_{i=1}^{N} V_{\alpha i, \beta j} = 0 \; , \tag{1.2.6.}$$

will effect the separation.

In equation (1.2.2.)

$$t_{\alpha i} = \sum_{j=1}^{N} \sum_{\beta} V_{\alpha i, \beta j} x_{\beta j} \quad , \tag{1.2.7.}$$

and on the further assumption that the transformation is linear so that

$$V_{\alpha i, \beta j} = \delta_{\alpha \beta} \tilde{V}_{ij} \quad , \tag{1.2.8.}$$

The equation (1.2.4b) is obtained at once using the chain rule with:

$$\tilde{G} = \tilde{V} \, m^{-1} \tilde{V}^T \quad , \tag{1.2.9}$$

where m is diagonal matrix (N by N) of the particle masses. G is (N-1) by (N-1).

The elements of the 3 by 3N matrix partition $\hat{m}$ in (1.2.3.) are seen, by comparison (1.2.2.) with (1.2.3.) to be

$$(\hat{m})_{\alpha, \beta j} = \delta_{\alpha \beta} m_j . M^{-1} . \tag{1.2.10}$$

For the transformation (1.2.2.) to be valid and useful it must be invertable, so that

$$(\underline{x}) = \left(\underline{T} \; \middle| \; \underline{V} \right) \left(\frac{X}{t} \right) , \tag{1.2.11.}$$

where the elements of the 3N by 3 matrix partition $\underline{T}$ are just

$$(\underline{T})_{\alpha i, \beta} = \delta_{\alpha \beta} . \tag{1.2.12.}$$

The precise form of the potential operator is found with the aid of (1.2.11.), and since all the potentials are inverse square in the space fixed frame, it is apparent that they depend only on the t_i when expressed in the new frame. The new frame has cartesian unit vectors parallel to the space fixed unit vectors, and moves with the system defined by the t_i. Because of (1.2.11) it is often convenient to think of the new frame centred on the centre of mass, but the origin is, of course, arbitrary providing that the frame moves with the t_i.

It is perfectly possible, in principle, to compute with $\hat{H}_I$ directly, but since H_I is rotationally invariant, it must contain the description of the rotations of the system as a whole. Since traditionally, one thinks of the energies associated with rotational motion as small compared unit the remaining internal motions (electronic and nuclear) there are clearly advantage to be gained by separation of the rotational motion so that the energy associated with it is not swamped by the other energies in calculation.

The transformations required to separate off rotation, involve a great deal of tedious and intricate algebra and here merely the results of the transformation are presented and explained. More details can be found in references 3 and 4.

The final internal Hamiltonian can be written as

$$\hat{K}_I = \hat{K}_1 + \hat{K}_2 + V(q), \tag{1.2.13.}$$

with

$$\hat{K}_1 = \tfrac{1}{2} \sum_{\alpha\beta} M_{\alpha\beta} \hat{L}_\alpha^T \hat{L}_\beta \frac{\hbar^2}{2} \sum_{k\ell=1}^{3N-6} G_{k\ell}^q \frac{\partial^2}{\partial q_k \partial q_\ell}$$

$$+ \frac{\hbar}{2i} \sum_\delta \sum_{k=1}^{3N-6} (W_{k\delta} \hat{L}_\delta^T \frac{\partial}{\partial q_k} + W_{k\delta} \frac{\partial}{\partial q_k} \hat{L}_\delta^T), \tag{1.2.14a}$$

$$\hat{K}_2 = \frac{\hbar}{2i} \sum_\delta \nu_\delta \hat{L}_\delta^T - \frac{\hbar^2}{2} \sum_k \tau_k \frac{\partial}{\partial q_k} . \tag{1.2.14b}$$

The jacobian for the transformation is, to within a constant factor,

$$|\underline{J}| = |\underline{D}|^{-1} |\underline{M}|^{-\frac{1}{2}} |\underline{G}^q - \underline{WM}^{-1}\underline{W}^T|^{-\frac{1}{2}} .$$

In the above equations

$$\underline{M} = \sum_{i,j=1}^{N-1} \widetilde{G}_{ij} \underline{\omega}^{iT} \underline{\omega}^j , \tag{1.2.15a}$$

$$\underline{G}^q = \sum_{i,j=1}^{N-1} \widetilde{G}_{ij} \underline{Q}^{iT} \underline{Q}^j , \tag{1.2.15b}$$

$$W = \sum_{i,j=1}^{N-1} \widetilde{G}_{ij} \underline{Q}^{iT} \underline{\omega}^j , \tag{1.2.15c}$$

$$\hat{L}_\alpha^T = \hat{L}_\alpha^F + \hat{\pi}_\alpha \tag{1.2.16.}$$

and the ν_δ and τ_k are rather complicated functions of the 3N-6 internal co-ordinates q_k . Formulae for them are given in reference 4

A brief discussion of the transformation will help to explain its status and motivate the definitions of the terms in (1.2.15) and (1.2.16) above.

It is assumed that it is possible to find 3N-6 internal co-ordinates q_k and 3 Euler angles ϕ_m, m = 1,2,3, to describe the motion of the system. In this case one may write

$$\underline{t}_i = \underline{C} \, \underline{z}_i , \tag{1.2.17.}$$

where $\underline{C}$ is an orthogonal matrix (with $|\underline{C}| = \pm 1$) made up from the Euler angles. The internal cartesian $\underline{z}_i$ are functions of the q_k alone. Since these are 3N-3 components $z_{\alpha i}$, not all of these can be independent. There must be three <u>embedding relations</u> among them, of the form

$$f_m(\underline{z}_i) = 0, \; m = 1,2,3,$$
(1.2.18.)

which fix the Euler angles in the system. The embedding relations are satisfied as identities when the $\underline{z}_i$ are expressed in a compatible set of q_k. If the $\underline{z}_i$ are expressed in terms of the $\underline{t}_i$ by (1.2.17) then the embedding relations provide the constraint conditions that define the elements of $\underline{C}$ (and hence the Euler angles) in terms of the $\underline{t}_i$.

The matrix $\underline{C}$ can of course be seen as the matrix that relates the space-fixed frame, into a frame which in some sense, rotates with the system, so that

$$\hat{\underline{\varepsilon}} = \hat{\underline{e}} \; \underline{C}.$$
(1.2.19.)

The sense in which it rotates with the system depends of course on the precise nature of the embedding conditions.

Simply by using the fact that there must be some embedding relation for a valid transformation, it is possible to show that

$$\frac{\partial \phi_m}{\partial t_{\alpha i}} = (\underline{C} \; \underline{\omega}^i \; \underline{D})_{\alpha m},$$
(1.2.20.)

where $\underline{\omega}^i$ can be shown to have elements that are functions of the q_k alone and $\underline{D}$ to have elements that are functions of the Euler angles alone.

The elements of $\underline{\omega}^i$ are defined by using the fact that

$$\frac{\partial}{\partial t_{\alpha i}}(\underline{C}^T \; \underline{C}) = 0,$$
(1.2.21.)

so that

$$\frac{\partial \underline{C}^T}{\partial t_{\alpha i}} \; (\underline{C}^T \underline{C}) = \underline{\Omega}^{\alpha i}$$
(1.2.22.)

where $\underline{\Omega}^{\alpha i}$ is skew symmetric. Choosing $\underline{\Omega}^{\alpha i}$ to have the form.

$$\underline{\Omega}^{\alpha i} = \begin{pmatrix} 0 & -\omega^{\alpha i}_z & \omega^{\alpha i}_y \\ \omega^{\alpha i}_z & 0 & -\omega^{\alpha i}_x \\ -\omega^{\alpha i}_y & \omega^{\alpha i}_x & 0 \end{pmatrix} \tag{1.2.23}$$

then defining the elements of $\underline{\omega}^i$ as

$$\omega^i_{\gamma\delta} = \sum_{\alpha} C_{\alpha\gamma}\omega^{\alpha}_{\delta} , \tag{1.2.24}$$

leads to the form (1.2.20). Of course the $\omega^{\alpha i}_{\gamma}$ and hence the $\omega^i_{\partial\delta}$ are defined only in a formal sense by the above relations. In any practical problem one must choose a set of embedding relations and hence get $\underline{C}$ as a function of the $t_{\alpha i}$ and actually calculate the $\omega^{\alpha i}_{\gamma}$.

Now writing

$$\frac{\partial}{\partial t_{\alpha i}} = \sum_{m=1}^{3} \frac{\partial\phi_m}{\partial t_{\alpha i}} \frac{\partial}{\partial\phi_m} + \sum_{k=1}^{3N-6} \frac{\partial q_k}{\partial t_{\alpha i}} \frac{\partial}{\partial q_k} , \tag{1.2.25a}$$

and using (1.2.19) and (1.2.17) yields

$$\frac{\partial}{\partial t_{\alpha i}} = \frac{i}{\hbar} \sum_{\beta} (\underline{C\omega}^i)_{\alpha\beta} \hat{L}^F + \sum_{k} (\underline{C}(\underline{Q}^i + \underline{\omega}^i\underline{N}))_k \frac{\partial}{\partial q_k} \tag{1.2.25b}$$

Here

$$\hat{L}^F_{\beta} = \frac{\hbar}{i} \sum_{\gamma} D_{\gamma\beta} \frac{\partial}{\partial\phi_{\beta}} , \tag{1.2.26}$$

$$(Q^i)_{\alpha k} = \frac{\partial q_k}{\partial z_{\alpha i}} , \tag{1.2.27}$$

$$(N)_{\delta k} = \sum_{\varepsilon} \sum_{i=1}^{N-1} \hat{z}^i_{\delta\varepsilon} \frac{\partial q_k}{\partial z_{\varepsilon j}} , \tag{1.2.28}$$

Where $\hat{\underline{z}}^i$ in (1.2.28) is a matrix constructed like $\underline{\Omega}^{\alpha i}$ in (1.2.22) but with z_{zj}, in place of $\omega^{\alpha i}_z$, z_{yj} in place of $\omega^{\alpha i}_j$ and so on.

It is a straightforward matter to show that $\hat{L}^F_{\alpha}$ is related to the total internal angular momentum $\hat{L}^I_{\gamma}$ expressed in terms of the $t_{\alpha i}$, by the relation

$$\hat{\underline{L}}^I = -\hat{\underline{CL}}^F \tag{1.2.29}$$

where $\underline{L}^I$ and $\hat{\underline{L}}^F$ are column matrices of the elements L_α^I and L_α^F. The elements of both $\hat{\underline{L}}^I$ and $\hat{\underline{L}}^F$ obey standard commutation rules

$$\frac{i}{\hbar}\hat{\pi}_\alpha = \sum_k N_{\alpha k} \frac{\partial}{\partial q_k} , \tag{1.2.30}$$

(1.2.25) can be re-written as

$$\frac{\partial}{\partial t_{\alpha i}} = \frac{i}{\hbar} \sum_\beta (\underline{C}\omega^i)_{\alpha\beta} \hat{\underline{L}}^T_\beta + \sum_k (\underline{C}\underline{Q}^i)_{\alpha k} \frac{\partial}{\partial q_k} , \tag{1.2.31}$$

and the kinetic energy operators in (1.2.14) follow on substituting (1.2.31) into (1.2.4b).

Now a definite choice of Euler angles does not in fact define the internal co-ordinates q_k uniquely. However it does put certain conditions on their choice. In fact by using the properties of the jacobian for the transformation and its inverse it can be shown that

$$\sum_{i=1}^{N-1} \hat{\underline{z}}^{iT} \underline{\omega}^i = \underline{E}_3 \tag{1.2.32}$$

$$\sum_{i=1}^{N-1} \hat{\underline{Q}}^{iT} \underline{\omega}^i = \underline{O}_{3N-6,3} \tag{1.2.33}$$

where

$$(\hat{\underline{Q}}^i)_{k\alpha} = \partial z_{\alpha i}/\partial q_k \tag{1.2.34}$$

These relations must be true, as identities where $\underline{z}^i$, $\underline{\omega}^i$ and $\hat{\underline{Q}}^i$ are expressed in a properly chosen set of internal co-ordinates. Eq. (1.2.33) can, of course, also be used as a constraint equation in choosing the internals, if $\underline{\omega}^i$ is known formally.

The above Hamiltonian contains all the particles, both electrons and nuclei, but a separation can be effected by choosing the $\underline{t}_i$ in the original transformation as

$$\underline{t}_i = \underline{x}_i - \underline{X}_m, \quad i = 1, N_n - 1 \tag{1.2.35}$$

$$\underline{t}_{i-1} = \underline{x}_i - \underline{X}_m, \quad i = N_n + 1, \dots N$$

where $\underline{X}_m$ is the centre of nuclear mass, and where the nuclei are assumed numbered $1,2...N_n$ and the electrons numbered N_n+1, N_n+2 up to N_1. The first N_n-1 $\underline{t}_i$, $i = N_n...N-1$ involve the electrons alone, relative to the centre-of-nuclear mass. If $3N_n-6$ q_k are then chosen from any the first N_n-1 $\underline{t}_i$ and the remaining $\underline{t}_i$ used directly as internal co-ordinates then $\hat{K}_1$ may be written

$$K_1 = \hat{K}_{11} + \hat{K}_{12} \tag{1.2.36}$$

Here, $\hat{K}_{11}$ has exactly the same form as $\hat{K}_1$ but with sums running up only to $3N_n-6$, and with $\underline{M}$, $\underline{G}$ and $\underline{W}$ functions of the $3N_n-6$ q_k associated with nuclei alone. Unfortunately however $\hat{\Pi}_\alpha$ still contains a contribution from the electronic variables. The operator $\hat{K}_{12}$ just involves the electronic variables and is

$$\hat{K}_{12} = \frac{-\hbar^2}{2\mu} \sum_{i=1}^{N_e} \nabla^2(i) - \frac{\hbar^2}{2M_n} \sum_{ij=1}^{N_e} \vec{\nabla}(i)\vec{\nabla}(j) \tag{1.2.37}$$

where $\mu^{-1} = (m_e^{-1}+M_n^{-1})$ with m_e the electron mass and M_n the total nuclear mass. The implied variables in (1.2.34) are just the N_e electronic variables.

It should be mentioned that it is often more convenient to keep all the particle indices $i = 1,2,...N$ explicit in the expression for the Hamiltonian until the final choice of embedded frame and a consequent choice of internal co-ordinates is made. In a case where all the indices are kept there are, of course, three relations among the $\underline{t}_i$ and six relations among the $\underline{z}_i$ so that the co-ordinates form a redundant set. In such a set there is, strictly, no inversion relation like (1.2.11), but it can be shown that this does not matter. Correct results are obtained simply by extending the sums like (1.2.13) and (1.2.28), that define terms in the Hamiltonians, from $i = 1$ to $N-1$ to $i = 1$ to N. In this case it turns out also that there is no need to actually calculate $\underline{G}$, its elements in the sums are always equivalent to $\delta_{ij} m_i^{-1}$. This observation is parallel to one long known for internal co-ordinates in the Eckart[5] Hamiltonian, where it is the custom to use a redundant co-ordinate set to describe, for example, the valence-angle-bends in methane, in order to preserve the symmetry properties of the wave function.

Finally, it should be pointed out that the jacobian for the transformation is, in general, a function of the ϕ_m and the q_k and that sometimes it is convenient to incorporate its square root, or part of it, into the wave function, thus 're-normalising' the Hamiltonian. The process is exactly analogous to that often performed when treating systems in spherical polar co-ordinates, where the original wave function $R(r)$ is replaced by $P(r) = rR(r)$, incorporating the square root of the r^2 part of the jacobian. An

appropriate change is then made to the kinetic energy operator re-
normalizing it so that its eigenfunction is now $P(r)$. The process
is described in detail by Kemble.[6]

1.3. SPECIFIC CHOICES FOR THE TRIATOMIC

At the start of our work it was our hope (as explained above)
to be able to tackle the triatomic problem without making the usual
separation of electronic and nuclear motion. To this end we
considered a triatomic in the co-ordinate system used by Bhatia and
Temkin [7] in a nuclear physics three body problem. The attractive-
ness of this system arises from the ease in which permutations of a
pair of identical particles may be represented in the internal
co-ordinate system. In this system the three nuclei define a plane
and one axis (the ε_x axis say) bisects the bond angle at the unique
centre. Denoting this bond angle by θ and the two bond lengths by
r_1 and r_2 and the lab-fixed variables of the identical nuclei by $\underline{x}_1$
and $\underline{x}_2$ and of the unique nucleus by $\underline{x}_3$ then

$$\underline{t}_1 = \underline{x}_1 - \underline{x}_3 \tag{1.3.1a}$$

$$\underline{t}_2 = \underline{x}_2 - \underline{x}_3 \tag{1.3.1b}$$

$$r_1 = |\underline{t}_1|, \quad r_2 = |\underline{t}_2|, \quad \cos\theta = |\underline{t}_1^T \underline{t}_2|/r_1 r_2$$

The embedding conditions are then (see (1.2.17) and (1.2.18))

$$\underline{t}_i = \underline{C}\underline{z}_i \tag{1.3.2}$$

such that

$$\underline{z}_1 = r_1 \begin{pmatrix} \cos\theta/2 \\ \sin\theta/2 \\ 0 \end{pmatrix} \qquad \underline{z}_2 = r_2 \begin{pmatrix} \cos\theta/2 \\ -\sin\theta/2 \\ 0 \end{pmatrix} \tag{1.3.3}$$

ensuring that the $\hat{\varepsilon}_x$ axis bisects the bond angle and that the nuclei
define a plane. That (1.3.3) provides two relations for (1.2.18) is
obvious, the third relation is obtained by noticing that

$$(r_2 \underline{z}_1 + r_1 \underline{z}_2)_y = 0$$

If these relations are differentiated with respect to $t_{\alpha i}$ then
an expression for the elements of $\underline{\Omega}^{\alpha i}$ for (1.2.22) can be obtained.
For example

$$\frac{\partial}{\partial t_{\alpha i}} (\underline{C}^T \underline{t}_1)_z = 0 \tag{1.3.4}$$

implies that

$$- \omega_y^{\alpha i} z_{xi} + \omega_x^{\alpha i} z_{yi} + C_{\alpha z} \delta_{1i} = 0 \tag{1.3.5}$$

and from the two remaining relations and the above the $\omega_\gamma^{\alpha i}$ can be determined.

A complete derivation of the Hamiltonian involves much intricate algebra, which is not much to the point here, since having obtained it in its final form it was found not possible to evaluate all the integrals that would be needed in any reasonable basis set. The work began here is continuing, but so far does not look too hopeful.

Attention was therefore turned to the traditional approach to vibration-rotation spectra, which was founded by Eckart[5] and given its modern form by Watson,[8] a form later confirmed by Louck.[9]

In this approach it is assumed that a solution to the clamped nucleus problem for electronic motion provides a set of equilibrium positions for the nuclei. In the rotating frame $\hat{\underline{\varepsilon}}$, these positions are denoted by the vectors

$$\overline{a}_i = \hat{\underline{\varepsilon}} \underline{a}_i, \quad i = 1, 2 \ldots N_n \tag{1.3.6}$$

where $\underline{a}_i$ are constant columns.

Because it is convenient in this approach to specify all the nuclear variables for the $\underline{a}_i$ it is customary to use the redundant set of frame-fixed cartesians

$$\underline{z}_i = \hat{\underline{\varepsilon}} \underline{z}_i \tag{1.3.7}$$

under the constraints that

$$\sum_{i=1}^{N_n} m_i \underline{z}_i = \underline{0} \tag{1.3.8}$$

The three constraints corresponding to (1.3.8) effectively remove the centre-of-mass motion and refer all variables to the centre of nuclear mass.

The three frame embedding conditions for the $\hat{\varepsilon}_\alpha$ are usually written as

$$\sum_{i=1}^{N_n} m_i (\vec{a}_i \times \vec{z}_i) = \vec{0} \tag{1.2.9}$$

but it is convenient for our purposes to re-write this as

$$\sum_{i=1}^{N_n} m_i \underline{A}^i \underline{z}_i = \underline{0} \tag{1.3.10}$$

where $\underline{A}^i$ is like $\hat{\underline{z}}^i$ in (1.2.28), but defined in terms of the $a_{\alpha i}$ rather than the $z_{\alpha i}$. Writing the $\underline{z}_i$ as in (1.2.17) and differentiating the embedding relation with respect to the $t_{\alpha i}$ and using (1.2.22) and (1.2.24) an equation for the $\underline{\omega}^i$ is obtained. This equation is

$$\underline{\omega}^i = m_i \underline{A}^i \underline{B}^{-1} \tag{1.3.11}$$

where

$$\underline{B} = \sum_{j=1}^{N_n} m_j \underline{A}^{jT} \hat{\underline{z}}^j \tag{1.3.12}$$

The internal co-ordinates are now chosen to generate a set of $\underline{z}_j$ that satisfy (1.3.8) and (1.3.9) as identities in this system. They are usually chosen in the form

$$q_\ell = \sum_{i=1}^{N_n} \sum_\alpha m_i^{\frac{1}{2}} \ell_{\alpha i \ell} (z_{\alpha i} - a_{\alpha i}) \tag{1.3.13}$$

where the $\ell_{\alpha i \ell}$ are constants. It is always possible to choose these constants such that

$$\sum_{i=1}^{N_n} \sum_\alpha \ell_{\alpha i k} \, \ell_{\alpha i \ell} = \delta_{k\ell} \tag{1.3.14}$$

while continuing to satisfy (1.3.8) and (1.3.9). If this choice is made then (1.3.13) is easily inverted to yield

$$(z_{\alpha i} - a_{\alpha i}) = m_i^{-\frac{1}{2}} \sum_k \ell_{\alpha i k} q_k \tag{1.3.15}$$

In practice it is the custom to arrive at the $\ell_{\alpha i \ell}$ in two

stages. A non-orthogonal set of internal co-ordinates that satisfy
the Eckart conditions is first found using the Wilson "s-vector"
method, and the $\ell_{\alpha i k}$ are then constructed from these co-ordinates
using the Wilson "F-G matrix" technique. The process is described
in such standard texts as Wilson, Decius and Cross,[10] but it
essentially involves diagonalising a quadratic approximation to the
potential. The resulting co-ordinates are often called (harmonic)
normal co-ordinates.

If such co-ordinates are used and if the part of the jacobian
that depends on them is incorporated into the wave function the
total operator becomes, after much algebra,

$$H = \frac{1}{2} \sum_{\alpha\beta} \mu_{\alpha\beta}(\Pi_\alpha - \pi_\alpha)(\Pi_\beta - \pi_\beta)$$

$$- \frac{\hbar^2}{2} \sum_{k=1}^{3N_n-6} \frac{\partial^2}{\partial q_k^2} - \frac{\hbar^2}{8} \, \mathrm{tr}\, \mu + V \tag{1.3.16}$$

Here conventional notation has been used so that Π_α is just $-L_\alpha^F$
(see 1.2.26) and $\underline{\mu}$ is just $\underline{M}$ (see 1.2.15a) with the ω^i given by
(1.3.11). The matrix $\underline{N}$ simplifies considerably in this scheme so
that the operator π_α (see 1.2.30) can be written as

$$\pi_\alpha = \frac{\hbar}{i} \sum_{k\ell} \epsilon_{k\ell}^\alpha \, q_k \, \frac{\partial}{\partial q_\ell} \tag{1.3.17}$$

with

$$\epsilon_{k\ell}^\alpha = \sum_{i=1}^{N_n} (\vec{\ell}_{ik} \times \vec{\ell}_{i\ell})_\alpha \tag{1.3.18}$$

1.4. CALCULATIONS ON CH_2^+ AND H_2O

In all the calculations described in this section the Eckart
Hamiltonian (1.3.16) was used, with V the potential energy function,
computed by ab-initio methods using the clamped-nucleus electronic
Hamiltonian. The method of solving the vibration problem was
essentially that of Whitehead and Handy[11] and can be summarised
as follows. Starting from some guess at the quadratic part of the
potential V a set of (harmonic) normal co-ordinates q_i were found,
each associated with an eigen-value λ_i. The approximate eigen-
solutions of the Eckart Hamiltonian were then written as the linear
combination

$$\Phi = \sum_{n_1 n_2 n_3} C(n_1, n_2, n_3) \Phi_{n_1 n_2 n_3}^{JK}(\bar{q}_1, \bar{q}_2, \bar{q}_3), \tag{1.4.1}$$

where

$$\phi^{JK}_{n_1 n_2 n_3} = \exp(-\tfrac{1}{2} \sum_{i=1}^{3} q_i^2) P_{JKM}(\phi_1, \phi_2, \phi_3) \prod_{i=1}^{3} H_{n_i}(\bar{q}_i) \quad (1.4.2)$$

Here $\bar{q}_i$ is the dimensionless normal co-ordinate given by $\bar{q}_i = 2\pi(\nu_i/h)^{\frac{1}{2}} q_i$, with $\nu_i = \lambda_i^{\frac{1}{2}}/2\pi$, $H_{n_i}(\bar{q}_i)$ is a Hermite polynomial of order n_i. P_{JKM} is a Jacobi polynomial of the Euler angles and by making the choice $M = 0$ (a choice to which the energy is indifferent) the polynomial reduces to a spherical harmonic. Results for the $J = 0$ state only are reported here, so the precise form of the angular functions is of no concern. In this work q_1 is the harmonic bend, q_2 the symmetric stretch and q_3 the assymetric stretch.

The choice of (1.4.1) for the trial wave-function generates a linear variational problem of a conventional type, precisely analogous to an electronic CI problem. The matrix elements involving the second derivatives of the normal co-ordinates may be evaluated analytically in this case, but those involving μ and the potential V must be done by quadrature. The quadrature scheme used in this work was the rather natural Gauss-Hermite scheme, with a given number of points along each dimensionless normal co-ordinate. Since the normal co-ordinates are known, once the number of points along each co-ordinate is chosen, then the nuclear geometry at which V must be computed, is known.

It is clear that the better the initial guess at the normal co-ordinates, the shorter the expansion (1.4.1) is. The shorter the expansion is, the lower the order of the Hermite polynomials need be and thus the fewer the points that are necessary to obtain an accurate quadrature scheme. Since the potential must be evaluated at each of the quadrature points and such an evaluation is costly, a premium is placed on a good first guess at the normal co-ordinates.

However to get a good set of normal co-ordinates inevitably involves a fair amount of initial trial calculations. In this work the trial calculation of V was carried out using the UHF method. The initial normal co-ordinates were determined on the basis of an assumed geometry by means of a conventional force-constant normal-co-ordinate program. UHF calculations were then carried out at the geometries dictated by the chosen quadrature scheme, initially 6 integration parts in each direction. The maximum value of n chosen in the vibrational function was 5. Full details of the calculations may be found in reference 12.

The results obtained from this initial calculation were recognisable in terms of the input functions in that, for example, one low lying eigen-vector had its main contribution from the simple A_1 bending function ($n_1 = n_2 = n_3 = 0$) and so on. On the basis of this assignment however it was clear that the results were inconsistent with the input. Thus the A_1 bending frequency appeared at about 850 cm^{-1} as compared with the harmonic input frequency of about 1350 cm^{-1}. Of course the fact that the output and input results are inconsistent does not make the output results invalid, but insofar as consistency indicates a good initial basis set and thus makes for a short expansion, consistency is desirable.

To try and achieve consistency a least squares fit was made to the UHF surface and new force constants were determined and the calculations repeated using normal co-ordinates appropriate to these force constants. But again the output was inconsistent with the input. This time however it was the identification of the eigen-vectors which proved the problem. Thus for example the state in which the harmonic ground state function ($n_1 = n_2 = n_3 = 0$) pre-dominated turned out to be the fourth-from lowest energy state and was much lower than the state in which the single bend predominated.

It is perhaps worthwhile pointing out that no such troubles were encountered in parallel calculations on H_2O and so it was felt that this behaviour was due in some way to the intrinsic 'floppiness' of CH_2^+ in its bending mode.

After much further numerical investigation, including a potential calculated from a large CI calculation (again for details se ref. 12 the source of the trouble was identified in the "Watson term", tr$\underline{\mu}$ in (1.3.16).

For the triatomic molecule defined in the x-y plane it is easy to show that μ has the form

$$\mu = \begin{pmatrix} & 0 \\ \underline{\mu}_{2 \times 2} & 0 \\ 0 \quad 0 & \mu_z \end{pmatrix} \tag{1.4.3}$$

where

$$\underline{\mu}_{2 \times 2} = \frac{M}{4m_1 m_2 m_3} \frac{1}{A^2} \begin{pmatrix} K_{xx} & K_{xy} \\ K_{yx} & K_{yy} \end{pmatrix}$$

and

$$\mu_z = \sum_{i=1}^{3} m_i (a_{xi}^2 + a_{yi}^2) / (\sum_{i=1}^{3} m_i (a_{xi} z_{xi} + a_{yi} z_{yi}))^2 \tag{1.4.5}$$

Here M is the total mass of the three particles, each of mass m_i and A is the area of the molecule given by

$$A = \tfrac{1}{2} \frac{\mu}{m_3} \begin{vmatrix} z_{x1} & z_{y1} \\ z_{x2} & z_{y2} \end{vmatrix} = \tfrac{1}{2} \frac{\mu}{m_3} |\underline{R}| \tag{1.4.6}$$

Table 1. The maximum values of $1/|\underline{R}|^2$ on the
sampling points of the Gauss Hermite grids

| Integration | Molecule | |
Scheme	H_2O	CH_2^+
6 4 4	0.416	4.172
6 6 4	0.565	9.273
8 6 6	0.839	1.870E + 3
8 8 6	1.169	1.165E + 4
10 8 8	2.075	1.778E + 4
10 10 8	3.185	4.539E + 5

where the $z_{\alpha i}$ are the cartesian components of the nuclear positions relative to the centre-of-nuclear mass. The elements of $K_{\alpha\beta}$ in (1.4.4) are given by

$$K_{\alpha\beta} = \sum_{i=1}^{3} m_i z_{\alpha i} z_{\beta i} \qquad\qquad (1.4.7)$$

As is easily seen, and as indeed is well known, as the molecule becomes linear A tends to zero so that $\underline{\mu}$ becomes strongly singular. Thus if in the course of integration $\underline{\mu}$ is sampled in a region near its singularity it is to be expected that the results of a calculation will be extremely sensitive to the precise details of the numerical integration scheme and of the basis. That this is what is happening here can be seen from Table 1 where the maximum value of $|\underline{R}|^{-2}$ is displayed as a function of the integration scheme used both for H_2O and CH_2^+.

Table 2. The Polynomial expansion of $|\underline{R}|$ for H_2O and CH_2^+

| Term | Coefficient | |
	H_2O	CH_2^+
1	−2.8132	−2.5042
$\bar{q}_1$	0.19623	0.58880
$\bar{q}_2$	0.40651	0.37655
$\bar{q}_1^{\,2}$	0.033761	0.029826
$\bar{q}_2^{\,2}$	−0.014687	−0.014019
$\bar{q}_3^{\,2}$	0.014278	0.011467
$\bar{q}_1 \bar{q}_2$	−0.014070	−0.038354

The precise form of $|\underline{R}|$ is shown in Table 2 as a polynomial in the scaled normal co-ordinates actually used, together with the equivalent polynomial for the H_2O calculations. Since the coefficent of the bending normal co-ordinate is roughly three times as big in CH_2^+ as it is in H_2O, it is seen that, all other things being equal, the singular region in CH_2^+ will be reached in about one third the extension of the bending co-ordinate as in H_2O.

It is thus reasonable to ask whether it is possible at all to describe CH_2^+ as a basically triangular molecule executing vibrations, since in large amplitude motion the Hamiltonian as conventionally formulated appears to break down. Such a description is of course possible for water, precisely because the vibrational wavefunctions vanish before the singular region in $\underline{\mu}$ is encountered, which is another way of saying that vibrations in H_2O are not of large amplitude. It is clear furthermore that none of the usual large amplitude reformulations of the conventional problem help here, precisely because none of them eliminate the singularity arising from the "Watson term".

One possible way to attempt to answer this question would be to attempt a trial expansion in terms of vibrational functions corresponding to "stiffer" bending modes, so that they vanish in the singular region of $\underline{\mu}$. However this approach, if it were to work, would involve extremely large expansions, and if it did not converge very well not much would be learned anyway.

However another method of "stiffening" the wave function seemed worth a try. Given any trial function, Φ, that gave good results in the problem neglecting tr $\underline{\mu}$, then form from it a new function

$$\Phi = f(|\underline{R}|)\Phi$$

where the function $f(|\underline{R}|)$ is chosen to vanish strongly in the region where $|\underline{R}|$ vanishes. A suitable function will be one which basically leaves unchanged the results of a problem which is ordinarily convergent, but forces convergence on a recalcitrant problem.

After quite a lot of numerical experiment it was found that $|\underline{R}|$ itself seemed a perfectly reasonable choice. In Table 3 the results of calculations on H_2O are shown as a function of basis size which confirm that this choice does not affect an already convergent problem. In Table 4 the results of using the technique on CH_2^+ are shown, for the Hamiltonians with and without the Watson terms. Of course in this Table no comparison with the pure Φ result is sensibly possible, but, for what it is worth the bending band origin calculated from Φ, simply goes on getting bigger as the basis set gets larger and as the number of integration points consequently increase.

Table 3. The convergence of the H_2O CI A1 fundamentals with an increase in the basis size

Basis	basis type Φ		basis type $|\underline{R}|\Phi$	
(n_1,n_2,n_3)	100	010	100	010
$<(5,5,3)$ [a] $\Sigma n_i \leqslant 5$	1630.53	3718.76	1630.81	3718.38
$<(6,6,3)$ [b] $\Sigma n_i \leqslant 6$	1630.56	3718.14	1630.41	3718.20
$<(7,7,3)$ [c] $\Sigma n_i \leqslant 7$	1630.44	3717.85	1630.46	3717.22

Integration schemes   (a)   12 12 12
                         (b)   10 10 8
                         (c)   12 12 12

Table 4. The convergence of the CH_2^+ CI 644 ($\nu_{bend} = 797$ cm^{-1}) Hermite polynomial generated surface A1 fundamentals. Basis $|\underline{R}|n_1n_2n_3$

Basis (n_1,n_2,n_3)	No Watson term		With Watson term	
	100	010	100	010
$\leqslant(4,3,3)$ [a]	955.8	2934.9	930.5	2935.1
$\leqslant(5,3,3)$ [b]	943.8	2934.7	909.7	2935.1
$\leqslant(6,3,3)$ [c]	945.1	2935.0	908.5	2835.4

(a)   Integration scheme 866
(b)   Integration scheme 1088
(c)   Integration scheme 1066

The results of Table 4 are physically reasonable, the predicted bending frequency of 908.5 cm^{-1} is reasonable enough and certainly not an intrinsically stiff bend. Furthermore the function $|\underline{R}|\Phi$ is not negligible when sampled by the integration scheme in the region where μ is large. This is shown by the rather large (35 cm^{-1}) lowering of the bond origin when the tr μ term is included in the Hamiltonian. Furthermore the lowering is stable under increase of the number of integration points and the convergence under basis set increase is not too bad.

Obviously there is a lot more work to be done before anything really definitive can be said about this approach, particularly in cases for which $J \neq 0$ so that μ enters the kinetic energy operator completely (rather than just via μ_z as in the $J = 0$ case). However the method looks promising and the results presented here are perhaps strong enough to enable it to be said that CH_2^+ is a basically triangular molecule, exhibiting large amplitude bending vibrations so that the traditional picture is probably reasonable, at least for $J = 0$ states.

However having said that it is clear that at least to some extent, the structure aspect of CH_2^+ had to be forced upon it by forcing a compact vibrational function on it. That this vibrational function was not rejected by the molecule in the course of computation seems to indicate some physical reality in this case, but it is by no means clear that such forcing is always possible. Interesting cases in which spectroscopic evidence makes it seem possible that it would not be are the NH_2 and NO_2 radicals. Thus the concept of molecular structure remains a slippery one even in these simple cases. That this is now coming to be recognized more widely can be seen by consulting the special issue of the Israel Journal of Chemistry (19 (1980)) in which a substantial proporton of the articles are devoted to a consideration of these problems.

ACKNOWLEDGEMENTS

The author would like to thank the Science Research Council for grants that supported this work. Also Dr N. C. Handy for letting us have a copy of his and Dr Whitehead's vibrational analysis program.

1.   G. M. Zhislin, Dokl. Akad. Nauk SSSR 117: 931 (1957).
2.   J. Uchiyama, Publ. RIMS, Kyoto Univ., A2: 117, 1966.
3.   B. T. Sutcliffe, Ch. 1, in "Quantum Dynamics of Molecules, R. G. Woolley, Ed., Plenum Press, N. York (1980).
4.   B. T. Sutcliffe, Israel J. Chem., 19: 220 (1980).
5.   C. Eckart, Phys. Rev., 47: 552 (1935).
6.   E. C. Kemble, "The Fundamental Principles of Quantum Mechanics", McGraw-Hill, New York (1937).

7. A. K. Bhatia and A. Temkin, Rev. Mod. Phys. Phys. 137A: 1335
    (1965).
8. J. K. G. Watson, Mol. Phys., 15: 479 (1968).
9. J. D. Louck, J. Molec. Spec., 61: 107 (1976).
10. E. B. Wilson, J. C. Decius and P. C. Cross, "Molecular
    Vibrations", McGraw-Hill, N. York (1955).
11. R. J. Whitehead and N. C. Handy, J. Mol. Spec., 55: 356 (1975).
12. R. Bartholomae, D. Martin and B. T. Sutcliffe, J. Mol. Spec.
    (accepted for publication, Nov., 1980).

POTENTIAL ENERGY SURFACES AND SOME PROBLEMS

OF ENERGY CONVERSION IN MOLECULAR COLLISIONS

Lutz Zülicke

Central Institute of Physical Chemistry
Academy of Sciences of G.D.R.
DDR-1199 Berlin-Adlershof

INTRODUCTION

The problem of energy conversion in molecular collisions has several aspects of obvious practical importance: Enhancement of the rates of specific reaction branches by excitation of specific reactant modes opens the possibility of selective activation using, e. g., laser or synchrotron radiation. Deposition of reaction energy in specific product modes is the basis of chemical laser action. Of particular interest is the combination of these two effects, i. e. energy channeling from reactants to products. Finally, nonreactive energy transfer between molecular species leads to activation and deactivation, stabilization and destabilization, it determines the distribution functions under nonequilibrium conditions and therefore plays a decisive role in all fields of gas dynamics and kinetics.

The aim of this lecture is to discuss the connection of energy conversion in molecular collisions with characteristic features of the potential energy surfaces (PES) which govern the processes. Particular emphasis will be given to the choice of the adiabatic separation in a semiclassical description. To give a full account of progress in this field is not possible within the frame of this lecture; therefore, we will present a coarse overview and illustrate some parts by examples.

SEPARATION AND COUPLING OF MODES

The concept of energy conversion is connected with the concept of approximate adiabatic separation of various modes of motion in a molecular system in bound or collision states. Let us consider a bimolecular collision of the type

$$K(k) + L(l) \rightarrow X(k') + Y(l') \tag{1}$$

where $K(k)$, $L(l)$, $X(k')$ and $Y(l')$ are atomic or molecular species in specific quantum states collectively denoted by k, l, k' and l', respectively.

For the elementary process (1) we can formulate a detailed rate equation for formation of the product species X in the state k',

$$(d/dt)\left[X(k')\right] = \sigma_{ab}(kl|E|k'l')u\left[K(k)\right]\left[L(l)\right] , \tag{2}$$

where E denotes the total energy, u is the relative (center of mass) translational energy, and a,b are indices characterizing the two scattering channels (left and right side of eq. (1)). The total cross section of the bimolecular collision process (1), $\sigma_{ab}(m|E|m')$ with $m \equiv (k,l)$, as defined by eq. (2), is related to the transition probability:

$$\sigma_{ab}(m|E|m') = \pi\, k_m^{-2} P_{ab}(m|E|m') \tag{3}$$

with $k_m = (\mu_m/\hbar)u_m$ where μ_m denotes the reduced mass of the colliding partners. This quantity $P_{ab}(m|E|m')$ obeying normalization and symmetry conditions, follows directly from the collision dynamics, it contains measurable information about the exchange of energy among the different modes of the system (characterized by quantum numbers) in going from channel a to channel b.

Phenomenological Description

According to experimental (mainly spectroscopic) evidence of characteristic frequencies (or velocities) lying in different widely spaced regions, we are used to separate phenomenologically several modes of motion: electronic, nuclear vibrational, rotational and translational. The energy E of the system is then approximately written as

$$E = U_{e,n} + E_{vib,nv} + E_{rot,nvj} + E_{tr} ; \tag{4}$$

to each of the corresponding groups of internal degrees of freedom (e, vib, rot), a set of quantum numbers is attributed. This
partitioning corresponds to a hierarchy of adiabatic separations
of the degrees of freedom into "fast" and "slow" subsystems:
(full set) → (electronic) + (nuclear), (nuclear) → (vibration)
+ (rotation, translation), (rotation, translation) → (rotation)
+ (translation). The criterion for the validity of an adiabatic
separation is the smallness of the ratio of characteristic times
τ_q and τ_Q of the fast and the slow subsystem which we describe
by coordinates q and Q, respectively:

$$1/\gamma = \tau_q/\tau_Q \ll 1 , \tag{5}$$

the quantity γ is called the Massey parameter. If this relation
is fulfilled, the motion of the fast subsystem (q) is only parametrically dependent on the coordinates (Q) of the slow subsystem,
and the latter moves under the influence of an average effective
potential caused by the fast subsystem's motion. Usually, relation (5) will be valid only in certain regions of the Q space.
In case that $\tau_q/\tau_Q \approx 1$, the adiabatic separation breaks down
and transitions between quantum states of the q subsystem occur.

In Fig. 1, the energy partitioning in the asymptotic reactant and product regions of a colliding system consisting of an
atom and a diatomic molecule, is shown schematically.

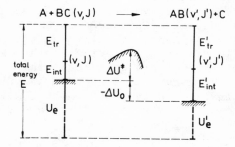

Fig. 1. Energy partitioning for reactants and products of an
A + BC → AB + C collision process. $E_{int} \approx E_{vib} + E_{rot}$
denotes the internal energy of the diatomic, $-\triangle U_0$ is the
reaction energy and $\triangle U^{\ddagger}$ an electronic potential energy
barrier separating reactants and products.

On the basis of this decomposition we may now discuss the
energy conversion problem. In general, exchange of energy between
different modes of motion requires some coupling of the modes.
There are two possibilities: (i) adiabatic coupling allows for
energy exchange between modes keeping fixed the quantum numbers
of those subsystems which are faster than that under consideration; (ii) nonadiabatic coupling connected with changes of the
quantum numbers of the faster subsystems.

Theoretical Approaches

The calculation of probabilities $P_{ab}(m|E|m')$ can be done
in several different ways: In a dynamical treatment one has to
solve the equations of motion of the particles involved; this
gives the most detailed description. Unfortunately, this approach
is limited to simple few-particle systems because of the enormous
number of open channels appearing with increasing number of degrees of freedom. By means of statistical methods, the dynamical
treatment can be circumvented so that the limitations become
less severe. Introducing information-theoretical concepts, certain aspects of the dynamical and statistical treatments can be
combined to a "semithermodynamical" approach; it provides the
desired data (probabilities) in an approximation which deviates
from the statistical limit only so far as to satisfy the most
relevant dynamical constraints of the process. Fundamentals of
these basic approaches to molecular collisions can be found in
recent monographs[1,2].

All these approaches take advantage of the concept of approximate separation of modes. This not only simplifies the calculations but also allows for a more transparent analysis of the
interplay of the various degrees of freedom in a colliding system. Because we are interested in detailed information on the
energetic aspects, we concentrate ourselves on a dynamical treatment of small systems. Furthermore, we use a simple semiclassical (or classical-path) description[3-5] in which the fast subsystem q is described quantum-mechanically and the slow subsystem
Q by classical trajectories. The choice of the q-Q separation is
dictated by the criteria of (i) well-localized q-Q coupling,
(ii) simplicity of the dynamical Q problem and (iii) justifica-
tion of applying classical mechanics to the Q motion. Having
decided about the q-Q subdivision of the degrees of freedom, the
total Hamiltonian $\hat{H}$ is written as

$$\hat{H} = \hat{T}_Q + \hat{H}_q(q,Q) \ , \qquad (6)$$

and the q problem is assumed to be solved in an approximate
way,

$$\hat{H}_q^0 \, \Phi_n^0(q;Q) = U_n^0(Q) \, \Phi_n^0(q;Q) \, , \tag{7}$$

with the Hamiltonian

$$\hat{H}_q^0 = \hat{H}_q - \hat{V} \, . \tag{8}$$

The Q subsystem is described by a classical multidimensional trajectory $Q(t)$ which is a solution of the classical equations of motion, the Hamiltonian function being $H(Q,P) = T(P) + \mathcal{U}(Q)$ with an appropriate potential $\mathcal{U}(Q)$ which, in the simple adiabatic approximation, is to be identified with $U_n^0(Q)$ for a given state n of the q subsystem.

In general, the fast subsystem q is identified with the electronic degrees of freedom (Φ_n^0 = electronic basis), the slow subsystem being the nuclear degrees of freedom. Sometimes, it will be appropriate to include into the fast subsystem also part of the nuclear degrees of freedom, e. g. the vibrational (Φ_n^0 = vibronic basis). In the latter case, we have the advantage to account for a larger part of the quantum effects in the semiclassical treatment and to have a smaller dimension of the dynamical Q problem. On the other hand, the calculation of the vibronic PES $U_n^0(Q)$ is in general much more difficult than that of the electronic PES and the Q dynamics and nonadiabatic coupling becomes more involved because of the large number of coupled states and the complicated crossing behaviour (overlapping crossing regions). Therefore, to use a vibronic representation needs drastic simplifications and will be preferable in special situations only. Such situations are met, for example, if (i) the nonadiabatic regions appear at large distances of the colliding partners so that structure and determination of vibronic PES are comparatively simple, (ii) the energies are sufficiently low so that we have small dynamic widths and, therefore, well-localized crossings.

Factors Determining Molecular Collisions

For an elementary process (1) proceeding from channel a, state m, to channel b, state m′, energetical and dynamical requirements are to be fulfilled[o]. The total energy of the system must be sufficiently high so that along some path from reactants to products the kinetic energy is non-zero everywhere. This condition is not sufficient, however. A reactive or inelastic process requires transfer of momentum between degrees of freedom (modes of motion), this is possible only if the modes are coupled together and if the interaction time is of the same order of magnitude as the reciprocal transition frequency. This coupling and resonance depends on the PES behaviour, the mass ratios of the atoms and the energy distribution among the reactant modes.

ENERGY CONVERSION IN MOLECULAR COLLISIONS

Using the considerations in the preceding section, we give in Table 1 a classification according to the choice of the basis and the process type (adiabatic-nonadiabatic, nonreactive-reactive) together with a brief review of problems studied by means of the different descriptions and approximations. Throughout the following discussion, we confine ourselves to atom-diatomic collisions.

Table 1. Classification of Description Schemes

Basis	q-Q Coupling		Problems
electronic	adiabatic	nonreactive	energy transfer: TV, RV, TRV, VV
		reactive	energy disposal
			energy consumption
			energy channeling
			branching ratios
	nonadia-	nonreactive	VE
	batic	reactive	
vibronic	adiabatic	nonreactive	TR
	nonadia-	nonreactive	VT, VE
	batic	reactive	intermediate energy conversion

Electronic Basis

In the first category of the scheme of Table 1, electronically <u>adiabatic</u> nonreactive processes, there are several powerful model theories for describing energy transfer between translational (T), vibrational (V) and rotational (R) modes as reviewed in the literature[7]. In the semiclassical picture adopted here, the conditions mentioned above are easily verified, and the dynamics is clearly understood. This holds also for a quan-

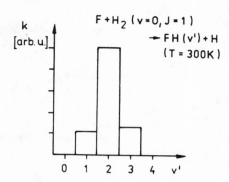

Fig. 2. Relative rates of formation of product vibrational states
v' of HF for reactant H$_2$ in state v=0, J= 1 (thermal distribution of collision energies, T = 300 K). Data taken
from Polanyi and Schreiber.[9]

tum-mechanical or quasiclassical (classical-limit) description
which can be carried out for simple systems without any serious
difficulty.

For electronically adiabatic reactive processes, we likewise have a rather detailed knowledge of questions like deposition of reaction energy into specific product modes, consumption
of reactant energy in endoergic processes, transformation of
specific reactant energy forms into specific product energy forms
and influence of the form of reactant energy on branching ratios.
In addition to systematic model studies of Polanyi and his
school[8], a number of realistic prototypes of systems has been
investigated both experimentally and theoretically. For exoergic
processes, the basic effect of transforming (electronic) reaction energy E_e into product vibrational excitation energy E'_{vib}
(EV conversion) is best illustrated by the F + H$_2$ → FH + H
reaction (Fig. 2). The average product vibrational excitation
energy $\langle E'_{vib} \rangle$ can be correlated with a characteristic feature
of the potential energy surface: the percentage of attractive
(during the approach of the reactants) and mixed (during the
tight interaction) release of the reaction energy. Concerning
the energetical and dynamical requirements of the F + H$_2$ reaction, in agreement with the general Polanyi rules translational
energy is effective in surmounting the small barrier of about
7 kJ/mol in the reactant valley (Fig. 3).

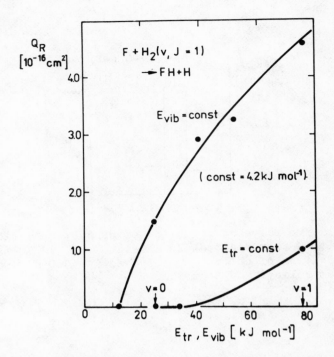

Fig. 3. Cross section of the F + H2(J=1) reaction in dependence
on the reactant relative translational and vibrational
energy. Data taken from Polanyi and Schreiber.[9]

        Concerning the pecularities of endoergic, electronically
adiabatic reactive processes, the He + H_2^+ → HeH^+ + H rearrangement has been studied in detail both experimentally and
theoretically.[10-12] In this case, the endoergicity barrier of
about 0.8 eV is located late in the product valley; therefore,
according to Polanyi's rules, vibrational energy should be more
effective in promoting the reaction than translational energy.
Trajectory calculations show this behaviour clearly. Only for
the strongly endoergic processes with $v \leq 2$, translational energy
favours rearrangement, for higher v it even hinders (Fig. 4).
As can be inferred from the angular distributions, for the
endoergic processes with $v \leq 2$ hard, near-central collisions make
the translational energy conversion possible which is necessary
for overcoming the potential barrier; accordingly, this results
in dominant backward and sideway scattering. For the higher vibrational states, the mechanism is changed and reaction comes

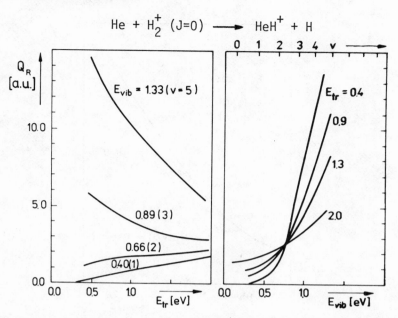

Fig. 4. Cross section of the He + H$_2^+$(J=0) reaction in dependence on the reactant relative translational and vibrational energy.[12]

prevailingly from glancing collisions so that we see mainly forward scattering.[12]

Whereas the effects of reactant translational and vibrational energy are comparatively well understood, there is far less information about the role of reactant rotational energy.[13] In a few well-documented examples, different influence of reactant rotational excitation is found and not in all cases a conclusive interpretation has been possible. Dynamically, the rotational effect is connected with the steric requirements of the reaction, the anisotropy of the potential energy surface; it occurs if the rotational period and the energy are of the same order of magnitude as the characteristic time and the energy differences, respectively, of other modes of motion. We found in the case of He + H$_2^+$ very strong rotational excitation to have approximately the same effect as vibrational excitation of the same energy (Fig. 5).

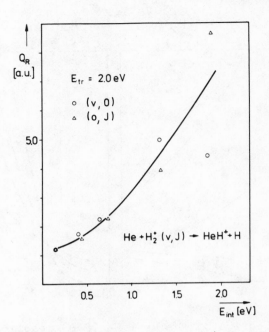

Fig. 5. Total cross section of the He + H$_2^+$ reaction in dependence on E$_{int}$ for pure vibrational and pure rotational
excitation.[12]

Electronically <u>nonadiabatic</u> processes have been treated so
far mainly within the semiclassical (classical-path) approach by
the surface-hopping trajectory (SHT) technique.[4] The simplest
situation we have probably in the case of harpooning processes
between an alkali atom (M) and a halogen molecule (X$_2$); Fig. 6
shows a schematic diagram of diabatic PES for the lowest covalent
and ionic electronic states. In Fig. 7 a branching trajectory
at the crossing seam of the (in this region) approximately parabolic covalent term with the nearly plane ionic term is schematically drawn. Calculations of several simple atom-diatomic collision problems, nonreactive as well as reactive, gave satisfying
results, the most detailed investigations being those for the
isotopic H$^+$ + H$_2$ reactions.[14]

The SHT method suffers from severe restrictions, for example
neglect of tunneling and interference effects, validity for a
certain energy range only etc. Nevertheless, the basic feature of
electronic-to-nuclear energy conversion is qualitatively accounted
for. The deficiencies of SHT can be partly overcome by generalizations like complex-trajectory approaches (classical S matrix)[15]
or other, more recent classical-limit methods[16]. Full quantum

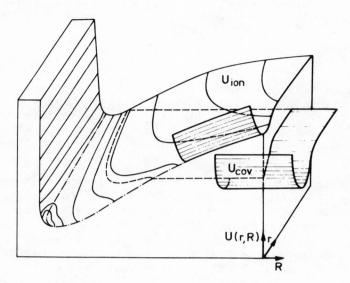

Fig. 6. Lowest covalent and ionic diabatic PES for a system
M + X_2 (schematically).[18]

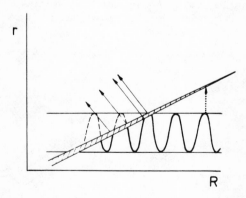

Fig. 7. Branching trajectory in an (R,r) plane. The hatched
stripe indicates the crossing seam of the parabolic
covalent term with the plane ionic term; arrows represent
trajectory branches on the ionic term. The outer right
dotted arrow corresponds to the real part of a complex
trajectory reaching the crossing seam in a classically
forbidden region.[19]

treatments of electronically nonadiabatic processes are yet at
the beginning.[17]

Vibronic Basis

Using a vibronic basis, it is in principle possible to take
into account some important quantum effects connected with vibrational motion. Vibronically <u>adiabatic</u> processes determine TR conversion.[7] In reactive collisions, they do not play any role in
general because vibronic adiabacity is hardly maintained in rearrangement.

Vibronically <u>nonadiabatic</u> nonreactive processes are the
elementary steps of TV and TRV conversions.[7] In the case of a
nondegenerate electronic state (Fig. 8a), a vibrational transition can be imagined as a tunneling process between the two nonintersecting adiabatic vibronic terms (Landau-Teller transition).
If we have a degenerate electronic state and the degeneracy is
removed by the intermolecular interaction, two mechanisms are
possible for TV conversion (Fig. 8b): the tunneling mechanism
without change of the electronic state and the crossing (or
pseudocrossing) mechanism with change of the electronic state,
the latter being described, in the simplest case, by the Landau-
Zener approximation. If the corresponding transition regions are
well separated, both contributions can be treated as independent;
they lead to different temperature dependence of the vibrational
deactivation probability. Using such an approach, it was possible
to explain qualitatively the order of magnitude as well as the

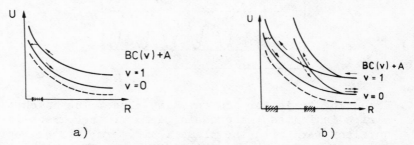

a)   b)

Fig. 8. Schematic diagrams (taken from Andreev and Nikitin[7b]) of
the lowest vibronic states of a system A + BC as function
of R(A-BC) for asymptotically a) nondegenerate electronic
state, b) degenerate electronic state. The hatched sections
on the abscissa indicate the nonadiabatic transition regions

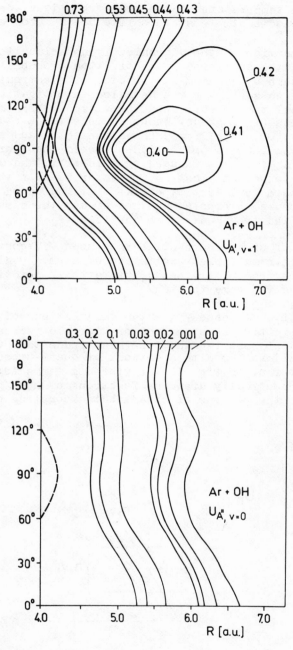

Fig. 9. Vibronic PES of the system $Ar(^1S) + OH(^2\Pi)$ in states
(A',v=1) and (A", v=0); energies in eV. The dashed curve
indicates the crossing.

variation with temperature of the mean vibrational deactivation
probability of some simple systems like O + N_2, for example. [7b,20]

On the dynamics of the vibronically nonadiabatic contribution
to the one-quantum collisional deactivation probability P_{10} of a
diatomic molecule by an atom in the case of electronic degeneracy,
not much is known so far. For a detailed investigation, it is necessary to calculate the corresponding vibronic PES in the energetically accessible regions. We began recently a study[21] of the
system $Ar(^1S) + OH(^2\Pi)$, Fig. 9 shows the third and second vibronic PES. The nonadiabatic transitions between the vibronic PES
(A',v=1) and (A'',v=0) are induced by spin-orbit and Coriolis interactions. As expected, most of trajectories with thermal energies are of a simple in-out type like that indicated in Fig. 8b.
It seems not shure at present whether or not the assumptions made
in the simplified treatment[20] are justified.

A number of other nonreactive processes (EV conversion:
quenching of intramultiplet excitation of atoms by diatomic molecules or the reverse VE process of vibrational deactivation) can
be described on the same footing. [7b]

For reactive processes, quantum calculations of close-coup-
ling type implicitly use some kind of vibronic representation. A
semiclassical (classical-path) treatment leads to a very complicated dynamical problem with a network of quasi-crossing or crossing vibronic terms. In Fig. 10 the vibronic terms for the M + X_2
system are schematically drawn as functions of $R(M-X_2)$ at large
R. In general, the regions of nonadiabatic coupling cannot be

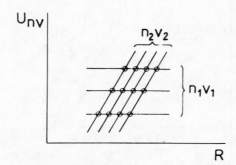

Fig. 10. Vibronic terms corresponding to the electronic terms
        of Fig. 6.[19] The circles indicate regions of quasicross-
        ing (in the adiabatic representation) or crossing (in a
        diabatic representation).

considered localized and nonadiabatic transitions are not independent if the static and dynamic widths of these regions are of
the same order of magnitude as the distance between two crossings.
If nonadiabatic transitions would be localized and independent,
some kind of SHT procedure could be applied, in the simplest case
a multiple (pseudo-) crossing approach [19,22-24] and in the limit
of quasicontinuous spectrum a semiclassical optical model[25]. A
situation appropriate for such treatment is met in the case of
harpooning processes where the crossings occur at large distances
and the dynamics can be treated as a three-step process: vibronically adiabatic nuclear motion, electron jump (described in a vibronic representation by the semiclassical optical model[25]) and
subsequent nuclear motion described in an electronic representation. In this way it has been possible[18,19] to explain properly
the conversion, connected with the electronic transition, of
electronic energy into vibrational energy of the molecular ion
X_2^-, followed by the vibrational deactivation of X_2^- during the
interaction with M^+ (Fig. 11), or by rearrangement to $MX + X$ or
by dissociation to $M^+ + X^- + X$.

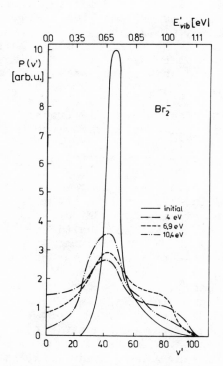

Fig. 11. Distribution of vibrational states v' of Br_2^- at several collision energies (normalized to the same area under
the curves); full curve: initial distribution immediately
after the electronic transition.

CONCLUSIONS

In a semiclassical treatment of molecular collisions, the description should be fitted to the problem by a proper choice of the adiabatic separation. Applying an appropriate representation, the treatment can be simplified and improved. Although quite complicated in general, a vibronic basis will be advantageous in special situations, as shown by the examples of the present paper. The connection between energy partitioning and conversion with characteristics of the PES is rather complicated, most certain information and generalizations have been achieved so far only for the case of electronically adiabatic processes.

REFERENCES

1. W. H. Miller, ed., "Dynamics of Molecular Collisions", Parts A and B (Modern Theoretical Chemistry, Vol. 1 and 2), Plenum Press, New York and London (1976).
2. R. B. Bernstein, ed., "Atom-Molecule Collision Theory. A Guide for the Experimentalist", Plenum Press, New York and London (1979).
3. R. N. Porter and L. M. Raff, Classical Trajectory Methods in Molecular Collisions, in: Ref. 1, Part B.
4. J. C. Tully, Nonadiabatic Processes in Molecular Collisions, in: Ref. 1, Part B.
5. E. E. Nikitin and L. Zülicke, "Selected Topics of the Theory of Chemical Elementary Processes" (Lecture Notes in Chemistry, Vol. 8), Springer-Verlag, Berlin/Heidelberg/New York (1978).
6. P. J. Kuntz, Features of Potential Energy Surfaces and Their Effect on Collisions, in: Ref. 1, Part B.
7. a) E. E. Nikitin, Theory of Energy Transfer in Molecular Collisions, in: "Physical Chemistry", Vol. VI A, Kinetics of Gas Reactions, F. Jost ed., Academic Press, New York/San Francisco/London (1974).
    b) E. A. Andreev and E. E. Nikitin, Transfer of Vibrational and Electronic Energy in Atomic and Molecular Collisions, Khimiya Plasmy 3:28 (1976) (in Russian).
8. J. C. Polanyi and J. L. Schreiber, The Dynamics of Bimolecular Reactions, in: "Physical Chemistry", Vol VI A, Kinetics of Gas Reactions, F. Jost ed., Academic Press, New York/San Francisco/London (1974).
9. J. C. Polanyi and J.L. Schreiber, The Reaction of $F + H_2 \rightarrow HF + H$. A Case Study in Reaction Dynamics, Faraday Disc. Chem. Soc. 62:267 (1977).
10. F. Schneider, U. Havemann, L. Zülicke, V. Pacák, K. Birkinshaw and Z. Herman, Dynamics of the Reaction $H_2^+(He;H)HeH^+$,

Comparison of Beam Experiments with Quasi-Classical Trajectory Studies, Chem. Phys. Letters 37:323 (1976).

11. W. N. Whitton and P. J. Kuntz, Trajectory Calculation of the Effectiveness of Reagent Vibration in the $H_2^+ + He \rightarrow HeH^+ + H$ or $He + H + H^+$ Reactions, J. Chem. Phys. 64:3624 (1976).

12. Ch. Zuhrt, F. Schneider, U. Havemann, L. Zülicke and Z. Herman, Dynamics of the Reaction $H_2^+(He;H)HeH^+$. Influence of Various Forms of Reactants Energy on the Total and Differential Cross Section, Chem. Phys. 38:205 (1979).

13. R. D. Levine and J. Manz, The Effect of Reagent Energy on Chemical Reaction Rates: An Information Theoretic Analysis, J. Chem. Phys. 63:4280 (1975).

14. J. R. Krenos, R. K. Preston, R. Wolfgang and J. C. Tully, Molecular Beam and Trajectory Studies of Reactions of $H^+ + H_2$, J. Chem. Phys. 60:1634 (1974).

15. W. H. Miller, The Classical S-Matrix in Molecular Collisions, Advan. Chem. Phys. 30:77 (1975).

16. W. H. Miller and C. W. McCurdy, Classical Trajectory Model for Electronically Nonadiabatic Collision Phenomena. A Classical Analog for Electronic Degrees of Freedom, J. Chem. Phys. 69:5163 (1978).

17. Z. H. Top and M. Baer, Incorporation of Electronically Nonadiabatic Effects into Bimolecular Reactive Systems, J. Chem. Phys. 66:1363 (1977); Chem. Phys. 25:1 (1977).

18. U. Havemann, L. Zülicke, A. A. Zembekov and E. E. Nikitin, Simple Semiclassical Optical Model of Harpooning Elementary Processes $K + Br_2$, Chem. Phys. 41:285 (1979).

19. A. A. Zembekov, E. E. Nikitin, U. Havemann and L. Zülicke, Dynamics of Harpooning Reactions, a Prototype of Chemi-Ionization Processes, Khimiya Plasmy 6:3 (1979) (in Russian).

20. E. E. Nikitin and S. Ya. Umansky, Effect of Vibronic Interaction upon the Vibrational Relaxation of Diatomic Molecules, Faraday Disc. Chem. Soc. 53:7 (1972).

21. Ch. Zuhrt, S. Ya. Umansky et al., to be published.

22. E. Bauer, E. R. Fisher and F. R. Gilmore, De-Excitation of Electronically Excited Sodium by Nitrogen, J. Chem. Phys. 51:4173 (1969).

23. E. A. Gislason and J. G. Sachs, Multiple-Crossing Electron Jump Model for Reactions of Metal Atoms with Diatomic Halogen Molecules, J. Chem. Phys. 62:2678 (1975).

24. Yu. N. Demkov and V. I. Osherov, Stationary and Nonstationary Quantum Problems Solvable by the Contour Integral Method, JETP 53:1589 (1967).

25. E. E. Nikitin, Optical Model for Spectator-Stripping Reactions, Chem. Phys. Letters 1:266 (1967).

DERIVATIVE COUPLING ELEMENTS IN ELECTRONICALLY ADIABATIC

REPRESENTATIONS AND THEIR USE IN SCATTERING CALCULATIONS

Bruce C. Garrett[*]

Battelle Columbus Laboratories
505 King Avenue
Columbus, OH 43201

Donald G. Truhlar

Department of Chemistry
University of Minnesota
Minneapolis, MN 55455

and Carl F. Melius

Sandia Laboratories
Livermore, CA 94550

INTRODUCTION

The Born-Oppenheimer electronically adiabatic basis provides the most convenient representation for obtaining molecular-structure and potential-energy-surface information for systems of chemical interest. In this representation the electronic Hamiltonian is diagonal, and the adiabatic energies may be defined and calculated accurately by the variational principle. For systems in which the adiabatic states are well separated in energy, the nuclear motion at chemical energies can be treated adequately using a single adiabatic potential energy surface. For systems in which the coupling of electronic states is important, the coupling can be included consistently in this

---

[*]Present address: Chemical Dynamics Corporation, 1550 West Henderson Road, Suite N-140, Columbus, OH 43220.

representation through matrix elements of the nuclear-motion derivative operators.

A diabatic representation is any representation in which the
electronic Hamiltonian is not diagonal. As compared to adiabatic
representations, suitably chosen diabatic basis sets may have the
advantage of providing representations that are more uncoupled for
high-energy collisions and narrowly avoided adiabatic curve crossings.
These representations can also be used to define potential energy surfaces which are more smoothly varying functions of internuclear coordinates than the adiabatic surfaces are. A further advantage of
diabatic representations is that they may be chosen so that the
coupling of electronic states is dominated by potential coupling
terms rather than nuclear-motion derivative terms.

Diabatic bases often provide the most convenient representation
for performing scattering calculations in which the electronic states
are coupled; however, as already mentioned, adiabatic bases are the
most convenient for obtaining accurate potential energy surfaces. A
review and discussion of selected aspects of the coupling of electronically adiabatic and diabatic states has been presented elsewhere.[1]
In this article we examine one of the fundamental problems of obtaining accurate derivative coupling terms in adiabatic representations —
the fact that the derivative coupling terms depend upon the location
of the origin of the electronic coordinates.[2] A further complication
is that for an arbitrary origin of the electronic coordinates the
derivative coupling terms do not in general vanish at large internuclear separations. Some methods for dealing with this long-range
coupling are discussed in this article.

To illustrate these considerations we consider atom-atom collisions; however, the fundamental problems are general ones, and
similar considerations are required for atom-molecule scattering.
The particular examples chosen for the present study are collisions
of ^{39}K with protium (^{1}H) and with a fictitious heavy isotope of
hydrogen (^{39}H). For these examples the derivative coupling terms
are examined, and scattering calculations for several different
choices of electronic origin are compared.

COUPLED-CHANNEL EQUATIONS

General Theory

A fundamental problem in defining the derivative coupling terms
is their dependence upon the choice of origin for the electronic
coordinates.[2] For collisions at energies of chemical interest of
systems with narrowly avoided adiabatic curve crossings, the origin

dependence of the coupling terms is expected to have only a small
effect on the calculated cross sections. However, for systems with
weak coupling or nonlocalized interactions or for high collision
energies, the origin dependence of the derivative coupling terms is
more important. Although the choice of origin has been discussed in
the context of high-energy ion-atom collisions,[3] the effects of the
origin dependence of the derivative coupling terms have not been
tested for collisions at chemical energies. In this section we write
the quantum mechanical coupled-channel equations for a general choice
of origin for the electronic degrees of freedom, and we discuss some
methods of correctly imposing asymptotic boundary conditions on the
solutions of these equations. Further discussion of these topics is
presented elsewhere.[1]

We consider a system of two nuclei A and B of masses m_A and m_B
respectively and N electrons of mass m_e, with the center of mass at
rest. Spin is neglected, the charges of A and B are assumed different, and ionization processes are excluded. We define the relative
internuclear coordinate $\underset{\sim}{R}'$ in a space-fixed system as the vector from
nucleus A to nucleus B. (We deonte coordinates in space-fixed
systems with primes, and we will denote the magnitude of a vector $\underset{\sim}{x}'$
by x'.) The vector from nucleus A to electron i is denoted $\underset{\sim}{r}'_{Ai}$.
The origin for electron i can be redefined to any point along the
internuclear axis by the definition

$$\underset{\sim}{r}'_{\eta i} = \underset{\sim}{r}'_{Ai} - n_i \underset{\sim}{R}' \tag{1}$$

where n_i is an arbitrary number. For $n_i = 0$ the origin of electron i
is at nucleus A, whereas for $n_i = 1$ the origin is at nucleus B. For
the set of coordinates $R', \{r'_{\eta i}\}_{i=1}^N$, which we will denote by the short-
hand $(\underset{\sim}{R}', \underset{\sim}{x}_\eta)$, the total Hamiltonian for the electronic and nuclear
degrees of freedom is

$$H = H_e - \frac{\hbar^2}{2\mu_{AB}} \nabla^2_{\underset{\sim}{R}'_\eta} - \sum_{i=1}^N \hbar^2 \left(\frac{1 - n_i}{m_A} - \frac{n_i}{m_B}\right) \underset{\sim}{\nabla}_{\underset{\sim}{R}'_\eta} \cdot \underset{\sim}{\nabla}_{\underset{\sim}{r}'_{\eta i}}$$

$$- \sum_{i,j=1}^N \frac{\hbar^2}{2} \left[\frac{(1 - 2n_i + n_i n_j)}{m_A} + \frac{n_i n_j}{m_B}\right] \underset{\sim}{\nabla}_{\underset{\sim}{r}'_{\eta i}} \cdot \underset{\sim}{\nabla}_{\underset{\sim}{r}'_{\eta j}} \tag{2}$$

where the electronic Hamiltonian is defined as

$$H_e = - \frac{\hbar^2}{2m_e} \sum_{i=1}^N \nabla^2_{\underset{\sim}{r}'_{\eta i}} + V(\underset{\sim}{x}'_\eta, \underset{\sim}{R}') \tag{3}$$

and

$$\mu_{AB} = \frac{m_A m_B}{m_A + m_B} \tag{4}$$

$\underset{\sim}{R}'_\eta$ is the same as $\underset{\sim}{R}'$, however $\underset{\sim}{\nabla}_{R'_\eta}$ denotes that $\{\underset{\sim}{x}'_{\eta i}\}_{i=1}^N$ are fixed during the variation of $\underset{\sim}{R}'$.

The Born-Oppenheimer electronically adiabatic space-fixed basis functions are the solutions of

$$H_e \, \phi_{n\alpha}^{Sa}(\underset{\sim}{x}'_\eta, \underset{\sim}{R}') = \varepsilon_\alpha^a(R) \, \phi_{n\alpha}^{Sa}(\underset{\sim}{x}'_\eta, \underset{\sim}{R}') \tag{5}$$

where η denotes the collection of n_i values, α is a collective index for the electronic quantum numbers, and $\varepsilon_\alpha^a(R)$ is the adiabatic energy which depends only upon the magnitude of R ($\equiv R'$) of $\underset{\sim}{R}'$.

The coupled-channel scattering equations can be obtained by expanding the total wavefunction Φ_{α_0} in the Born-Oppenheimer basis functions

$$\Phi_{\alpha_0} = (1/R') \sum_\alpha \phi_{n\alpha}^{Sa}(\underset{\sim}{x}'_\eta, \underset{\sim}{R}') \, g_{n\alpha\alpha_0}^a(\underset{\sim}{R}') \tag{6}$$

and substituting this expansion into the Schroedinger equation where the Hamiltonian is given by equation (2). The index α_0 indicates the initial quantum state of the system. The coupled equations are then given by

$$[-\frac{\hbar^2}{2\mu_{AB}} \nabla_{R'}^2 + \varepsilon_\alpha^a(R) - E] \, g_{n\alpha\alpha_0}^a(\underset{\sim}{R}') = \sum_\beta \, [\frac{\hbar^2}{2\mu_{AB}} \, (<\phi_{n\alpha}^{Sa}|\nabla_{R'_\eta}^2|\phi_{n\beta}^{Sa}>_{x'_\eta}$$

$$+ 2<\phi_{n\alpha}^{Sa}|\underset{\sim}{\nabla}_{R'_\eta}|\phi_{n\beta}^{Sa}>_{x'_\eta} \cdot \underset{\sim}{\nabla}_{R'}) + \hbar^2 \sum_{i=1}^N \, (\frac{1-n_i}{m_A} - \frac{n_i}{m_B})$$

$$\times \, (<\phi_{n\alpha}^{Sa}|\underset{\sim}{\nabla}_{r'_{\eta i}} \cdot \underset{\sim}{\nabla}_{R'_\eta}|\phi_{n\beta}^{Sa}>_{x'_\eta} + <\phi_{n\alpha}^{Sa}|\underset{\sim}{\nabla}_{r'_{\eta i}}|\phi_{n\beta}^{Sa}>_{x'_\eta} \cdot \underset{\sim}{\nabla}_{R'})$$

$$+ \frac{\hbar^2}{2} \sum_{i,j=1}^N \, (\frac{1-2n_i - n_i n_j}{m_A} + \frac{n_i n_j}{m_B}) <\phi_{n\alpha}^{Sa}|\underset{\sim}{\nabla}_{r'_{\eta i}} \cdot \underset{\sim}{\nabla}_{r'_{\eta j}}|\phi_{n\beta}^{Sa}>_{x'_\eta}]$$

$$\times \, g_{n\beta\alpha_0}^a(\underset{\sim}{R}') \tag{7}$$

where the subscript on a matrix element denotes the variables integrated over. Equation (7) is general for any choice of origins for
the electronic degrees of freedom. The terms on the right hand side
of equation (7) are the nonadiabatic coupling terms; and we are free
to use any choices of origin for these terms. A convenient choice is
that which eliminates the cross terms between the nuclear and electronic kinetic energy. This choice is the center of mass of the
nuclei (CMN) and is specified by setting all n_i to the value

$$C = \frac{m_B}{m_A + m_B} \tag{8}$$

The first-derivative coupling term for any arbitrary choice of origin
is related to the coupling term for the origin at the CMN by

$$\langle \phi_{C\alpha}^{Sa} | \nabla_{R_C'} | \phi_{C\beta}^{Sa} \rangle_{x_C'} = \langle \phi_{n\alpha}^{Sa} | \nabla_{R_\eta'} | \phi_{n\beta}^{Sa} \rangle_{x_\eta'} + \sum_{i=1}^{N} (C - n_i)$$

$$\times \langle \phi_{n\alpha}^{Sa} | \nabla_{r_{\eta i}'} | \phi_{\nu\beta}^{Sa} \rangle_{x_\eta'} \tag{9}$$

We emphasize that the origin used in defining the first-derivative
coupling matrices should be consistent with the origin used for the
other terms in the coupled equations.

The coupled equations (7) are useful for displaying how the
choice of electronic origin affects the various coupling terms; however, this form of the equations is impractical for computations.
Computationally convenient scattering equations can be obtained using
a body-fixed coordinate system (R, x_η) in which the z-axis lies along
the internuclear axis. (Body-fixed coordinates are denoted without
primes.) The total scattering wavefunction is expanded in coefficients of the irreducible representations of the rotation group and
the Born-Oppenheimer basis functions.[4] The coefficients of this
expansion are the radial wavefunctions for relative motion of the
nuclei. The details of this procedure as well as the dependence of
the body-fixed coupled-channel radial equations upon electronic origin
are given elsewhere.[1] We will present here approximate equations in
which the mass polarization term [i.e., the cross terms in the electronic kinetic energy, the last term of equation (7)] and the angular
coupling terms are neglected. The effect of neglecting the angular
coupling terms for collisions at thermal energies requires further
study; in the present article we are not interested in obtaining
totally converged cross sections for a specific problem but instead
are interested in examining the importance of the origin dependence
of the derivative coupling elements in physically realistic model
systems. By neglecting angular coupling we neglect coupling Σ states

to Π states in atom-atom collisions; then the body-fixed radial equations for Σ states, with the electronic origin at the CMN, are given by

$$[- \frac{\hbar^2}{2\mu_{AB}} \frac{d^2}{dR^2} + \frac{\hbar^2 \ell(\ell+1)}{2\mu_{AB} R^2} - E] \chi^a_{C\alpha\alpha_0}(R) + \sum_\beta [H^a_{\alpha\beta}(R) + 2F^a_{C\alpha\beta}(R) \frac{d}{dR}$$

$$+ G^a_{C\alpha\beta}(R)] \chi^a_{C\beta\alpha_0}(R) = 0 \tag{10}$$

where

$$H^a_{\alpha\beta}(R) = \langle \phi^a_{C\alpha} | H_e | \phi^a_{C\beta} \rangle_{\chi} = \varepsilon^a_\alpha(R) \delta_{\alpha\beta} \tag{11}$$

$$F^a_{C\alpha\beta}(R) = - \frac{\hbar^2}{2\mu_{AB}} \langle \phi^a_{C\alpha} | (\frac{\partial}{\partial R})_{\chi} | \phi^a_{C\beta} \rangle_{\chi} \tag{12}$$

$$G^a_{C\alpha\beta}(R) = - \frac{\hbar^2}{2\mu_{AB}} \langle \phi^a_{C\alpha} | (\frac{\partial^2}{\partial R^2})_{\chi} | \phi^a_{C\beta} \rangle_{\chi} \tag{13}$$

and χ is χ_C. We have denoted the body-fixed electronic wavefunctions as $\phi^a_{C\alpha}(\chi,R)$, a special case of $\phi^a_{\eta\alpha}(\chi_\eta,R)$.

The simple form of equation (10) is a result of taking the origin of the electronic degrees of freedom to be the center of mass of the nuclei; this removes the nuclear-electronic derivative cross terms. The analog of equations (10) for a different choice of electronic origin can be obtained by re-introducing the inconvenient nuclear-electronic derivative cross terms and transforming the nuclear first- and second-derivative coupling terms to the new origin. For example, for a new electronic origin somewhere else on the internuclear line, the required relation for the nuclear first-derivative term is

$$\langle \phi^a_{C\alpha} | (\frac{\partial}{\partial R})_{\chi} | \phi^a_{C\beta} \rangle_{\chi} = \langle \phi^a_{\eta\alpha} | (\frac{\partial}{\partial R})_{\chi_\eta} | \phi^a_{\eta\beta} \rangle_{\chi_\eta}$$

$$+ \sum_i (C - \eta_i) \langle \phi^a_{\eta\alpha} | (\frac{\partial}{\partial z_{\eta i}}) | \phi^a_{\eta\beta} \rangle_{\chi_\eta} \tag{14}$$

where $z_{\eta i}$ is the internuclear-axis-component of $\chi_{\eta i}$.

In this article, rather than actually calculating the matrix elements $G^a_{\eta\alpha\beta}(R)$ from the electronic wavefunctions, we assume they are given by the relation

$$G_{\eta}^{a}(R) = \frac{dF_{\eta}^{a}}{dR} - \frac{2\mu_{AB}}{\hbar^2} [F_{\eta}^{a}(R)]^2 \tag{15}$$

This equation holds rigorously for an infinite basis set, but we apply it even though our matrices are only of order 3. Tests of this assumption have been made for the KH system;[5] they show that it makes a negligible difference in the transition probabilities in low-energy collisions.

Infinite-Range Coupling

An important aspect of the coupled equations (10) is that the coupling does not necessarily go to zero for large internuclear separations. For example, consider the collision of two atoms in which all electronic excitations and deexcitations occur on one nucleus A. For the electronic origin located at A the first-derivative coupling term is

$$F_{0\alpha\beta}^{a}(R) = - \frac{\hbar^2}{2\mu_{AB}} \lim_{\delta \to 0} \frac{1}{\delta} (<\phi_{0\alpha}^{a}(x_0,R) | \phi_{0\beta}^{a}(x_0,R+\delta)>_{x_0} - \delta_{\alpha\beta}) \tag{16}$$

and this vanishes at large R by the orthonormality of the atomic orbitals on A. For the electronic origin at the CMN, however, we get

$$F_{C\alpha\beta}^{a}(R) = - \frac{\hbar^2}{2\mu_{AB}} \lim_{\delta \to 0} \frac{1}{\delta} (<\phi_{C\alpha}^{a}(x,R) | \phi_{C\beta}^{a}(x,R+\delta)>_{x_0} - \delta_{\alpha\beta}) \tag{17}$$

In (16) the nuclear displacement is carried out by moving B with A fixed, but in (17) the displacement must be carried out with the CMN fixed. Thus, at large R, the matrix element on the right side of (16) tends to an overlap of displaced atomic orbitals, and $F_{\eta\alpha\beta}^{a}(R)$ does not tend to $\delta_{\alpha\beta}$.

The infinite-range coupling of electronic states presents a problem in applying asymptotic boundary conditions to the scattering wavefunction. Bates and McCarroll[6] have observed that imposing the usual scattering boundary conditions in the (R,x) coordinates using the usual adiabatic basis functions $\phi_{C\alpha}^{a}(x,R)$ neglects the asymptotic motion of the electrons with respect to the CMN. Therefore they proposed including electron translation factors in the adiabatic basis to account for this asymptotic behavior. For some transitions, the infinite-range coupling can be eliminated by applying the asymptotic

boundary conditions in a coordinate system in which electronic origin
are at a nucleus and the internuclear distance is replaced by the dis
tance between the centers of mass of the two separated atoms.[1,7,8]
Unfortunately a coordinate system involving the atomic centers of mas
makes the electronic part of the problem so much more difficult that
it is apparently out of the question as a general computational technique. A more practical, but still rigorously correct way of treatin
the infinite-range coupling problem is by a diagonalization method.[9]
In this method the second order coupled equations are transformed
into a set of equivalent first-order equations that can be diagonalized.

In this article we examine less rigorous, but easier ways to
eliminate the infinite-range coupling. In the first method, suggested by Chen et al.,[10] we use equations (10) but we replace $F^a_{C\alpha\beta}(R)$
by $F^a_{C\alpha\beta}(R) - F^a_{C\alpha\beta}(\infty)$. Although Chen et al.[10] tried to justify this
by a semiclassical analysis, the method has been criticized by
Thorson and coworkers.[11] Notice that since we always assume (15) for
$G^a_{\eta}(R)$, it will vanish asymptotically if $F^a_{\eta}(R)$ vanishes.

In the second method, we retain the convenient form of equation
(10), with no nuclear-electronic derivative cross terms, but we
replace $F^a_{C\alpha\beta}(R)$ by $F^a_{0\alpha\beta}(R)$ and $G^a_C(R)$ by values calculated using $F^a_0(R)$
and (15). $F^a_0(R)$ is computed with the electronic origin on K. In thi
article we only consider excitations and de-excitations of K by
ground-state H; thus an electronic origin on K makes the nuclear deri
vative coupling terms vanish at infinity. Notice the nature of the
inconsistency involved in this method, namely it corresponds to using
the CMN as electronic origin for the electronic-nuclear derivative
cross terms but using the K as origin for the nuclear derivative
coupling elements. This procedure is essentially the same as the one
adopted by some workers for the semiclassical coupled equations for
high-energy, ion-atom scattering problems,[3] but for the quantal case
it can only by justified on an ad hoc basis.

The third method is also an analog for the quantal coupled-
channels equations of a method proposed for the semiclassical equations for high-energy scattering. This method, proposed by one of
the authors and Goddard (MG),[12,13] is applicable to systems with only
one active electron. In it, η becomes a function of α, β, and R,
and $F^a_{C\alpha\beta}(R)$ is replaced by $F^a_{\eta_{\alpha\beta}(R)\alpha\beta}(R)$ [$\equiv F^a_{MG\alpha\beta}(R)$], again without
re-introducing the electronic-nuclear derivative cross terms. $G^a_{MG}(R)$
is assumed to be given by (15) using $F^a_{MG}(R)$. Each $\eta_{\alpha\beta}(R)$ is chosen
to make $F^a_{\eta_{\alpha\beta}(R)\alpha\beta}(R)$ vanish at $R = \infty$, even for systems in which electronic excitation can occur at both nuclei and for systems in which
charge exchange channels are open. Rather than specify $\eta_{\alpha\beta}(R)$, it
is more convenient to specify the position $z_{\alpha\beta}(R)$ of the new electronic origin by its value in the CMN coordinate system. In this
coordinate system, the MG origin is

$$z_{\alpha\beta}(R) = \frac{<abs[\phi_{C\alpha}^a(\underset{\sim}{x},R)]|z|abs[\phi_{C\beta}^a(\underset{\sim}{x},R)]>_{\underset{\sim}{x}}}{<abs[\phi_{C\alpha}^a(\underset{\sim}{x},R)]|abs[\phi_{C\beta}^a(\underset{\sim}{x},R)]>_{\underset{\sim}{x}}} \qquad (18)$$

where abs(f) denotes the absolute value of f. Because of the complicated dependence of this electronic origin upon the internuclear coordinate it would be difficult to derive the appropriate coupled equations consistent with this choice of origin.

CALCULATIONS

The adiabatic potential curves for the $^1\Sigma^+$ states of KH were calculated by a one-electron model[14,15] for alkali hydrides using effective core potentials to represent K^+ and H. Further details of these calculations are presented elsewhere.[5] In review, the adiabatic energies are obtained by adding one-electron eigenvalues calculated with effective core potentials representing K^+ and H, to the KH^+ core energy. The KH^+ core energy is approximated by a calculation of KH^+ employing only a single H 1s basis function. This is called method 2H in previous work.[5,15] The calculated adiabatic potential curves $\varepsilon_\alpha^a(R)$ are plotted in Fig. 1 for the three lowest $^1\Sigma^+$ curves of KH. Also shown are the RKR curves[16,17] for the X and A states.

The first-derivative coupling matrices $F_{\eta\alpha\beta}^a(R)$ were calculated from the one-electron wavefunctions for two choices of electronic origin by using equations (16) and (17) with $\delta = 0.001\ a_0$. For electronic origins on the internuclear line the first-derivative coupling matrix depends linearly on the choice of origin. Because of this linear relationship the first-derivative coupling matrix element $F_{\eta\alpha\beta}^a(R)$ for any other choice of origin could be and was obtained as a linear combination of $F_{0\alpha\beta}^a(R)$ and $F_{C\alpha\beta}^a(R)$. For discussion purposes we define the mass-independent matrix

$$\underset{\sim}{f}_\eta^a(R) = -(2\mu_{AB}/\hbar^2)\ \underset{\sim}{F}_\eta^a(R) \qquad (19)$$

The derivative coupling terms $f_{\eta\alpha\beta}^a(R)$ are plotted in Figs. 2-4 for four choices of origin: at the K, at the H, at the CMN for $^{39}K^1H$, and at the MG origin. Recall that the MG origin depends on α, β, and R. The CMN for $^{39}K^{39}H$ is the same as the geometrical center of the nuclei (GCN), i.e., the bond midpoint. Although the coupling matrices are not shown for the GCN origin, their shape is easily visualized since they are exactly halfway between the results shown for the origins at K and at H.

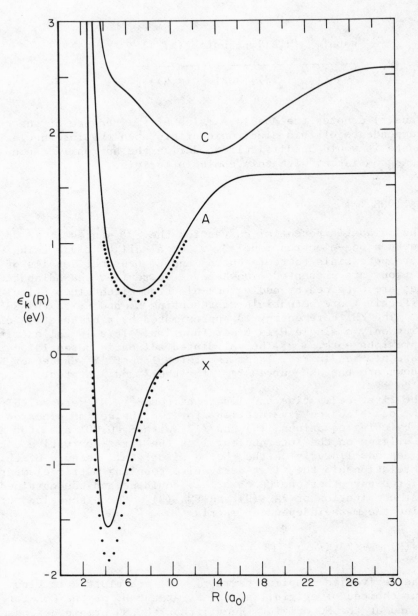

Fig. 1. Adiabatic potential energy curves as a function of internuclear distance for the three lowest $^1\Sigma^+$ states of KH. The
curves are the results of an ab initio pseudopotential calculation as obtained by method 2H. The points are spectroscopic RKR values for the X and A potential curves.

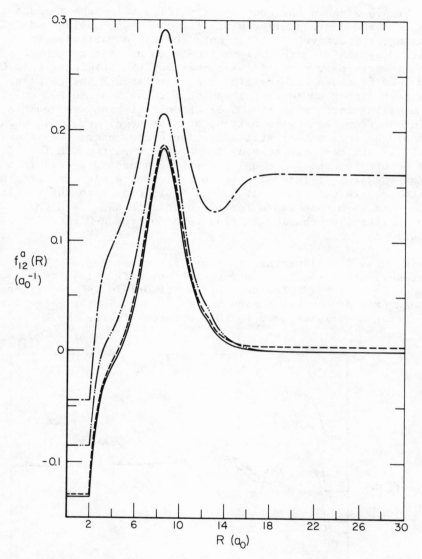

Fig. 2. First-derivative coupling term between the X and A state of
KH as a function of internuclear distance. The four curves
are for four different choices of the origin for the electronic coordinate. The solid curve is for the electronic
origin at the K^+ core, the dashed curve is for the electronic
origin at the center of mass of the K^+ core and ^1H atom, the
long-short dashed curve is for the electronic origin at the
^1H atom, and the dash-dotted curve is for the Melius-Goddard
prescription for the electronic origin.

Figure 2 shows the coupling $f^a_{\eta 12}(R)$ between the X and A states.
Figures 3 and 4 are similar plots for the X–C coupling and A–C
coupling, respectively. First note the large variation in $f^a_{\eta}(R)$
obtained by shifting the origin from K to H. Because the CMN for
$^{39}K^1H$ is very near K the $f^a_{\eta}(R)$ curves for the origin at this CMN are
very near to those with origin at K. At large R the coupling for the
origin at either CMN does not vanish for the X–A and A–C coupling.
The coupling terms with the MG electronic origins are qualitatively
different from those with origins at K or at either CMN. When one or
both of the interacting states has ionic character, the value of
$z_{\alpha\beta}(R)$ is shifted more towards the H core than the CMN for $^{39}K^1H$.
This occurs for the X–A coupling between 4 and 12 a_0 and in this
region $f^a_{MG12}(R)$ is intermediate to the results with the origin at K
and those with the origin at H, but it is closer to the results for
the origin at H than those for the origin at the $^{39}K^1H$ CMN. This
same qualitative behavior is also seen for the A–C coupling between
8 and 10 a_0.

Figures 2–4 illustrate an important general property of the
first-derivative coupling matrices. Notice from (14) that the difference between $F^a_{\eta}(R)$ for two choices of origin is proportional to
the electric dipole transition matrix. Thus, in the large-R limit,
the different origins lead to the same results for the X–C coupling

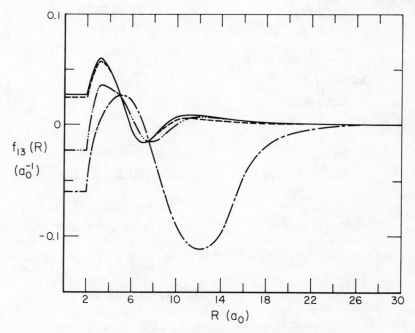

Fig. 3.   Same as Fig. 2 except for coupling between the X and C states
          of KH as a function of internuclear distance.

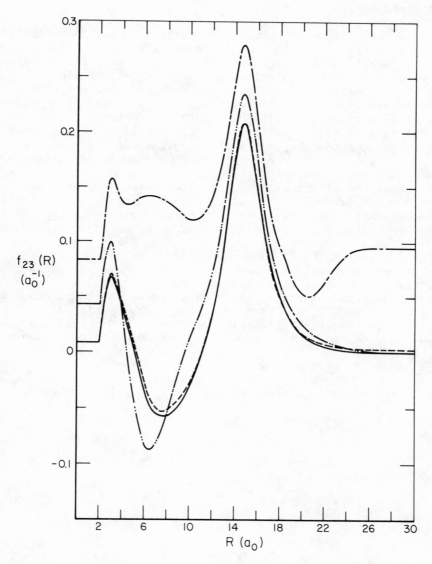

Fig. 4.   Same as Fig. 2 except for coupling between A and C states
of KH as a function of internuclear distance.

because the 4^2S-5^2S transition is forbidden, but the origin choice does matter for the X-A and A-C couplings because the A state dissociates to the 4^2P state which is connected to both 4^2S and 5^2S by the electric dipole operator.

It is interesting to examine how quantum scattering calculations of electronic transition probabilities are altered by changes in the electronic origin of the derivative coupling terms. If the calculations were done correctly including a complete set of electronic states and all kinds of coupling terms with a consistent choice of electronic origin, then the results would be independent of the choice of electronic origin. We wish to examine the effect of using one set of radial equations, namely equations (10)-(13), but with several different choices for the derivative coupling terms as discussed above. We neglect angular coupling and the mass polarization term; however, the model problem studied here still allows a realistic assessment of the effect of using coupling terms with arbitrary choice of electronic origin upon the calculation of transition probabilities. All the derivative coupling matrices we use vanish asymptotically; thus the boundary conditions can be applied by standard methods.

The scattering equations were solved using a R-matrix propagation method described in detail elsewhere.[5] In all cases we use a basis of the three lowest electronically adiabatic $^1\Sigma^+$ states: state 1 is the X state which dissociates to $K(4^2S)$, state 2 is the A state which dissociates to $K(4^2P)$, and state 3 is the C state which dissociates to $K(5^2S)$. The excitation energies of the 4^2P and 5^2S states are 1.61 and 2.55 eV, respectively. We neglect spin-orbit coupling.

We consider three processes:

$$K(4^2S) + H(1^2S) \rightarrow K(4^2P) + H(1^2S) \qquad\qquad (1 \rightarrow 2)$$

$$K(4^2S) + H(1^2S) \rightarrow K(5^2S) + H(1^2S) \qquad\qquad (1 \rightarrow 3)$$

and

$$K(4^2P) + H(1^2S) \rightarrow K(5^2S) + H(1^2S) \qquad\qquad (2 \rightarrow 3)$$

RESULTS

$^{39}K^1H$

We calculated the cross sections for two total energies, 1.6327 eV and 2.7212 eV. At the lower energy, two channels are open, and at the higher energy, three. Since 2/3 of the collisions of $K(4^2P)$ with $H(1^1S)$ occur in the Π manifold, and only 1/3 in the Σ manifold, we multiplied $\sigma_{2\rightarrow3}$ by 1/3. With this factor the calculated

Table 1. Cross sections (a_0^2) for excitation processes in
$^{39}K + {}^1H$ collisions

Electronic origin	E = 1.6327 eV	E = 2.7212 eV		
	$\sigma_{1 \to 2}$	$\sigma_{1 \to 2}$	$\sigma_{1 \to 3}$	$\sigma_{2 \to 3}$
at K	1.60(-5)[a]	9.24(-4)	1.73(-6)	4.06(-1)
at CMN	1.59(-5)	8.88(-4)	1.73(-6)	3.91(-1)
Melius-Goddard	1.22(-5)	8.71(-4)	1.70(-6)	4.04(-1)

[a]Numbers in parentheses are powers of ten.

cross sections satisfy detailed balance including the threefold
degeneracy of P states but neglecting spin-orbit coupling. The
factors of 1/4, because only 1/4 of all collisions occur in the
singlet manifold, are not included. Thus, as in reference 5, the
results are cross sections for singlet collisions only. The results
are given in Table 1.

As expected for $^{39}K + {}^1H$, where the CMN lies so close to K, the
cross sections for these two origins are very similar. The largest
difference for the four cases in Table 1 is 4%. In contrast the
physically motivated MG procedure leads to a much larger difference
in one case; the difference is 30-31%. It is very interesting that
alternative methods of coping with the formal problem of infinite-
range nonadiabatic coupling terms can lead to such a difference even
for the integral cross section of this reasonably simple case. We
note that the difference between the results calculated with different origins is greatest for the case where the initial and final
states are connected by the greatest optical oscillator strength.
In general one expects[18] the effects to be largest in such cases for
the reasons already dicussed in connection with Figs. 2-4.

Instead of comparing the cross sections we could have compared
the s-wave transition probabilities. (For the probabilities we do
not include the 1/3 factor for multiple potential energy curves.)
This comparison is shown in Table 2. The differences are larger
than those in Table 1. This could be expected since the inelastic
transition probabilities are oscillatory functions of orbital angular
momentum ℓ for this example. Thus there is a certain amount of cancellation of errors in the integral cross sections. One would expect
less cancellation in the differential cross sections.

Table 2. Inelastic s-wave transition probabilities for
excitation processes in $^{39}K + {}^1H$ collisions

Electronic origin	E = 1.6327 eV	E = 2.7212 eV		
	$P_{1 \to 2}$	$P_{1 \to 2}$	$P_{1 \to 3}$	$P_{2 \to 3}$
at K	1.56(−6)	1.84(−7)	5.70(−10)	2.15(−3)
at CMN	1.57(−6)	2.22(−7)	4.06(−10)	2.17(−3)
Melius−Goddard	1.24(−6)	6.65(−7)	2.05(−9)	2.07(−3)

$^{39}K + {}^{39}H$

In many systems, e.g., collisions of K with a particle heavier
than itself in the absence of narrowly avoided curve crossings, we
would expect a larger difference between the results calculated with
origin at one of the nuclei and those calculated with the origin at
the CMN. To illustrate this effect of the mass of the collision
partner, we calculated the s-wave transition probabilities for the
same potential curves but with the masses of both particles taken as
38.964 amu. The $f_{\eta}^a(R)$ values were also the same for the cases of
origin at K or MG origin, but for the CMN origin, we must now use
the results of origin at the GCN. The results we obtained for the
s-wave transition probability $P_{1 \to 2}$ are shown as a function of energy
in Fig. 5.

For all three origins, the transition probability oscillates as
a function of energy. In general, the oscillations of all three
cases are reasonably well in phase (the most significant exceptions
being for the energy range 2.0-2.3 eV). The magnitudes of the transition probabilities at the maxima of the oscillations are in worse
agreement than the phases of the oscillations. In general the
results for the origin at K and those for the MG origin agree better
with each other than either agrees with the GCN origin. A more
detailed comparison of these results for energies near the maxima
in the oscillatory curve is given in Table 3. The results for the
K and MG choices agree within a factor of 2 or better for all cases
in the table, but those for the K and GCN choices differ by more than
a factor of 2 in half the cases. Overall the K and GCN results
differ by from 6% to a factor of 3.8 in thirteen of the cases, and
in one case the difference is a factor of 190. Of course the percentage deviations may be much larger if we make the comparisons near

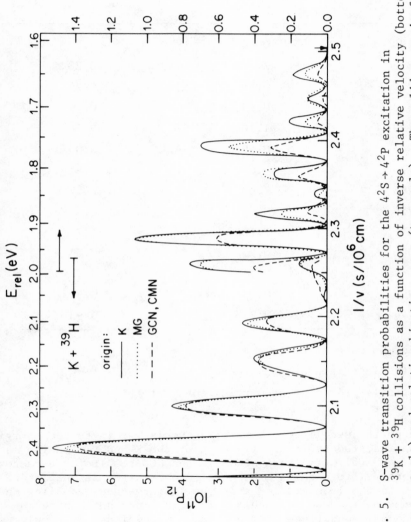

Fig. 5. S-wave transition probabilities for the $4^2S \rightarrow 4^2P$ excitation in $^{39}K + ^{39}H$ collisions as a function of inverse relative velocity (bottom scale) and relative kinetic energy (top scale). The solid curve is for electronic origin at K; the dashed curve is for electronic origin at the center of mass of the nuclei, which in this case is the same as the geometrical center of the nuclei, and the dotted curve is for the Melius-Goddard prescription for the electronic origin.

Table 3.   Ratios of s-wave inelastic transition probabilities
           for excitations from ground to first excited state
           in ^{39}K + ^{39}H collisions using different electronic
           origins

Energy (eV)	$P_{1 \to 2}(MG^a)/P_{1 \to 2}(K^b)$	$P_{1 \to 2}(GCN^c)/P_{1 \to 2}(K^b)$
1.6490	0.64	0.27
1.6898	1.21	0.28
1.7225	0.58	0.26
1.7633	0.73	0.42
1.8123	1.16	0.56
1.8449	0.59	0.0053
1.8776	0.66	0.31
1.9266	1.00	0.57
1.9756	0.92	0.43
2.0409	1.16	1.77
2.1062	0.95	0.73
2.1769	0.94	0.92
2.2994	0.98	0.94
2.3946	0.95	0.93

[a] Melius-Goddard electronic origin.
[b] Electronic origin at K nucleus.
[c] Electronic origin at geometric center of the nuclei.

the minima of the oscillatory curve. Returning again to the energies
near the maxima, the typical difference of the K and MG results is
25% but the typical difference of the K and GCN results is a factor
of 2-3.   These differences are especially noteworthy because it has
been stated in the literature[19] that the infinte-range coupling due
to the expansion in Born-Oppenheimer functions not satisfying the
correct boundary conditions does not cause significant difficulties
in low-energy collisions.*   More recently a contrary opinion has

_____
*Consistent with this widely held opinion, the infinite-range coupling
 effect has also been neglected in low-energy atom-molecule collision
 studies.[20]

been stated in the literature,[21] but to our knowledge the present calculations are the first numerical test of this effect for low-energy
collisions.

SUMMARY

We have performed numerical comparisons of three different ways
of treating the first-derivative coupling terms in the quantum mechanical coupled-channel equations for atom-atom scattering in an adiabatic representation. We have shown that different ways of handling
the formal problems with the derivative coupling operators can lead
to errors of 25% to a factor of 2 or 3 in computed transition probabilities and inelastic cross sections. We even found a case where
the predictions of two of the methods differ by a factor of 190 even
for an energy near a maximum in the oscillatory curve of transition
probability vs. energy.

ACKNOWLEDGMENT

The authors are grateful to Mike Redmon for a critical reading
of the manuscript. This work was supported in part by the National
Science Foundation through grant no. CHE772-27415 and by the
Air Force Office of Scientific Research, United States Air Force
(AFSC), under contract no. F49620-79-C-0050. The United States
Government is authorized to reproduce and distribute reprints for
government purposes notwithstanding any copyright notation hereon.

REFERENCES

1. B. C. Garrett and D. G. Truhlar, The coupling of electronically
       adiabatic states in atomic and molecular collisions, in:
       "Theoretical Chemistry: Advances and Perspectives", Vol. 6A,
       D. Henderson, ed., Academic Press, New York (1981), p. 215.
2. D. R. Bates, H. S. W. Massey, and A. L. Stewart, Inelastic col-
       lisions between atoms. I. General theoretical considerations,
       Proc. Roy. Soc. Lond., Ser. A 216:437 (1953).
3. See, e.g., R. D. Piacentini and A. Salin, Molecular treatment
       of the He^{2+}-H collisions, J. Phys. B 7:1666 (1974); C. Harel
       and A. Salin, Charge exchange in collisions of highly ionized
       ions and atoms, J. Phys. B 10:3511 (1977).
4. R. T Pack and J. O. Hirschfelder, Separation of rotational coor-
       dinates from the N-electron diatomic Schrödinger equation,
       J. Chem. Phys. 49:4009 (1968).
5. B. C. Garrett, M. J. Redmon, D. G. Truhlar, and C. F. Melius,
       Ab initio treatment of electronically inelastic K + H colli-
       sions using a direct integration method for the solution

of the coupled-channel scattering equations in electronically
adiabatic representations, J. Chem. Phys. 74: in press.

6. D. R. Bates and R. McCarroll, Electron capture in slow colli-
    sions, Proc. Roy. Soc. Lond., Ser. A 245:175 (1958).

7. M. H. Mittleman and H. Tai, Low energy atom-atom scattering:
    Corrections to the He-He interaction, Phys. Rev. A 8:1880
    (1973).

8. W. R. Thorson and J. B. Delos, Theory of near-adiabatic colli-
    sions. II. Scattering coordinate method, Phys. Rev. A 18:135
    (1978).

9. A time-independent quantum mechanical formulation proposed by
    M. E. Riley is presented in reference 1. A semiclassical
    formulation is given by S. K. Knudson and W. R. Thorson,
    Lyman -α excitation and resonant charge exchange in slow
    H^+-H(1s) collisions, Can. J. Phys. 48:313 (1970).

10. J. C. Y. Chen, V. H. Ponce, and K. M. Watson, Translational
    factors in eikonal approximation and their effect on channel
    couplings, J. Phys. B 6:965 (1973).

11. V. SethuRaman, W. R. Thorson, and C. F. Lebeda, Impact ionization
    in the proton-H-atom system. V. Final cross section calcula-
    tions, Phys. Rev. A 8:1316 (1973); W. R. Thorson and J. B.
    Delos, Theory of near-adiabatic collisions. I. Electron trans-
    lation factor method, Phys. Rev. A 18:117 (1978).

12. C. F. Melius and W. A. Goddard III, The theoretical description
    of an asymmetric nonresonant charge transfer process:
    Li + Na^+ $\rightleftarrows$ Li^+ + Na, the two state approximation, Chem. Phys.
    Lett. 15:524 (1972).

13. C. F. Melius and W. A. Goddard III, Charge-transfer process
    using the molecular-wavefunction approach: The asymmetric
    charge transfer and excitation in Li + Na^+ and Na + Li^+, Phys.
    Rev. A 10:1541 (1974).

14. C. F. Melius, The H^+ + H^- neutralization process using the mole-
    cular wavefunction approach, Bull. Amer. Phys. Soc. 19:1199
    (1974).

15. C. F. Melius, R. W. Numrich, and D. G. Truhlar, Calculations of
    potential energy curves for the ground states of NaH^+ and
    KH^+ and Π state of NaH and KH, J. Phys. Chem. 83:1221 (1979).

16. R. W. Numrich and D. G. Truhlar, The mixing of ionic and covalent
    configurations for NaH, KH, and MgH^+. Potential energy curves
    and couplings between molecular states, J. Phys. Chem. 79:2745
    (1975).

17. R. W. Numrich and D. G. Truhlar, Detailed study of the inter-
    action of covalent and ionic states in collisions of Na and
    K with H, J. Phys. Chem. 82:168 (1978).

18. E. J. Shipsey, J. C. Browne, and R. E. Olson, Theoretical charge-
    exchange cross sections for B^{+3} + He and C^{+4} + He collisions,
    Phys. Rev. A 15:2166 (1977).

19. See, e.g., S. Geltman, "Topics in Atomic Collision Theory",
    Academic Press, New York (1969), section 21; J. E. Bayfield,

E. E. Nikitin, and A. I. Reznikov, Semiclassical scattering matrix for two-state exponential model, Chem. Phys. Lett. 19:471 (1973).

20. See, e.g., P. L. DeVries and T. F. George, Quantum mechanical theory of a structured atom-diatom collision system: A + BC($^1\Sigma$), J. Chem. Phys. 67:1293 (1977); F. Rebentrost and W. A. Lester, Jr., Nonadiabatic effects in the collision of F(^{2}P) with H$_2$($^1\Sigma_g^+$). III. Scattering theory and coupled-channel computations, J. Chem. Phys. 67:3367 (1977).

21. A. Riera and A. Salin, Limitations on the use of the perturbed stationary-state methods for the description of slow atom-atom collisions, J. Phys. B 9:2877 (1976).

DYNAMICAL CALCULATIONS ON TRANSPORT PROCESSES

Dennis J. Diestler and John W. Allen

Department of Chemistry
Purdue University
West Lafayette, Indiana 47907 USA

INTRODUCTION

Measurements of relaxation processes can be conveniently classified as either frequency-resolved or time-resolved. The frequency-resolved measurements are generally expressible in terms of the Fourier transform of the time-correlation function (TCF) of a characteristic dynamical variable.[1] Thermal transport coefficients are given as the zero-frequency Fourier component of the appropriate TCF. For example, the self-diffusion coefficient D can be written as

$$D = \frac{1}{3} \int_0^\infty dt \ \langle \vec{v}(0) \cdot \vec{v}(t) \rangle_{.eq} , \qquad (1)$$

where $\vec{v}$ is the velocity of the diffusing "tagged" particle. In (1) the angular brackets $\langle ... \rangle_{eq}$ denote an ensemble average over the *equilibrium* distribution function f_{eq}, i.e.

$$\langle X \rangle_{eq} = \int d\underset{\sim}{r} \int d\underset{\sim}{p} \ f_{eq}(\underset{\sim}{r},\underset{\sim}{p}) \ X(\underset{\sim}{r},\underset{\sim}{p})$$

Spectra and scattering cross sections can generally be cast as nonzero-frequency Fourier components of the TCF's of appropriate matrix elements of the transition operator. As an example, the absorption spectrum can be expressed as

$$I_a(\omega) \ \alpha \ \int_{-\infty}^\infty dt \ \exp(-i\omega t) \langle \vec{\mu}(0) \cdot \vec{\mu}(t) \rangle ,$$

where $\vec{\mu}$ is the dipole-moment operator of the system and $<...>$ denotes an average over the distribution function describing the initial state of the system.

In time-resolved experiments one measures the expectation value of the relevant dynamical variable. For instance, the excess energy $\delta\epsilon$ contained in a specific mode of the system, is given by

$$<\delta\epsilon(t)>_{ne} = \int d\underset{\sim}{r}\int d\underset{\sim}{p} \ f_{ne}(\underset{\sim}{r},\underset{\sim}{p})\delta\epsilon[\underset{\sim}{r}(t),\underset{\sim}{p}(t)] \ . \tag{2}$$

In (2) f_{ne} is the initial nonequilibrium distribution function and

$$\delta\epsilon = \epsilon(t) - <\epsilon>_{eq} \ .$$

If the system is initially displaced only slightly from equilibrium, one can show[2] that

$$<\delta\epsilon(t)>_{ne} \simeq <\delta\epsilon(t)\delta\epsilon(0)>_{eq} \ .$$

Thus, in this case energy relaxation can be described in terms of the decay of correlations in $\delta\epsilon$ in the equilibrium system. It is convenient to define a characteristic energy-relaxation time by

$$\tau_\epsilon = \int_0^\infty dt \ \bar{C}_\epsilon(t) \ , \tag{3}$$

where

$$\bar{C}_\epsilon(t) \equiv <\delta\epsilon(t)\delta\epsilon(0)>/<\delta\epsilon(0)\delta\epsilon(0)> \tag{4}$$

is the *normalized* TACF. Of course, (3) and (4) can be generalized to an arbitrary dynamical variable A.

Information on energy relaxation is also contained in spectra. In the case of a dilute solution of two-level molecules in an "inert" solvent, the absorption spectrum is a Lorentzian centered at the (shifted) transition frequency and having a width (FWHH) Γ given by[3]

$$\Gamma = T_2^{-1} = T_1^{-1} + T_2'^{-1} \tag{5}$$

where

$$T_1^{-1} = \frac{1}{2}(T_{1i}^{-1} + T_{1f}^{-1})$$

and

$$T_{1i}^{-1} = \frac{1}{2}\hbar^{-2} \sum_{f \neq i} \int_{-\infty}^{\infty} dt \ \exp(i\omega_{fi}t)<V_{if}(0)V_{fi}(t)>_B^{\circ} \qquad (6a)$$

$$T_2'^{-1} = \frac{1}{2}\hbar^{-2} \int_{-\infty}^{\infty} dt \ <\Delta_{if}(0) \ \Delta_{if}(t)>_B^{\circ} \qquad (6b)$$

$$V_{if} \equiv <i|V|f> \qquad (6c)$$

$$\Delta_{if} = V_{ii} - V_{ff} \qquad (6d)$$

$$\omega_{fi} = (\varepsilon_f^M - \varepsilon_i^M)/\hbar \qquad (6e)$$

In (6) ε_i^M is the energy of the unperturbed (isolated) molecule in
state i; V is the *assumed weak* coupling between the molecule and
solvent; $<...>_B^{\circ}$ signifies an ensemble average over the states of
the unperturbed solvent. From (5) we see that the total rate of dephasing T_2^{-1} comprises two contributions: (1) T_1^{-1}, the average
rate of energy relaxation, due to inelastic transitions that alter
the populations of the molecular states i and f; $T_2'^{-1}$, the rate
of "pure dephasing," which arises from elastic transitions leaving
the populations of states i and f unchanged. According to (6),
both energy-relaxation and pure dephasing rates are expressible as
Fourier components of the TACF of appropriate matrix elements of
the molecule-solvent coupling V.

In summary, the description of a relaxation process, either
in the frequency domain or the time domain, involves the determination of an appropriate time-autocorrelation function. We shall
turn next to this task, restricting attention to systems in *equilibrium*, or at least sufficiently near equilibrium that the linear-
response approximation holds. Also, we shall limit the consideration
to *classical* systems.

II. DEFINITION AND PROPERTIES OF CLASSICAL TIME-AUTOCORRELATION
    FUNCTIONS

For a classical N-particle system Liouville's operator can be
written in anti-Hermitian form as

$$\hat{L} = \sum_{j=1}^{N} [(\partial H/\partial \vec{p}_j) \cdot \partial/\partial \vec{r}_j - (\partial H/\partial \vec{r}_j) \cdot \partial/\partial \vec{p}_j] ,$$

where $\vec{r}_j$ and $\vec{p}_j$ respectively denote the position and (conjugate)
momentum of the j-th particle and H is the Hamiltonian. We shall

assume that all dynamical variables are explicit functions only of coordinates and momenta and that they are infinitely differentiable with respect to both $\vec{r}_j$ and $\vec{p}_j$. This restriction eliminates impulsive interactions and implies that the TACF is regular at the origin. Now the equilibrium TACF of a given variable, say A, can be cast variously as

$$C_A(t) = <A(0)A^*(t)>$$

$$= <A(0)[\exp(\hat{L}t)A^*(0)]>$$

$$= \sum_{n=0}^{\infty} (-1)^n a_n t^{2n}/(2n!) \tag{7}$$

where

$$a_n \equiv <(\hat{L}^n A)(\hat{L}^n A^*)> \tag{8}$$

is the 2n-th moment of the Fourier transform of $C_A(t)$, which is proportional to

$$\tilde{C}_A(\omega) = \int_0^{\infty} dt \cos(\omega t) C_A(t) .$$

Note that the subscript "eq" on the angular brackets has been dropped to simplify notation; henceforth, <...> will denote the equilibrium ensemble average.

By means of projection-operator techniques,[2,4,5] one can derive the following generalized Langevin equation (GLE) for the time evolution of an *arbitrary* dynamical *vector* $\underline{A}$:

$$\dot{\underline{A}} = i \underline{\Omega} \cdot \underline{A} - \int_0^t dt' \underline{K}(t') \cdot \underline{A}(t-t') + \underline{f}(t) , \tag{9}$$

where $\underline{A}$ denotes the column vector of dynamical variables (functions of $\vec{p}_j$ and $\vec{r}_j$). In (9) $\underline{\Omega}$ is the "frequency matrix," defined by

$$i\underline{\Omega} \equiv [<\hat{L}\underline{A} \otimes \underline{A}^T>] \cdot [<\underline{A} \otimes \underline{A}^T>]^{-1} ;$$

$\underline{K}$ is the "damping matrix," or "memory matrix," defined by

$$\underline{K}(t) \equiv [<\underline{f}(t) \otimes \underline{f}(0)^T>] \cdot [<\underline{A} \otimes \underline{A}^T>]^{-1} ;$$

the so-called "random force" $\underline{f}$ is given by

$$\underline{f}(t) \equiv \exp[t(1-\hat{P})\hat{L}](1-\hat{P})\underline{A}$$

The symbol $\underline{A} \otimes \underline{B}^T$ represents a direct product of column vector $\underline{A}$ with row vector $\underline{B}^T$, T denoting the transpose; $\hat{P}$ is a projection operator defined by its action on an arbitrary dynamical vector $\underline{X}$ as

$$\hat{P}\underline{X} \equiv <\underline{X} \otimes \underline{A}^T> \cdot [\underline{A} \otimes \underline{A}^T]^{-1} \cdot \underline{A} ,$$

i.e. $\hat{P}$ projects any vector onto the dynamical subspace spanned by the elements of $\underline{A}$. If $\underline{A}$ consists of simply one variable A, then (9) reduces to

$$A(t) = i\Omega A(t) - \int_0^\infty dt' K(t') A(t-t') + f(t) \tag{10}$$

Upon multiplying both members of (10) from the right by A* and averaging the resulting equation over the equilibrium ensemble, we obtain

$$\dot{C}_A(t) = - \int_0^t dt' K(t') C_A(t-t') . \tag{11}$$

It is worth mentioning that (11) is *exact* in that no approximations beyond the validity of classical mechanics were made in reaching it. However, it should also be emphasized that (11) is deceptively simple in appearance; all of the complicated many-body effects are buried in the memory function, which is itself a TACF of the "random force" f evolving under the projected Liouville operator. In fact, we have really just transformed the problem of determining $C_A(t)$ into that of finding $K(t)$.

A variety of approaches to the determination of C_A have been taken, ranging from the construction of modelling functions for the memory function (or for $C_A(t)$ itself) to the numerical solution of the classical equations of motion for the full N-particle system (i.e. molecular dynamics). In the remainder of this article we wish to present a rather unusual treatment of relaxation processes via time-correlation functions. Our procedure is rooted essentially in a combinatorial analysis of the TACF. We shall first outline the basic formalism. We then show how it leads to useful systematic approximations to the TACF and to the corresponding transport coefficients.

III. COMBINATORIAL ANALYSIS OF TIME-AUTOCORRELATION FUNCTIONS

We define the set of dynamical variables

$$B_j(t) = \hat{L}^j A(t), \quad j = 0,1,2,\ldots,$$

where A is the variable of chief interest, i.e. the one whose TACF we ultimately desire. We next transform from the set $\{B_j\}_{j=0}^{\infty}$ to the Schmidt-orthogonalized set[6-10] $\{Z_j\}_{j=0}^{\infty}$, given explicitly by

$$Z_o = B_o = A \tag{12a}$$

$$Z_n = B_n - \sum_{k=0}^{n-1} <B_n Z_k^*><Z_k Z_k^*>^{-1} Z_k \tag{12b}$$

It is easy to see that

$$<Z_k Z_j^*> = \delta_{kj} <Z_k Z_k^*> . \tag{13}$$

Moreover, under the restriction of infinite differentiability introduced earlier, it can be shown that

$$\hat{L} Z_k = Z_{k+1} - \lambda_k Z_{k-1} ,$$

where the *lowering coefficient* λ_k is given by

$$\lambda_k = <Z_k Z_k^*>/<Z_{k-1} Z_{k-1}^*>, \ k \geq 1$$

Thus, we describe the dynamical evolution of A in the basis $\{Z_j\}$ in which $\hat{L}$ is resolved into "raising" and "lowering" operators as

$$\hat{L} = \hat{R} + \hat{L} \tag{14}$$

defined by their respective action on Z_j as

$$\hat{R} Z_j = Z_{j+1}, \ j \geq 0 \tag{15a}$$

$$\hat{L} Z_j = - \lambda_j Z_{j-1}, \ j \geq 1 \tag{15b}$$

$$\hat{L} Z_o = 0 \tag{15c}$$

We shall now proceed to derive an explicit relation between the moments a_n and lowering coefficients λ_k. From (8), (12a) and (14) we have

$$a_n = (-1)^n <Z_o (\hat{L}+\hat{R})^{2n} Z_o^*> . \tag{16}$$

Note that in writing (16) we have also used the anti-Hermiticity of $\hat{L}$, i.e. the property

$$<X(\hat{L}Y)> = - <(\hat{L}X)Y> ,$$

where X and Y are arbitrary dynamical variables. Now observe that $(\hat{L}+\hat{R})^{2n}$ is a sum of 4^n terms, each consisting of a product of 2n factors of $\hat{R}$ and $\hat{L}$. Each term operates upon Z_o^* to yield $Z_\ell^*(0 \leq \ell \leq 2n)$ multiplied by a constant. Hence, we can write

$$\hat{L}^{2n}Z_o^* = (\hat{L}+\hat{R})^{2n}Z_o^* = \sum_{\ell=0}^{2n} h_\ell^{(n)} Z_\ell^* , \qquad (17)$$

where $h_\ell^{(n)}$ is a sum of simple products of lowering coefficients. Substituting (17) into (16), we obtain

$$a_n = (-1)^n h_o^{(n)} <Z_o Z_o^*> , \qquad (18)$$

where we have also invoked orthogonality of the Z_k's [see (13)].

It is clear from (18) that we need find only the coefficient $h_o^{(n)}$ in order to establish the relation between the a_n and the λ_k. We observe next that terms proportional to Z_ℓ^* in (17) must arise from sequences of $(\hat{L}+\hat{R})^{2n}$ that have ℓ more factors of $\hat{R}$ than of $\hat{L}$. This follows easily from relations (15). Thus, $h_o^{(n)}$ must derive from sequences having *equal* numbers (n) of factors of $\hat{R}$ and $\hat{L}$. Therefore, $h_o^{(n)}$ is a sum of products of $(-1)^n$ and n factors of lowering coefficients λ_k, with the index k ranging from 1 to n at most. It can be shown[11] that the total number of sequences of $(\hat{R}+\hat{L})^{2n}$ that contribute to $h_o^{(n)}$ is

$$K_{2n} \equiv (2n)!/[n!(n+1)!] .$$

Moreover, all such contributing sequences must terminate with $\hat{L}$ (i.e. the leftmost factor of the sequence is $\hat{L}$). Were the terminal factor $\hat{R}$, then the final resulting Schmidt variable Z_k^* would possess an index k>0. Indeed, since the index must be 0, the terminal factor of $\hat{L}$ must operate upon Z_1^*, thus yielding a final factor of $-\lambda_1$.

Now let us define the m-th *stage* of a contributing sequence as the operation with the m-th factor of $\hat{L}$ giving rise to the m-th factor of λ_k. From the preceding discussion it should be clear that λ_1 appears at the n-th (and last) stage. At the (n-1)-th stage either λ_1 or λ_2 appears. Thus, we can write tentatively

$$a_n = <Z_o Z_o^*> \lambda_1 \sum_{k_2=1}^{2} \lambda_{k_2} \times [\text{additional terms}]$$

If λ_1 appears at stage (n-1) then at stage (n-2) either λ_1 or λ_2

can appear. On the other hand, if λ_2 appears at stage $(n-1)$, then only λ_1, λ_2 or λ_3 can appear at stage $(n-2)$. Thus, we have

$$a_n = <Z_o Z_o^*> \lambda_1 \sum_{k_2=1}^{k_1+1} \lambda_{k_2} \sum_{k_3=1}^{k_2+1} \lambda_{k_3} \times [\text{additional terms}]$$

Extending this reasoning through the n successive stages, we have finally the general relation between the a's and λ's:

$$a_n = <Z_o Z_o^*> \sum_{k_1=1}^{1} \lambda_{k_1} \sum_{k_2=1}^{k_1+1} \lambda_{k_2} \sum_{k_3=1}^{k_2+1} \lambda_{k_3} \cdots \sum_{k_n=1}^{k_{n-1}+1} \lambda_{k_n} \tag{19}$$

Using (19), we calculate the first four moments of the normalized TACF $[\bar{C}_A = C_A/<A^2>]$ as

$$\bar{a}_1 = \lambda_1$$

$$\bar{a}_2 = \lambda_1(\lambda_1+\lambda_2)$$

$$\bar{a}_3 = \lambda_1(\lambda_1+\lambda_2)^2 + \lambda_1\lambda_2\lambda_3$$

$$\bar{a}_4 = \lambda_1(\lambda_1+\lambda_2)^3 + 2\lambda_1^2\lambda_2\lambda_3 + 2\lambda_1\lambda_2^2\lambda_3 + \lambda_1\lambda_2\lambda_3\lambda_4$$

IV. DYNAMICAL EMBEDDING

A somewhat more computationally viable relation between the a's and λ's can be derived from the observation that all the information required to compute the TACF of $\dot{A} = \hat{L}A$, i.e. $C_{\dot{A}}(t)$, is already contained in the TACF of A itself. This is easily seen by examining the following sequence:

$$C_{\dot{A}} = \sum_{n=0}^{\infty} (-1)^n a_n^{(1)} t^{2n}/(2n)!$$

$$= \sum_{n=0}^{\infty} (-1)^n a_{n+1} t^{2n}/(2n)!$$

$$= - d^2 C_A/dt^2 .$$

We say that $C_A^\bullet$ is "dynamically embedded" in C_A, in as much as the moments of $C_A^\bullet$, $\{a_n^{(1)}\}_{n=0}^\infty = \{a_n\}_{n=1}^\infty$ form a subset of those of $C_A[\{a_n\}_{n=0}^\infty]$. By a rather lengthy analysis it can be shown[11] that the lowering coefficients associated with $C_A^\bullet$ and C_A are connected by the relations

$$\lambda_{2n-1}^{(1)} \lambda_{2n}^{(1)} = \lambda_{2n}^{(0)} \lambda_{2n+1}^{(0)} \tag{20a}$$

$$\lambda_{2n}^{(1)} + \lambda_{2n+1}^{(1)} = \lambda_{2n+1}^{(0)} + \lambda_{2n+2}^{(0)} , \tag{20b}$$

where the superscripts 0 and 1 respectively refer to C_A and $C_A^\bullet$.

The embedding procedure can be immediately generalized. Consider the TACF of $\hat{L}^k A$:

$$C_{\hat{L}^k A}(t) = \sum_{n=0}^\infty (-1)^n a_n^{(k)} t^{2n}/(2n)!$$

$$= \sum_{n=0}^\infty (-1)^n a_{n+k} t^{2n}/(2n)!$$

$$= (-1)^k d^{2k} C_A / dt^{2k} .$$

The generalization of (20) is then

$$\lambda_{2n-1}^{(k)} \lambda_{2n}^{(k)} = \lambda_{2n}^{(k-1)} \lambda_{2n+1}^{(k-1)} \tag{21a}$$

$$\lambda_{2n}^{(k)} + \lambda_{2n+1}^{(k)} = \lambda_{2n+1}^{(k-1)} + \lambda_{2n+2}^{(k-1)} . \tag{21b}$$

The normalized moments of the k-th embedded TACF are given by

$$\bar{a}_{n+1}^{(k)} = \prod_{\ell=k}^{n+k} \lambda_1^{(\ell)} = a_{n+1}^{(k)}/C_{\hat{L}^k A}(0) \tag{22}$$

Equations (21) and (22) constitute a convenient algorithm for generating the lowering coefficients of a given TACF from its moments and *vice versa*. The procedure is nicely illustrated with the Gaussian $C_G(t) = \exp(-t^2/2)$ as a model TACF. From the Maclaurin expansion the moments of C_G are easily found to be $\bar{a}_n^{(0)} = a_n = (2n-1)!!$ Then from (22) the $\lambda_1^{(\ell)}$ are calculated (up to $\ell=5$) and entered into Table 1 as the first column. Clearly, $\lambda_1^{(\ell)}$ is the first moment of the ℓ-th embedded TACF derived from C_G.

Table 1.   Embedding table for
           the Gaussian
           $C_G(t) = \exp(-t^2/2)$

n \ ℓ	1	2	3	4	5	6
0	1	2	3	4	5	6
1	3	2	5	4	7	
2	5	2	7	4		
3	7	2	9			
4	9	2				
5	11					

Now from (21b) we have

$$\lambda_2^{(\ell-1)} = \lambda_1^{(\ell)} - \lambda_1^{(\ell-1)} \quad, \quad \ell = 1,2, \ldots 5,$$

which generates the second column of the Table.   Next, we rearrange (21a) to obtain

$$\lambda_3^{(\ell-1)} = \lambda_2^{(\ell)} \, \lambda_1^{(\ell)} \, / \lambda_2^{(\ell-1)} \quad, \quad \ell = 1,2,3,4,$$

which gives us the third column.   Then we return to (21b) to compute

$$\lambda_4^{(\ell-1)} = \lambda_3^{(\ell)} + \lambda_2^{(\ell)} - \lambda_3^{(\ell-1)} \quad, \quad \ell = 1,2,3. \tag{23}$$

Eq. (23) yields the fourth column.   Continuing in this fashion alternately using (21a) and (21b), we eventually fill in the entire embedding table.   Then the lowering coefficient $\lambda_j^{(0)}$ of C_G is simply the j-th entry of the first ($\ell=0$) row.

Of course, the above procedure can be "reversed" to calculate the first column of the table, and from (22) the moments, if we are given the first row, i.e. the lowering coefficients.

V.   BOUNDS ON TIME-CORRELATION FUNCTIONS VIA THE INFINITE-ORDER
     MACLAURIN EXPANSION

The results of Sections III and IV can be employed to answer
a very important question, namely: Precisely how much can we state
about the time-dependence of the TACF, given only the first n
moments? Platz and Gordon[12] have offered an answer by approximating
the TACF

$$C(t) = \sum_{i=1}^{n} \rho_i \cos(\omega_i t) \tag{24}$$

as the Fourier transform of a non-negative spectrum

$$\tilde{C}(\omega) = \sum_{i=1}^{n} \rho_i \delta(\omega-\omega_i) \tag{25}$$

that "optimizes the correlation" in the sense of giving optimal
bounds on the exact TACF $C_A(t)$ for suitably chosen positive weights
ρ_i and frequencies ω_i. The 2n unknowns $\{\rho_i\}$ and $\{\omega_i\}$ are determined
by an optimization procedure constrained to reproduce the first n
even moments $\{a_k\}_{k=0}^{n-1}$ of the exact TACF, i.e.

$$a_k = \sum_{i=1}^{n} \omega_i^{2k} \rho_i \, , \quad k = 0,1, \ldots n-1 \, . \tag{26}$$

Unfortunately the rather involved optimization procedure obscures
somewhat the mechanism by which information contained in the known
moments is "communicated" to the higher moments.

Now using the embedding relations (21) we can compute the first
n lowering coefficients corresponding to the known (first n) moments.
We then approximate the higher moments as

$$a_k' = a_k(\lambda_1,\lambda_2, \ldots \lambda_n,0,0 \ldots 0), \quad k \geq n,$$

which we calculate via the embedding relations. Since the λ's are
all positive, it follows from (19) that a_k' is a lower bound on a_k.

We next extend the optimization procedure[12] by requiring that
the additional n *approximate* moments $\{a_k'\}_{k=n}^{2n-1}$ satisfy the same
constraints as the first n, namely

$$a_k' = \sum_{i=1}^{n} \omega_i^{2k} \rho_i, \quad k=n, n+1, \ldots 2n-1 \, . \tag{27}$$

Together (26) and (27) constitute a set of 2n equations in 2n
unknowns, which has a unique solution if, and only if, $\text{Det}(\underline{\underline{\omega}}) \neq 0$,
where $(\underline{\underline{\omega}})_{ki} = \omega_i^{2k}$, k=0,1, ... n-1. Since $\underline{\underline{\omega}}$ is a matrix of the
Vandermonde type, $\text{Det}(\underline{\underline{\omega}})$ is nonzero if, and only if, $\omega_i \neq \omega_j$,
$i \neq j$.[13] It can be shown[11] that this condition is always satisfied
by classical systems and, further, that the resulting solution of
(26) and (27) yields optimal bounds on $C_A(t)$. Thus, augmentation
of (26) with (27) yields a procedure equivalent to that of Platz
and Gordon.[12]

We can now write an infinite partial Maclaurin expansion for
$C_A(t)$ as

$$C_M'(t) = C_M(t) + \sum_{k=n}^{\infty} (-1)^k a_k' t^{2k}/(2k)!$$

where

$$C_M(t) \equiv \sum_{k=0}^{n-1} (-1)^k a_k t^{2k}/(2k)!$$

Since (26) and (27) solve the same problem as (24) and (25), C_M' is
the *exact* Maclaurin expansion of C(t) given by (24). Thus
$|C_M'(t)| \leq 1$ for all t and, as a partial summation of full Maclaurin
series, C_M' is a better representation of $C_A(t)$ than is the truncated
expansion C_M.

This is illustrated in Fig. 1, where we plot the single-
particle velocity TACF for an isotopic impurity of mass 2m in an
otherwise pure infinite, one-dimensional harmonic chain of atoms
of mass m. We chose this model because it is exactly solvable.[14]
The first 21 (even) moments were used to calculate the various
approximations to $C_A(t)$. The "partial" moments a_k' were evaluated
from the embedding relations. The summations required in computing
C_M and C_M' were carried out directly, 55 partial moments being used
to calculate C_M'.

We observe that the optimization technique of Platz and
Gordon,[12] which must somehow solve a system of n equations [namely
(26)] in 2n unknowns, effectively builds in the additional constraints (27). The key point is that the moments are not independent quantities. Knowledge of the first n is carried forward into
all the higher ones, as (19) clearly indicates. In other words,
the partial moments $a_k'(k \geq n)$, which are lower bounds on the exact
moments, reflect all the information that we have from the first
n moments. Including this knowledge in the infinite partial
Maclaurin series C_M' gives a better estimate of $C_A(t)$ than simply
truncating the exact series C_M at the n-th term.

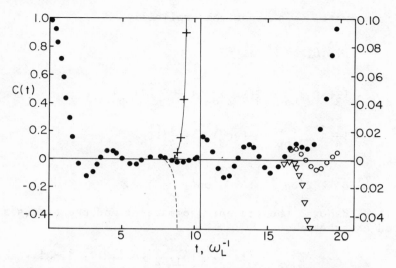

Fig. 1: $C_M'(t)$ and $C_M(t)$ for the infinite one-dimensional harmonic chain with a single isotopic impurity of mass twice that of the host atoms. ●, C_M' (21 moments); —, C_M (21 moments); △, C_M' (20 moments); --- C_M (20 moments); ◯, exact result $J_1(2\omega_L t)/(\omega_L t)$ (Ref. 14). Note shift in scale at $\omega_L t = 10.4$

VI. ZASSENHAUS APPROXIMATIONS TO THE TIME-CORRELATION FUNCTION

The resolution (14) of $\hat{L}$ into a simple sum of two operators suggests the possibility of an approximation to $C_A(t)$ based on the general Zassenhaus formula[15,16,17]

$$e^{t(\hat{A}+\hat{B})} = e^{t\hat{A}} e^{t\hat{B}} \prod_{n=2}^{\infty}{}' \ e^{t^n \hat{C}_n} ,$$

where $\hat{A}$ and $\hat{B}$ are any two operators and the $\hat{C}_n$ are derived from $\hat{A}$ and $\hat{B}$ through a set of recursive relations. The prime on Π' signifies an ordered product of the factors following, i.e. the factors occur in order of increasing n from left to right. Now taking $\hat{A} = \hat{R}$ and $\hat{B} = \hat{L}$, we can cast the "propagator" $\exp(\hat{L}t)$ as

$$e^{t\hat{L}} = e^{t(\hat{R}+\hat{L})} = e^{t\hat{R}} e^{t\hat{L}} \prod_{n=2}^{\infty}{}' \ e^{t^n \hat{C}_n(\hat{R},\hat{L})} \tag{28}$$

The first several $\hat{C}_n$'s are given explicitly as

$$\hat{C}_2 = -\frac{1}{2}[\hat{R},\hat{L}]$$

$$\hat{C}_3 = -[\frac{1}{3}\{\hat{R}\hat{C}_2\}_{op} + \frac{1}{3}\{\hat{L}\hat{C}_2\}_{op} + \frac{1}{3!}\{\hat{R}\hat{L}^2\}_{op}]$$

$$\hat{C}_4 = -[\frac{1}{4}\{\hat{R}\hat{L}\hat{C}_2\}_{op} + \frac{1}{4}\{\hat{R}\hat{C}_3\}_{op} + \frac{1}{4}\{\hat{L}\hat{C}_3\}_{op}$$

$$+ \frac{1}{8}\{\hat{C}_2\hat{C}_2\}_{op} + \frac{1}{4!}\{\hat{R}\hat{L}^3\}_{op}] ,$$

where $[\hat{A},\hat{B}]$ denotes the ordinary commutator and the braces signify a nested commutator

$$\{\hat{A}_1\hat{A}_2\hat{A}_3 \ldots \hat{A}_k\}_{op} \equiv [[[\ldots [\hat{A}_1,\hat{A}_2],\hat{A}_3], \ldots],\hat{A}_k] .$$

Substituting (28) into (7) and recalling (12a), we can write the TACF as

$$C_A(t) = <Z_o[e^{t\hat{R}}e^{t\hat{L}} \prod_{n=2}^{\infty} e^{t^n\hat{C}_n}]Z_o^* > . \tag{29}$$

Observe that

$$[e^{t\hat{L}} \prod_{n=2}^{\infty} e^{t^n\hat{C}_n}]Z_o^* = \sum_{k=0}^{\infty} f_k(t) Z_k^* ,$$

where $f_k(t)$ is a function only of t. Operating upon Z_k^* with $e^{t\hat{R}}$, where

$$e^{t\hat{R}} \equiv 1 + t\hat{R} + t^2\hat{R}^2/2 + \ldots t^n\hat{R}^n/n! \ldots , \tag{30}$$

raises the index k, except for the first term. On account of the orthogonality of the Schmidt variables [see (13)] only the first term of the expansion (30) contributes to the TACF and we can accordingly simplify (29) to

$$C_A(t) = <Z_o[e^{t\hat{L}} \prod_{n=2}^{\infty} e^{t^n\hat{C}_n}]Z_o^*>$$

$$= f_o(t) <Z_oZ_o^*>$$

$$= f_o(t) C_A(0) .$$

Now the relevant part of the propagator may be written more explicitly as

$$
e^{t\hat{L}} \prod_{n=2}^{\infty}{}' \; e^{t^n \hat{C}_n} = e^{t\hat{L}} e^{t^2 \hat{C}_2} \{ 1 + \sum_{k=1}^{\infty} \sum_{j=3}^{\infty} t^{kj} (\hat{C}_j)^k / k!
$$

$$
+ \sum_{k_1=1}^{\infty} \sum_{k_2=1}^{\infty} \sum_{j_1=3}^{\infty} \sum_{j_2=4}^{\infty} t^{k_1 j_1 + k_2 j_2} (\hat{C}_{j_1})^{k_1} (\hat{C}_{j_2})^{k_2} / (k_1! k_2!)
$$

$$
\ldots + \sum_{\underline{k}_n} \sum_{\underline{j}_n} t^{\underline{k}_n \cdot \underline{j}_n} \prod_{\ell=1}^{n} (\hat{C}_{j_\ell})^{k_\ell} / \prod_{\ell=1}^{n} k_\ell! + \ldots \} \tag{31}
$$

In (31) $\underline{k}_n$ is an n-component vector, each component an integer between 1 and ∞; the same is true of $\underline{j}_n$ except for the restriction $2 < j_1 < j_2 < j_3 \ldots < j_n$, which allows for the fact that the product $\prod_{n=2}^{}{}'$ is ordered.

There is literally an infinity of ways in which (31) can be truncated to yield approximations to the full propagator. Further, there is the potential problem that the chosen truncation may not converge, even in infinite order, i.e. even if an infinite number of terms are retained in (31). To resolve this problem, we allow only those truncations that when summed give a function whose Maclaurin expansion agrees with the exact expansion through a specified order in the time. Thus a truncation that gives $C_A(t)$ exactly through order t^6 is

$$
e^{t\hat{L}} e^{t^2 \hat{C}_2} \{ 1 + t^3 \hat{C}_3^{(1)} + t^4 [\hat{C}_4^{(0)} + \hat{C}_4^{(2)}]
$$

$$
+ t^5 \hat{C}_5^{(1)} + t^6 [\hat{C}_6^{(0)} + \hat{C}_3^{(-1)} \hat{C}_3^{(1)} / 2] \} \tag{32}
$$

where $\hat{C}_j^{(k)}$ denotes the component of $\hat{C}_j$ that operates upon z_ℓ^* to yield $z_{\ell+k}^*$, i.e. $\hat{C}_j^{(k)}$ raises the index of the Schmidt variable by k. The choice represented by (32) contains the least number of operators and is exact to the specified order in t. This approach has a certain mathematical integrity in that a sequence of such truncations, indexed by the order of "exactness" defined above, yields a sequence of approximations that converges to $C_A(t)$ wherever Maclaurin's expansion converges.

A little algebraic manipulation yields the following expression for $C_A(t)$ to order t^4:

$$\bar{C}_A(t) = C_A(t)/C_A(0) = [1 + (6\lambda_1^2 - 3\lambda_1\lambda_2)t^4/4!]$$

$$x \exp(-\lambda_1 t^2/2) - [(8\lambda_1^2 - 4\lambda_1\lambda_2)t^4/4!]$$

$$x \exp[-(\lambda_2 - \lambda_1)t^2/2] . \tag{33}$$

Note that the Gaussian factors in (33) are due to the fact that Z_k is an eigenfunction of the commutator $\hat{C}_2$, i.e.

$$\hat{C}_2 Z_k = -\frac{1}{2}[\hat{R},\hat{L}]Z_k = - [(\lambda_{k+1} - \lambda_k)/2]Z_k .$$

Moreover, the Gaussians imply that (33) is essentially an *infinite-order partial* summation of Maclaurin's expansion. We expect (33) to be of practical value only if $\lambda_{k+1} > \lambda_k$, so that the finite truncations of the Zassenhaus formula that we have defined decay at long time. However, even if this inequality is not satisfied, (33) cannot be a worse representation of $C_A(t)$ than the Maclaurin expansion to the same order.

We have recently applied[10] the Zassenhaus approximation to an analysis of computer-simulation data[18] on the normalized single-particle velocity TACF ϕ for a Lennard-Jones liquid. As a special case of (11), we have

$$\dot{\phi} = - \int_0^t dt' K(t') \phi(t-t') , \tag{34}$$

where A is taken to be the velocity $\vec{v}$ of a single "typical" molecule. By virtue of the universal character of the analysis of TACF's presented in Section III, we conclude that the "memory function" (MF) $K(t')$, which is the TACF of the "random force," can be expressed in terms of lowering coefficients, say μ_j, which are analogous to the λ_j. Substituting the Maclaurin expansions for ϕ and K (in terms of their respective lowering coefficients λ_j and μ_j) into (34) and equating coefficients of like powers of t on either side of the resulting equation, we obtain

$$\mu_1 = \lambda_2, \ \mu_2 = \lambda_3, \ \mu_3 = \lambda_4, \ \cdots \ . \tag{35}$$

Indeed, relation (35) can be generalized[19] to

$$\mu_j^{(N)} = \lambda_{j+N} ,$$

where $\mu_j^{(N)}$ are the lowering coefficients associated with the MF of the N-Schmidt-variable GLE. Now using (33) and (35), we can write the following Zassenhaus approximation for the MF to order t^4:

$$K^{(4)}(t) = e^{-\lambda_2 t^2/2} + g(t) \lambda_2 (\lambda_3 - 2\lambda_2) t^4/4! ,$$ (36)

where

$$g(t) = g^{(4)}(t) \equiv 4e^{-(\lambda_3-\lambda_2)t^2/2} - 3e^{-\lambda_2 t^2/2}$$

On the basis of previous considerations we conclude that expression (36) is exact to $O(t^4)$.

In analyzing their molecular-dynamics data, Levesque and Verlet[18] assumed the following form for the MF

$$K_{LV} = e^{-B_o t^2/2} + A_o e^{-\alpha_o t} t^4/\lambda_1 ,$$ (37)

where A_o, B_o and α_o are determined by a least-squares fit of the computed ϕ and the lowering coefficient

$$\lambda_1 \equiv \langle \dot{\vec{v}} \cdot \dot{\vec{v}} \rangle / \langle \vec{v} \cdot \vec{v} \rangle = \rho \int d\vec{r}\ g(r)\ \nabla_{\vec{r}}^2\ V(r)/3m$$ (38)

is obtained by performing the integration indicated in (38) numerically. Clearly, on account of the isotropy of the fluid and the spherical symmetry of the intermolecular potential $V(r)$ the integral is one dimensional. (In (38) $g(r)$ is the radial (pair) distribution function, m is the molecular mass and ρ is the number density).

The similarity between expressions (36) and (37) is striking. Indeed, if we require that $K^{(4)}$ and K_{LV} agree to $O(t^4)$, we find that

$$B_o = \lambda_2$$

$$A_o = \lambda_1 \lambda_2 (\lambda_3 - 2\lambda_2)/4!$$

Thus, K_{LV} has the form (36) with $g^{(4)}$ replaced by

$$g_{LV} \equiv e^{-\alpha_o t} .$$

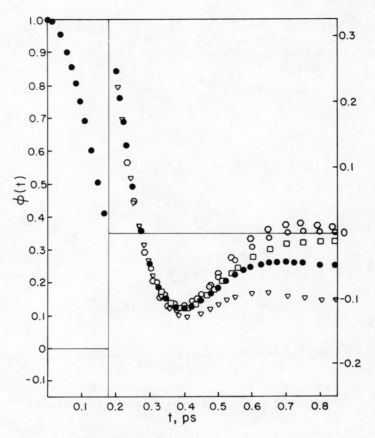

Fig. 2: Plots of single-particle velocity TACF for LJ(12,6) liquid
($\rho^* = 0.85$, $T^* = 0.76$) determined by solving (34) with MF
given by (36) with various possible g's. ●, g_{LV} (the
simulated result). O, g_s [$\beta = \beta^{(4)}$]; ∇, g_s [$\beta = \beta_{LV}$];
□, g_s [β chosen to give best fit of ϕ through first
minimum]; ◯, $g^{(4)}$. Note change of scale in ordinate at
$t \simeq 0.18$ psec. For $t < 0.18$ psec all approximations
agree within indicated limits.

A bothersome feature of g_{LV} is that it gives rise to odd powers in
t in the MF. Nevertheless, for the interval of time of principal
interest (.3 psec $\lesssim$ t $\lesssim$ 1.5 psec) the exponential describes well
not only the negative minimum in ϕ that sets in at high density,
but also the negative plateau beyond $t \simeq$ .6 psec. (see Fig. 2).
Thus, it seems necessary to retain the exponential in order to
realize a good description of ϕ over the range of t of interest.

Evenness in t and exponential behavior can be nicely reconciled by employing the function

$$g_s(t) = \text{sech}(\beta t) .$$

The parameter β can be estimated in a variety of ways, e.g. by requiring the Maclaurin expansions of $g^{(4)}$ and g_s to agree to order t^2, which yields

$$\beta^{(4)} = (4\lambda_3 - 7\lambda_2)^{1/2}$$

Another possibility is matching g_{LV} and g_s in the limit of large t, which gives

$$\beta_{LV} = \alpha_o .$$

Using the finite-difference procedure of Mountain,[20] we have solved (34) numerically for $\phi(t)$ corresponding to various approximate MF's. Taking a step size of .01 (in the reduced units of Ref. 18), we estimate that ϕ is precise to within 1% up to .5 psec and to 5-10% up to 1.0 psec. In Fig. 2 we plot ϕ for the thermodynamic state $\rho^* = 0.85$ and $T^* = 0.76$ with the MF given by (36). (The reduced units used here and later are defined by $\rho^* \equiv \rho/\sigma^3$, $T^* \equiv kT/\epsilon$, where σ and ϵ are the Lennard-Jones potential parameters.) Plots corresponding to four different versions of g are shown: $g^{(4)}$; g_{LV}; g_s with $\beta = \beta^{(4)}$; g_s with $\beta = \beta_{LV}$. It appears that the best agreement with the simulated data ($g=g_{LV}$) is obtained with $g_s(\beta = \beta^{(4)})$. This result is quite encouraging because it means that knowledge of just the short-time ($t \leq .5$) behavior of K(t') is sufficient to yield a good approximation of ϕ up to about 1 psec.

VII. APPROXIMATIONS TO THE CLASSICAL TRANSPORT INTEGRAL: SELF-DIFFUSION

In this Section we shall derive a systematic approximation to the zero-frequency component of the classical transport integral

$$\tilde{C}_A(0) = \int_o^\infty dt\ \bar{C}_A(t) . \tag{39}$$

Our starting point is the continued-fraction representation of the Laplace transform of $\bar{C}_A(t)$, which can be written in terms of the lowering coefficients as

$$\hat{C}_A(s) \equiv \int_o^\infty dt\ \exp(-st)\ \bar{C}_A(t) \tag{40a}$$

$$= \frac{1}{s} + \frac{\lambda_1}{s} + \frac{\lambda_2}{s} + \cdots, \quad s>0 \tag{40b}$$

Expression (40b) follows easily from Mori's[6] formalism by noting that Mori's Δ_j^2 is identical to λ_j and that Mori's ω_j vanishes for classical equilibrium systems. It can also be derived[19] via a set of linear fractional transformations[21] resulting from the Laplace transform of the GLE in the infinite-variable Schmidt basis.

If we ignore for the moment the restriction s>0 in (40), we conclude that

$$\tilde{C}_A(0) = \hat{C}_A(0) . \tag{41}$$

Then setting s=0 in (40b), we obtain

$$\hat{C}_A(0) = 1/\lambda_1/\lambda_2/ \ldots, \tag{42}$$

which "simplifies" to the ill-conditioned result

$$\hat{C}_A(0) = (\lambda_2\lambda_4\lambda_6 \ldots)/(\lambda_1\lambda_3\lambda_5 \ldots) \tag{43}$$

Let us tentatively interpret (43) to be given by the "odd" approximant of (42), i.e.

$$[\hat{C}_A(0)]_{odd} = \lim_{j \to \infty} (\lambda_2\lambda_4\lambda_6 \cdots \lambda_{2j-2})/(\lambda_1\lambda_3\lambda_5 \cdots \lambda_{2j-1}) .$$

Now since λ_k's have units of $(time)^{-2}$ and $\hat{C}_A(0)$ has units of time [see (39)], the "odd" approximant must have units of $(time)^2$. Now let us take (43) to be given by the "even" approximant, i.e.

$$[\hat{C}_A(0)]_{even} = \lim_{j \to \infty} (\lambda_2\lambda_4\lambda_6 \cdots \lambda_{2j})/(\lambda_1\lambda_3\lambda_5 \cdots \lambda_{2j-1}) ,$$

which is clearly unitless. If we insist on using the approximants to evaluate $\tilde{C}_A(0)$, this dimensional analysis suggests that we consider the geometric mean of an "odd" approximant and the succeeding "even" one, i.e.

$$\hat{C}_A(0) = \{[\hat{C}_A(0)]_{odd}[\hat{C}_A(0)]_{even}\}^{1/2} = \lim_{j \to \infty} h_j \lambda_{2j}^{-1/2}$$

where h_j is defined by

$$h_j \equiv (\lambda_2\lambda_4\lambda_6 \cdots \lambda_{2j})/(\lambda_1\lambda_3\lambda_5 \cdots \lambda_{2j-1}).$$

Alternatively, we could consider the mean of an "even" approximant and the succeeding "odd" one to get

$$\hat{C}_A(0) = \lim_{j \to \infty} h_j \; \lambda_{2j+1}^{-1/2}$$

It can be shown[22] that a sufficient condition for the sequences

$$\hat{C}_A^{(2n)}(0) = h_n \; \lambda_{2n}^{-1/2} \tag{44a}$$

$$\hat{C}_A^{(2n+1)}(0) = h_n \; \lambda_{2n+1}^{-1/2} \tag{44b}$$

to converge to the same limit (provided that limit is nonzero and finite) $\hat{C}_A$ is that

$$\lim_{n \to \infty} \lambda_{n+1}/\lambda_n = 1 \tag{45}$$

Intuitively, (45) seems like an eminently reasonable requirement, although clearly not the only mathematically possible one.

To examine the utility of formulas (44), we have again turned[22] to the computational experiments of Levesque and Verlet[18] on the Lennard-Jones liquid. We shall focus on the diffusion coefficient, which can be expressed as

$$D = (kT/m) \int_o^\infty dt \; \phi(t) \; , \tag{46}$$

i.e. as the zero-frequency Fourier component of the single-particle velocity TACF. From (44) and (46) we have for the first four mean approximants to D

$$D_1 = \eta \; \lambda_1^{-1/2} \tag{47a}$$

$$D_2 = \eta(\lambda_2/\lambda_1)\lambda_2^{-1/2} \tag{47b}$$

$$D_3 = \eta(\lambda_2/\lambda_1)\lambda_3^{-1/2} \tag{47c}$$

$$D_4 = \eta(\lambda_2\lambda_4/\lambda_1\lambda_3)\lambda_4^{-1/2} \; , \tag{47d}$$

where $\eta \equiv kT/m$. The values of the lowering coefficients are estimated from the simulated data[18] in the manner discussed in Section

VI. Except at high density ($\rho^* \gtrsim 0.8$) and low temperature ($T^* \sim 1.0$), i.e. near the triple point, D_2 and D_4 give the best agreement (to within 15%) with the simulated value D_{LV}. The odd-indexed approximants work best at the triple point, where they are significantly better than the even-indexed ones. As T^* increases, the even-indexed approximations become superior.

A very interesting feature of the sequence (47) is the way in which the functional dependence of D upon the thermodynamic state develops. Levesque and Verlet[18] summarize their data (for $\rho^* \gtrsim 0.65$) with the following "empirical" expression:

$$D_{LV} = 0.006423T^* \rho^{*-2} - .028\rho^* + 0.0222 \tag{48}$$

The leading term of (48) accounts for about 90% of D_{LV}. We have observed the state dependences of the lowering coefficients to be roughly

$$\lambda_1 \alpha\ T^{*1/2} \rho^{*2}$$

$$\lambda_i \alpha\ T^* \ , \ i = 2,3,4 \tag{49}$$

Recalling that η is proportional to T^*, one sees from (47) and (49) that

$$D_1 \alpha\ T^{*3/4} \rho^{*-1}$$

$$D_i \alpha\ T^* \rho^{*-2} \ , \ i = 2,3,4 \ .$$

Thus, the primary functional dependence of D [see (48)] seems to have "set in" by D_2. Since, with D_3 and either D_2 or D_4, one can obtain estimates within 85 or 90% of D_{LV} and since the $T^* \rho^{*-2}$ term of (48) is sufficient to come within 90% of D_{LV}, it is clear that the thermodynamic state dependence of D can be well understood from that of the first four lowering coefficients [or, equivalently, the first four moments of $\phi(\omega)$].

VIII. ENERGY TRANSPORT IN FLUIDS

Buoyed by the success of formulas (44) in estimating the diffusion coefficient, we should like to ascertain whether they are equally useful in estimating other characteristic microscopic times, in particular those associated with energy-relaxation processes. The simplest example is single-particle (translational) kinetic-energy relaxation, for which we can define the time constant [see (3)]

$$\tau = \int_0^\infty dt \ \bar{C}_{KE}(t) \qquad\qquad (50)$$

where

$$\bar{C}_{KE}(t) = <[p^2(0) - <p^2>][p^2(t) - <p^2>]>/$$

$$[<p^4> - <p^2>^2]$$

is the normalized single-particle kinetic-energy TACF. Clearly
τ given by (50) is the analog of D given by (46). Similarly,
the mean approximants to τ are given by expressions analogous to
(47). Here we shall estimate τ from simply the first mean approximant, namely

$$\tau_1 = [\lambda_1^{(KE)}]^{-1/2}$$

It is straightforward to show that $\lambda_1^{(KE)}$ is related to the first
lowering coefficient λ_1 of the velocity TACF ϕ by

$$\lambda_1^{(KE)} = 2\lambda_1$$

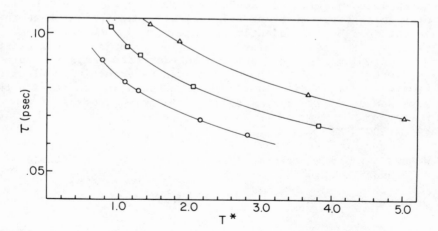

Fig. 3:   Single-particle kinetic-energy relaxation time as a
          function of thermodynamic state for the Leonard-Jones
          fluid. Δ, $\rho^* = 0.65$; $\square$, $\rho^* = 0.75$; $\bigcirc$, $\rho^* = 0.85$.

Therefore,

$$\tau \sim (2\lambda_1)^{-1/2} \tag{51}$$

The kinetic-energy relaxation time for the Lennard-Jones fluid in various thermodynamic states has been calculated from the data of Levesque and Verlet[18] using (51). The results, given in Fig. 3, are in accord with one's expectations that the rate of relaxation should increase with increasing temperature at fixed density and should increase with increasing density at fixed temperature. Also, the order of magnitude of τ of about 10^{-13} sec seems reasonable and is in rough agreement with the results of other molecular-dynamics calculations.[23]

While these results are encouraging, it is still necessary to evaluate several higher approximants τ_i in order to ascertain whether convergence has been achieved. Unfortunately, the higher lowering coefficients associated with $\bar{C}_{KE}$ are not as simply related to those of ϕ as is $\lambda_1^{(KE)}$.

IX.   SUMMARY AND CONCLUSION

We have presented a dynamical treatment of irreversible processes that derives essentially from a combinatorial approach to the classical moment problem. Given that many experimental measurements on relaxation processes, e.g. scattering cross sections, absorption and emission spectra and thermal transport coefficients, are expressible in terms of the Fourier transform of the equilibrium TACF of a characteristic dynamical variable, and further that these measurements can be conveniently characterized in terms of the moments of the Fourier transform, it follows that new insight into the moment problem may yield new insight into the relaxation process being measured. Indeed, we have presented here results reflecting some progress along these lines. We are continuing to develop the formalism and to apply it to specific relaxation processes.

X.   ACKNOWLEDGEMENT

We are pleased to acknowledge the support of the National Science Foundation.

REFERENCES

1.   D. A. McQuarrie, "Statistical Mechanics," Harper and Row, New York (1976).
2.   See, for example, J. T. Hynes and J. M. Deutch, Nonequilibrium Problems - Projection Operator Techniques, in: Vol. 11B, "Physical Chemistry, an Advanced Treatise," H. Eyring, D. Henderson, and W. Jost, eds., Academic, New York (1975).

3. See, for example, D. Kivelson and K. Ogan, "Spin Relaxation in Terms of Mori's Formalism," Adv. Mag. Res. 7: 71 (1974).

4. R. Zwanzig, Statistical Mechanics of Irreversibility, in: "Lectures in Theoretical Physics," Vol. 3, W. E. Brittin, B. W. Bowns and J. Downs, eds., Interscience, New York (1961).

5. H. Mori, "Transport, Collective Motion, and Brownian Motion," Prog. Theoret. Phys. 33: 423 (1965).

6. H. Mori, "A Continued-Fraction Representation of the Time-Correlation Functions," Prog. Theoret. Phys. 34: 399 (1965).

7. M. Dupuis, "Moment and Continued-Fraction Expansions of Time-Autocorrelation Functions," Prog. Theoret. Phys 37: 502 (1967).

8. F. Lado, "Density Autocorrelation Function in a Classical Fluid from Initial Correlations," Phys. Rev. A2: 1467 (1970).

9. M. Weinberg and R. Kapral, "Continued-Fraction Description of Collective Motion in Simple Fluids," Phys. Rev. A4: 1127 (1971)

10. J. W. Allen and D. J. Diestler, "On the Calculation of Classical Time-Autocorrelation Functions: Resolution of the Liouville Operator," J. Chem. Phys. 73: 4597 (1980).

11. J. W. Allen and D. J. Diestler, unpublished.

12. O. Platz and R. G. Gordon, "Rigorous Bounds on Time-Dependent Correlation Functions," Phys. Rev. Lett. 30: 264 (1973).

13. V. I. Smirnov, "Linear Algebra and Group Theory," Dover, New York (1970).

14. R. J. Rubin, "Statistical Dynamics of Simple Cubic Lattices. Model for the Study of Brownian Motion," J. Math. Phys. 1: 309 (1960).

15. R. M. Wilcox, "Exponential Operators and Parameter Differentiation in Quantum Physics," J. Math Phys. 8: 962 (1967).

16. W. Magnus, "On the Solution of Differential Equations for a Linear Operator," Comm. Pure and Appl. Math. 7: 649 (1954).

17. W. Magnus, A. Karrass and D. Solitar, "Combinatorial Group Theory," Dover, New York (1976).

18. D. Levesque and L. Verlet, "Computer 'Experiments' on Classical Fluids. III. Time-Dependent Self-Correlation Functions," Phys. Rev. A2: 2514 (1970).

19. J. W. Allen and D. J. Diestler, "Analytic Characteristics of Time-Autocorrelation Functions," Phys. Rev. A, submitted for publication.

20. R. D. Mountain, "Single-Particle Motions in Liquids: Qualitative Features of Memory Functions," J. Res. Nat. Bur. Stds. 78A: 413 (1974).

21. H. S. Wall, "Continued Fractions," Van Nostrand, New York, 1948.

22. J. W. Allen and D. J. Diestler, "Universal Formulas for Classical Transport Integrals: The Zero-Frequency Component," J. Chem. Phys. 73: 6316 (1980).

23.   G. Harp and B. J. Berne, "Linear- and Angular-Momentum Auto-
      correlation Functions in Diatomic Liquids," J. Chem. Phys.
      49: 1249 (1968).

A COMMENT ON DYNAMICAL CHAOS IN CLASSICAL AND

QUANTUM MECHANICAL HAMILTONIAN SYSTEMS

Stuart A. Rice

The Department of Chemistry and
The James Franck Institute
The University of Chicago
Chicago, Illinois 60637

I. INTRODUCTION

The development, via laser technology, of new and very selective methods for preparing nonstationary states of molecules has stimulated interest in the dynamics of intramolecular relaxation processes. The earliest theoretical discussions of intramolecular energy transfer are now more than fifty years old,[1] and deal with its influence on the rate of thermal unimolecular reactions. The theory assumes, and experiment confirms,[2] that for the vast majority of unimolecular reactions induced by collisional excitation the rate of intramolecular energy transfer greatly exceeds the rate of reaction, so the latter depends only on the energy of the molecule. However, because rapid intramolecular energy transfer is assumed to occur, the theory does not address questions such as:
(i) What conditions must the molecular parameters, e.g. anharmonicities, vibration-rotation couplings, ..., satisy in order to make intramolecular energy transfer more rapid than reaction?
(ii) Do the conditions defined by the answer to (i) describe almost all molecules? Do these conditions depend on the excitation process?
(iii) Are there special initial states for which intramolecular energy transfer is slower than reaction?
(iv) If the answer to (iii) is "yes", how can these states be prepared?
(v) If the answer to (iii) is "yes", and the answer to (iv) is known, how does the outcome of the reaction depend on the initial state?

A principal tool in the new theoretical approaches to the study of intramolecular energy transfer has been the construction of anal-

ogies between the behavior of classical and quantum mechanical model
systems. It is now well established, from analytic and numerical
studies, that the motion determined by a nonintegrable classical
mechanical Hamiltonian is quasi-periodic for low energy and chaotic
for high energy.[3,4] A variety of subjective criteria suggest that similar behavior occurs for the motion determined by a quantum mechanical Hamiltonian, but the lack of a suitable definition of chaos for
a quantum mechanical system makes the inference a very primitive,
and possibly incorrect, characterization of the motion. In any
event, it is the onset of chaos which is now identified with the
onset of rapid intramolecular energy transfer. This paper discusses
a few aspects of dynamical chaos in classical and quantum mechanical
Hamiltonian systems, with attention focussed on what is meant by
chaos and how it is measured. The contents of the paper overlap
but are not identical with the contents of the talk I gave at the
Bielefeld conference: How Stationary are the Internal States of a
Molecule? In particular, I have omitted any description of the
background classical mechanics,[10] KAM [11] theorem, Nordholm-Rice [5]
calculations, etc., all of which have been reviewed elsewhere.[1,12,13]
The work which is described was carried out in collaboration with
Dr. Ronnie Kosloff.[14,15]

II. CHAOTIC MOTION: BACKGROUND INFORMATION

We begin our discussion with a summary of some of the characteristics of motion in a classical mechanical system. In particular, we note that:
(i) When the equation of motion is separable, the trajectories are
quasi-periodic and are restricted to lie on a torus in the phase
space of the system. When the equation of motion is not separable
some of the trajectories are quasi-periodic, but not all; these
other trajectories are not restricted to lie on a torus although
they, of course, do lie on the energy surface. [4,10]

(ii) The Kolmogorov- Arnold-Moser (KAM) theorem establishes that
most of the quasi-periodic trajectories of some unperturbed system
survive under sufficiently small perturbation, but there is also
generated a set of chaotic trajectories. The relative weight of
the chaotic trajectories grows as the energy increases, and they
eventually fill all of the accessible phase space.[11,16]
(iii) The onset of chaotic behavior is relatively sharp, but continuous, at an energy E_c. Below E_c almost all trajectories are
quasi-periodic, but there are some chaotic trajectories. Above E_c
most trajectories are chaotic, but there is also a dense set of
quasi-periodic trajectories. The quasiperiodic trajectories embedded in the chaotic domain above E_c do not reflect global invar-
iants of the system because their properties, e.g. the periods of
the orbits, are extremely sensitive to very small changes in the
initial conditions. [16,17]
(iv) Consider two trajectories corresponding to infinitesimally

different initial conditions. In the quasiperiodic motion domain,
$E < E_c$, the distance between these trajectories grows linearly with
time, whereas in the chaotic motion domain, $E > E_c$, it grows exponentially with time. [18]
(v) The Kolmogorov entropy of the system, which can be thought of
as a measure of the chaos generated by the motion, is zero in the
domain of quasi-periodic motion and positive in the domain of chaotic
motion. Moreover, the value of the Kolmogorov entropy is related to
the average over the phase space of the characteristic e-folding time
for exponential growth of the distance between initially close trajectories. [11,19-21]

Consider now the quantum mechanical description of a system which
undergoes the classical mechanical quasi-periodic-to-chaotic-motion
transition. Is there a change in the character of the stationary
states of this system at E_c? If the answswer to this question is yes,
what are the properties of the stationary states for $E > E_c$?

In order to address the questions just posed we must select a
criterion which signals the transition to chaotic behavior. Nordholm
and Rice [5] suggested that a state of the system is "ergodic" if an
excitation initially not uniformly distributed over the energy surface becomes uniformly distributed as $t \to \infty$. They showed, for several examples which undergo the classical mechanical quasi-periodic-
to-chaotic motion transition, that if the eigenstates are represented as a superposition of harmonic oscillator basis states, then
there is a "KAM-like" change in the amplitudes of the contributing
basis functions at about E_c. Stratt, Handy and Miller [7] repeated
these calculations using the analogue of natural orbital basis states
to examine the distribution of projections of amplitudes of the eig-
en states; they reach the same conclusion as do Nordholm and Rice
on the existence of a "KAM-like" transition in the quantum mechanical system. Stratt, Handy and MIller also follow up a suggestion
made by Pechukas, [22] and suggest that the nodal pattern of a wavefunction corresponding to $E > E_c$ is much more "irregular" than the
pattern when $E < E_c$. Pomphrey, [8] and later Marcus and coworkers, [6]
suggest that for $E > E_c$ the eigenvalues of the system become very
sensitive to such variations. The behavior of the nodal pattern of
the wavefunction as a function of energy, and of the eigenvalue
sensitivity to variation in coupling as a function of energy, both
confirm the findings of Nordholm and Rice.

Yet all of the criteria thus far introduced to categorize the
behavior of quantum mechanical systems are subjective. For example,
although the projections of the wavefunction on to harmonic oscillator or natural oscillator basis functions show a "KAM-like" transition, the wavefunction itself and the corresponding Wigner function appear to be regular deep in the classically chaotic domain, [9]
and it is possible to invent separable systems for which the nodal
pattern is as "regular" or "irregular" as desired. Furthermore, the
evidence from semiclassical quantum theory suggests that only a sub-

set of quasi-periodic trajectories embedded in the domain of chaotic trajectories correlate with eigenstates of the system.[23] Despite the fact that the measure of these particular embedded quasi-periodic trajectories is so very small compared to the measure of all trajectories, they appear to exhaust the eigenvalue spectrum. What then is the meaning of the "KAM-like" behavior of the projections of a full system eigenvector on to some given set of basis vectors?

In order to be able to compare classical mechanical and quantum mechanical chaotic motion, we must introduce some common measure of the chaos. At the qualitative level, the most striking feature of motion in the classically chaotic domain is the exponential divergence of the separation of two trajectories started with infinitesimally different initial conditions. However, because there is no quantum mechanical analogue to the classical mechanical trajectory, we cannot use this criterion outside the domain of classical mechanics. On the other hand, we have already noted that the Kolmogorov entropy of a classical mechanical system is related to the average rate of divergence of initially adjacent pairs of trajectories,[21] so if the Kolmogorov entropy can be generalized to the quantun mechanical domain we would have a means of comparing the behavior of a given system when described alternatively by classical and quantum mechanics.

In Section IX of this paper we develop a generalization of the Kolmogorov entropy suitable for quantum mechanical systems, and examine its behavior in the cases that the eigenvalue spectrum is discrete or continuous. We also examine the quantum dynamics of two models [24,25] previously proposed as examples of systems which undergo a transition to quantum mechanical chaotic motion. We shall show that when the eigenvalue spectrum is discrete both the classical and quantum mechanical Kolmogorov entropy are zero and the corresponding system dynamics is quasi-periodic. In addition, we shall show that the models previously used to demonstrate quantum mechanical chaos have been incompletely analyzed; when exact solutions for the dynamics of these models are obtained the behavior agrees completely with the predictions based on the Kolmogorov entropy. Some implications of these results for the study of intramolecular energy transfer are discussed in the last section of this paper.

III. THE KOLMOGOROV ENTROPY OF A CLASSICAL MECHANICAL SYSTEM

In this section we consider some of the properties of the Kolmogorov entropy of a classical mechanical system. Our intent is to provide only such information as aids understanding of the extension to quantum mechanical systems made in Section IX. [26]

The classical mechanical description of a system we use is based on a phase space Γ and a time evolution operator $\hat{U}_t$ defined on that phase space. Given the Hamiltonian equations of motion, Liouville's theorem guarantees that a mapping of the phase space into itself,

$$\Pi_t = \hat{U}_t \Pi_o ,$$

(3.1)

is measure preserving. However, the "shape" of a volume element in phase space can be, and usually is, grossly distorted by successive mappings. Thus, although the motion of each representative point is determined uniquely by the initial conditions and the equations of motion, the relative motion of two points initially close together can appear to be erratic. In a sense, evolution which generates chaotic components of the relative motion can be thought of as a source of random noise. The Kolmogorov entropy is a function defined for the purpose of classifying the extent to which the relative motion is chaotic; it is based on an evaluation of the average amount of uncertainty in the relative location of phase points generated by motion of the system.

To implement the notion that uncertainty is associated with some aspects of the relative motion of points in phase space, we define a partition $\mathbb{P}^{(0)}$ of Π. This partition consists of a collection of non-empty, nonintersecting sets $\Omega_i(0)$ that completely cover Π. The volume of phase space associated with $\Omega_i(0)$ is

$$W(\Omega_i^{(0)}) = \int_{\Omega_i^{(0)}} f \, d\Pi$$

(3.2)

where f is the density of phase points in Π. The evolution operator $\hat{U}_t$ maps the partition $\mathbb{P}(0)$ into a new partition $\mathbb{P}(1)$ such that

$$\Omega_i^{(1)} = \hat{U}_t \Omega_i^{(0)}$$

(3.3)

We associate an information entropy with the partition $\mathbb{P}(0)$ by use of the definition

$$h(\mathbb{P}^{(0)}) \equiv - \sum_i W(\Omega_i^{(0)}) \ln W(\Omega_i^{(0)}).$$

(3.4)

Since $\hat{U}_t W(\Omega_i(0)) = W(\Omega_i^{(0)})$ by Liouville's theorem, we have $h(\mathbb{P}(0)) = h(\hat{U}_t \mathbb{P}(0))$, i.e. the information entropy is conserved under the motion of the system.

Consider, now, two partitions of the phase space Π, namely $\mathbb{P}^{(1)}$ and $\mathbb{P}(2)$. From these two partitions we generate the product partition $\mathbb{P}^{(2)} \vee \mathbb{P}^{(1)}$ which consists of all intersections $\Omega_i^{(2)} \cap \Omega_j^{(1)}$. The joint information entropy of the product partition is

$$h(\mathbb{P}^{(2)} \vee \mathbb{P}^{(1)}) = - \sum_{i,j}' \pi_{ij} \ln \pi_{ij}$$

(3.5)

where

$$\pi_{ij} \equiv W(\Omega_i^{(2)} \cap \Omega_j^{(2)}).\tag{3.6}$$

We now define the conditional information entropy

$$h(\mathbb{P}^{(2)}|\mathbb{P}^{(1)}) \equiv -\sum_{i,j} \pi_{ij} \ln(\pi_{ij}/W(\Omega_j^{(1)})),\tag{3.7}$$

which refers to the overlap between the partitions $\mathbb{P}^{(2)}$ and $\mathbb{P}^{(1)}$. Note that when $\mathbb{P}^{(2)} = \mathbb{P}^{(1)}$, $h(\mathbb{P}^{(2)}|\mathbb{P}^{(1)}) = 0$. Therefore, if $\mathbb{P}^{(2)} = \hat{U}_t \mathbb{P}^{(1)}$, we can take $h(\mathbb{P}^{(2)}|\mathbb{P}^{(1)})$ to be a measure of the distortion in the shape of the volume element generated by evolution under the equation of motion. Proceeding in an analogous fashion, we define the joint partition $\mathbb{P}_n$ as the __product__ of all the partitions produced by evolution for n time steps under the equation of motion,

$$\mathbb{P}_n \equiv \mathbb{P}^{(0)} \vee \hat{U}_t \mathbb{P}^{(0)} \vee \hat{U}_t^2 \mathbb{P}^{(0)} \vee \cdots \vee \hat{U}_t^n \mathbb{P}^{(0)},\tag{3.8}$$

and we define the average entropy per time step for a given partition $\mathbb{P}^{(0)}$ and evolution operator $\hat{U}_t$ by the limiting process

$$h(\mathbb{P}^{(0)}, \hat{U}_t) \equiv \lim_{n \to \infty} \frac{1}{n} h(\mathbb{P}_n).\tag{3.9}$$

If $\hat{U}_t$ is the identity operator, $\hat{U}_t^n \mathbb{P}^{(0)} = \mathbb{P}^{(0)}$ and since $h(\mathbb{P}^{(0)})$ is finite, taking the limit in (9) gives $h(\mathbb{P}^{(0)}, \hat{U}_t) = 0$. At the other extreme, if all the successive partitions are independent, it can be shown that $h(\mathbb{P}^{(0)}, \hat{U}_t) = h(\mathbb{P}^{(0)})$.

The average entropy per time step can also be represented in the form [27]

$$h(\mathbb{P}^{(0)}, \hat{U}_t) = \lim_{n \to \infty} h(\mathbb{P}^{(n)}|\mathbb{P}^{(0)} \vee \mathbb{P}^{(1)} \vee \cdots \vee \mathbb{P}^{(n-1)})\tag{3.10}$$

which can be utilized as follows: if, under a long succession of mappings, the partition $\mathbb{P}^{(n)}$ cannot be completely inferred from the previous n-1 partitions, then $h(\mathbb{P}^{(0)}, \hat{U}_t) > 0$ and the motion of the system is considered chaotic; if the converse is true $h(\mathbb{P}^{(0)}, \hat{U}_t) = 0$ and the motion of the system is quasiperiodic. The Kolmogorov entropy is designed to represent only the chaotic properties of the motion generated by the evolution operator $\hat{U}_t$. Given all possible partitions of the phase space $\mathbb{\pi}$, the Kolmogorov entropy is defined by

$$h_k \equiv \sup_{\mathbb{P}} h(\mathbb{P}, \hat{U}_t).\tag{3.11}$$

That is, by that partition which maximizes the average entropy per time step for given $\hat{U}_t$.

The Kolmogorov entropy of a classical mechanical system can be shown to have the following properties:
(i) h_K is invariant to an isomorphism of the evolution operator $\hat{U}_t$, so that if desired an analogous motion on a different space can be used to calculate its value. [11, 19, 20]
(ii) It is found that [11, 19, 20]

$$h_K(\hat{U}_t^n) = n\, h_K(\hat{U}_t),$$

(3.12)

from which one can draw two important inferences. First the Kolmogorov entropy per time step is invariant to the time scale and, second, if the motion is periodic for a finite number of time steps, n, the Kolmogorov entropy is zero. The first inference follows from division of both sides of (3.12) by n, while the second is a consequence of the fact that for a periodic evolution operator $\hat{U}_t m = 1$ for some m, and every partition is invariant under the identity operator $\hat{1}$.
(iii) In a mixing system, for which $h_K > 0$, a small region of phase space of volume W will become uniformly distributed with e-folding time $-\ln W/h_K$. To be more specific, the exponential rate of divergence of pairs of initially adjacent trajectories is measured by the Liapanov characteristic number $\chi(\Gamma)$, where Γ is a point in the phase space Π. A remarkable theorem by Piesen[21] relates the Ljapunow number to the Kolmogorov entropy:

$$h_K(\hat{U}_t) = \sum_i \int \chi_i(\Gamma)\, d\Pi.$$

(3.13)

The sum in (3.13) is over all vectors along the tangent mapping of the evolution operator for which $\chi(\Gamma) > 0$; Piesen shows that there can be no more than N-1 of these vectors for a system with N degrees of freedom.[21] Eq. (3.13) establishes a link between the mechanical notion of diverging trajectories and the entropic measure of the chaotic properties of the system evolution under the given equation of motion.

IV.   AN ASIDE ON THE ONSET OF CHAOS

Several methods have been proposed for locating the onset of chaotic motion, of which one is particularly relevant to the arguments advanced in this paper. Brumer and Duff,[28] Toda,[29] and Cerjan and Reinhardt [30] (BDTCR) assume that the existence of a local instability in the relative motion of a pair of trajectories which were initially close generates a global, long-time, instability; the local instability is found by use of a stationary point analysis of the nonlinear dynamics of the relative motion of the neighboring trajectories. In contrast, by construction, the Kolmogorov entropy

characterizes the global dynamics of a system. Yet, Brumer and
Duff's numerical calculations show that there is a correlation between the rate of adjacent trajectory divergence in the neighborhood
of a local instability and the global rate of adjacent trajectory
divergence. It appears to be the case that the onset of chaotic
motion is closely connected with the penetration of the trajectory
into a region of space in which it is unstable with respect to small
changes in initial conditions, even if that region is of limited
size. We shall return to this comment later. Although it cannot be
generally true that the stationary points of the difference flow control the global dynamics, the BDTCR method successfully predicts the
onset of chaotic behavior in several systems.

Kosloff and Rice[14] have proposed an improved local criterion for
the transition from quasiperiodic to chaotic motion in classical
mechanical Hamiltonian systems. Their approach addresses the consequences of the following observation:[31] When two diverging trajectories are on the same line of flow in phase space, they eventually cover the same region of phase space, and the motion is not
mixing; it is only neighboring trajectories that diverge in the
direction perpendicular to the flow in phase space which contribute
to chaotic motion. The BDTCR method does not take into account the
consequences of this observation.

The Kosloff-Rice analysis of the onset of chaotic motion under
a given Hamiltonian resembles that of BDTCR in its use of linearized
equations of motion for the difference between two trajectories. By
modifying a method developed by Gutzwiller[32] for the analysis of
semiclassical quantization, they identify the directions of flow in
phase space, and then use the fact that divergent motion in the direction of the flow is independent of that perpendicular to this flow.
By construction, the Kosloff-Rice criterion gives qualitative information which further categorizes the trajectories that satisfy the
BDTCR condition.

Consider the trajectory represented in Fig. 1; the analysis is
simplified if we change to a new set of coordinates which moves with
the trajectory. In these new coordinates q_1 is in the direction of
the flow and q_2 is perpendicular to the flow. Therefore, from Hamilton's equations of motion, with the conventional notation, we find

$$\dot{q}_1 = -\frac{\partial H}{\partial p_1} ; \quad \dot{q}_2 = 0. \tag{4.1}$$

The motion under the Hamiltonian H is now represented by the action
function $S(q',q'',t)$, which is defined by the partial differential
equations

$$H\left(\frac{\partial S}{\partial q_1}, q'\right) = E, \tag{4.2}$$

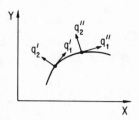

Fig. 1. A Coordinate system which moves with the trajectory.

$$H\left(\frac{\partial S}{\partial q''}, q''\right) = E,$$

(4.3)

where q' and q'' are on the system trajectory and t=t'' -t'. Differentiating (4.2) with respect to q'' and (4.3) with respect to q' we obtain

$$\sum_j \frac{\partial H}{\partial p_j'} \frac{\partial^2 S}{\partial q_i'' \partial q_j'} = 0,$$

(4.4)

$$\sum_j \frac{\partial H}{\partial p_j''} \frac{\partial^2 S}{\partial q_i' \partial q_j''} = 0,$$

(4.5)

which, after introduction of (4.1) leads to the decoupling condition

$$\frac{\partial^2 S}{\partial q_i'' \partial q_i'} = \frac{\partial^2 S}{\partial q_i' \partial q_i''} = 0.$$

(4.6)

We now seek the linearized mapping which transforms the initial difference between two trajectories into the final difference. This mapping is defined by

$$\begin{pmatrix} \xi'' \\ \eta'' \end{pmatrix} = M \begin{pmatrix} \xi' \\ \eta' \end{pmatrix}$$

(4.7)

where

$$\xi \equiv q - q^o \; ; \; \eta \equiv p - p^o.$$

(4.8)

Given that

$$p' = \frac{\partial S(q', q'', t)}{\partial q'}, \quad p'' = \frac{\partial S(q', q'', t)}{\partial q''}, \tag{4.9}$$

we find

$$\eta' = \frac{\partial^2 S}{\partial q' \partial q'} \xi' - \frac{\partial^2 S}{\partial q' \partial q''} \xi''$$

$$\eta'' = \frac{\partial^2 S}{\partial q'' \partial q'} \xi' + \frac{\partial^2 S}{\partial q'' \partial q''} \xi'' \tag{4.10}$$

For fixed t we then obtain the linearized mapping

$$M = \begin{pmatrix} -ba & -b^{-1} \\ b' - cb^{-1}a & -cb^{-1} \end{pmatrix} \tag{4.11}$$

where

$$a \equiv \frac{\partial^2 S}{\partial q' \partial q'}; \quad b \equiv \frac{\partial^2 S}{\partial q' \partial q''}; \quad c \equiv \frac{\partial^2 S}{\partial q'' \partial q''}. \tag{4.12}$$

A stability analysis of the motion examines the eigenvalues of the matrix (4.11). Because the mapping (4.11) is area preserving the product of the eigenvalues is one. If the eigenvalues are on the unit circle the motion is stable; otherwise the motion is unstable and we find diverging trajectories. For a two dimensional system the eigenvalues are

$$\lambda_{1,2} = \frac{-(a+c) \pm \left[(a+c)^2 - 4b^2\right]^{1/2}}{2b} \tag{4.13}$$

Two trajectories will diverge when the square root term of equation (4.13) is real.

   In order to obtain a local criterion for the divergence of trajectory pairs we use an expression for the action S(q',q'',t) which is valid for short times, namely

$$S(q', q'', t) = \sum_j m \frac{(q_j' - q_j'')^2}{2\Delta t} + V\left(\frac{q' + q''}{2}\right) \Delta t. \tag{4.14}$$

Taking the derivative of S and inserting that derivative back in equation (4.14) we obtain the following condition for the divergence of trajectory pairs:

$$m \, \frac{\partial^2 V}{\partial q_{\perp}^2} \leq 0.$$

(4.15)

Note that the derivative is taken in the direction perpendicular to the motion. Condition (4.15), considered as a criterion for the onset of chaotic behavior is (as in the BDTCR method) local in character, and cannot account for the influence of global symmetries on the behavior of the system. One such example comes to mind immediately: if the potential is separable this analysis will not reveal this property, hence can then lead to an incorrect prediction of the onset of chaotic behavior. For this reason we suggest that the Kosloff-Rice analysis of the onset of chaotic motion be applied only to irreducible potential surfaces.

V. AN EXAMPLE: THE HENON-HEILES SYSTEM

Kosloff and Rice have applied the analysis of Section IV to the Henon-Heiles system,[33] which has the Hamiltonian:

$$H = \frac{1}{2} \left(p_x^2 + p_y^2 + x^2 + y^2 \right) + x^2 y - \frac{1}{3} y^3.$$

(5.1)

Using condition (4.15), we calculate d^2V/dZ^2, where Z is perpendicular to an arbitrary direction of flow (described by the angle θ, defined relative to a symmetry axis of the potential surface). The direction Z can be related to x and y by

$$Z = -x \sin\theta + y \cos\theta$$

(5.2)

so that

$$\frac{\partial^2 V}{\partial Z^2} = 1 - 2y \cos 2\theta - 2x \sin 2\theta.$$

(5.3)

The critical energy for the onset of chaos is determined by the boundaries of the region where d^2V/dZ^2 changes sign from positive to negative; these are found by minimization of d^2V/dZ^2 with respect to θ. We find the critical energy locus on the Henon-Heiles potential surface to be

$$x^2 + y^2 = \frac{1}{4}.$$

(5.4)

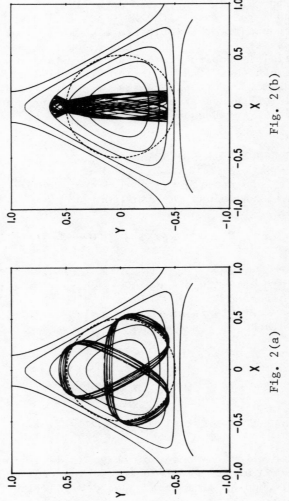

Fig. 2(a)   Fig. 2(b)

Two high energy trajectories, each of which shows a high degree
of correlation in the motion. In this and all the succeeding
figures the dashed circle is the BDTCR critical energy locus for
the onset of chaotic motion. Fig. 2a, E = 1.o5; Fig. 2b, E = .618.
All energies are measured relative to the dissociation energy.

Eq. (5.4) is just the BDTCR condition for the onset of chaos. However, the Kosloff-Rice analysis requires both that (5.4) be satisfied and that the trajectory cross the critical locus at an angle
such that it samples a region where $d^2V/dZ^2<0$. We show in Figs. 2a
and 2b two high energy trajectories which represent quasiperiodic
motion on the Henon-Heiles surface even though the critical energy
locus is crossed. Figs. 3a, 4a and 5a illustrate chaotic trajectories on the Henon-Heiles surface, and Figs. 3b, 4b and 5b show the
domains where d^2V/dZ^2 is negative for these trajectories. The results displayed show clearly that the regions where $d^2V/dZ^2>0$ are
only a small fraction of the available surface, and that the sizes
of these regions depends on the energy of the system.

We note, in passing, that the Kosloff-Rice method of inferring
the onset of chaotic motion correctly predicts that mixing flow cannot occur in a one dimensional system, since in that case the motion
is always in the direction of the flow.

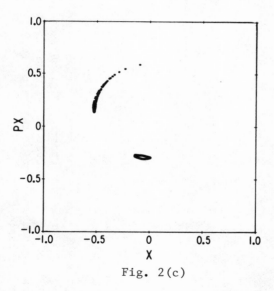

Fig. 2(c)

The surface of section for the
trajectory shown in Fig. 2a. Note
that this surface of section, and
those in Figs. 3c, 4c and 5c, is
a projection on the x, p_x plane,
which is different from the y, p_y
projection used in Ref. 17.

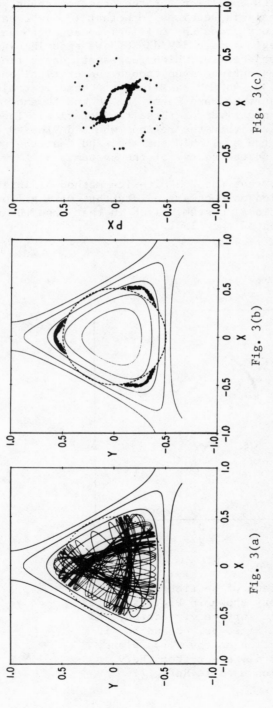

Fig. 3(a)

Fig. 3(b)

Fig. 3(c)

A chaotic trajectory which was used to calculate the average entropy per digit from the macro-state occupation number representation of the trajectory. (E=.702). Note that the trajectory gets locked into relatively long lived quasi-periodic motion.

The regions in which the trajectory of Fig. 3a encounters negative curvature perpendicular to its direction.

The surface of section for the trajectory shown in Fig. 3a.

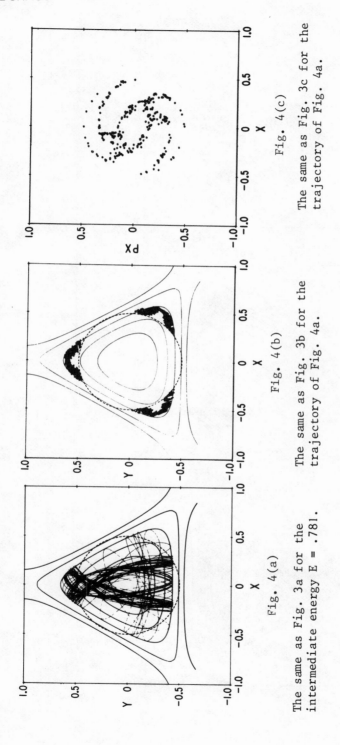

Fig. 4(a)

The same as Fig. 3a for the
intermediate energy E = .781.

Fig. 4(b)

The same as Fig. 3b for the
trajectory of Fig. 4a.

Fig. 4(c)

The same as Fig. 3c for the
trajectory of Fig. 4a.

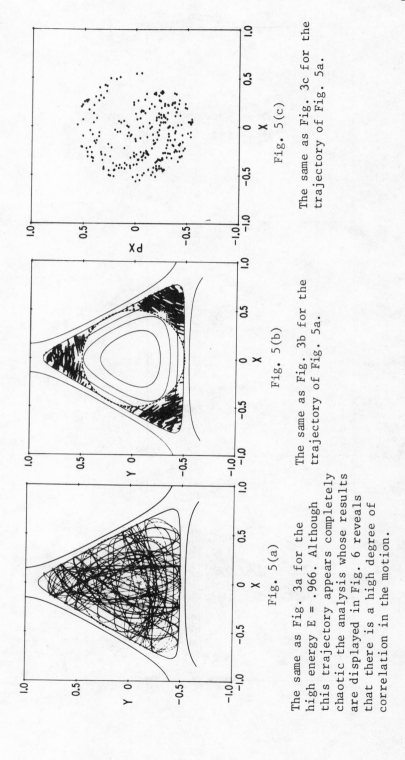

Fig. 5(a)

Fig. 5(b)

Fig. 5(c)

The same as Fig. 3a for the
high energy E = .966. Although
this trajectory appears completely
chaotic the analysis whose results
are displayed in Fig. 6 reveals
that there is a high degree of
correlation in the motion.

The same as Fig. 3b for the
trajectory of Fig. 5a.

The same as Fig. 3c for the
trajectory of Fig. 5a.

VI. MORE ON THE LOCAL CRITERION FOR THE ONSET OF CHAOS

Why should a local criterion ever, except by accident, correctly predict the onset of chaotic motion in a Hamiltonian system? We cannot answer this question, but there is a hint that suggests a route for investigation.

It has been known for many years that geodesic flow on a compact manifold with negative curvature is ergodic and mixing.[34,35] This behavior follows from the characteristic that on a manifold with negative curvature adjacent geodesics diverge exponentially, and this leads to chaos in the corresponding geodesic flow. The statements just made refer to manifolds with negative curvature everywhere, whereas the potential energy surfaces used to characterize intramolecular motion have both positive and negative curvature. Nevertheless, it is reasonable to expect that if a potential energy surface has regions in which the curvature perpendicular to the trajectory is negative, and these regions are not too small, the exponential divergence of adjacent trajectories generated within any one region will have the consequence that incoming trajectories that pierce its boundary at nearby points become outgoing trajectories randomly distributed in direction. The smaller the region of the prescribed negative curvature, the less likely it will be that the exiting trajectories will be randomly distributed in direction. It seems at least plausible that the areas with the prescribed negative curvature for the Henon-Heiles potential are "large enough," but we have no means of bounding the required size. What is needed is a generalization of the theorems of Hopf [34] and Hedlund [36] to the case that the manifold has some regions where the curvature is not negative, at least in the sense of establishing a bound for the minimum measure for the regions of negative curvature required to achieve ergodicity and mixing.

If we accept the notion that ergodicity and mixing can be generated when the measure of the regions with negative curvature perpendicular to the trajectory is greater than zero but less than one, the average measure of these regions provides a natural cell size for coarse graining the phase space. We define a coarse grained density by

$$\varphi(m) \equiv \frac{1}{W(m)} \int \cdots \int_{m^{th} cell} f \, d\pi \qquad (6.1)$$

where $W(m)$, the volume of the mth cell, is chosen to be the average measure of the regions with the prescribed negative curvature and, for simplicity, independent of m. We now define a coarse grained entropy by

$$\bar{S} \equiv -\sum_m{}' \varphi(m) \ln \varphi(m) W \tag{6.2}$$

$$= -\int f \ln \varphi(m) \, d\Pi, \tag{6.3}$$

from which we obtain

$$\frac{d\bar{S}}{dt} = -\sum_m{}' \frac{\partial \varphi(m)}{\partial t} \ln \varphi(m), \tag{6.4}$$

where we have used the time invariance of the normalization of the distribution function. We can calculate $\partial\varphi(m)/\partial t$ by combining (6.1) and the Liouville equation; this gives

$$\frac{\partial \varphi(m)}{\partial t} = \frac{1}{W} \int \cdots \int_{m^{th}\,cell} \frac{\partial f}{\partial t} \, d\Pi \tag{6.5}$$

$$= -\frac{1}{W} \int \cdots \int_{\sigma(m)} f \, \underset{\sim}{v} \cdot d\underset{\sim}{\sigma} \tag{6.6}$$

where $\sigma(m)$ is the surface of the mth cell and $\underset{\sim}{v}$ is the flow velocity of the phase fluid. Substitution of (6.6) into (6.4) then gives

$$\frac{d\bar{S}}{dt} = \frac{1}{W} \sum_m \int \cdots \int_{\sigma(m)} f \, \underset{\sim}{v} \cdot d\underset{\sim}{\sigma} \ln \varphi(m). \tag{6.7}$$

Since f and $\underset{\sim}{v}$ are pointwise continuous across cell boundaries, a non-vanishing contribution to the sum in (6.7) occurs only when $\varphi(m)$ changes discontinuously at a boundary. If two cells, m and m', have a mutual interface $[\sigma(m),\sigma(m')]$, then the contribution to $d\bar{S}/dt$ from that interface is

$$\int \cdots \int_{[\sigma(m),\,\sigma(m')]} f \, \underset{\sim}{v} \cdot d\underset{\sim}{\sigma} \ln (\varphi(m)/\varphi(m')), \tag{6.8}$$

where the positive direction of flow is from m towards m'.

Eq. (6.8), when combined with the notion that a region of the prescribed negative curvature with nonzero measure generates chaotic trajectories, leads to an interesting conclusion. Given that adjacent trajectories that enter such a region diverge exponentially, we expect that equal sized regions elsewhere in the phase space are crossed by trajectories that are uniformly distributed with respect to direction-- we need only sample enough trajectories for this to be the case. Then the flow across each of the coarse grained cell boundaries is like a diffusion process, and we expect the net flow to be from a region where φ is large to a region where φ is small. If so, the contribution from each interface given by (6.8) is positive, and we conclude that

$$\frac{d\overline{S}}{dt} \leq 0,$$

(6.9)

equality holding only when the distribution is uniform in the coarse grained sense. Note that in this plausibility argument it is not necessary that the trajectory on a potential energy surface have negative curvature everywhere perpendicular to its direction, but only that the regions of negative curvature which exist be "large enough" that they serve as "sources" of random trajectories which chaotically traverse the remainder of the surface. Clearly, a verification of the plausibility argument depends on extending the Hopf-Hedlund theorem, as already mentioned above.

VII. AN UPPER BOUND FOR THE KOLMOGOROV ENTROPY

Earlier we noted that numerical calculations show that there is a correlation between the rate of divergence of adjacent trajectories in the neighborhood of a local instability and the global rate of divergence of adjacent trajectories. This finding suggests that we seek a relationahip between the local criterion for onset of chaotic motion and the Kolmogorov entropy. This section duscusses one such possible relationship.

We have found that, using the Kosloff-Rice criterion, there are only a few small regions on the potential energy surface that contribute to generating mixing motion of the trajectory. Suppose we accept this as generally valid; then the following formal transformation can be used to relate the global and local descriptions of the mixing flow. Let each of the regions in which condition (4.15) is satisfied be considered to define a "state" of the system. We then abstractly represent the motion defined by the trajectory as a set of transitions from one "state" to another. Thus, the motion corresponding to any particular trajectory is mapped into a sequence of numbers, X_1, X_2, ..., X_j, ..., characterizing the states through which the trajectory passes. By virtue of our assumption that entry into the region which defines a "state" generates exponential divergence of adjacent trajectories and thereby chaotic motion and mixing, the evolution of a small neighborhood centered around the initial

conditions of our particular trajectory must map in a stochastic manner on to a succession of "states." We infer, then, that the evolution of such a neighborhood in phase space can be described by a stochastic process which is characterized by a probability measure of
the sequences of successive "states" it covers, $P(X_1, \ldots, X_n)$. Of
course, the number of "states" n through which the trajectory passes
increases indefinitely as $t \to \infty$. The average information entropy per
step of the stochastic sequence representing the trajectory is

$$h_{p^{\circ}q^{\circ}}(P(X_1, \ldots, X_n)) = -\sum_{\{X_1, \ldots X_n\}} P(X_1, \ldots, X_n) \ln P(X_1, \ldots, X_n),$$

(7.1)

with the usual definition [27]

$$h_{p^{\circ}q^{\circ}}(P(X_1, \ldots, X_n))$$
$$= -\sum_{\{X_1, \ldots X_n\}} P(X_1, \ldots, X_n) \ln P(X_1, \ldots, X_n).$$
(7.2)

Note that (7.1) defines a conditional information entropy, dependent
on the initial conditions. To obtain the Kolmogorov entropy, we average $h_{p^0 q^0}$ over all points in phase space,

$$h_K = \int_{\Pi} h_{p^{\circ}q^{\circ}} \, d\Pi.$$

(7.3)

While direct calculation of h_K is difficult, an upper bound can be
found easily by calculating the average information entropy of a finite sequence of "states". To obtain an upper bound it is sufficient
to neglect correlations of order longer than n in the sequence defining $P(X_1, \ldots, X_n)$ or, alternatively, to assume that the correlations
between "states" decay in time more rapidly than the time required
for n steps. Then an estimate for the probability measure of the
finite sequence, $P(X_1, \ldots, X_n)$, can be obtained by use of the ergodic
hypothesis. The theorems of information theory guarantee that this
procedure is convergent from above as n increases. [27]

Consider again the Henon-Heiles system, and the trajectories
shown in Figs. 2-5. The trajectory of Fig. 2a has a transition probability that favors the initial "state" and, as a consequence, has a
low average information entropy per step. Similarly, the precessing
trajectory shown in Fig. 2b is very highly correlated, since, in our
language, the initial "state" recurs every three steps; the average
information entropy per step is very low. On the other hand, the
trajectory of Fig. 5 looks as if there is no correlation between
the successive "states." For this trajectory, an upper bound to the
Kolmogorov entropy is

$$h_K < \ln 3.$$

(7.4)

The Kosloff-Rice [14] method of obtaining an upper bound to the Kolmogorov entropy, which is an extension of Shimada's method, [37] can be
applied to a variety of Hamiltonian systems. For example, whenever
the motion is dominated by attractors in the phase plane, so that the
trajectory weaves about in these attractors and occasionally undergoes a transition from one to another, the dynamics can be mapped
onto an equivalent system in which each attractor represents a
"state", and an occupation number locates the trajectory vis a vis
the several "states" of the system. Shimada introduced this idea in
an elegant analysis of the chaotic behavior of the Lorenz attractor.[37]
Note, however, that if the motion between attractors is correlated,
the upper bound found for the Kolmogorov entropy by neglect of correlations can be very far from the actual value. We address this
problem in the next section.

VIII.   CORRELATIONS ALONG A TRAJECTORY CONNECTING MACROSTATES

     We now investigate, following Kosloff and Rice,[14] the decay of
correlation in the successive occupation of a set of macrostates tra-
versed by one trajectory. To make the approach we use as clear as
possible, we treat in detail the particular case of trajectories on
the Henon-Heiles potential surface. As shown in Figs. 3b, 4b and 5b,
on this surface we find three regions in which neighboring trajectories diverge. Using the mapping described in Section VII, when a trajectory enters one of these regions the system is said to be in a
macrostate labelled by the index 1, 2 or 3, as the case may be. A
given trajectory was followed long enough that a sequence of macrostates represented by a number with 10,000 digits was produced. To
analyze the correlation in successive occupation of the macrostates
1,2,3 the numerical representation of the course of the trajectory
was subdivided into smaller sequences of N digits, and the frequency
of occurrence of each N digit number in the parent "trajectory number"
was computed. After normalization by the frequency of occurrence calculated for the case of complete lack of correlation, the entropy of
the N digit sequence was calculated using Eq. (7.2). The average
entropy per digit, calculated as just described, is plotted in Fig.6
for the sequence length range N = 2 to N = 8 for three energies in
the chaotic domain.

     We have also examined the correlation in the successive occupa-
tion of a set of macrostates using a different scheme to represent
the trajectory. In this scheme, if a trajectory returned in one
"step" to the macrostate from which it started the transition was
represented by the digit 1; if not, it was represented by the digit
0. Thus, the temporal evolution of the trajectory, with respect to
penetration of the regions designated, was represented as a long binary number. This number was then divided in sequences of length N,
and the entropy per digit associated with those sequences computed as

described in the preceding paragraph; the results are shown in Fig.6, for values of N up to N = 11 for three energies in the chaotic domain.

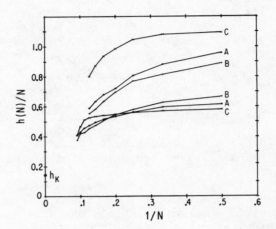

Fig. 6. The average information entropy per unit digit length as a function of N. for three different energies, calculated using the macrostate occupation number representation of the trajectory (see text). The lines marked by A are the lower energy E = .7o2. The lines marked by B are the intermediate energy E = .781. The lines marked by C are the high energy E = .966. The upper group of lines is obtained from the three state representation, and the lower group from the two state representation, of the trajectory.

We also show in Fig. 6 the value of the Kolmogorov entropy for the Henon-Heiles system, computed directly by Benettin et al. [31] For the highest energy they consider, the estimated value of the Kolmogorov entropy is 0.14, far below the values we find from the analysis of limited sequences of successive transitions between macrostates visited by a trajectory. Indeed, the most striking feature of the calculations displayed in Fig. 6 is the very slow convergence to the asymptotic value of the Kolmogorov entropy. We note that neither the 8 digit sequence in the three macrostate representation, nor the 11 digit sequence in the binary transition representation, give estimates of the Kolmogorov entropy at all close to the value 0.14. Using simple plotting schemes, it appears that the rate of convergence of the calculation of the Kolmogorov entropy with respect to the length of the sequence of occupation of macrostates is slower than $(\ln N)^{-1}$ We believe this result to be of considerable importance, since it shows very clearly that the correlation in occupation of macrostates visited by a trajectory is very long lived.

In contrast, when two neighboring trajectories are compared with respect to the ordering in the sequence of macrostates each visits, the cross correlation decays much more rapidly than does the auto-correlation along one trajectory. Of course, the precise lifetime of the cross correlation between trajectories depends on how nearly the same are their initial conditions. Nevertheless, it is clear that even in the chaotic domain the short time dynamics represented by a single trajectory retains much of the character of quasiperiodic motion.

IX.   THE KOLMOGOROV ENTROPY OF A QUANTUM MECHANICAL SYSTEM

The first step in the extension of the Kolmogorov entropy concept to quantum mechanical systems is the definition of a quantum mechanical analogue of the partition $I\!P$ defined on the phase space $I\!\Gamma$. A few obvious but necessary remarks are a preamble to our argument.

A quantum mechanical description of a system is based on a Hilbert space $I\!H$, a positive density operator with unit trace $\hat{\rho}$, and a set of operators $\{\hat{A}\}$ on the Hilbert space defined to represent physical observables. The density operator represents the state of the system, and the expectation value of an observable A is just $\mathrm{tr}(\hat{\rho}\,\hat{A})$. It is convenient to think of the evolution of the system under the equation of motion as a trace preserving positive mapping of density operators into density operators. Then the state of the system at time t_2 is obtained from the state at time t_1 by the transformation

$$\hat{\rho}(t_2) = tr_1\,(\,\hat{U}_{12}\,\hat{\rho}\,(t_1)\,),$$

(9.1)

where $\hat{U}_{12}$ is the evolution superoperator and tr_1 is the trace operation with respect to $\hat{\rho}(t_1)$. An operator $\hat{A}$ admits a spectral resolution into a set of orthogonal projection operators,

$$\hat{A} = \sum_i \alpha_i \, \hat{P}_{A_i}.$$

(9.2)

Given the set of projection operators $\hat{P}_{A_i}$ we can construct a resolution of the identity operator in the form

$$\hat{1} = \sum_i \hat{P}_{A_i}. \tag{9.3}$$

The set of projection operators $\{\hat{P}_{A_i}\}$ describes all the distinct outcomes that can be realized when the expectation value of $\hat{A}$ is observed.

We now define the set of orthogonal projection operators $\{\hat{P}_{A_i}\}$ which resolve the identity operator to be the quantum mechanical analogue of the partition P on the classical mechanical phase space Π. Furthermore, we define the entropy associated with an observable A [38] in a state k as the amount of uncertainty in $\hat{\rho}$ resolved by the measurement of A, i.e.

$$h(\hat{\rho}, \hat{A}) \equiv - \sum_k p_k \ln p_k \tag{9.4}$$

where

$$p_k \equiv tr(\hat{\rho}\,\hat{P}_{A_k}). \tag{9.5}$$

In addition to an analogue of a partition on the phase space, we must also define an average entropy per unit time. To do so, we must construct the joint probability of obtaining from a series of measurements of A the outcome α_1 at t_2, α_2 at t_2, ..., α_n at t_n. Imagine an ensemble of identical systems in a state defined by the density operator $\hat{\rho}$. At each of the times t_1, t_2, ..., t_n a sample is extracted from the ensemble and the observable A measured; after each extraction of a sample the remainder of the ensemble is allowed to evolve undisturbed. The Hilbert space on which the joint probability of obtaining α_1 at t_1, α_2 at t_2, ... is defined as the tensor product space

$$|H^{(1)} \otimes |H^{(2)} \otimes \cdots \otimes |H^{(n)} \tag{9.6}$$

on which we can define the density operator

$$\hat{\rho}_n \equiv \hat{\rho} \otimes \hat{U}\hat{\rho} \otimes \cdots \otimes \hat{U}^n \rho. \tag{9.7}$$

The combined density operator $\hat{\rho}_n$ describes the correlated state of the system for all the times t_1, t_2, ..., t_n. Given (9.7), the required joint probability is defined by

$$P(A^{(1)} = \alpha_1, \ldots, A^{(n)} = \alpha_n)$$
$$= tr_{1,2,\ldots,n}(\hat{P}_{A_1^{(1)}} \otimes \hat{P}_{A_2^{(2)}} \otimes \cdots \otimes \hat{P}_{A_n^{(n)}} \hat{\rho}_n), \tag{9.8}$$

where the $A^{(\ell)}$ refer to the observable A at time t_ℓ. Note that the projection operators in the argument of the trace of Eq. (9.8) are at different times, hence they commute, and their ordering is not important.

We can now define the average information entropy associated with measuring the observable A as

$$h(\hat{A}, \hat{\rho}) \equiv \lim_{n \to \infty} h(\hat{A}_n, \hat{\rho}_n)$$

(9.9)

where

$$\hat{A}_n \equiv \hat{A}^{(1)} \otimes \hat{A}^{(2)} \otimes \cdots \otimes \hat{A}^{(n)}.$$

(9.10)

From the set of all possible operators corresponding to observables we choose that one which maximizes $h(\hat{A}, \hat{\rho})$. Thus, our quantum mechanical generalization of the Kolmogorov entropy is defined by

$$h_K(\hat{\rho}) \equiv \sup_A h(\hat{A}, \hat{\rho}).$$

(9.11)

This definition of h_K is quite general; note that it does not depend on how chaos is generated [34] and, since it is based on a classical probability space, most of the theorems which describe the behavior of the classical mechanical Kolmogorov entropy apply equally well to the quantum mechanical extension (9.11). In particular Eq. (3.12) remains valid for both the quantum mechanical and classical mechanical Kolmogorov entropy.

We now note that our definition (9.11) implies there is an important difference between the behavior of classical mechanical and quantum mechanical systems with respect to the onset of chaotic motion. For, if the spectrum of a unitary evolution operator is discrete, the Kolmogorov entropy of the corresponding motion is zero. Because of its importance, we will sketch the proof of this statement.

If the evolution operator has a discrete spectrum, it can be expanded in terms of a set of orthogonal projectors in operator space. We write, for one unit time interval T,

$$\hat{U}_T = \sum_j \hat{g}_j e^{i \Delta_j T}$$

(9.12)

where $\hat{g}_j$ is the amplitude and $\Delta_j T$ the phase of projector j. Clearly, the evolution operator corresponding to a time interval nT is

$$\hat{U}_T^n = \sum_j \hat{g}_j e^{i \Delta_j n T},$$

(9.13)

and it is possible to find a number n such that $\hat{U}\frac{n}{T}$ is as close to
the identity operator as desired. Then, using Eq. (3.12), the Kolmogorov entropy is zero.

The inference that the Kolmogorov entropy is zero when the evolution operator has a discrete spectrum is true in both quantum mechanics and classical mechanics. Nevertheless, there are important
differences between classical and quantum mechanical motion on the
same potential surface. Thus, if we consider motion of a bounded
system, the spectrum of the quantum mechanical evolution operator
is always discrete, whereas that of the classical mechanical evolution operator is discrete in the quasi-periodic domain and continuous
in the chaotic domain. We note that in this case of bounded motion,
there will usually be a regime in which classical mechanical trajectories diverge and the corresponding Kolmogorov entropy is positive,
while at the same energy the quantum mechanical Kolmogorov entropy
is zero. Therefore, for bounded motion on the same potential surface, chaotic classical mechanical motion does not have a quantum
mechanical analogue. Of course, in the limit that $\hbar \to 0$ the quantum
mechanical evolution operator's spectrum can become continuous even
for bounded motion, and in that case the limiting classical mechanical motion is chaotic. If when $\hbar \to 0$ the spectrum of the evolution
operator remains discrete, the limiting classical mechanical motion
is quasi-periodic.

X.   COMPARISON OF THE CLASSICAL AND QUANTUM MECHANICAL KOLMOGOROV
     ENTROPIES

In this section we discuss some aspects of the relationship between the classical mechanical and quantum mechanical concepts involved in defining the Kolmogorov entropy. Given the nature of these
concepts, we show how the von Neumann axiomatic formulation of quantum mechanics [40] leads to our formulation of the Kolmogorov entropy.

Consider, first, the prescription used to define the classical
mechanical Kolmogorov entropy, namely the successive partitioning
of the phase space Π. Recall that the microstate of a classical
mechanical system can be represented by a point Γ in the space Π,
and the evolution of the microstate of the system by the trajectory
$\Gamma(t)$ that satisfies the equation of motion for the given initial
conditions. Any particular partition $\mathcal{P}^{(0)}$, which consists of a
collection of nonempty, nonintersecting sets that completely cover
Π, also defines a set of macrostates of the system, each macrostate consisting of all the microstates whose representative points
lie within one of the sets of points of Π generated by that particular partition. Thus, a partition $\mathcal{P}^{(0)}$ of the phase space Π can
be thought of as a specification of the set of all possible outcomes
of an experiment which, at time t_0, determines the macrostate of the
system. Put another way, the possible outcomes of an observation of
the macrostate of the system are mapped, by the definition of the

partition $\mathit{I\!P}^{(0)}$, into the corresponding sets of points which represent those macrostates.

The partitioning of the phase space $\mathit{I\!\Gamma}$ leads to the consideration of a classical (Kolmogorov) probability space. This probability space consists of a triple ($\mathit{I\!\Gamma}, \Omega, \mu$), where the elements $\mathit{I\!\Gamma}$ correspond to the possible outcomes of a random experiment, the sets in Ω correspond to random events and the measure $\mu(A)$ for $A \in \Omega$ gives the probability that the random event occurs. Given the probability space ($\mathit{I\!\Gamma}, \Omega, \mu$), a random variable corresponds to a measurable quantity for the random experiment. If B is some subset of the possible outcomes, and f is a random variable, $f^{-1}(B)$ is the event that f has a value in B and $\mu_f \equiv \mu[f^{-1}(B)]$ is the probability of that event. Clearly, μ_f is the distribution of f. The expectation value of f is defined in the usual way as $\int f d\mu$. In classical probability theory, successive observations do not interfere with one another and the corresponding random variables commute. Thus, in the product partition defined in Eq. (3.8), the ordering of the component partitions is irrelevant.

We now consider the prescription proposed for the definition of the quantum mechanical Kolmogorov entropy. Our construction is based on the axiom: [40-43]

The probability that an observable A has a value in a (Borel) set B when the system is in the state described by ψ is $<\hat{P}_A(B)\psi, \psi>$, where $\sum_B \hat{P}_A(B) = \hat{1}$ is the resolution of the identity operator for $\hat{A}$ (see Eqs. (9.2) and (9.3)). Thus $<\hat{P}_A(B)\psi, \psi>$ is the distribution of the observable A in the state ψ, and comparison of the classical and quantum mechanical definitions shows that observables correspond to random variables, and that projection operators $\hat{P}_A(B)$ correspond to events. It can be shown that any finite number of observables can be treated as random variables if and only if they commute.[44] We may in general interpret the set $\{\hat{P}\}$ of orthogonal projections on a Hilbert space $\mathit{I\!H}$ as a set of quantum mechanical events. Note that there is a one-to-one correspondence between orthogonal projections and closed subspaces of $\mathit{I\!H}$, so that said set of closed subspaces can also be considered quantum mechanical events. Although the preceding has been stated for the case that the system is in a pure state represented by ψ, the argument is easily extended to the case that the state of the system is represented by the density operator $\hat{\rho}$. To do so we need only note that the mapping $\hat{P} \longmapsto <\hat{P}\psi, \psi>, \hat{P} \in \{\hat{P}\}$, which replaces the classical probability measure μ when the system is in a pure state, is itself replaced by $\hat{P} \longmapsto tr(\hat{\rho}\hat{P})$ when the system is in the mixed state represented by the density operator $\hat{\rho}$.

In summary, if $\mathit{I\!H}$ is a Hilbert space, $\{\hat{P}\}$ the set of orthogonal projections on $\mathit{I\!H}$, and $\hat{\rho}$ a density operator with unit trace, then the triple ($\mathit{I\!H}, \{\hat{P}\}, \hat{\rho}$) is the quantum probability counterpart to the classical probability space (Π,Ω,μ). In terms of this counter-part we have the strong theorem: [43]

If (Π,Ω) is a measurable space, there exists a Hilbert space $I\!H$ and a σ-isomorphism T_σ from Ω to the set $\{\hat{P}\}$ of orthogonal projections on $I\!H$. For any probability measure μ on (Π,Ω) there exists a density operator $\hat{\rho}$ such the $\mu(\Omega) = \mathrm{Tr}(T_\sigma(\Omega)\hat{\rho})$ for every $\Omega \in \Pi$. For any class of random variables $\{f_\alpha\}$ on (Π_1,Ω) there exists a class of observables $\{A\}$ such that $\hat{P}_{A_\alpha}(E) = T_\sigma[f_\alpha^{-1}(E)]$ for every E contained in the Borel subsets of Π.

Examination of our procedure for the construction of the quantum mechanical Kolmogorov entropy shows that it is in one-to-one correspondence with the classical mechanical Kolmogorov entropy. In particular, the definition displayed in Eq. (9.8) guarantees the commutation of the projection operators, hence permits treatment of the observables as random variables.

The definition of the quantum mechanical Kolmogorov entropy which we propose differs somewhat from those introduced by Connes and Stormer, Emch and Lindblad. [39] Connes and Stormer, and also Emch, construct a noncommutative counterpart of the classical partition; in our construction, by virtue of the sampling method by which measurements are made, measurements at different times do not interfere with each other. It is only when this condition is met that a classical probability space can be defined. Lindblad, and also Emch, uses an external bath to generate a stochastic force on the system, which force plays a role in the definition of the entropy. The lattice structure of his definition resembles ours, but there is no reference made to the observables of the system. In view of these remarks, we conclude that our definition of the quantum mechanical Kolmogorov entropy is in closer correspondence with the classical Kolmogorov entropy than those previously proposed, and is more general.

XI. TWO MODELS COMPARING CLASSICAL AND QUANTUM MECHANICAL CHAOS

The conclusion that the Kolmogorov entropy of a quantum mechanical system with discrete energy level spectrum is zero conflicts with the suggestion of a "KAM-like" transition in the projection of the system wavefunction onto a set of basis states, and also with the results of the analyses of two model systems designed for the purpose of comparing the nature and onset of classical mechanical and quantum mechanical chaotic motion. In this section we reexamine the behavior of the model systems proposed by Berry, Balazs, Tabor and Voros [24] (BBTV) and by Casati, Chirikov, Izraelev and Ford [25] (CCIF); we reserve for later discussion our reinterpretation of the results of the calculations of Nordholm and Rice, [5] Stratt, Handy and Miller, [7] and Marcus. [6]

A. The BBTV MOdel

Berry, Balazs, Tabor and Voros [24] have proposed a model Hamiltonian, which defines a class of one dimensional systems, for which
it is possible to study both the area preserving mappings on the
classical mechanical phase space and the corresponding quantum mechanical evolution generated by unitary transformations of the system
wavefunction (called a quantum mapping by BBTV). This Hamiltonian
is periodic, with period T, and its time average describes a particle of mass μ moving in a potential V(q). They write

$$\hat{H}(\hat{q},\hat{p},t) = \frac{\hat{p}^2}{2\mu\gamma} \qquad (0 < t < \gamma T)$$
$$\hat{H}(\hat{q},\hat{p},t) = \frac{V(q)}{1-\gamma} \qquad (\gamma T < t < T)$$

$$(11.1)$$

and

$$\overline{H}(\hat{q},\hat{p}) = \frac{\hat{p}^2}{2\mu} + \hat{V}(\hat{q}).$$

$$(11.2)$$

Notice that when γ is close to unity the Hamiltonain (11.1) is that
of a free particle which is periodically perturbed with a potential
$\hat{V}(\hat{q})\delta_p$ (t/T), where δ_p(t/T) is a periodic delta function which "turns
on" for an infinitesimal interval every T seconds. Given this par-
ticular perturbation, the energy is alternately purely kinetic or
purely potential. The quantum mapping operator, which describes
the action of $\hat{H}$ for interval t = 0 to t = T, is found to be

$$\hat{G}(\psi) = \hat{U}_T \psi \hat{U}_T^\dagger,$$

$$(11.3)$$

$$\hat{U}_T = exp\left(-\frac{iT}{\hbar}\hat{V}(\hat{q})\right) exp\left(-\frac{iT}{2\hbar\mu}\hat{p}^2\right).$$

$$(11.4)$$

The classical mechanical motion of the system is described by the
same Hamiltonian with operators $\hat{q}$ and $\hat{p}$ replaced by the corresponding
conjugate variables. The mapping of the phase space generated by
this Hamiltonian for the time interval t = nT to t = (n + 1) T is

$$q^{(n+1)} = q^{(n)} + \frac{p^{(n)}}{\mu} T$$

$$(11.5)$$

$$p^{(n+1)} = p^{(n)} - TV'(q^{(n+1)})$$

$$(11.6)$$

where $V'(q)$ is the derivative of V with respect to q.

We consider the case that $\hat{V}(\hat{q})$ is quadratic in $\hat{q}$,

$$\hat{V}(\hat{q}) = \tfrac{1}{2}\mu\omega^2\hat{q}^2, \tag{11.7}$$

for which the classical map $X^{(n+1)} = \textbf{M} X^{(n)}$ is linear,

$$\textbf{M} = \begin{pmatrix} 1 & T/\mu \\ -\mu\omega^2 T & 1-\omega^2 T \end{pmatrix}. \tag{11.8}$$

A calculation of the effect of the operation of $\hat{G}$ on $\hat{p}$ and $\hat{q}$ leads to the same result as the classical mechanical mapping (11.8). We now determine the spectrum of the evolution operator $\hat{G}$. The first step is to calculate the eigenvalues and eigenoperators of $\textbf{M}$, we find

$$\hat{G}\,\hat{I}_{1,2} = \lambda_{1,2}\hat{I}_{1,2} \tag{11.9}$$

where

$$\hat{I}_{1,2} = \hat{q} + \beta_{1,2}\hat{p} \tag{11.10}$$

and

$$\lambda_{1,2} = -1 - \tfrac{1}{2}\omega^2 T^2 \pm i\omega\left(1 - \tfrac{1}{4}\omega^2 T^2\right)^{1/2}, \tag{11.11}$$

$$\beta_{1,2} = \pm\frac{i}{\mu\omega}\left(1 - \tfrac{1}{4}\omega^2 T^2\right)^{1/2}. \tag{11.12}$$

We now define the operator $\hat{N} = \hat{I}_1\hat{I}_2$, which is an invariant of motion, as can be seen from the relation

$$\begin{aligned}
\hat{G}\hat{N} &= \hat{U}_T\hat{I}_1\hat{I}_2\hat{U}_T^+ \\
&= \hat{U}_T\hat{I}_1\hat{U}_T^+\hat{U}_T\hat{I}_2\hat{U}_T^+ = \lambda_1\lambda_2\hat{I}_1\hat{I}_2.
\end{aligned} \tag{11.13}$$

Because it is an invariant of the motion, $\hat{N}$ plays a role similar to that played by the Hamiltonian in the sense that $\hat{N}$ commutes with $\hat{U}_T$ and therefore $\hat{N}$ and $\hat{U}_T$ have a common set of eigenfunctions. Clearly, when the spectrum of $\hat{N}$ is continuous, the spectrum of $\hat{G}$ will also be continuous. Writing out $\hat{N}$ explicitly we find:

$$\hat{N} = \hat{q}^2 + \frac{1 - \omega^2 T^2/4}{\mu^2\omega^2}\,\hat{p}^2. \tag{11.14}$$

Note that $\hat{N}$ resembles the Hamiltonian of a harmonic oscillator with the modified frequency

$$\omega' = \frac{\mu\omega^2/2}{1-\omega^2T^2/4} \cdot$$

(11.15)

Now the spectrum of an oscillator is discrete if $\omega' > 0$, in which case we recover the BBTV condition of stability. But if $\omega' < 0$, the harmonic potential is inverted (has negative curvature), in which case the spectrum is continuous and the mapping is unstable. Thus, there appears to be complete correspondence between the classical mechanical and quantum mechanical mappings and onset of chaotic motion. The motion becomes mixing for the same values of T and ω, just where the spectrum of the system becomes continuous. However, this correspondence may be misleading for three reasons:

1) It is hard to imagine a physical process in which the motion consists alternately of segments with purely kinetic and purely potential energy. For, there is no zero order Hamiltonian which corresponds to such a motion, and therefore it is difficult to consider energy changes in the system. A more realistic model is one in which a free particle with the zero order Hamiltonian

$$\hat{H}_o = \frac{\hat{p}^2}{2\mu}$$

(11.16)

is perturbed by the periodic time dependent potential

$$\hat{V}(t) = \mu\omega^2\hat{q}^2\delta_p(t/T).$$

(11.17)

The equation of motion of this system can be solved exactly. One finds that the motion is stable for all values of T when $\omega > 0$, and that the spectrum of the system is discrete.

2) In order for the motion in the system to be mixing, it is necessary that the evolution operator have a continuous spectrum. In a quantum mechanical description this condition implies that the motion is unbounded, and therefore we can not define a compact probability space. In classical mechanics the difficulty posed by unbounded motion is overcome by defining a modified mapping with periodic boundary conditions,[16] e.g. a mapping of the unit square into itself. This procedure produces a compact space which preserves the ergoidc characteristics of the full dynamics. Because of the compactness of the space, the Kolmogorov entropy can then be calculated. For the linear map (11.8) the Kolmogorov entropy is found to be positive.[16] Of course, if we change the boundary conditions of the quantum mechanical system we also change the dynamics; under cyclic boundary conditions the spectrum is always discrete, hence the mixing properties of the motion are lost.

3) The BBTV model is closer to a classical mechanical description

than is at first apparent from the formalism. For, the definition
of the Hamiltonian (11.1) implies that at the same time t either the
potential or the kinetic energy is zero, so the commutator of the
potential and kinetic energy operators vanishes. In this sense, the
model omits an essential feature of the quantum dynamics of all ordinary systems.

B. The CCIF MOdel: The Periodically Perturbed Pendulum

The system discussed by Casati, Chirikov, Izraelev and Ford[25]
has the classical mechanical Hamiltonian

$$\hat{H} = \frac{p_\theta^2}{2\mu \ell^2} - (\mu \ell^2 \omega_o^2 \cos \theta) \, \delta_p (t/T),$$

(11.18)

where θ is the angular displacement and p_θ the corresponding angular
momentum, μ, ℓ, and ω_0 are the mass, length and small amplitude frequency of the pendulum, and $\delta_p(t/T)$ is, as before, a periodic delta
function. Let $\theta^{(n)}$ and $p_\theta^{(n)}$ be the values of θ and p_θ just prior
to the nth application of the perturbing force. Direct integration
of the equations of motion for θ and $\dot{p}_\theta$ yields

$$\tilde{p}^{(n+1)} = \tilde{p}^{(n)} - K \sin \theta^{(n)},$$

(11.19)

$$\theta^{(n+1)} = \theta^{(n)} + \tilde{p}^{(n+1)},$$

(11.20)

where $\tilde{p}^{(n)} \equiv p_\theta^{(n)} T/\mu \ell^2$ and $K \equiv (\omega_0 T)^2$. In the limit $K \to 0$ the trajectories of this system are periodic, since the system is then an
ordinary conservative, gravitational pendulum. When K is small most
of the trajectories are quasi-periodic, as implied by the KAM theorem. However, when K >> 1 the trajectories are chaotic and the motion of $\tilde{p}$ is diffusive.

To solve the equivalent quantum mechanical problem CCIF note
that the system can be regarded as a perturbed rigid rotor. It is,
therefore, convenient to expand the wavefunction for the perturbed
system in free rotor basis functions

$$\psi(\theta, T) = \frac{1}{2\pi} \sum_{n=-\infty}^{\infty} A_n(t) e^{in\theta}.$$

(11.21)

Integration of the quantum mechanical equation of motion for one
period T defines a unitary transformation that relates the wavefunctions at times t and t + T. In the momentum representation this
transformation is given by

$$A_n(t+T) = \sum_{r=-\infty}^{\infty} A_r(t) b_{n-r}(k) e^{-ir^2\tau/2}$$

(11.22)

where

$$b_s(k) = b_{-s}(k) \equiv i^s J_s(k),$$

(11.23)

$$k \equiv \frac{\mu l^2 \omega_o^2 T}{\hbar},$$

(11.24)

$$\tau \equiv \frac{\hbar T}{\mu l^2},$$

(11.25)

and $J_s(k)$ is a Bessel function of the first kind. CCIF have numerically integrated the quantum mechanical equation of motion for many
periods T. The results of these calculations show an evolution in
time which is significantly different from the time evolution of the
corresponding classical mechanical system. Details of the behavior
are discussed in Ref. 25, and we here single out only two points.
First, the classical mechanical motion is parameterized by the single
variable $K = k\tau$, whereas the quantum mechanical motion depends separately on k and τ. Second, and more striking, for particular values
of τ the quantum mechanical motion is deterministic even though the
classical motion is chaotic for the corresponding value of $k\tau$, e.g.
when $\tau = 4\pi$ the motion is strictly periodic. These findings have a
simple explanation in terms of the differences between the quantum
mechanical and classical mechanical descriptions of the system.

The model under consideration has two periodicities that generate constraints on the quantum dynamics not present in the classical
dynamics of the system. First, the periodicity of the potential in
θ , just as in the case of motion in a regular lattice, leads to a
constraint on the relationship between p_θ and θ and restricts changes
in p_θ to multiples of the basic wave vector of the system. In our
case, when the coordinate is an angle, that basic wave vector corresponds to an angular momentum

$$p_o = \hbar.$$

(11.26)

Second, the periodicity of the potential in time leads to a constraint on energy transfers, all of which must be a multiple of the
unit energy

$$E_o = \frac{h}{T} .$$

(11.27)

The existence of the unit momentum and unit energy can also be thought of as the consequence of satisfying the uncertainty relations for a potential with the angular and temporal periodicities of our model.

Let $p_\theta^{(n)}$ and $E^{(n)}$ be the angular momentum and energy prior to the nth application of the perturbing force. Since the angular momentum and energy can change only by integer amounts, we have

$$\frac{1}{2m\ell^2}(m_1 p_o + p_\theta^{(n)})^2 = m_2 E_o + E^{(n)},$$

(11.28)

where m_1 and m_2 are integers. Eq. (11.28), which plays the role of a selection rule, greatly restricts the response of the system to the perturbation. Indeed, for each initial state characterized by $p_\theta^{(n)}$ and $E^{(n)}$ there is a particular set of final states, characterized by those integer values of m_1 and m_2 that satisfy Eq.(11.28). The composition of this set of final states, that is the distribution with respect to m_1 and m_2, is very sensitive to the values of $p_\theta^{(n)}$ and $E^{(n)}$, and also to the value of T. In this sense the quantum mechanical solution retains the sensitivity to initial conditions characteristic of the corresponding classical solution in the chaotic domain. Nevertheless, the eigenvalues of energy and angular momentum are well defined and simply related, so the corresponding motion is not well described if called chaotic. The importance of Eq. (11.28) is readily illustrated by interpretation of the observations that when $\tau = 4\pi$ the motion of the system is periodic and the average energy grows proportional to t^2. For, using Eqs. (11.25), (11.26) and (11.27), we find that when $\tau = 4\pi$, $E_0 = p_0^2/2 \, \mu\ell^2$, i.e. unit energy and angular momentum are transferred under the purturbation.

The result obtained above recalls the observation that semiclassical quantization appears to satisfactorily reproduce the eigenvalues of a system even when the energy corresponds to a classical trajectory which is deep in the chaotic region. That is, the condition (11.28) is analogous to the requirement that semiclassical quantization pick out a subset of the quasi-periodic orbits embedded in the chaotic domain, and that the vast majority of the orbits (which are chaotic) are irrelevant to the quantum dynamics.

XII. DISCUSSION

Part of the interest in the nature of dynamical chaos in Hamiltonian systems arises from the association of chaos with rapid intramolecular energy transfer. To what extent do the results discussed in this paper contribute to an understanding of intramolecular energy transfer?

Consider first classical mechanical dynamical chaos. We have
argued that it is only those regions of the potential energy surface
beyond the critical energy locus and for which the curvature perpendicular to a trajectory is negative that contribute to generating
chaos. If, as seems likely, the Henon-Heiles potential energy surface is typical in that regions with the required curvature occupy
only a small fraction of the energetically accessible surface, it is
possible to imagine trajectories that never encounter the conditions
that generate chaos. When this occurs, there are domains of stability with respect to small changes in the initial conditions of those
trajectories. Note that the dense set of periodic trajectories embedded in the chaotic domain found by Helleman and Bountis [45] shows
great sensitivity of the periods to changes in initial conditions.
Thus, although these trajectories are periodic they do not avoid the
negative curvature regions under discussion. Note also that on the
Henon-Heiles surface the trajectory shown in Fig. 2 is quasiperiodic
even though it enters regions where the surface curvature perpendicular to the trajectory is negative. The trajectories that never encounter the conditions that generate chaos, if they exist, will
correspond to quasiperiodic motions of the system, and these motions
do not exchange energy with other motions of the system. Obviously,
the embedded periodic motions also do not exchange energy with other
motions of the system. On the other hand, it is presumed that a
Fourier analysis of a chaotic trajectory will have contributions
from all the system motions, hence such a trajectory corresponds to
motion in which there is sharing of energy between all the modes of
the system. Although the presumption concerning the sharing of
energy is valid as $t \to \infty$, our analysis of the correlation in occupation of macrostates visited sequentially by one trajectory
strongly suggests there is a time scale on which the energy remains
localized in some subset of the modes of motion of the system. If
other processes are fast enough to intercept the system before the
energy distribution has become randomized, it does not matter that
the asymptotic state corresponds to delocalization of the system
energy; in such a case the system behaves as if the energy is localized. The results we have do not establish that it is possible to
find such interceptor processes, but they do suggest that any discussion of intramolecular energy exchange and chaos must explicitly
recognize the existence of different time scales for different competing processes, and the dynamical consequences that flow from
those differences.

The behavior of a quantum mechanical system with respect to the
onset of chaos is different from that of a classical mechanical system. We have shown that a generalization of the concept of Kolmogorov entropy to quantum mechanics leads to the conclusion that the
Kolmogorov entropy of a bounded system (which has a discrete spectrum) is zero. The classical mechanical Kolmogorov entropy is the

average e-folding time for the exponential divergence of neighboring
trajectories; the quantum mechanical Kolmogorov entropy is not so
simply interpreted. Nevertheless, the fact that it is possible for
the classical mechanical motion under a given Hamiltonian to be
chaotic, in which case the Kolmogorov entropy is necessarily positive, while the quantum mechanical motion under the same Hamiltonian
is characterized by a Kolmogorov entropy equal to zero, illustrates
dramatically the importance of the interference effects which are
characteristic of quantum mechanics. Clearly, this result also implies that quantum mechanical dynamical chaos and classical mechanical dynamical chaos are not identical. Yet, as pointed out in Section II, numerical investigations of model systems, taken together
with the qualitative definitions of chaos previously adopted, [5,7]
imply that the onset of chaotic motion is similar in the classical
mechanical and quantum mechanical descriptions of a given model system. We now attempt to resolve this apparent inconsistency of interpretation.

Consider again the observation, based on the Helleman-Bountis[45]
analysis of the Henon-Heiles system, that there is a dense set of
quasi-periodic trajectories embedded in the chaotic motion domain,
and also the inference, based on numerical calculations, that semiclassical quantization of a subset of these quasiperiodic trajectories exhausts the true eigenvalue spectrum of the system. [23]

The inference that only the quasiperiodic trajectories correlate
with the eigenstates of the system has been proposed many times. An
excellent suggestion (not a proof) as to why this is so has been advanced by Freed; [46] he showed, from a careful examination of the
semiclassical approximation to the Green's function of the system,
that interference effects destroy the contributions to the action
of all but the quasiperiodic trajectories.

Consider, now, a system with two degrees of freedom. Helleman
and Bountis characterize the embedded quasi-periodic trajectories by
the ratio of frequencies, σ , which must be a rational fraction, and
the Poincaré recurrence time of the system T_r. For a given value of
σ the initial conditions that generate embedded quasiperiodic trajectories are a smooth function of T_r, but T_r changes discontinuously as a function of σ. Therefore, a small change in initial conditions, which generates a small change in σ , leads to a dramatic
change of quasiperiodic trajectory topology. It is to be expected
that the accommodation of such dramatic changes in topology requires
that an embedded trajectory of the type under discussion will span
the full phase space in one or a few directions. In contrast, prior
to the onset of chaotic motion the closed trajectories sample only
a very limited region of the phase space. If a quantum mechanical
description of such a system is based on expansion of the system
wavefunction in localized basis functions then, inevitably, more
basis functions are needed to represent the wavefunction in the

classical chaotic domain (just to span the necessary region of the phase space) than in the classical quasi-periodic domain. This particular representation of the wavefunction thus leads to a "KAM-like" transition in the amplitudes of the basis functions, but does not contradict the fundamental result that so long as the spectrum is discrete the motion of the quantum mechanical system is not chaotic.

The preceding remark is a simple version of a more formal analysis which we now sketch. Note that our definition of the quantum mechanical Kolmogorov entropy is based on the use of a sampling measurement procedure. Given this measurement procedure it is possible to assert that the unsampled bulk of the ensemble evolves undisturbed, i.e. for the purpose of our analysis we can consider that the motion of the system is not perturbed by the measurement process. We chose the sampling measurement procedure for just this reason, since in this case the quantum mechanical and classical mechanical concepts of observation are similar. A different measurement procedure leads to a significantly different picture of the evolution of the quantum mechanical system.

Suppose that the expectation value of the operator $\hat{A}$ is measured at the times t_1, t_2, ..., t_n on the entire ensemble of systems. Given this procedure, the evolution of the system is altered by the measurement process. Indeed, it is now necessary to consider the system, by virtue of the coupling to the measurement apparatus, to be open, and the assumption that the evolution operator is unitary is no longer valid. Von Neuman [40] proposed that after the expectation value of $\hat{A}$ has been determined the system is described by the reduced density operator

$$\hat{\rho}' = \sum_k p_k (\hat{P}_k \hat{\rho} \hat{P}_k).$$ (12.1)

In the measurement procedure now under discussion this reduction is accomplished successively at t_1, t_2, ..., generating density operators $\hat{\rho}'$, $\hat{\rho}''$, ... It has been shown by Kraus [47,48] that successive measurements of the expectation value of $\hat{A}$ in the fashion just described, with free evolution of the ensemble between measurements, generates a semigroup evolution, which is the quantum mechanical analogue of a Markov process. In this case the unitary operator of Eq. (9.1) is replaced by

$$\hat{G}(\psi) = \sum_k p_k \hat{U}_T (\hat{P}_k \psi \hat{P}_k) \hat{U}_T^\dagger.$$ (12.2)

Our formulation of the quantum mechanical Kolmogorov entropy can equally well be based on the destructive measurement procedure just described; the only change required is the replacement of the unitary evolution operator of Eq. (9.1) by the semigroup evolution

operator of Eq. (12.2). However, when the Kolmogorov entropy is defined with respect to the semigroup evolution operator (12.2), a
stochastic element which has no classical mechanical analogue is introduced in the analysis. To illustrate this point, consider a
linear array of Stern-Gerlach analyzers, each designed to measure the
z-component of the spin of a particle; a beam of particles is passed
from analyzer to analyzer in sequence. Suppose that between each
pair of analyzers there is a constant magnetic field in the x-direction. The effect of this field is to rotate the particle's spin
direction in the yz plane prior to its admission to the next Stern-
Gerlach analyzer. It is readily seen that each analyzer acts on the
whole ensemble of systems (the beam that passes through). The magnetic field can be adjusted to rotate the spin to any angle θ in the
yz plane, and the correlations between successive measurements of
the z component of the particle spin depend on the angle θ. In the
experiment described, the Kolmogorov entropy associated with the
evolution of the system can vary from zero when $\theta = 0$ to ln2 when
$\theta = \frac{\pi}{2}$, i.e.

$$h_K = s\left(\sin^2 \frac{\theta}{2}\right) \tag{12.3}$$

where

$$s(x) \equiv -x \ln x - (1-x) \ln (1-x). \tag{12.4}$$

It is also important to note that, given the destructive measurement procedure involving the whole ensemble of systems, altera-
tion of the interval between measurements also alters the ensemble's
evolution in time. In particular, in the limit that the measurement
of the expectation value of $\hat{A}$ is made continuously, the ensemble of
systems and measuring apparatus pass into a stationary state, i.e.
evolution in time ceases.   [49,50]

We now ask if, for a given potential energy surface, the existence of classical mechanical chaotic motion has any significance
for the interpretation of the stochastic element introduced by the
destructive measurement process of quantum mechanics. To answer
this question we imagine constructing a nondestructive measurement
of some property of the system out of local observables in phase
space; if this is possible the local measurement operator commutes
with the effective evolution operator. Now, if the motion of the
system is separable, a local measurement of the expectation value
of an operator associated with one degree of freedom does not interfere with the evolution of other degrees of freedom. On the other
hand, if the motion of the system is not separable, any measurement
of the expectation value of an operator associated with only one
degree of freedom does interfere with the evolution of the other
degrees of freedom. We can use this observation as follows: the
calculations of Nordholm and Rice, [5] and of Stratt, Handy and
Miller, [7] demonstrate that in the region where the classical mech-

anical motion is chaotic the eigenstates have global character, where by global character we mean that the wavefunction spans a large region of the phase space, so that correlations between motions in different portions of the phase space cannot be ignored. We conclude that in the domain where the classical motion is chaotic it is not possible to construct a nondestructive local measurement of a property of the corresponding quantum mechanical motion. Since the projection of the system wavefunction onto basis states can be thought of as a measurement of the amplitudes of localized functions, the existence of a correlation between the onset of classical mechanical chaotic motion and "KAM-like" behavior of the amplitudes of the basis functions is to be expected. Nevertheless, it remains the case that so long as the spectrum of the system is discrete the Kolmogorov entropy is zero and the quantum mechanical motion is not chaotic.

Consider a comparison of the decay of wave packets constructed, respectively, from a superposition of states of a separable system which in the classical limit has quasi-periodic trajectories, and from a superposition of states of a nonseparable system which in the classical limit has chaotic trajectories. In each case the spectrum of the system is assumed to be discrete. We now note that the decay of a wave packet depends only on its spectral content. Then if the spectra of the two systems have similar distributions of states for similar values of the energy, we predict that the rates of decay of the two wave packets will be similar despite the gross difference in behavior of the classical limit trajectories. The calculations of Brumer and Shapiro [51] provide convincing evidence of the validity of this statement. Note that the behavior described derives from the nature of the wave packet, specifically its spectral content. Our use of the Kolmogorov entropy permits this behavior to be anticipated by virtue of the categorization of the consequences of the nature of the spectrum of the system with respect to quasi-periodic and chaotic motion.

Thus far in the discussion of the Kolmogorov entropy of a quantum mechanical system we have focussed attention on the case for which the evolution operator has a discrete spectrum. It can be shown that a necessary but not sufficient condition that the Kolmogorov entropy of a system be positive is that the spectrum of the evolution operator be continuous.[19,20] Given the relationship between the properties of the Kolmogorov entropy and the evolution operator, it is interesting to note that all derivations of the so called Master Equation of which we are aware introduce at some stage of the analysis an approximation, or a limiting condition, which has the effect of making the system's spectrum continuous.[52] It is always assumed in these derivations that a continuous spectrum provides both the necessary and sufficient conditions for irreversible behavior. Yet, the mathematical theory of mixing motion imposes further conditions, and it remains possible that there exist examples for which some types of wave packet initial states do not uni-

formly sample the full phase space even when the system's spectrum
is continuous. This caveat is of considerable importance to the
study of intramolecular relaxation processes. It is reasonable to
expect the dynamics of a coupled system of many nonlinear oscillators, such as a molecule, to have different kinetic behavior in
different energy regions. The obvious extremes are complete dynamical reversibility described by a unitary evolution operator built
from the discrete bound state spectrum, and irreversible decay of
an initial state described by a relaxation operator determined from
the Master Equation for the diagonal elements of the density matrix.
In between these extremes is a domain in which there might be interesting quantum stochastic dynamics, in the sense that the evolution
operator defines a semi-group and a continuous positive mapping of
the density operator, but otherwise all interference effects are
retained. [53] We speculate that in this quantum stochastic domain
there can be special wave packet excitations which do not satisfy
the criteria for quasi-ergodic motion.

Although the preceding paragraphs have emphasized how classical
and quantum mechanical dynamical chaos differ, it is important to
recognize that there are also strong similarities between the behavior of classical and quantum mechanical systems with respect to the
probability of energy flow between local states, particularly so
when the local states are nearly equal amplitude distributions of
some set of basis states. [54] For this case all amplitude interference effects appear to wash out. In contrast, if the local states
correspond to superpositions of basis states in which one amplitude
dominates the mixture, the quantum and classical mechanical probabilities for energy flow between them have residual differences of
important magnitude, and the dephasing of the wave packet retains
some of the consequences of amplitude interference. Even though the
asymptotic state of a system corresponds to a delocalized energy
distribution, the time scale for attaining that distribution can be
long enough to permit interception by competing processes, particularly when interference effects influence the dephasing of the prepared wave packet.

On the basis of the observations described above, we speculate
that it will generally be found to be the case that some of the
pathways to dynamical chaos have correlations that die out slowly.
If this speculation is correct, and if we learn how to excite wave
packets which have the requisite properties, even though chaos and
intermolecular energy flow are related it remains possible, in principle, to find interceptor processes that permit exploitation of
some of the properties of energy localization in a molecule.

XIII. ACKNOWLEDGEMENTS

The new work reported in this paper has been carried out in collaboration with Dr. Ronnie Kosloff.

This research has been supported by grants from the Air Force Office of Scientific Research (AF 10 AFOSR 80-0004) and the National Science Foundation (NSF CHE78-01573).

REFERENCES

1.   O. K. Rice and H. C. Ramsperger, J. Am. Chem. Soc. 49, 1617
        (1927); L. S. Kassel, J. Phys. Chem. 32, 225 (1928).
2.   See, for example, W. Forst, Theory of Unimolecular Reactions,
        Academic Press, New York (1973); P. J. Robinson and K. A.
        Holbrook, Unimolecular Reactions, Wiley, New York (1972).
3.   G. Casati and J. Ford, Editors, Stochastic Behavior in Classi-
        cal and Quantum Hamiltonian Systems, Springer-Verlag,
        Berlin (1979).
4.   S. Jorna, Editor, Topics in Nonlinear Dynamics, AIP Conference
        Proceedings No. 46, New York (1978).
5.   K. S. J. Nordholm and S. A. Rice, J. Chem. Phys. 62, 157 (1975).
6.   D. W. Noid, M. L. Koszykowski, M. Tabor and R. A. Marcus,
        J. Chem. Phys. 72, 6169 (1980).
7.   R. M. Stratt, N. C. Handy and W. H. Miller, J. Chem. Phys. 71,
        3311 (1980).
8.   N. Pomphrey, J. Phys. B7, 1909 (1974).
9.   J. S. Hutchinson and R. E. Wyatt, Chem. Phys. Lett. 72, 378
        1980).
10. V. I. Arnold and A. Avez, Ergodic Problems of Classical Mech-
        anics, Benjamin, New York (1974).
11. A. N. Kolmogorov, Dokl. Akad. Nauk. SSSR 124, 774 (1959)
12. See, for example, S. A. Rice in Quantum Dynamics of Molecules,
        Ed. R. G. Wooley, Plenum Press, New York (1980) p. 257.
13. S. A. Rice, Adv. Chem. Phys. in press.
14. R. Kosloff and S. A. Rice, J. Chem. Phys. in press.
15. R. Kosloff and S. A. Rice, J. Chem. Phys. in press.
16. B. V. Chirikov, Phys. Repts. 52, 263 (1979).
17. J. Ford, Adv. Chem. Phys. 24, 155 (1973).
18. G. Benettin, C. Froeschle and J. P. Scheidecker, Phys. Rev.A19,
        2454 (1979); G. Benettin, L. Galgani and J. M. Strelcyn,
        Phys. Rev. A14, 2338 (1976).
19. V. A. Rokhlin, Uspenhi Mat. Nauk 22, 1 (1967) (Russian Math.
        Surveys 22, 1 (1967)); V. A. Rokhlin, Izv. Akad. Nauk SSSR
        Ser. Mat. 25, 499 (1961); (Amer. Math. Soc. Transl. (2)39,
        1 (1964)).
20. Ya. Sinai, Izv. Akad. Nauk SSSR Ser. Mat. 25, 899 (1961), (Amer.
        Math Soc. Transl. (2)39, 83 (1964)). Ya. Sinai Izv. Akad.
        Nauk SSSR Ser. Mat. 30, 15 (1966), Amer. Math. Soc. Transl.
        (2)68, 34 (1967).

21. Ya. B. Pesin, Uspekhi Mat. Nauk 32:4, 55 (1977), Russian Math.
    Surveys 32:4, 55 (1977)).
22. P. Pechukas, J. Chem. Phys. 57, 5577 (1972).
23. R. T. Swimm and J. B. Delos, J. Chem. Phys. 71, 1706 (1979).
24. M. V. Berry, M. L. Balazs, M. Tabor and A. Voros, Ann. Phys.
    122, 26 (1979).
25. G. Casati, B. V. Chirikov, F. M. Izrealev and J. Ford, in
    Stochastic Behavior in Classical and Quantum Hamiltonian
    Systems. (Springer-Verlag Berlin, 1979)p. 334.
26. For details see A. Wehrl. Rev. Mod. Phys. 50, 221 (1978).
27. A. Ash, Information Theory, (Wiley-Interscience, New York,1965).
28. J. W. Duff and P. Brumer, J. Chem. Phys. 65, 3566 (1977);
    J. W. Duff and P. Brumer, J. Chem. Phys. 67, 4898 (1977);
    P. Brumer, J. Comp. Phys. 14, 391 (1973); P. Brumer, Adv.
    Chem. Phys. (in press).
29. M. Toda, Phys. Lett. A48, 335 (1974).
30. C. Cerjan and W. P. Reinhardt, J. Chem. Phys. 71, 1819 (1979).
31. G. Benettin, L. Galagani, and J. M. Strelcyn, Phys. Rev. A14,
    2338 (1976); G. Benetting, G. Froechle, and J. P. Schiedecher,
    Phys. Rev. A19, 2454 (1979); Ya. A. Pesin, Dokl, Akad. Nauk,
    SSSR 226, 774 (1976). Y. A. Pesin, Uspekhi Mat Nauk 32:4 55
    (1977), (Russian Math Surveys 32:4, 55 (1977)).
32. M. Gutziller, J. Math. Phys. 12, 343 (1979).
33. M. Henon and C. Heils, Astron. J. 69, 73 (1964).
34. E. Hopf, Ber. Verh. Sachs. Akad. Wiss. Leipzig 91, 261 (1939).
35. Ya. Sinai, Dokl. Akad. Nauk. SSSR 131, 752 (1960); D. V.
    Ansonov and Ya. Sinai, Usp. Mat. Nauk. 22, 107 (1967).
36. G. A. Hedlund, Ann. Math. 35, 787 (1934); Am. J. Math. 62, 233
    (1940); J. C. Oxtoby and S. M. Ulam, Ann. Math. 42, 874
    (1941).
37. I. Shimada, Prog. Theor. Phys. 62, 61 (1979).
38. R. S. Ingarden, Acta. Phys. Pol. 43, 3 (1973).
39. There are other definitions of the quantum mechanical Kolmogorov
    entropy. See, for example: A. Connes and E. Stormer, Acta
    Math. 134, 289 (1975); G. Emch, Comm. Math. Phys. 49, 191
    (1976), and G. Lindblad, Comm. Math. Phys. 65, 281 (1979).
40. J. von Neumann, Mathematical Foundations of Quantum Mechanics,
    Princeton University Press, Princeton, N. J. 1968).
41. H. Krips, J. Math. Phys. 18, 1015 (1977); Foundations of Physics
    4, 181, 381 (1974).
42. A. Gleason, J. Math. Mech. 6, 885 (1957).
43. S. Gudder, Stochastic Methods in Quantum Mechanics (North Hol-
    land, New York, 1979).
44. E. Nelson, Dynamical Theories of Brownian Motion, Princeton
    University Press, Princeton, N. J., 1967).
45. R. Hellman and T. Bountis, in Stochastic Behavior in Classical
    and Quantum Hamiltonian Systems (Springer-Verlag, Berlin,
    1979)) p. 353.
46. K. F. Freed, Disc. Faraday Soc. 55, 68 (1973).
47. K. Kraus, Ann. Phys. 64, 311 (1971).

48. K. Matsuno, J. Math. Phys. 16, 2368 (1975).
49. R. Kosloff, Adv. Chem. Phys. in press.
50. B. Misra and E. C. G. Sudarshan, J. Math. Phys. 18, 756 (1977).
51. P. Brumer and M. Shapiro, Chem. Phys. Lett. 72, 528 (1980).
52. K. J. Kay, J. Chem. Phys. 61, 5205 (1974); W. M. Gelbart,
    S. A. Rice and K. F. Freed, J. Chem. Phys. 57, 4699 (1972);
    L. Van Hove, Physica 21, 517 (1955); G. P. Berman and G. M.
    Zaslavsky, Physica 91A, 450 (1978).
53. G. Lindblad, Comm. Math. Phys. 48, 119 (1976).
54. K. G. Kay, J. Chem. Phys. 72, 5955 (1980).

THEORETICAL ANALYSIS OF EXPERIMENTAL PROBES OF

DYNAMICS OF INTRAMOLECULAR VIBRATIONAL RELAXATION[*]

Karl F. Freed [a] and Abraham Nitzan [b]

[a] The James Franck Institute and
Department of Chemistry
The University of Chicago
Chicago, Illinois 60637

[b] Institute of Chemistry
Tel Aviv University
Ramat-Aviv, Tel Aviv, Israel

INTRODUCTION

The phenomenon of intramolecular vibrational relxation is postulated to play a central role in the description of unimolecular reaction processes of polyatomic molecules. For instance, the famous RRKM theory is generally presented as being predicated on the assumption that vibrational energy is rapidly randomized among the different vibrational degrees of freedom on time scales which are rapid compared to the decomposition times of these molecules.[1,2] A number of different theoretical approaches have been undertaken to understand better the phenomena of vibrational energy scrambling in molecules. Other talks at this conference discuss the transition from quasiperiodic to stochastic behavior in classical mechanical descriptions of vibrational energy in molecules with the hopes that in an as yet undefined fashion this is somehow relevant to the description of energy randomization processes occurring in real molecules under experimental conditions. This lecture is concerned with providing a theoretical basis for understanding recent experiments on intramolecular vibrational relaxation.

It is not as widely recognized that the popular RRKM-type theories of unimolecular decomposition can be formulated with the totally

* Supported, in part, by NSF Grant CHE80-23456 and the U.S.-Israel Binational Science Foundation.

opposite hypothesis that energy is not randomized on the decomposition time scales. [3]   Rather, the collisional preparation of the ener-
gized states of the system is sufficiently chaotic to produce excited
molecular levels with a statistical distribution among these states.
These eigenstates are highly anharmonic motions of the system which
may under some conditions appear local mode in nature or be of a more
global variety akin to some normal modes of vibration. In chemical
activation experiments the initial state is not a statistical distribution. [4] This initial state can be taken to be a nonstationary superposition of a large number of vibrational eigenstates. Then the
observed energy flow in chemical activation experiments [4] can be
viewed [3] as the natural nonrandom time evolution of this prepared
wave packet in which the energy is initially localized in one end of
the molecule. By invoking the assumption of a zero rate of intramolecular vibrational relaxation along with the random lifetime hypothesis of the traditional RRKM formulations, we arrive at the identical mathematical representation of unimolecular decay theory. The
collisional preparation of the system leads in a first approximation
to a statistical distribution among the available vibrational eigenstates. The importance of further studies of intramolecular vibrational relaxation is highlighted by the fact that the kinetically
extreme models of complete and of no vibrational energy relaxation
lead to the same RRKM theory.

Intramolecular vibrational relaxation obviously can not occur
between the sparsely spaced vibrational energy levels in a diatomic
molecule. On the other hand, vibrational relaxation is definitely
a commonplace occurrence in solids. Hence, intramolecular vibrational relaxation must occur in sufficiently large polyatomic molecules
under the appropriate experimental circumstances, and it should be
absent in sufficiently small polyatomic molecules. There is also a
situation intermediate between the two extremes in which there is a
transition from the small molecule, no vibrational energy randomization limit to the large molecule case with intramolecular energy relaxation. It is this intermediate case and its relationship to recent experiments on intramolecular vibrational energy flows [5] which
is the central focus of this talk. The work by Mukamel [6] considers
a more general and sophisticated description of intramolecular vibrational relaxation processes with more emphasis on the large molecule
limit where intramolecular vibrational relaxation must occur.

Let us begin by considering the values of intramolecular vibrational relaxation rates that have been inferred from a number of recent sophisticated experiments using such techniques as picosecond
spectroscopy and laser studies in molecular beams. The systems,
their total vibrational energy content and the estimated intramolecular vibrational relaxation rates are presented in Table 1.[5,7-17,4]
An examination of the table displays the perplexing tendency to have
very long intramolecular vibrational relaxation times for some systems containing very high vibrational energies and high vibrational

TABLE 1

Molecule and Electronic State	No. of Atoms	Excess Vibrational Energy cm	Estimated IVR Lifetimes sec	Ref.	Comments
Tetracene (S_1)	30	$>10^4$	$>5 \times 10^{-9}$	7	from excess E dependence of electronic radiationless relaxation
Pentacene (S_1)	36	$>10^3$	$>5 \times 10^{-9}$	7	
Pentacene (S_0)	36	1.9×10^4	$>10^{-6}$	8	$S_0 - S_1$ lineshape following internal conversion. Beam
Naphthalene (T_1)	18	1.1×10^4	$>10^{-6}$	9	T - T absorption following inter-system crossing.
Naphthalene (S_0)	18	$8000 - 20,000$	$>10^{-13}$	11	overtone linewidths:2'K
Naphthalene (S_1/S_2)	18	435	1.2×10^{-11}	5	overtone linewidths supersonic beam; excess energy is measured from the S_1 origin
		3069	3.3×10^{-12}		
		3760	3.4×10^{-12}		
		4057	3.2×10^{-12}		
		4296	2.2×10^{-12}		
		5205	1.2×10^{-12}		
Naphthalene $(D_8)(S_1/S_2)$	18	3449	2.7×10^{-12}		
		3768	2.3×10^{-12}		
		4640	1.2×10^{-12}		
Benzene (S_0)	12	$8000 - 20,000$	10^{-13}	12	overtone lineshapes; room temperature
Coumarine 6 (S_0)	43	5950	$4 \pm 1 \times 10^{-12}$	13	IR-visible double resonance
Dimethyl POPOP (S_1)	50	3000	2×10^{-12}	14	time evolution of fluorescence spectrum
$F_2 \!-\! F \!-\! F \!-\! F_2$ H_2 D_2 (S_0)	16	3×10^5	10^{-12}	4	pressure dependence of competing thermal reactions

(continued)

TABLE 1 (Cont'd)

Molecule and Electronic State	No. of Atoms	Excess Vibrational Energy cm^{-1}	Estimated IVR Lifetimes sec	Ref.	Comments
p-difluorobenzene (S_1)	12	2190	10^{-12}	15	quenching of redistributed fluorescence
SF_6 (S_0)	7	1000–3000	$1-30 \times 10^{-12}$	16	recovery of saturation in IR excitations
		2000	$>>10^{-6}$	17	IR-IR double resonance
		≥ 1000	$1- 5 \times 10^{-9}$	17	IR-IR double resonance
		1 quantum in			
methylbenzene (S_1)	15	$6b_0^1$	$>3 \times 10^{-6}$	5	from relative intensities of direct and redistributed fluorescence
		12_0^1	$>2 \times 10^{-6}$		
		$18a_0^1$	$>3 \times 10^{-7}$		
n(t) Propylbenzene (S_1)	21	$6b_0^1$	$>2 \times 10^{-6}$	5	
		12_0^1	1×10^{-8}		
		$18a_0^1$	5×10^{-9}		
n(t) Hexylbenzene (S_1)	30	$6b_0^1$	$<2 \times 10^{-9}$	5	
		12_0^1	$<1 \times 10^{-9}$		
Naphthalene (S_1)	18	1×10^4	$>10^{-8}$	10	from excess E, dependence of electronic radiationless relaxation

densities of states, whereas a number of other systems have very short intramolecular vibrational relaxation lifetimes under circumstances in which the density of states is significantly lower. This is in marked conflict with the intuitive concept that the rate of intramolecular vibrational relaxation should grow significantly with increases in the vibrational density of states. It should be noted that some of the experiments in Table 1 or the interpretations of these experiments have been called into question, and the Table is presented here to show the confusing status of the literature concerning experiments to directly probe intramolecular vibrational relaxation rates.

Some of the important reasons for the huge variance of literature values for intramolecular vibrational relaxation rates are as follows:
(a) Intramolecular vibrational relaxation processes are generally undefined as dynamical processes with specified initial and final states.
(b) When defined, the given initial states are often not those that are prepared under the experimental circumstances.
(c) Many arguments concerning intramolecular vibrational relaxation are based on non-measurable quantities and therefore lie outside the scope of proper quantum mechanical descriptions.
(d) Some estimates of intramolecular vibrational relaxation rely on nondynamical measurements. Rather sophisticated understanding of intramolecular vibrational relaxation is required to enable the extraction of intramolecular vibrational relaxation rates from experiments which do not directly view a time varying vibrational energy content in particular vibrational states.
For example, descriptions of RRKM theory generally do not specifically define the initial states of the system. The kinetically equivalent formulation with or without intramolecular vibrational relaxation result, in part, from utilizing different representations for the initial states of the system.

On the other hand, a good deal of the new sophisticated experimental results can be simply explained by using a generalization of known material in the theory of electronic relaxation processes. The researchers in the field of electronic relaxation theory have kept this material highly secret. But rather than the customary method of having the theoretical developments classified, the researchers have kept the material secret by presenting a number of detailed review articles [18-20] on the subject which have yet to permeate to workers in the field of intramolecular vibrational energy randomization. In 1976 there already appeared two applications of the results of electronic relaxation theory to descriptions of intramolecular vibrational relaxation processes.[21,22] Freed [21] discussed the effects of intramolecular vibrational relaxation processes on the observed vibrational energy dependence of electronic relaxation processes in tetracene and pentacene. [7] Mukamel and Jortner [22] introduced the con-

cepts of electronic relaxation processes into a description of the
intramolecular vibrational relaxation occurring in multiphoton decomposition processes.

It is therefore appropriate to reiterate the correspondence between the two phenomena. The electronic relaxation theory is presented in terms of an initial state ϕ_0 before excitation, a zeroth
order state ϕ_s which is prepared by the optical excitation, and a
dense manifold of final levels $\{\phi_\ell\}$ into which the state ϕ_s can decay.
The same cast of characters with slightly changed names also arises
in a description of intramolecular vibrational relaxation processes
as presented below in Fig. 1. The theory of electronic relaxation
processes separates the behavior into categories associated with the
small, intermediate and large molecule limits.[18-20] By the strong
similarity of the energy level diagram appropriate to intramolecular
vibrational relaxation processes, it is also useful to introduce
these limiting cases for the vibrational problem. [18,21,22]

Before we turn to a brief review of those known results of electronic relaxation processes which are of relevance to the description
of intramolecular vibrational relaxation, it is useful to pause for
a moment to examine the definition or meaning of relaxation. The concept of relaxation is generally associated with the loss or irreversible decay of some physically measurable quantity. For instance,
population decay (T_1) and phase memory decay (T_2) are two well understood concepts which are studied in the wide variety of two-level
systems. A problem arises when in some instances clever experiments
are able to reverse what appears to be an irreversible decay. The
case of spin echos is an example where there is apparently a rapid
irreversible loss of spin magnetization which can be re-formed, in
part, through a particular pulse sequence. This example serves to
demonstrate the care that must be exercised to distinguish between
population and phase memory decays.

It should be emphasized that the quantum mechanical description
of isolated bounded systems yields a discrete energy level spectrum.
Then there is rigorously no possibility for irreversible decay, so
no vibrational energy relaxation is, in fact, mathematically possible
for an isolated, bound polyatomic molecule. Thus, it is necessary
to discuss questions of practical or apparent decay which occurs on
the time scales of real experiments. [18,25]As described below, this
leads to an apparent isolated molecule "decay" phenomena which has
some analogies with dephasing processes and which has been referred
to as "an intramolecular dephasing process." This has caused some
confusion with researchers who are familiar with the ordinary concepts of dephasing, so for the purposes of this lecture, we refer to
this apparent intramolecular relaxation as "unphasing."

II. THE SMALL, LARGE, AND INTERMEDIATE CASES OF

INTRAMOLECULAR VIBRATIONAL RELAXATION

After excitation the experiments begin with the vibrational energy in some zeroth order vibrational states corresponding to some
nonlinear anharmonic vibrations or some local mode-type vibrations,
etc. The mode of preparation of the system is not described yet, but
it is an important consideration to determine whether the designated
initial state of the system is one which can be prepared in realistic
experiments. At some subsequent time a measurement is performed which
finds the state of the system with the vibrational energy in other
modes, bonds, etc. The situation can be depicted by the energy level
scheme in Figure 1. ϕ_0 represents the state of the system before
preparation of the excited vibrational state. The preparation technique leads to the excitation of the vibrational degrees of freedom
and to the production of the zeroth order state ϕ_s. ϕ_s is coupled
to a dense manifold of vibrational levels $\{\phi_\ell\}$ through the intramolecular $v_{s\ell}$ which is responsible for the intramolecular vibrational relaxation. Interrogation of the time dependence of the system

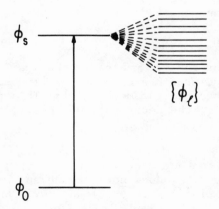

Fig. 1. Zeroth order energy level diagram. ϕ_0 is a ground state vibration level populated before the excitation process. ϕ_s is
a zeroth order excited vibrational level for which the excitation process has nonzero matrix elements for transitions
from ϕ_0. The manifold of zeroth order vibrational levels
$\{\phi_\ell\}$ have no matrix elements for excitation from ϕ_0. The
$\{\phi_\ell\}$ are coupled to ϕ_s by the matrix element $v_{s\ell}$; this coupling is represented by dashed lines.

finds that at subsequent times the molecule is in some linear superposition of the states ϕ and $\{\phi_\ell\}$. It is this observed time dependent
behavior of the system which is interpreted as involving an "intramolecular vibrational relaxation process."

As noted above, the energy level diagram in Figure 1 is identical to that which has extensively been studied for electronic relaxation. [20-22] The electronic relaxation theory delineates three limiting cases with qualitatively different behavior. The exact parallel
of the energy level diagrams implies that an identical situation occurs for the case of intramolecular vibrational energy relaxation.
[18,21,23] Thus, we anticipate observing three limiting cases of intramolecular vibrational energy relaxation as follows: (a) In the
small molecule limit the time dependence of the system is described
in terms of the individual vibrational eigenstates which need not be
harmonic oscillator states. There is no vibrational energy relaxation. Modes of preparation of the system generally lead to the preparation of individual vibrational eigenstates, or, perhaps, because
of accidental degeneracy, a pair of vibrational eigenstates. (b) In
the opposite large molecule limit the $\{\phi_\ell\}$ manifold is very dense
and irreversible intramolecular vibrational relaxation occurs. A
quantification of the description "very dense" is discussed below.
(c) In the intermediate case the $\{\phi_\ell\}$ manifold is rather dense, but
the apparent "decay" is not an irreversible one. This is the situation in which the "unphasing" processes occur.

A. Conditions for Limiting Cases

The criteria for the occurrence of the small, medium and large
limits of intramolecular vibrational relaxation can be expressed in
terms of a single physical parameter. [18,21] This parameter is represented in terms of Γ_ℓ, the decay rate of the $\{\phi_\ell\}$ and

$$\epsilon_\ell = \rho_\ell^{-1} \tag{1}$$

the mean spacing between adjacent (in energy) $\{\phi_\ell\}$ levels. ρ_ℓ is the
density of $\{\phi_\ell\}$ levels. Γ_ℓ does not include any contribution from
intramolecular vibrational energy relaxation. The small molecule
limit is governed by the condition,

$$\Gamma_\ell / \epsilon_\ell \ll 1 \text{ , small molecule limit.} \tag{2}$$

Here the system is most simply described in terms of the vibrational
eigenstates $\{\psi_n\}$ which diagonalize the vibrational Hamiltonian. Excitation prepares the system initially in either a single pure vibrational eigenstate ψ_n or a linear superposition of eigenstates $\{\psi_n\}$.
There is no possibility for irreversible intramolecular vibrational
relaxation; the dynamics is governed by the simple evolution of the
individual vibrational eigenstates. On the other hand, the large
molecule limit is governed by the opposite extreme of

$$\Gamma_\ell / \varepsilon_\ell \gg 1 \text{ , statistical limit.} \tag{3}$$

Here we may have irreversible electronic relaxation. Note that condition (3) is the same one obtained by Kosloff and Rice [24] from a quantum mechanical extension of the Kolmogorov entropy concept.

The conditions in (2) and (3) for the small and large molecule limits, and perforce for the lack or presence of irreversible intramolecular vibrational relaxation, are somewhat mathematical constraints since they do not include the important physical considerations of (a) the method of preparation of the system, and (b) the question of practical irreversibility, i.e., whether the system behaves in a fashion displaying apparent intramolecular vibrational relaxation on the time scales of real experiments.

B. Preparation of the Initial Excited State

Let us consider first the question of the nature of the preparation of the initial state before intramolecular vibrational relaxation. Let τ_p be the duration of the pulse of exciting radiation. Under normal circumstances we have the uncertainty limits $\tau_p \Delta\omega_p \gtrsim 1$ with $\Delta\omega_p$ the frequency spread of the incident light. In the small molecule limit of condition (2) if the pulse duration satisfies the inequality,

$$\hbar / \tau_p \sim \Delta\omega_p \ll \varepsilon_0 \lesssim |E_n - E_m| , \tag{4}$$

where ε_0 is the minimum spacing between neighboring energy levels, the incident radiation can excite only individual vibration eigenstates. Simple single exponential decay of the excited states ensues with no intramolecular vibrational relaxation. If, on the other hand, the pulse is shortened so that the duration satisfies the condition,

$$\hbar / \tau_p \sim \Delta\omega_p \gtrsim \varepsilon_0 \lesssim |E_n - E_m| \tag{5}$$

then this radiation must coherently excite a linear combination of the vibrational eigenstates $\{\Psi_n\}$. The state of the system after the excitation pulse (at t = 0) is then

$$\Psi(0) = a_s \phi_s + \sum_\ell b_\ell \phi_\ell \equiv \sum_n c_n \Psi_n . \tag{6}$$

(It is not always possible to separate the dynamics into excitation and subsequent decay, but for our purposes this idealization is sufficient. The time dependent decay displayed by the initial state (6) can be rather complicated. There is no line broadening from intramolecular vibrational relaxation even with the presence of a number of terms in the summation in (6) provided that the small molecule limit of (2) is still obeyed. However, there is the possibility of observing "practical" intramolecular vibrational relaxation when starting with the initial state (6) if the experiments are performed on certain time scales.

In the large molecule limit of (3) any pulse must excite a linear combination of the zeroth order levels $\{\phi_s, \phi_s\}$ of a form like (6). Here irreversible intramolecular vibrational relaxation does contribute to the line broadening because of the large number of levels contributing to the sums in (6). It is, therefore, possible in principle, to study the line broadening and sharp spectral structure under super-high resolution to distinguish between the statistical limit and the intermediate case of intramolecular vibrational relaxation. [23] However, this process is technically complicated by trivial sources of broadening such as those produced by rotational bands, Doppler broadening, etc., which serve to wash out structure.

C. Practical Irreversibility

A central concept in studying intramolecular relaxation processes is the concept of practical irreversibility.[18,25] Given an experimental time scale τ_{obs} the relevant questions is whether the system appears to decay on time scales comparable with τ_{obs} . The answer is determined [18,25] by the magnitude of the parameter $[\Gamma_\ell + \hbar/\tau_{obs}]/\varepsilon_\ell$. The system behaves as if it were in a small molecule limit for times $t \approx t_{obs}$ if the condition

$$[\Gamma_\ell + \hbar/\tau_{obs}]/\varepsilon_\ell \ll 1, \text{ small molecule,} \qquad (7)$$

is satisfied. Condition (7) implies that for $t \lesssim \tau_{obs}$ there is no intramolecular vibrational relaxation.

Molecules in the small molecule limit can behave as if they were in the statistical limit for timescales τ_{obs} if

$$[\Gamma_\ell + \hbar/\tau_{obs}]/\varepsilon_\ell \gg 1, \text{ apparent statistical limit for}$$
$$t \lesssim \tau_{obs}, \qquad (8)$$

is obeyed. Then the system appears to undergo an intramolecular vibrational relaxation process, but there is no additional line broadening due to this apparent intramolecular vibrational relaxation. Molecules conforming to both the small molecule limit (2) and the apparent or practical statistical limit (8) provide the intermediate case of intramolecular vibrational relaxation, a situation which we have termed the "too many level small molecule limit" because practical excitation techniques produce a superposition of too many individual levels to ever wish to resolve all of them. Thus, short, broad, pulsed excitation as depicted in Fig. 2 leads to a linear superposition of a number of vibrational eigenstates. The initial state of the system is approximately given by the nonstationary state ϕ_s when the pulse is short enough so that there is coherent excita-

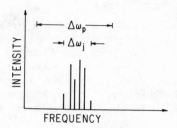

Fig. 2.   A schematic representation of the absorption spectrum to
the set of vibrational eigenstates $\{\psi_n\}$ associated with a
given ϕ_s. The spread of these levels is $\Delta\omega_j$ which is dependent on the coupling strength $v_{s\ell}$. The absorption intensity
for each ψ_n is proportional to $|<\phi_s|\psi_n>|^2$. If the excit-
ing light has $\Delta\omega_p$ of (4) greater than $\Delta\omega_j$, then this excitation process coherently excites the zeroth order level ϕ_s.

tion of all of the vibrational eigenstates whose parentage is the
zeroth order level ϕ_s. This initial state is written as

$$\Psi(0) \overset{\cdot}{\simeq} \phi_s$$
$$= \sum_n C_{ns}\psi_n \tag{9}$$

The subsequent time evolution of this state is simply given by

$$\Psi(t) = \sum_n C_{ns} \exp[-i E_n t/\hbar - \Gamma_n t/2]\psi_n. \tag{10}$$

The probability $P_s(t)$ that the system be found in the initial zeroth
order state ϕ_s at time t is then given by the lengthy expression

$$P_s(t) = |<\phi|\Psi(t)>|^2 = \sum_n |C_{ns}|^4 \exp(-\Gamma_n t)$$
$$+2 \sum_{n \neq n'} |C_{ns}|^2 |C_{n's}|^2 \exp[-(\Gamma_n + \Gamma_{n'})t/2] \cos(\omega_{nn'}t) \tag{11}$$

where $\omega_{nn'} = (E_n - E_{n'})/\hbar$ is the transition frequency between the levels n and n'.

D. Time Evolution and "Unphasing"

To study the detailed time evolution of the system [26] in the interesting intermediate case, it is useful to separately consider the two summations in (11). The term involving a single summation over n represents the incoherent superposition of the decays of individual vibrational eigenstates, whereas the summation over $n \neq n'$ represents the "coherent" contribution due to interference between pairs of vibrational eigenstates. For the initial time t = 0, we have the relation

$$\cos(\omega_{nn'}t)\Big|_{t=0} = 1.$$

Thus, all of the vibrational eigenstates $\{\psi_n\}$ in (10) are in phase at t = 0. However, for t = 0 the individual $\{\psi_n\}$ in the coherent term of (11) are random functions of the vibrational eigenstate indices n and n'. In the too many level small molecule limit the average of this $\cos(\omega_{nn'}t)$ term over the contributions n and n' levels is found [26,18,23] to behave at short times as the exponential decay,

$$\cos(\omega_{nn'}t) > \; \dot{\simeq} \; \exp(-\Delta t), \tag{12}$$

where the "unphasing" time Δ^{-1} is given by the golden rule-like expression

$$\Delta \; \dot{=} \; \frac{2\pi}{\hbar} \sum_\ell | v_{s\ell}|^2 \, \delta(E_s - E_\ell). \tag{13}$$

The incoherent contribution to (11) can be seen to exhibit a slow decay rate with a magnitude given by an average value $\bar{\Gamma} \simeq <\Gamma_n>$ whereas the faster coherent decay is with an average rate $\bar{\Gamma} + \Delta$. Thus, a double exponential decay of the resonance emission is anticipated from the initially excited zeroth order level ϕ_s. (In certain circumstances a triple exponential decay is to be expected.[27])

The general equation (11) is rather complicated with a summation over many contributing vibrational eigenstates. It may be studied further by numerical simulations. [28] Here we consider the popular egalitarian model [26,23] in which it is assumed that there are N coupled $\{\phi_\ell\}$ levels to each ϕ_s level. The zeroth order ϕ_s level is taken to be equally distributed among the individual vibrational eigenstates $\{\psi_n\}$ leading to the relation

$$|C_{ns}|^2 = (N + 1)^{-1} \text{ for all n.} \tag{14}$$

Given this egalitarian model, the decay of ϕ_S in (11) simplifies considerably to the result

$$P_s(t) = \begin{cases} \exp[-(\Gamma+\Delta)t], & \text{short } t, \\ (N+1)^{-1}\exp(-\Gamma t), & \text{long } t. \end{cases} \tag{15}$$

Eq. (15) displays the short and long time decay characteristics described physically above. It also links the relative intensities of the short and long time components to the number of effectively coupled $\{\phi_\ell\}$ levels. Note that in the statistical limit of $N \to \infty$ only the short time decay component persists. The decay rate of (13) then becomes the golden rule approximation to the intramolecular vibrational relaxation rate of the statistical limit molecule.

III. APPLICATION OF GENERAL PRINCIPLES TO INTRAMOLECULAR VIBRATIONAL RELAXATION IN REAL MOLECULAR SYSTEMS.

We consider first the absorption spectrum anticipated for molecules corresponding to the intermediate case of intramolecular vibrational relaxation. This discussion then naturally leads into a description of the expected emission spectrum and the time evolution of the emission spectra. It is convenient to introduce the predicted absorption and emission spectra in a stepwise fashion. Each step is rather simple and straightforward but the final results could appear counter-intuitive if just presented directly. Hence, for clarity the full arguments are summarized.

A. Absorption Spectrum

Let us begin with a zeroth order description in terms of harmonic oscillator states with anharmonic couplings between these zeroth order basis functions. In a supersonic nozzle the molecules can initially be taken to be in the vibrationless level $|0\rangle$ of the ground electronic state $|g\rangle$. The rotational degrees of freedom have some slight excitation, but for simplicity these are omitted here. In a zeroth order harmonic oscillator description there can be optical absorption from the ground vibronic state to vibronic components of the excited electronic state $|e\rangle$ in which there are excitations in a set of optically active vibrations a. For simplicity, Fig. 3 depicts a single quantum number n for the progression (s) in this optically active mode. In addition, in the zeroth order harmonic description the remaining vibrations $\{b\}$, the "bath modes," cannot become excited in the electronic transition g $\leftrightarrow$ e. Zeroth order allowed absorption processes are indicated by the arrows in Fig. 3. The excited electronic state also has vibronic components in which the $\{b\}$ vibrations are excited. These vibronic components cannot be reached by absorbtion from the ground vibronic state.

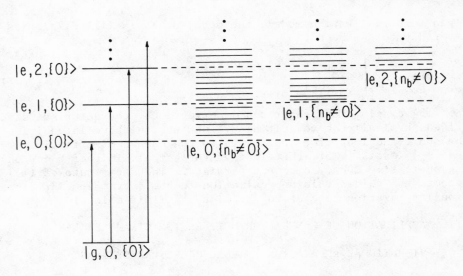

Fig. 3. Pictorial representation of the zeroth order absorption spec-
trum from the ground state vibrationless level $|g,0,\{0\}>$ to the excited
state levels $|e, n_a,\{0\}>$. n_a <u>sequentially</u> labels the zeroth order vi-
brational levels and generally involves a number of zeroth order vibrational modes. The remaining "bath modes" are characterized by the
zeroth order quantum numbers $\{n_b\}$. The excited state zeroth order
levels, having $n_b \neq 0$ for some bath modes, cannot be reached by exci-
tation from $|g,0,\{0\}>$. There are coupling matrix elements between
the $|e,n_a, \{0\}>$ and the $|e, n_a', \{n_b \neq 0\}>$ which convert this simple
zeroth order spectrum into the observed absorption spectrum. In the
intermediate case each zeroth order $|g, 0, \{0\}> \rightarrow |e, n_a,\{0\}>$ vibronic
transition is converted to a series of absorption peaks as represented in Fig. 2.

Anharmonic interactions mix the $\{n_b\}= 0$ levels with those having
$\{n_b\} \neq 0$. This mixing alters the absorption spectrum considerably
from the zeroth order one. In the small molecule limit and the too
many level small molecule limit as a special case, the anharmonic interactions are diagonalized by utilizing the vibrational eigenstates
$\{\Psi_{n_a j}\}$. The subscript n_a designates the parentage in the zeroth
order harmonic basis and the source of the absorption intensity from
g. These vibrational eigenstates are represented in the zeroth
order harmonic basis as

$$\Psi_{n_a j} = a_{n_a j} |e,n_a,\{0\}> + \sum_{\substack{n_a' < n_a \\ \{n_b\} \neq 0}} b_{n_a j, n_a', \{n_b\}} |e,n_a',\{n_b\}>.$$

$$(16)$$

Here the $|e, n_a, \{0\}\rangle$ component provides the source of all the absorption intensity from $|g, 0, \{0\}\rangle$, while the $|e, n'_a, \{n_b\}\rangle$ correspond to the $\{\phi_\ell\}$ and do not provide any absorption intensity from $|g, 0, \{0\}\rangle$. The latter, however, are very important in determining the nature of the emission spectrum.

The energies of the vibrational eigenstates are distributed about the zeroth order energy $E_{n_a, \{0\}}$ of the optically active zeroth order harmonic oscillator states $|e, n_a, \{0\}\rangle$, having the energies

$$E_{n_a j} = E_{n_a, \{0\}} + \delta_j \tag{17}$$

where δ_j is the energy shift associated with $\psi_{n_a j}$. Thus, the absorption spectrum, as depicted in Fig. 2 is expected to display a number of vibronic absorption lines for the $\{\psi_{n_a j}\}$ with intensities which are proportional to the population of the optically active harmonic states in the $\{\psi_{n_a j}\}$, i.e. proportional to $|a_{n_a j}|^2$. The positions of these vibronic absorption lines are governed by the transition energy from $|g, 0, \{0\}\rangle$ to the excited levels $\psi_{n_a j}$ with energies (17). When the couplings $v_{\ell s}$ are weak, the spread $\Delta \delta_j$ of these states may be rather narrow, producing an apparent narrow absorption line if resolution is insufficient and/or trivial sources of broadening mask the structure.

Note that in the intermediate case the widths of the individual vibronic absorption components are smaller than the separations between the individual vibronic lines. Thus, the intermediate case displays a structured absorption spectrum with no spectral broadening induced by the intramolecular vibrational relaxation. On the other hand, in the statistical limit, the spacing between the vibronic lines is smaller than their natural widths, so a single broadened absorption line appears, perhaps with some residual structure in this statistical limit. This absorption feature then has an intramolecular vibrational relaxation induced broadening $\Delta \omega_j$ as in Fig. 2. A similar absorption feature is obtained for each vibrational eigenstate which has as its parentage a given zeroth order optically active harmonic state. The situation becomes somewhat more complicated when the absorption features due to different zeroth order optically active harmonic levels begin to overlap with each other. Here we have considered only the limiting case when these absorption features are well separated from each other.

The observed spectrum even in supersonic nozzle spectrometers is complicated by the presence of some rotational excitation which leads to a rotational bandwidth for the individual vibronic transitions. This can smear out the structure expected in the intermediate case before the molecule truly conforms to the statistical limit. Thus, care must be taken in considering the role of rotations and other sources of spectral broadening before ascribing the apparent lack of structure in the absorption spectrum to the occurrence of the statistical limit of intramolecular vibrational relaxation.

B. The Emission Spectrum

The structure of the emission spectrum follows similarly. First, consider the expected behavior in terms of the zeroth order states. This is depicted schematically in Fig. 4. Fig. 4(a) represents the emission spectrum from the optically active zeroth order harmonic levels. The emission spectrum from each of the $|e,n_a\{0\}>$ zeroth order states involves a progression in mode(s) a. This is the same structure as in pure resonance fluorescence from the zeroth order optically active harmonic states. The emission spectrum from a zeroth order harmonic state with the remaining b modes excited $|e,n_a,\{n_b\}>$ is represented schematically in Fig. 4(b). Again there is a Franck-Condon resonance fluorescence type progression in mode a, but the quantum numbers $\{n_b\}$ remain unchanged in the $e \to g$ transition. Since the b modes may suffer very small frequency changes in the transition between excited and ground electronic states, the emission from $|e, n_a,\{n_b\}>$ has a spectrum which is shifted from that of $|e,n_a,\{0\}>$ by

$$\delta_j^{e0} = \sum_b n_b (\omega_b^e - \omega_b^g) \qquad (18)$$

where ω_b^e and ω_b^g are the frequencies of mode b in $|e>$ and $|g>$, respectively. Thus, apart from the small shifts (18), the emission spectrum from $|e,n_a,\{n_b\}>$ displays a pattern identical in form to that from $|e,n_a\{0\}>$.

When we introduce the anharmonic couplings between the optically active and the $\{n_b\} \neq 0$ zeroth   order vibronic states, the final state energies are shifted as in (17) but with slightly different values δ_j^e which depend on n_a. The final pattern is as follows: It is clear that the $|e,n_a\{0\}>$ term in (16) gives rise to a contribution to the emission spectrum from $\{\psi_{n_a}\}$ corresponding to resonance fluorescence to $|g,n_a,\{0\}>$. The term in $|e,n_a,\{n_b\}>$ in (16) provides contributions of the form of resonance fluorescence to $|g,n_a,,\{n_b\}>$ which are shifted by an energy δ_j^e from the simple resonance fluorescence of the zeroth order harmonic states. Since δ_j^e varies with the value of n_a and n_b, the "relaxed" spectrum, arising from the bath modes with $\{n_b\} \neq 0$, is distributed over a range in energies. This spectrum is depicted schematically in Fig. 5. The shortest wavelength peak can only arise from emission from the excited level $|e,n_a\{0\}>$ to the ground state vibronic level $|g,0,\{0\}>$. This corresponds to sharp zeroth order resonance fluorescence. The remaining bands in the spectrum in Fig. 5 can have sharp components associated with the unrelaxed resonance fluorescence emission from excited levels having $\{n_b\} = 0$, along with the broad background of "relaxed

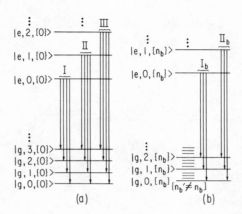

Fig. 4.   Schematic display of the zeroth order unrelaxed (a) and re-
          laxed (b) emission spectra. Each of the $|e,n_a,\{0\}>$ yield
          Franck-Condon progressions in the zeroth order emission spec-
          trum to $|g,n_a,\{0\}>$. Sequences I, II, and III arise for
          $n_a=$ 0,1, and 2, respectively. Likewise, the "relaxed" zer-
          oth order states $|e, n_a, \{n_b\}>$ produce Franck-Condon emission
          patterns to the ground state levels $|g,n_a,\{n\}>$. Note that
          $\{n_b\}$ is unchanged in the transition and that the transition
          energies depend on $\{n_b\}$.   Thus, the emission spectra I_b,
          II_b, etc. are slightly shifted in frequencies from their un-
          relaxed parents I and II, respectively. Adding the contri-
          butions to the zeroth order emission spectra from the dif-
          ferent accessible $|e,n_a,\{n_b\}>$ gives the broad "relaxed"
          background upon which the sharp "unrelaxed" peaks sit. The
          resultant experimental spectrum includes the effects of the
          small shifts δ_j of (17), but the basic pattern given in
          Fig. 5 resembles the superposition of relaxed and unrelaxed
          zeroth order emission spectra.

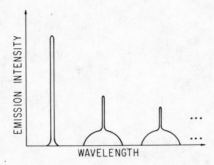

Fig. 5.   A qualitative representation of the emission spectrum. The
          sharp features emerge from the unrelaxed zeroth order
          emission, while the broad ones come from the relaxed contri-
          butions. The actual spectrum need not be symmetrical as
          represented. The shortest wavelength peak must arise solely
          from the unrelaxed contribution with parentage in the
          $|e,n_a,\{0\}\rangle \rightarrow |g,0,\{0\}\rangle$ transition. All longer wavelength
          features may display both relaxed and unrelaxed contribu-
          tions.

emission" from levels containing $\{n_b\} \neq 0$. Such a pattern occurs for
each value of n_a.

IV.   DYNAMICS OF INTRAMOLECULAR VIBRATIONAL RELAXATION

The term relaxation denotes a loss with time of energy, population, phase, etc. Thus, the optimal method for following a relaxation process implies a consideration of the time dependence of the
property or quantity whose relaxation is being studied. For intramolecular vibrational relaxation it is natural to investigate the
time dependence of the emission characteristics of the system, including both the unrelaxed and relaxed spectral components.

Consider the intermediate case where the pulse duration and frequency spread is given by (5). This implies that the initial state
of the system after excitation can be taken to be ϕ_S as in (9). The
choice of this initial condition is not a necessity, and the dynamics
of intramolecular vibrational relaxation in the intermediate case
depends intimately on the nature of the initially prepared state.
This dependence is perhaps best elucidated by a consideration of numerical simulations of the intramolecular vibrational relaxation dynamics using random matrix methods or molecular models. [28] For our
purposes this initial state choice is sufficient to display the sal-
ient physical features of the dynamics of intramolecular vibrational
relaxation. The time dependent intensity of the unrelaxed emission
is proportional to $P_S(t)$ as defined in (11). The corresponding dynamic relaxed intensity is proportional to $P_2(t) = \sum_{\ell} P_{\ell}(t)$ where

$$P_{\ell}(t) = |\langle \Psi(t) | \phi_{\ell} \rangle|^2. \tag{19}$$

The value of $P_S(t)$ using the egalitarian model is already presented
in (15). In the weak coupling limit Δ is given by (13) where ϕ_S is
the state $|e, n_a, \{0\}\rangle$ and $\{\phi_{\ell}\}$ are the states $|e, n_a', \{n_b\}\rangle$ with $n_{a'} < n_a$.
This value for Δ is obtained by virtue of the initial state (9). For
different choices of the initial state the unphasing rate Δ is
altered in general.

Within the egalitarian model the individual $P_{\ell}(t)$ are fluctuations and vanish on average. The quantity $P_2(t)$ may be evaluated
using random matrix methods. [23] However, if we focus on the total
relaxed emission which is proportional to $P_2(t)$, sum rule methods may
be utilized to calculate $P_2(t)$.[23] The resultant total intensity of
the relaxed emission at time t is then found [23] to be proportional
to

$$P_2(t) = \begin{cases} \exp(-\bar{\Gamma}t)[1-\exp(-\Delta t)], & \text{short } t \\ N(N+1)^{-1} \exp(-\bar{\Gamma}t), & \text{long } t \,. \end{cases} \tag{20}$$

Eqs. (15) and (20) provide the dynamical evolution of the unrelaxed
and the relaxed components, respectively, of the emission spectrum
for the special case of the initially excited states (9). The unrelaxed emission of (15) has a short rapid decay followed by a weaker
long time decay. The total relaxed emission intensity has a rising
time component associated with the "unphasing" rate Δ ; the unphasing
is followed by a long time relaxation with a time scale governed by
the decay rates of the individual $\{\phi_S, \phi_{\ell}\}$ levels in the absence of
any intramolecular vibrational relaxation. Eqs. (15) and (20) are
in accord with the predictions of a simple kinetic model of a single
level ϕ_S decaying into a set of levels $\{\phi_{\ell}\}$ with a decay rate Δ
where both the ϕ_S and ϕ_{ℓ} have decay rates to other sets of levels
given by average values $\bar{\Gamma}$. Nevertheless it should be stressed that
this decay is not an irreversible decay since, in principle, a se-

quence of pulses of radiation after time t = 0 could convert the system back to the state ϕ_S at some subsequent time, albeit with diminished total population because of the decay processes inherent in the decay rates $\{\Gamma_s, \Gamma_\ell\}$. In addition, when the system is prepared in the ground electronic state and is truly rotationally cold, then there is the possibility of observing quantum beat phenomena in both the relaxed and unrelaxed spectral dynamics. [23] In this case the kinetic model reproducing (15) and (20) would provide entirely falacious results. These beat phenomena arise from the explicit cosine terms in (11), etc., whose presence is generally washed out when there is an averaging over a large number of levels such as those present when there is rotational excitation in the initial ground state $|g,0,\{0\}\rangle$.[29]

The dynamics simplifies in the statistical limit of $N \to \infty$. The unrelaxed emission decays on a time scale determined by the decay rate equal to the sum of the intramolecular vibrational relaxation rate Δ plus all other decay rates $\overline{\Gamma}$, whereas the total relaxed emission is governed by the rise time associated with Δ^{-1} and the overall decay rate $\overline{\Gamma}$. It is in this large N limit that the relaxation process becomes truly irreversible.

The intermediate case total quantum yields for the relaxed and unrelaxed emission are obtained as

$$\Phi_{\text{unrelaxed}} = \frac{\gamma^R}{\overline{\Gamma}+\Delta} + \frac{\gamma^R}{N\overline{\Gamma}}$$

$$\Phi_{\text{relaxed}} = \frac{\gamma^R \Delta}{\overline{\Gamma}(\overline{\Gamma}+\Delta)} - \frac{\gamma^R}{N\overline{\Gamma}} .$$

(21)

Here γ^R is the average radiative decay rate of $\{\phi_S, \phi_\ell\}$, and the first terms of (21) are the statistical limiting values ($N \to \infty$).

V. APPLICATION TO EXPERIMENTS ON THE DYNAMICS OF INTRAMOLECULAR

VIBRATIONAL RELAXATION IN LARGE MOLECULES

Our analysis of the nature of the absorption and emission spectra, anticipated in cases of intramolecular vibrational relaxation, has already been utilized by Levy and coworkers [30] to explain the observed spectra in phthalocyanine molecules in a supersonic nozzle beam. This model was then elaborated by Smalley and coworkers in their discussion of their experiments on alkyl substituted benzenes. [5,6] The clever idea behind the Smalley group experiments is to vary the length of the alkyl chain to alter the density ρ_ℓ of the states $\{\phi_\ell\}$ and thereby to change the rate of intramolecular vibrational relaxation. The experiments are performed in a supersonic beam, and the incident radiation excites the alkyl benzene to the first excited singlet state S_1. Smalley and coworkers find that at low vibrational energies there are sharp absorption lines with linewidths much less than 1 cm^{-1}. Since the maximum frequency shifts of individual eigenstate δe_j of (17) are comparable to the coupling strength $v_{\ell s}$, this implies a small coupling $v_{\ell s}$. The experiments focus on the 6b, 12, and 18a ring vibrations which are found to be unshifted in going from the benzene absorption spectrum to the alkyl benzene absorption spectrum. [5] These vibrations have nodes at the ring carbon to which alkyl chain is connected. Hence, the interpretation is that excitation of these vibrations leads to excitation of ring modes which are essentially identical to those present in the parent benzene and which are unaffected by the presence of the alkyl chain in the absorption spectrum. The total emission spectrum is used to evaluate the relaxed and unrelaxed quantum yields. The dynamical evolution of the relaxed and unrelaxed spectra are followed by using a pulse with duration approximately 4 nsec and a frequency width of 0.3 cm^{-1}. The lifetime $\bar{\Gamma}$ of the excited S_1 states is approximately 80 nsec. As a function of the number of carbon atoms in the alkyl chain n, it is found that the relaxed quantum yield is negligible for n = 1 while the unrelaxed quantum yield becomes negligible for n = 6, 7. The relaxed quantum yield increases as the length of the alkyl chain increases, i.e., as the density of bath states ρ_ℓ increases, while the unrelaxed yield decreases as n increases.

Given the above discussions, it would then be expected that the alkyl benzenes behave in the small molecule limit of intramolecular vibrational relaxation for n = 1 and they progressively tend towards the intermediate or even statistical limits for n = 6 or 7 for the higher energy vibrations. Thus, based on the relaxed and unrelaxed yields, it should be possible to observe a double exponential decay of the relaxed and unrelaxed yields for intermediate values of n as described in (15) and (20). However the Smalley group experiments [5] observe only a single exponential decay of both the relaxed and unrelaxed spectrum with identical relaxation rates. There is no dynamical manifestation of the intramolecular vibrational relaxation despite the expectation from the relative relaxed and unrelaxed quantum yields that the rate of intramolecular vibrational relaxation should increase with n in an observable fashion on the experimental timescale.

There are two possible reasons why the experiments have failed to observe the expected double exponential behavior of the relaxed and unrelaxed emission.[23] Both possibilities can be tested by additional experiments which should be of considerable value in resolving this question and thereby in extending our understanding of intramolecular vibrational relaxation processes:

(1) The first possibility is that the excitation pulse is not sufficiently short to coherently excite the zeroth order level $|e, n, \{n_a=0\}> \equiv \phi_s$. Rather, a superposition of only a few of the vibrational eigenstates is initially prepared, and the summation over

$n(\equiv j$ now) in (6) involves too few terms to observe a short time unphasing. This possibility is readily tested by utilizing shorter excitation pulses. If this enables the observation of the expected
double exponential decay, then it would be of interest to study longer
time scale processes and to coherently excite an initial state other
than ϕ_S of (9) to see how the time dependence of the relaxed and unrelaxed emission spectra vary as the initial state is ranged over
different groups of vibrational eigenstates whose parantage resides
in a common ϕ_S.

(2)   The second possibility arises because the coupling between
the ring modes and the alkyl chain bath modes is channeled through
the bond between the ring carbon and its neighboring alkyl chain carbon.   (Other interactions may involve the hydrogen atoms on the first
alkyl carbon atom or the hydrogen in the ortho ring position, but the
basic argument remains the same.)   Thus, we take the intramolecular
coupling V to involve a bond oscillator coupling. This V is then
independent of n.   In the egalitarian model the total coupling
strength V is equally spread between the N strongly coupled $\{\phi_\ell\}$
levels,

$$<\phi_s|V|\phi_\ell> \approx (N+1)^{-\frac{1}{2}} V. \tag{22}$$

The effective density of bath levels increases with the number of
strongly coupled levels as

$$\rho_\ell^{EFF} = \rho_\ell^0 (N+1). \tag{23}$$

Using (22) and (23) in the golden rule rate expression (13) for the
intramolecular vibrational relaxation rate we obtain the results,

$$\Delta_{IVR} \rightarrow \frac{2\pi}{\hbar} \frac{|V|^2}{(N+1)} \rho_\ell^0 (N+1) = \frac{2\pi}{\hbar} |V|^2 \rho_\ell^0 , \tag{24}$$

which displays the intramolecular vibrational relaxation rate in
zeroth order to be independent of the alkyl chain length n. [23] Hence,
it would be necessary to employ shorter time scales to observe the
intramolecular vibrational relaxation rate.

Another simple test of the theory would be provided by experiments on dialkylated benzenes. [23] These molecules should likewise
leave some of the ring modes unshifted in the absorption spectrum
because these ring vibrations have a pair of nodes. The chain lengths
in the alkyl benzene and the dialkyl substituted benzene can be chosen
such that both molecules have approximately the same density ρ_ℓ of
bath states $\{\phi_\ell\}$.   However, the total coupling strength changes by
approximately a factor of two in going from the monoalkyl to the
dialkyl benzenes. Hence, the rates of intramolecular vibrational
relaxation should be approximately quadrupled by the dialkyl substi-

tution at constant density of states. The ratio of relaxed to unrelaxed yields should be increased accordingly.

VI. SUMMARY

A good deal of confusion has been present in the literature on
intramolecular vibrational relaxation because of the vaguely defined
nature of this process. We have provided simple quantitative definitions of intramolecular vibrational relaxation as a dynamical process associated with the irreversible or apparent change in vibrational state dynamics. The strong parallel with electronic relaxation processes has enabled the transcription of results from the vast
literature on the electronic relaxation problem 18-20,25   to the
case of intramolecular vibrational relaxation.21-23,6 We distinguish
between the small, intermediate and large molecule limits of intramolecular vibrational relaxation. In the small molecule limit there
is no intramolecular vibrational relaxation. Most forms of optical
excitation can excite either a single vibrational eigenstate or in
the case of accidental degeneracies a pair of vibrational eigenstates.
The large molecule limit of intramolecular vibrational relaxation is
characterized by a large density of vibrational levels such that the
mean spacing between vibrational levels considerably exceeds the
widths of these levels due to all relaxation processes other than
intramolecular vibrational relaxation.

We focus on the dynamics of intramolecular vibrational relaxation
in the interesting intermediate case. Here the intramolecular vibrational relaxation is not a truly irreversible decay, but it behaves
in a similar fashion kinetically when an average is performed over a
sufficient number of states to remove any quantum mechanical interference effects such as quantum beat phenomena. The intramolecular
vibrational relaxation in this intermediate case limit is then seen as
a purely intramolecular phase process, termed "unphasing," in which
there is a coherent excitation of a group of vibrational eigenstates.
As time evolves, this group of vibrational eigenstates become out of
phase with each other leading to the apparent intramolecular vibrational relaxation. The dynamics of the relaxed and unrelaxed emission spectra are described in the intermediate case using a simple
egalitarian model. Additional discussion has been presented elsewhere involving random matrix models of this dynamics.23 These present the same qualitative picture.

An analysis is provided of the Smalley group experiments on alkyl
benzene emissions where the emission spectra has components which can
be assigned to the unrelaxed and relaxed spectra of intramolecular
vibrational relaxation. However, the expected double exponential
decay of these two spectral components is not observed, and possible
reasons for their nonobservance are given.23 These suggest new experiments which should refine our conceptual and theoretical under=
standing of intramolecular vibrational relaxation phenomena.

REFERENCES

1. P. J. Robinson and K. A. Holbrook, Unimolecular Reactions, Wiley, New York, 1972.
2. I. Oref and G. S. Rabinovitch, Acc. Chem. Res. 12, 166 (1979) and references therein.
3. K. F. Freed, Faraday Discuss. Chem. Soc. 67, 231 (1979).
4. J. D. Rybrant and B. S. Rabinovitch, J. Chem. Phys. 54, 2275 (1971); J. Phys Chem. 75, 2164 (1971).
5. J. B. Hopkins, D. E. Powers and R. E. Smalley, J. Chem. Phys. 71, 3886 (1979); 72, 2905 (E) (1980); (b) J. B. Hopkins, D. E.Powers, S. Mukamel, and R. E. Smalley, J. Chem. Phys. 72, 5049 (1980); (c) J. B. Hopkins, D. E. Powers and R. E. Smalley, J. Chem. Phys. 73, 683 (1980).
6. S. Mukamel and R. E. Smalley, J. Chem. Phys. 73, 4156 (1980); S. Mukamel, Solid-State Sci. 18, 237 (1980).
7. S. Okajima and E. C. Lim, Chem. Phys. Lett. 37, 403 (1976).
8. R. K. Sander, B. Soep and R. N. Zare, J. Chem. Phys. 64, 1242 (1976).
9. B. Soep, C. Michel, A. Tramer, and L. Lindqvist, Chem. Phys.2, 293 (1973).
10. J. C. Hsieh, C. S. Huang and E. C. Lim, J. Chem. Phys. 60, 4345 (1974).
11. J. W. Perry and A. H. Zewail, J. Chem. Phys. 70, 582 (1979); Chem. Phys. Lett. 65, 31 (1980).
12. R. G. Bray and M. J. Berry, J. Chem. Phys. 71, 4909 (1979).
13. J. P. Maier, A. Seilmeir, A. Laubereau and W. Kaiser, Chem. Phys. Lett. 46, 527 (1977).
14. B. Kopainsky and W. Kaiser, Chem. Phys. Lett. 66, 39 (1979).
15. R. A. Coveleskie, D. A. Dolson and C. S. Parmenter, J. Chem. Phys. 72, 5774 (1980).
16. H. S. Kwok and E. Yablonovitch, Phys. Rev. Lett. 41, 745 (1978).
17. T. F. Deutch and S. J. Brueck, Chem. Phys. Lett. 54, 258 (1978); D. S. Frankel, J. Chem. Phys. 65, 1696 (1976).
18. K. F. Freed, Topics Appl. Phys. 15, 23 (1976); Acc. Chem. Res. 11, 74 (1976); Adv. Chem. Phys. 42, 207 (1980).
19. S. Mukamel and J. Jortner, in Excited States, ed. E. C. Lim, Academic Press, New York, 1977.
20. P. Avouris, W. M. Gelbart and M. A. El-Sayed, Chem. Rev. 77, 793 (1977).
21. K. F. Freed, Chem. Phys. Lett. 42, 600 (1976).
22. S. Mukamel and J. Jortner, J. Chem. Phys. 65, 5204 (1976).
23. K. F. Freed and A. Nitzan, J. Chem. Phys. 73, 4765 (1980).
24. S. A. Rice, this volume.
25. K. F. Freed, J. Chem. Phys. 52, 1345 (1970).
26. A. Nitzan, J. Jortner and P. Rentzepis, Proc. R. Soc. (London) A37, 367 (1972); A. Frad, F. Lahmani, A. Tramer and C. Tric, J. Chem. Phys. 60, 4419 (1974); R. van der Werf and J. Kommandeur, Chem. Phys. 16, 125 (1976).

27. K. F. Freed (unpublished); G. Atkinson (private communication); S. Leach (private communication).
28. B. Carneli, I. Scheck, A. Nitzan and J. Jortner, J. Chem. Phys. 71, 1928 (1980); W. M. Gelbart, D. F. Heller and M. L. Elert, Chem. Phys. 7, 116 (1975).
29. A. Villaeys and K. F. Freed, Chem. Phys., 13, 271 (1976).
30. P. S. H. Fitch, C. A. Haynam and D. H. Levy, J. Chem. Phys. 73, 1064 (1980);(in press).

SOME KINETIC AND SPECTROSCOPIC EVIDENCE ON INTRAMOLECULAR

RELAXATION PROCESSES IN POLYATOMIC MOLECULES

Martin Quack

Institut für Physikalische Chemie der
Universität, Tammannstr. 6
3400 Göttingen, West Germany

ABSTRACT

The description and definition of intramolecular vibrational
relaxation processes is discussed within the framework of the quantum mechanical and statistical mechanical equations of motion. The
evidence from quite different experimental sources is summarized
under the common aspect of vibrational relaxation. Although much of
the evidence remains ambiguous, there is good indication that a
localized vibrational excitation relaxes typically in 0.1 to 10 ps,
which is long compared to the classical vibrational period but
short compared to many optical and reactive processes.

1. INTRODUCTION

The collisional relaxation of polyatomic molecules in a thermal environment is a well defined and reasonably well understood
process, which is of great importance for chemical reactions under
thermal conditions.[1,2] On the other hand, relatively little is
known about intramolecular vibrational relaxation of isolated molecules and this phenomenon requires careful definition in the
first place. A process may be considered to be intramolecular if
there is negligible collisional or radiative interaction with the
environment during the time scale of interest (here usually less
than a nanosecond). Vibrational relaxation is defined to be the
irreversible decay (on the time scale of interest) of a localized
vibrational excitation in a molecule, characterized semiclassically
by vibrational amplitude in a certain bond or by vibrational energy
in a part of the molecule. Rotational degrees of freedom may be
included. When using such a definition one might note that the

493

rigorous treatment of the time dependence of every experimental observable requires a specific theoretical description. However, the
observable time dependences in various experiments may have a
common mechanistic background in one physical phenomenon in an
approximate sense. Thus it is the purpose of this paper to discuss
the implications of different experiments with respect to the rate
and the mechanistic role of intramolecular vibrational relaxation,
which is important for the chemical kinetics of molecules both under isolated and collisional conditions[1], for photochemistry[3] and
for the interpretation of frequency resolved and time resolved
spectroscopy in the IR[4] and UV[5].

2. STATISTICAL MECHANICS OF INTRAMOLECULAR RELAXATION

Time dependent processes in isolated molecules can be described by the time dependent Schrödinger equation

$$i \hbar \dot{\Psi} = \hat{H} \Psi \tag{1}$$

Neglecting weak interactions and radiative decay (or including the
states of the radiation field) H is hermitian and the solutions of
Eq. (1) are oscillatory, satisfying time reversal symmetry[6]. This
would appear to preclude relaxation phenomena. We therefore devote
this section to a brief discussion of the foundations of relaxation
in isolated molecules. The formulation is such that it allows directly for explicit numerical computations.

2.1. The Quantum Mechanical Equations of Intramolecular Motion

We begin by rewriting Eq. (1):

$$i \underline{\dot{b}} = (H_o + V) b = H b \tag{2}$$

The dot denotes derivation with respect to time, b is the amplitude
vector in the basis Ψ_k of eigenstates of $\hat{H}_o$, H_o being the corresponding diagonal matrix in frequency units (i.e. "$\hbar = 1$") and V
is the off diagonal coupling matrix. The choice of the basis
is dictated by the observable under consideration. In the context
of vibrational relaxation one possible choice (not the only one) is
the basis which makes H_o diagonal in local vibrational coordinates.
The solution of Eq. (2) with time independent H is given by Eq. (3):

$$b (t) = U (t, t_o) b (t_o) \tag{3a}$$

$$U (t, t_o) = \exp \left[- i H (t - t_o) \right] \tag{3b}$$

The time evolution matrix solves furthermore the Liouville-von Neumann equation for the time dependent density matrix P:

$$P(t) = U(t, t_o) P(t_o) U^+(t, t_o) \tag{4}$$

and the Heisenberg equation of motion for the representative matrix Q (in the basis ψ_k) of any observable, in particular coordinates and momenta of the atoms:

$$Q(t) = U^+(t, t_o) Q(t_o) U(t, t_o) \tag{5}$$

H being given, Eqs. (3) - (5) constitute the most general solution of a well defined mathematical formulation of the physical problem of intramolecular motion. The solution is simple, in principle, but becomes difficult if the order of the matrices becomes large (infinite). It is not clear a priori how from these dynamical equations an essentially statistical phenomenon such as relaxation may emerge. We shall discuss this first with a simple mathematical analogy, because this general problem in Eqs. (3) - (5) is too complex to be fully treated in a small space.

2.2. Relaxation and Equipartition in Number Theory

Consider the sequence of digits in the decimal representation of the number e:

$$e = \sum_{n=0}^{\infty} (n!)^{-1} = 2.71828182845904523 \ldots \tag{6}$$

Eq. (6) constitutes a well defined mathematical problem, simple in principle, but difficult to solve if (infinitely) many digits are required. One may ask now a more coarse grained question, for example, what is the relative frequency p of even (p_g) and odd (p_u) digits in this sequence as a function of the total number $N = N_g + N_u$ of decimal places considered:

$$p_g = 1 - p_u = N_g/N \tag{7}$$

This is shown in Fig. 1 for small N. After an initial relaxation phenomenon equipartition, $p_u \simeq p_g \simeq 1/2$, is attained. A mathematical proof of this equipartition for a general large N is difficult and does not exist for e, but it is known from number theory that almost all real numbers show this "normal" behaviour of equipartition[7]. A simple probability argument favours equipartition in e.

We repeat the basic ideas of the statistical consideration, which will also be used below: (1) A coarse grained quantity is considered (p_g, p_u), which does not contain the full mathematical

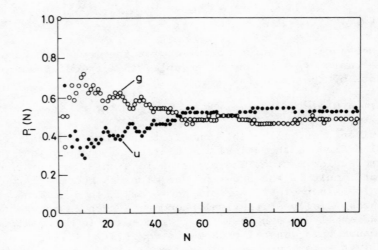

Fig. 1. Equipartition in e (see text)

information. (2) An (infinite) set of objects (the set of real numbers) sharing certain common properties is considered. (3) It is
proven that "almost all" members of the set show a certain behaviour
with respect to the coarse grained quantity. (4) It is argued by
probability and after inspection that one member (for instance e)
of this set also shows this behaviour. It is clear that new, simple
mathematical structures arise from new (coarse grained) questions.

2.3. The Pauli Equation[8] for Intramolecular Relaxation

We introduce coarse grained level populations p_K for approximately isoenergetic states characterized by the same quantum
number K in a particular local vibration (separable in H_o) but arbitrary quantum numbers for the other degrees of freedom:

$$p_K = \sum_k{}' p_{k\,(K)} = \sum_{k\,=x+1}^{x+N} b_{k(K)}\, b^*_{k(K)} \tag{8a}$$

Hereafter $\sum'$ indicates sums over indices of states in one level,
capital indices being used for levels. The number of states in one
level, N_K, is supposed to be large, say typically between 100 and
10^{20} in molecular problems. V may be considered to be off diagonal
with respect to the levels. Using Eq. (3) one obtains (setting
$t_o = 0$ for brevity of notation):

$$p_K(t) = \sum_k{}' \sum_J \sum_j{}' |U_{kj}|^2\, |b_j(0)|^2 \tag{8b}$$

$$+ \sum_k \sum_l \sum_{j\neq l} U_{kj}\, U^*_{kl}\, b_j(0)\, b^*_l(0)$$

The terms of the first sum in Eq. (8b) are necessarily all real, positive. On the other hand, the second multiple sum "almost always" (in the sense of section 2.2) contributes negligibly, since many $b_j(0)$ are supposed to be different from zero and have uncorrelated "irregular" phases, thus:

$$p_K(t) = \sum_J \sum_j{}' \sum_k{}' |U_{kj}|^2 \, p_j(0) \tag{9}$$

Only the coarse grained level population $p_J = \sum' p_j$ may be considered to be measurable. One can show[9] that the same probability argument leading from Eq. (8) to Eq. (9) allows one to replace Eq. (9) by Eq. (10) (in essence setting $p_j(0) = N_J^{-1} \, p_J(0)$ at all $t = 0$):

$$p_K(t) = \sum_J p_J(0) \, Y_{KJ}(t) \tag{10a}$$

$$Y_{KJ}(t) = N_J^{-1} \sum_j{}' \sum_k{}' |U_{kj}|^2 \tag{10b}$$

Eq. (10b) may have a more complicated general form, however always with the matrix $Y(t)$ depending only upon time differences as does $U(t)$, Eq. (3b). This fact can be used in the following way:

$$p(t) = Y(t) \, p(0) \tag{11}$$

$$p(t_1+t_2) = Y(t_1+t_2) \, p(0) \tag{12a}$$

$$= Y(t_2) \, p(t_1) \tag{12b}$$

$$= Y(t_2) \, Y(t_1) \, p(0) \tag{12c}$$

$$Y(t_1+t_2) = Y(t_2) \, Y(t_1) = Y(t_1) \, Y(t_2) \tag{13}$$

The only function satisfying Eq. (13), subject to wide conditions, is the exponential function with a time independent matrix K:

$$Y(t) = \exp(Kt) \tag{14}$$

By derivation of Eq. (14) one obtains the customary differential form of the Pauli equation[8]:

$$\dot{p} = K \, p \tag{15}$$

In contrast to the oscillatory solutions of Eqs. (1) - (3), Eqs. (14) and (15) have decaying solutions (relaxation). For sufficiently short times one can expand the exponential function and one obtains K by using Eq. (10b) in connection with Eq. (16) and first order perturbation theory[6,9,10]:

$$Y(t) = 1 + K t + \ldots \tag{16}$$

$$K_{KJ} = 2 \pi \ |V_{KJ}|^2 / \delta_K \tag{17}$$

$|V_{KJ}|^2$ stands for the average square coupling matrix element between the states in levels K and J and δ_K is the average frequency separation of states in level K (energies from H_o, replace $1/\delta_K$ by the density ρ_K in the case of a continuum). The last two steps imply several conditions upon the properties of H which are discussed in detail elsewhere[9,10].

The essential part of the derivation of the Pauli equation closely follows the reasoning of section 2.2: (1) A coarse grained quantity is considered (p_K, Eq. (8)). (2) A set of matrices H (sharing certain dynamical properties not discussed in detail here) is considered. (3) It is shown (by consideration of the phases in Eq. (8b)) that almost all members of the set lead to Eq. (14). (4) It is argued by probability that one particular member of the set (say H for a given molecule) also gives Eqs. (14) and (15). An essential part of this derivation, obviously similar to section 2.2, is the occurence of exceptions, which are rare and therefore improbable, but which might be selected theoretically more easily than experimentally. It should also be mentioned that the derivation of the Pauli equation can equivalently start with either Eqs. (3), (4) or (5)[9,10]. Under appropriate conditions, the Pauli equation has to be replaced by other equations, for instance for the decay of a single, pure state or for groups of states that are more weakly or more strongly coupled than allowed by the conditions for the Pauli equation (e.g. cases A, C, and D of ref 10, see also refs 11 - 14). The relaxation phenomenon is thus more general than the Pauli equation, which merely constitutes one important example.

Fig. 2 illustrates, how relaxation behaviour emerges from the solutions of the Schrödinger Eq. (2). In the example two groups of states with 59 states in each level are coupled by an appropriate matrix H, with the level population $p_1(0) = 1$ (but many states populated at t = 0). The points in Fig 2a indicate time dependent level populations from exact solutions of the Schrödinger equation. The lines are solutions of the Pauli equation, which is seen to be a good approximation because of coarse graining. Coarse graining is, however, not sufficient for Pauli behavior, as is illustrated by Fig 2b, where oscillatory level populations are shown for a similar spectrum with 59 states in each level. Very special ("improbable") assumptions must be made concerning the phases of the initial state vector in order to obtain such oscillations. With an arbitrary state vector b (0) and the same matrix H as in Fig. 2b one obtains almost always the relaxation shown in Fig. 2c. Fig. 2b is given here only to show the possibility of exceptions.

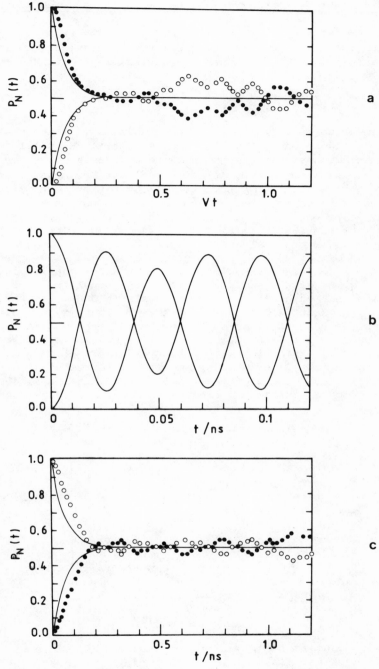

Fig. 2. Time dependent level populations in a two level system
(see detailed explanations in the text)

2.4. Equipartition, Microcanonical Equilibrium, and Temperature

For hermitian H, U is unitary and it follows from Eq. (16) that as t → ∞ one has a stationary level population corresponding to equipartition or microcanonical equilibrium as a consequence of the Paulie equation:

$$p_K / p_L = \delta_L / \delta_K = \wp_K / \wp_L \tag{18}$$

A requirement for Eq. (18) is that the matrix K is irreducible or that one considers one irreducible block of K, as we shall do hereafter (this amounts to assuming that the constants of the motion have been taken into account in advance[15]). It has been pointed out[10] that more generally one may have relaxation without equipartition, but we discard this case here. Also, Fig 2b illustrates the possibility of "improbable" situations with oscillatory level populations which are not considered here.

Microcanonical equilibrium leads to observable consequences discussed below. The nature of the microcanonical equilibrium is shown in Fig. 3 for the model example of the populations of the hypothetical harmonic C-C stretching levels in C_2H_6 at a total energy E_{tot} of about 41000 cm^{-1} (points, M). The very definition of levels implies that this energy is not exactly defined say to within $\Delta E < 1000$ $cm^{-1} \ll E_{tot}$. It is interesting to compare the microcanonical distribution with a thermal distribution, which would be a straight line in this logarithmic representation. Indeed, for low excitations the microcanonical distribution is closely approximated by a thermal distribution and this is just an example

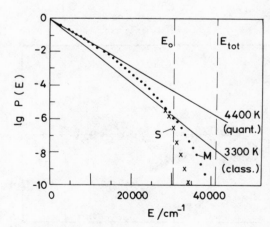

Fig. 3. Microcanonical (M, points) and canonical distributions (4400 K and 3300 K) for the C-C vibration in a model of C_2H_6. S is a steady state distribution with reaction (see section 3.3., threshold energy E_o). See also text.

for the fact that at equilibrium a "small subsystem" (C-C) is canonically distributed in a "large" microcanonical system (C_2H_6). However, at high excitations (important for reactive processes) the
thermal distribution fails completely. One must thus distinguish
two different regimes for the validity of the thermal approximation
and we shall come back to this point in section 3.3.

2.5. Redistribution or Relaxation?

The question may arise whether the observed consequences of
intramolecular rate processes indicate relaxation (corresponding
to Figs. 2a and c and possibly microcanonical equilibrium) or
rather redistribution, possibly including periodic exchange of
probability as in Fig. 2b. This question can be illustrated with
the chemical activation system investigated by Rynbrandt and Rabinovitch[16]:

$$CF_2 \text{———} CF \text{——} CF \text{——} CF_2 \quad \xrightarrow{(M)} \text{Stabilization}$$

with ring 1 (CH$_2$) and ring 2 (CD$_2$):

$$\xrightarrow{\quad} \text{Elimination of } CF_2 \text{ from 1}$$
$$\xrightarrow{\quad} \text{Elimination of } CF_2 \text{ from 2} \tag{19}$$

The initial energy may be localized in ring 1 or ring 2 depending
on the previous formation of the intermediate by addition of CH_2
or CD_2 to the appropriate monocyclic, unsaturated compound. By
assuming a conventional kinetic scheme with relaxation of the localized energy until microcanonical equilibrium between the two
rings is attained (Fig 2a!) a rate coefficient for relaxation of
about 10^{12} s^{-1} was found by measuring the relative yields in the
three channels of Eq. (19) as a function of [M][16]. This interpretation has been accepted by most workers in the field. Recently,
however, it has been suggested[17], that these results could as well
or possibly more adequately be described by assuming an oscillatory
redistribution of the energy between the two rings (cf. Fig. 2b!).
The frequency of the oscillation would then be related to the above
mentioned rate. This kind of interpretation, although possible in
principle, can be discarded on the probability grounds advanced in
section 2.2 and 2.3, and after inspection of the dynamical properties
of the system. Due to coarse graining the real experimental system
is theoretically expected to be characterized by a relaxation phenomenon as assumed by Rynbrandt and Rabinovitch. Otherwise extremely
unlikely assumptions would have to be made concerning the phases of
the initial state and the properties of the hamiltonian of this
system.

3. THE EVIDENCE ON INTRAMOLECULAR VIBRATIONAL RELAXATION

We summarize here the results of a variety of experiments, all bearing on intramolecular relaxation. We shall briefly discuss the main experimental idea and present a few typical examples, not an exhaustive review.

3.1 Equipartition in Bimolecular Reactions with Metastable Intermediates

In reactions with metastable intermediates, Eq. (20), one has competition between decomposition of the intermediate and vibrational relaxation to microcanonical equilibrium, if applicable:

$$X+RY \ (E_{to}, \ E_{io}) \rightarrow XRY^* \longrightarrow RX+Y \ (E_t, \ E_i) \tag{20}$$

The microcanonical equilibrium is reflected by restricted equipartition in the product (chemical channels) and product energy (physical channels) distributions after decomposition. The statistical models for scattering often use such (restricted) equipartition assumptions[14]. If the lifetime of the intermediate is known or can be estimated, the observation of the failure or success of the equipartition assumption contains information about the rate of vibrational relaxation. The experiment by Rynbrandt and Rabinovitch discussed in section 2.5 is the most splendid example of the use of such an idea under bulk conditions, indicating relaxation times of the order of a ps.

More recently, molecular beam experiments have appeared measuring the amount of equipartition reflected by the product translational energy distributions $P(E_t)$ of the decomposition products in reactions of halogen atoms with unsaturated halo-hydrocarbons and other reactions described by Eq. (20)[18,19,3,1]. Although some of the earlier evidence seemed to indicate that vibrational relaxation was sometimes slow, a more recent theoretical evaluation has shown that the available experimental results are consistent with vibrational relaxation in picoseconds[20]. A simpler and more fundamental example in this area is the experimental and theoretical (classical trajectory) study on the exchange reaction H^++D_2 $HD+D^+$ [21]. This work is also consistent with equipartition and fast relaxation in less than a ps.

3.2. Recombination Reactions in the Pressure Fall-Off and Unimolecular Reactions

Among the oldest experiments concerning intramolecular relaxation processes are studies of recombination reactions according to the Lindemann mechanism[22,23], which we write here in an extended form (still a mechanistic oversimplification):

$$X + Y \rightleftharpoons X \overset{(*)}{-} Y \tag{21a}$$

$$X \xrightarrow{(*)} Y+M \quad \rightleftarrows \quad XY+M \tag{21b}$$

$$X \xrightarrow{(*)} Y \quad \rightleftarrows \quad (XY)^* \tag{21c}$$

$$(XY)^* +M \rightleftarrows \quad XY+M \tag{21d}$$

The steps (21a) and (21b) are the same for diatomic XY and polyatomic XY. The intramolecular transfer of energy, step (21c), and the subsequent stabilization (21d) is typical for polyatomic systems. If the step (21c) were slow, polyatomic recombination, for example $CH_3 + CH_3$ [27], would be similar to diatomic recombination, for example $H + H$ [8]. In fact, dramatic differences are observed which indicate fast transfer of excitation, Eq. (21c), on the picosecond time scale. This qualitative conclusion does not depend upon model assumptions, say from RRKM theory. We refer to the literature on unimolecular reactions[23] and note that claims[17] that unimolecular reactions can be understood without assuming fast intramolecular transfer of excitation do not appear to be well founded. It is true, however, that because of the collisional environment the nature of the fast intramolecular transfer of excitation (relaxation or periodic) and quantitative data on the rates are not known from such studies.

More recent evidence from unimolecular dissociation after multiphoton excitation of polyatomic molecules under beam conditions[24] indicates fast intramolecular relaxation at least in nanoseconds. The first direct spectroscopic measurement of unimolecular isomerization lifetimes in the reaction Cycloheptatriene $\rightarrow$ Toluene under essentially collision free conditions at speciefied energies after optical excitation[25] is consistent with lifetimes calculated from statistical unimolecular rate theory[23,26]. This is indirect evidence that intramolecular vibrational relaxation is fast on the nanosecond time scale, but not without ambiguity.

3.3. Theory of Lifetimes in Unimolecular Reactions

The common statistical theories of lifetimes in unimolecular reactions (for example RRKM theory[26]) rest on the assumption that intramolecular vibrational relaxation is fast on the time scale of chemical reaction and that microcanonical equilibrium is maintained in the activated species. This results in an equation for the lifetime τ as a function of the energy E[26,29]:

$$\tau(E) = h \, \rho(E) / W(E) \tag{22}$$

$W(E)$ is the "number of states of the activated complex" and $\rho(E)$ the density of metastable states in the activated species. According to Eq. (22) the lifetimes decline rapidly as the energy increases beyond the threshold for reaction. Also, τ is predicted to be a

sensitive function of the number of intramolecular degrees of freedom. There is a large body of indirect and some direct evidence
that with "reasonable" assumptions on the activated complex (or
W(E)), Eq. (22) gives a quantitative description of the lifetimes
of metastable molecules and ions[1,3,14,23]. This is sometimes taken
as evidence for both fast relaxation and complete microcanonical
equilibrium on the time scale of reaction (typically between μs
and ps in current experiments). This conclusion is only partly
justified.

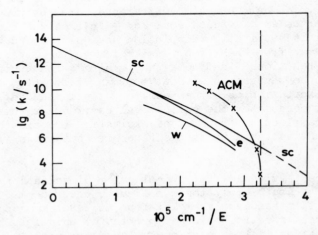

Fig. 4. Specific microcanoncial rate constants $K(E, J=0)$ for the
        decomposition $C_2H_6 \rightleftarrows 2CH_3$ according to various models in
        an Arrhenius-like diagram (see detailed explanations in
        the text).

More generally an inequality can be derived for average lifetimes from scattering theory[30,14] (see also refs. 31 and 32):

$$\tau \, (E, J, \ldots) \geqslant \quad h \, \varrho \, (E, J, \ldots) \, / \, W(E, J, \ldots) \tag{23}$$

Here the dependence on good quantum numbers corresponding to the
constants of motion (total angular momentum J, projection M, parity, etc.) is indicated explicitly. $W(E, J)$ is now the number of
dynamically accessible scattering channels[14]. Some quantitative
estimates for the deviation from the equality in Eq. (23) can be
obtained with a simple Pauli equation model of intramolecular energy transfer and reaction[9] (see also refs. 3, 12, 33). Let us consider as an example the chemical activation:

$$H + C_2H_5 \; \rightleftarrows \; C_2H_6{}^* \; \rightleftarrows \; 2CH_3 \tag{24}$$

Three major phases may be distinguished in this reaction (in the absence of collisions): (1) Formation of the C-H bond, (2) transfer of excitation to other vibrations including the reactive and initially unexcited C-C vibration (relaxation), (3) decomposition in the reactive C-C bond with a steady population of the C-C reactive states. The deviation of the steady state populations from microcanonical equilibrium leads to a lengthening of lifetimes. This deviation is shown in Fig. 3 for a particular model with finite relaxation rates (S for steady state, M for microcanonical equilibrium). Fig. 4 shows the corresponding rate constants $k(E, J=0) = 1/\tau$ $(E, J=0)$ in an Arrhenius like representation as a function of $1/E \propto 1/T$. The function labeled e is based on a simple quasi-harmonic potential model with exact statistics and microcanonical equilibrium (sc is the "semiclassical approximation" [23,26], not of interest here). W indicates the result of the exact steady state solution of the Pauli equation for the same quasi-harmonic model. As expected, the rates at high E are substantially lower than the equilibrium rates from Eq. (22). One might therefore think that this effect should be easily seen experimentally. However, there is a great uncertainty in the calculated absolute rates because of the lack of knowledge of the potential properties. This is shown by also reproducing the result from a more realistic anharmonic potential model (ACM[34]) which uses the equality in Eq. (23) (i.e. equilibrium). Clearly, the effects of different potentials are more important than the effects of assumptions on intramolecular relaxation rates and there appears to be little hope to get quantitative information on relaxation rates from lifetime studies in the ns-range, although measurements over a larger time range might reveal some trends. Fig. 4 also nicely illustrates the deviations from the straight line "Arrhenius" behaviour, which would be expected if the C-C vibrational levels were distributed canonically (see also Fig. 3). Therefore the "temperature" of the C-C distribution is not useful as a quantitative concept in the context of such a reactive process. Nevertheless it is qualitatively useful for understanding why larger molecules have longer liftimes than smaller ones with the same microcanonical energy above the threshold for reaction: The temperature of the corresponding "canonical" distribution in the reactive oscillator is lower.

3.4. Time Resolved Spectroscopy

A straightforward technique of measuring molecular relaxation times consists in probing spectroscopic properties as a function of time after local vibrational excitation. The actual experimental observable is then a time dependent transition moment, scattering cross section etc., which often can be associated with the time dependence of local vibrational amplitude. Two such experiments in vapour phase systems with the necessary high temperature resolution have been published recently, estimating relaxation times of less than a few ps[35]. In this connection, one should also mention the advent of new, sensitive techniques, capable of high time resolution, such as vapour phase CARS[36]. However, great care must be taken in identifying the observed time dependences with vibrational relaxation, since other origins are possible.

3.5. The Structure of Resonance Fluorescence Spectra

High resolution resonance fluorescence spectra of polyatomic molecules in the vapour phase have been obtained during the last decade (single vibronic level fluorescence, SVLF[37-43]). There might be some hope to obtain information on intramolecular vibrational relaxation times without time resolved measurements from the observation[38,40,42,46], that often there appear to be two spectrally different components of this fluorescence, one with resolved vibrational structure (S-component) and one unresolved broad background emission (U-component, this, too, is resonance fluorescence in the absence of external perturbations). Consider the following kinetic scheme (after optical excitation in the electronically excited state $*$):

$$A_i^* \; \underset{k_{ir}}{\overset{k_{ri}}{\rightleftharpoons}} \; A_r^* \; \overset{k_F}{\longrightarrow} \; A + h\nu_U \tag{25}$$
$$A_i^* \; \overset{k_F}{\longrightarrow} \; A + h\nu_S$$

If k_F is known, and assuming from the Pauli equation $k_{ri}/k_{ir} = \rho_r/\rho_i \gg 1$, one can obtain the relaxation rate constant k_{ri} from the evaluation of this kinetic scheme after measuring the fluorescence quantum yields for the S- and U-components. Indeed, the evaluation of a similar kinetic scheme was first proposed and used to obtain the first fast proton transfer reaction rate constants for electronically excited species in solution[47]. Weller's interpretation of the condensed phase experiments was quite straightforward. Several difficulties arise, however, with the corresponding scheme (25) for isolated molecules[40]: (1) The U-component may be partly due to sequence congestion. This difficulty has been overcome recently by measuring spectra of cold molecules in jets[42,43,45]. (2) The U-component may stem from a relative rotation of the normal coordinates in the two electronic states (Duschinski effect[44]). (3) The simple Pauli equation scheme may not be applicable (in fact it is not applicable in most cases of low excitation, where even in a statistical description case C[10], for instance, would be more adequate). In the absence of time resolved experiments, these difficulties make us prefer a time independent (scattering) description of the two components of the resonance fluorescence[40]. Nevertheless, time dependent descriptions have been suggested repeatedly[5,38,45]. Although this is not always unreasonable, the numbers quoted for relaxation times (in the nanosecond range for alkylbenzenes[45]) or the apparent "absence of relaxation" should be interpreted with caution, since they provide

neither upper nor lower bounds for the vibrational relaxation because of the difficulties (2) and (3) mentioned above. An interesting, new idea, closely related to the Rynbrandt-Rabinovitch experiment discussed in section 2.5, was the recent measurement of the two
fluorescence components (S and U) in competition with electronic
quenching[46]. At high quenching gas pressures relaxation rate constants k_{ri} 10^{11} s^{-1} were obtained for p-difluorobenzene. Of
course, in such an experiment collisional perturbations may not be
negligible.

3.6. Bandshapes of Vibrational Absorption

Often, but not necessarily always, local vibrational excitation
in a large molecule is associated with large transition moments in
the infrared: It is a local excitation which acts as chromophore[4].
One can also associate the "absorbing states" with eigenstates of
a zero order hamiltonian (local mode hamiltonian[48]). The coupling
with the other, optically inactive vibrational zero order states
will lead to a vibrational band structure or an apparent vibrational
broadening in the case of a high density of states. For a Lorentzian band shape the width Γ can be associated with an exponential
decay (rate constant k) after a hypothetical (or experimental)
short time, coherent, local excitation:

$$k = \Gamma / \hbar \qquad\qquad (26)$$

Large widths Γ can be easily measured and provide precisely the
required information on the rate of intramolecular vibrational relaxation. The situation is similar to line broadening by predissociation[49] and electronic relaxation[50]. A less helpful analogy exists
also to spin relaxation in magnetic resonance[51] (here the orders of
magnitude of the relevant parameters are quite different from the
vibrational problem). The large bandwidths $\Gamma \simeq 100$ cm^{-1} in high
overtones (5-7)of the C-H stretching vibration in the vapour spectra of benzene were interpreted by Bray and Berry[52] as being due to
vibrational relaxation with $k = 2 \cdot 10^{13}$ s^{-1} (see also refs. 53 and
54).

When evaluating vibrational bandshapes in this manner, one must
make sure that the observed widths are due to homogeneous vibrational structure. Difficulties arise from the rotational band structure
and from the inhomogeneous vibrational structure[4]. The rotational
structure can be taken into account by numerical simulation, if the
appropriate rotational constants are known (often not the case). A
less ambiguous alternative is to study the Q-brach structure of parallel bands of symmetric tops with a heavy frame (without vibrational structure the Q-branch would be quite sharp). In the case of
$(CF_3)_3$ C-H, for instance, this procedure has provided the first unambiguous vibrational width ($\Gamma = 3$ cm^{-1} for C-H) of a fundamental

transition in the IR[4]. An appreciable fraction of this was shown to
be homogeneous, giving $k = 5 \cdot 10^{11}$ s^{-1}. The homogeneous vibrational structure can quite generally be identified by measuring the
temperature dependence of the spectrum, ideally until T = 0K. A realistic alternative for small k is a time resolved experiment[35].

A systematic study of the isolated C-H chromphore in the molecules of the ideally suited series $C_nF_mX_e$-H with n up to 10 provided similar values ($\Gamma \leqslant 20$ cm^{-1} for the fundamentals and systematically larger values for the first and second overtones[4],[55]). A
very interesting situation arises in the "small" molecule $^{12}CF_3H$.
The C-H stretching fundamental is known to be strongly perturbed
by Fermi resonances, one of which occurs with the C-H bending overtone over a separation of 325 cm^{-1}. This resonance is also prominent
in C-H stretch overtones[56]. We have recently measured the fundamental at very high resolution, obtaining a complete vibrational band
structure of close resonances with separations as small as 0.14 and
0.6 cm^{-1} and extremely interesting rotational structure[4]. For isolated Fermi resonances the splittings would correspond to periodic
exchange of excitation with periods of 240 ps and 55 ps. However,
the multiple nature of this wide resonance structure really indicates the pre-stages of a very fast relaxation in less than a ps,
with fast recurrence, of course. CF_3H may be an extreme example,
but multiple resonances in small molecules are more abundant than
is often thought. They are of great importance for the dynamics of
molecular multiphoton excitation[10],[13],[24]. At low internal energy it
becomes meaningless to talk about the physical phenomenon of relaxation, which requires a high density of states. Still, strong anharmonic couplings are known to occur even at the zero point level as
has been discussed by E. Heller and coworkers for the famous CO_2-
Fermi resonance[57].

4. CONCLUSION, CRITIQUE, OUTLOOK

The evidence suggests that in ordinary tightly bound molecular
systems an initial vibrational excitation decays with a relaxation
time of the order of 0.1 to 10 ps. One must bear in mind that for a
meaningful definition of such a time dependence in discrete spectra
a sufficient energy bandwidth and therefore total excitation energy
is necessary[58]. It is not so well established, whether the decay
leads to microcanonical equilibrium (in those cases, where the general statistical mechanical conditions would allow this to be defined). There have been some reports on slow vibrational relaxation
(μs or slower). We cannot discuss these here in detail, because of
the limited space available, but we note the distinction between the
observation of a slow intramolecular process and its identification
with vibrational relaxation. If the average density of vibrational
states ρ is high, slow vibrational processes with periods $\tau \geqslant h\rho$
are expected. Isomerizations are well known to occur on long time

scales. The high degeneracy associated with optical isomers[59] (or alternatively the positive and negative parity states) may lead to extremely slow motions of a "vibrational" type (but not vibrational relaxation!). Vibrational exchange between van der Waals dimers is expected to be slow, as well as the even slower exchange of vibrational energy between molecules on the earth and the moon, respectively, connected by the gravitational interaction. Finally, other molecular degrees of freedom, such as nuclear spin, are expected to couple only on long time scales, which has interesting consequences for chemical reactions including intramolecular rearrangements[15]. Therefore, various phenomena may be associated with slow processes in molecules, but probably not with vibrational relaxation as defined here. It is more difficult to evaluate the relevance of classical trajectory results for models of benzene[60] and water[61], indicating possibly slow vibrational relaxation. Experimentally, such a slow relaxation remains to be discovered.

Acknowledgement: The author is greatly indebted to J. Troe for his generous support. The kind hospitality and the support from J. Hinze and the Zentrum für Interdisziplinäre Forschung as well as financial support from the Fonds der chemischen Industrie and the Deutsche Forschungsgemeinschaft (SFB 93 Photochemie mit Lasern) are gratefully achnowledged.

REFERENCES

1. M. Quack and J. Troe, in "Gas Kinetics and Energy Transfer" 2:175 (1977) (P.G. Ashmore and R.J. Donovan eds.) The Chemical Society, London.
2. M. Kneba and J. Wolfrum Ann. Rev. Phys. Chem. 31:47 (1980)
3. S.A. Rice, in "Excited States" (E.C. Lim ed.), Academic Press, New York (1975)
4. H.R. Dübal and M. Quack, Chem. Phys. Lett. 72:342 (1980), and to be published
5. C. Tric, Chem. Phys. 14:189 (1976); F. Lahmani, A. Tramer, J. Chem. Phys. 60:4431 (1974)
6. A. Messiah, "Mécanique Quantique", Dunod, Paris (1969); P.O. Löwdin, Advan. Quantum Chem. 8:323 (1967)
7. G.H. Hardy, and E.M. Wright, "An Introduction to the Theory of Numbers", Oxford University Press, London (1958)
8. W. Pauli, in "Probleme der Modernen Physik", P. Debeye ed., Hirzel, Leipzig (1928)
9. M. Quack, to be published
10. M. Quack, J. Chem. Phys. 69:1281 (1978)
11. R. Zwanzig, Physica (Utrecht), 30:1109 (1964)
12. R. Ramaswam y, S. Augustin, and H. Rabitz, J. Chem. Phys. 69:5509 (1978)
13. B. Carmeli, I. Schek, A. Nitzan, and J. Jortner, J. Chem. Phys. 72:1928 (1980); S. Mukamel, J. Chem. Phys. 71:2012 (1980)

14. M. Quack and J. Troe, in "Theoretical Chemistry, Advances and Perspectives", 6B:199 (1981), D. Henderson ed., Academic Press, New York

15. M. Quack, Mol. Phys. 34:477 (1977); and to be published

16. J.D. Rynbrandt, and B.S. Rabinovitch, J. Phys. Chem. 75 : 2164 (1971)

17. K. Freed, Far. Disc. Chem. Soc. 67:231 (1979)

18. K. Shobatake, Y.T. Lee, and S.A. Rice, J. Chem. Phys. 59:1435 (1973)

19. R.J. Buss, and Y.T. Lee, J. Phys. Chem. 83:34 (1979); R.J. Buss, M.J. Coggiola, and Y.T. Lee, Far. Disc. Chem. Soc. 67:221 (1979)

20. M. Quack, Chem. Phys. 51:353 (1980)

21. Ch. Schlier, atthis conference; D. Gerlich, U. Nowotny, Ch. Schlier, and E. Teloy, Chem. Phys. 47:245 (1980)

22. F.A. Lindemann, Trans. Far. Soc. 17: 598 (1922)

23. J. Troe, in "Physical Chemistry an Advanced Treatise", Vol. 6 B, W. Jost ed., Academic Press, New York (1975)

24. P.A. Schulz, A.S. Sudbo, D.J. Krajnovitch, H.S. Kwok, Y.R. Shen, and Y.T. Lee, Ann. Rev. Phys. Chem. 30:379 (1979)

25. H. Hippler, K. Luther, J. Troe, and R. Walsh, J. Chem. Phys. 68:323 (1978)

26. R.A. Marcus, and O.K. Rice, J. Phys. Colloid. Chem. 55:894 (1951)

27. H.E. van den Bergh, A.B. Callear, and R.J. Norstrom, Chem. Phys. Lett. 4: 101 (1969); K. Glänzer, M. Quack, and J. Troe, Symp. Int. Comb. 16:949 (1977)

28. J. Troe, Ann. Rev. Phys. Chem. 29:223 (1978); H.O. Pritchard, in "Reaction Kinetics", Vol. 1, P.G. Ashmore ed. the Chemical Society, London (1975)

29. H.M Rosenstock, M.B. Wallenstein, A.L. Wahrhaftig, and H. Eyring, Proc. Nat. Acad. Sci. USA, 38:667 (1952)

30. M. Quack, and J. Troe, Ber. Bunsenges. Phys. Chem. 79:469 (1975)

31. K. G. Kay, J. Chem. Phys. 64:2112 (1976)

32. W. H. Miller, at this conference, B.A. Waite, and W.H. Miller, J. Chem. Phys. 73:3713 (1980)

33. G.R. Fleming, O.L.J. Gijzeman, and S.H. Lin, J. Chem. Soc. Far. Trans. II, 70:37 (1974)

34. M. Quack, and J. Troe, Ber. Bunsenges. Physik. Chem. 78:240 (1974)

35. B. Kpainsky and W. Kaiser, Chem. Phys. Lett. 66:39 (1979); J. P. Maier, A. Seilmeier, A. Lauberau, and W. Kaiser, Chem. Phys. Lett. 46:527 (1977)

36. T.J. Aartsma, W.H. Hesselink, and D.A. Wiersma, Chem. Phys. Lett. 71:424 (1980); K. Luther, and W. Wieters, J. Chem. Phys. 73:4132 (1980)

37. C.S. Parmenter, and M.W. Schyler, J. Chem. Phys. 52:5366 (1970)

38. H.F. Kemper and M. Stockburger, Ber. Bunsenges. Phys Chem. 72:1044 (1968)

39. M. Stockburger, in "Organic Molecular Photophysics", J. Birks
     ed., Wiley, New York (1973)
40. M. Quack and M. Stockburger, J. Mol. Spectrosc. 43:87 (1972)
41. D.A. Chernoff and S.A. Rice, J. Chem. Phys. 70:2511 (1979)
42. D.E. Powers, J.B. Hopkins, and R.E. Smalley, J. Chem. Phys.
     72:5721 (1980)
43. D.H. Levy, at this conference and Ann. Rev. Phys. Chem. 31:197
     (1980)
44. J.B. Coon, R.E. De Wames, and C.M. Loyd, J. Mol. Spectrosc.
     8:285 (1962)
45. J.B. Hopkins, D.E. Powers, S. Mukamel, and R.E. Smalley, J.
     Chem. Phys. 72:5049 (1980)
46. R.A. Coveleskie, D.A. Dolson, and C.S. Parmenter, J. Chem. Phys.
     72:5774 (1980)
47. A. Weller, Ber. Bunsenges. Phys. Chem. 56:662 (1952); Z. Physik.
     Chem. 15:438 (1958)
48. B.R. Henry and W. Siebrand, J. Chem. Phys. 49:5369 (1968); B.
     R. Henry, Acc. Chem. Res. 10:207 (1977)
49. G. Herzberg, "Molecular Spectra and Molecular Structure" Vol.
     III, van Nostrand, New York (1966)
50. J. Jortner, S.A. Rice, and R.H. Hochstrasser, Adv. Photochem.
     7:149 (1969)
51. A. Abragam, "Principles of Nuclear Magnetism", Oxford U.P.,
     London (1961)
52. R.G. Bray, and M.J. Berry, J. Chem. Phys. 71:4909 (1979)
53. D. Heller, at this conference
54. S. Mukamel, at this conference
55. In this series of experiments we had also measured the CH fun-
     damental and first overtone of pentafluorobenzene. In the fun-
     damental at a resolution of 0.06 cm^{-1} there was only broad
     structure, therefore our interpretation is the same as for
     $(CF_3)_3$H and the other molecules in this series[4]. D. Heller[53] re-
     ported a spectrum of the fundamental of pentafluorobenzene
     with "sharp rotational" structure, which is not in good agreement
     with neither our experimental results nor our interpretation
     of the vibrational structure
56. H.J. Bernstein and G. Herzberg, J. Chem. Phys. 16:30 (1948)
57. E.J. Heller, at this conference; E.J. Heller, E.B. Stechel,
     and M.J. Davis, J. Chem. Phys. 00:000 (1980)
58. W. Heisenberg, "Physikalische Prinzipien der Quantentheorie;,
     BI, Mannheim (1958)
59. P. Pfeifer, at this conference
60. P.J. Nagy and W.L. Hase, Chem. Phys. Lett. 54:73 (1978)
61. R.T. Lawton and M.S. Child, Il Nuovo Cimento 00:000 (1980)
     (Abstracts from the European Conference on the Dynamics of
     Excited States, Pisa, 1980)

VIBRATIONAL ENERGY RELAXATION FROM CH-STRETCHING MODES

OF SMALL MOLECULES IN THE LIQUID

Sighart F. Fischer, Wolfgang Kaiser and
Renate Zygan-Maus

Physik-Department
Technische Universität München
D-8046 Garching, W. Germany

INTRODUCTORY REMARKS

In this contribution we will be concerned with the vibrational relaxation of polyatomic molecules in the liquid state. Compared to the gas phase in the collision-free regime the situation seems much more complicated, since little is known about the interaction of a polyatomic molecule imbedded in a solution like CCl_4. We will see that this is not necessarily so.

Experimentally it is possible to observe the population decay on the pico-second time scale.[1] We will be especially concerned with the energy decay of CH-stretching modes. In this case most of the energy goes into vibrational energy of the same molecule. The medium picks up relatively little translational energy. It only helps to induce the transition. In some cases this can be verified experimentally by observing the population of the final states. These are typically overtones or combination tones of bending modes.[2] The preparation of the initial state is usually different in the liquid than in the gas phase and requires special considerations.

We have to distinguish between the initially prepared states, the final states, and the coupling mechanism. In the ideal situation an eigenstate of the isolated molecule is excited and the interaction with the medium can be handled as weak perturbation. The final states are other eigenstates, for example overtones of lower frequency modes, together with the continuum of translational and rotational states. It is important for the theory that the eigenstates include the so-called Fermi resonance mixing which is due to intramolecular anharmonic coupling between symmetry-like neighboring

states. In liquids the transitions are collision-induced, but most
of the vibrational energy release goes into vibrational energy of
the same molecule. If many eigenstates of the molecule fall within
the width of the excitation pulse, the initial state has to be represented as a superposition of eigenstates. If these states form a
dense manifold, we are dealing with the so-called 'statistical limit'. The initial state is in this limit the oscillator-carrying
state. Contrary to electronic relaxation processes it is not obvious
how to construct the interaction between the oscillator-carrying
state and the dense manifold. For electronic excitation one can
argue that the initial state is a Born-Oppenheimer state or a pure
spin state leaving the Non-Born-Oppenheimer terms and the spin-orbit
coupling as interaction. For vibrational states a Born-Oppenheimer
separation is in general not useful unless there is one high-frequency mode coupled to many low-frequency modes. We will see that
the 'statistical limit' is not directly applicable for the small
molecules to be studied here.

When only a few states are coherently excited, we are dealing with
the intermediate coupling situation. In this case a double exponential decay is predicted. The fast decay is due to destructive interference, which may include quantum beats followed by a slow exponential decay representative of the population lifetime of the individual states.

       For molecules in the liquid state the interaction with the sur-
rounding medium causes rapid dephasing and destructive interference
within the time of excitation, which is 5 ps in the experiments discussed here. Therefore we can identify the observed exponential decay with the population lifetime of the individual state or a group
of states. If the energy exchange between several vibrational states
is faster than the duration of the excitation pulse, an energy redistribution can take place before the onset of the decay process.
That means, several states may have been pumped directly or indi-
rectly and the observed decay may be lengthened since a reservoir
or several states has to be depopulated.

EXPERIMENTAL RESULTS

       We want to select a few experimental results[3,4] which point
very clearly at a relation between the lifetime of the CH-stretching
mode and its Fermi resonance mixing with combination tones in the
3000 cm^{-1} energy regime. As measure of the Fermi resonance mixing
we can define the intensity ratio R from the infrared spectrum
between the final state and the initial state. Strictly speaking
this applies only if the intensity of the final state is only due
to the intensity borrowing from the CH-stretching mode. For overtones or higher-order combination tones of bending modes this

assumption applies actually quite well since these modes can be described in first approximation by harmonic oscillators. A further measure of the Fermi resonance mixing is the shift in energy of the final state from its value predicted within a harmonic model. In the following we want to discuss these factors.

We start with the chloroform molecule ($CHCl_3$). This molecule is an excellent example to demonstrate the importance of the Fermi resonance. Since there is only one CH bond, we have only one CH-stretching mode at ν_1 = 3016 cm^{-1} (see Fig. 1). The overtone of the bending mode is at 2398 cm^{-1} far apart from the stretching mode ν_1. Therefore, one would expect a small transition rate into the overtone of the bending mode and a long lifetime for ν_1. The observed time of 1.6 ps is, however, relatively short (see Fig. 2). The key to the problem can be found in the infrared spectrum showing a line at 2970 cm^{-1}, a shoulder at 3070 cm^{-1}, and additional peaks around 2900 cm^{-1}. These bands must be due to combination tones. Tentatively, we assign the A_1-component of the $\nu_2+3\nu_5$ states to the main peak at

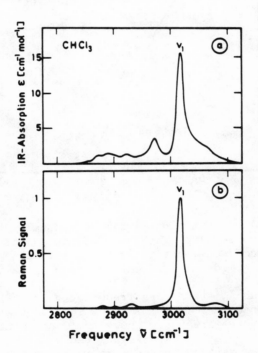

Fig. 1. Infrared (a) and Raman (b) spectrum of chloroform around the CH-stretching mode ν_1. The band at 2970 cm^{-1} points to a Fermi resonance between a combination mode and the ν_1-mode. The bands around 2900 cm^{-1} and the shoulder at 3070 cm^{-1} suggest additional Fermi resonances with the ν_1-stretching mode.

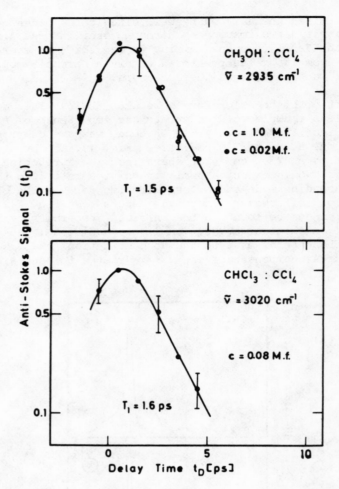

Fig. 2. Anti-Stokes scattering signal versus delay time of the
         probing pulse. Top: The decay of the CH-stretching mode
         at 2935 cm^{-1} of mechanol is measured. Bottom: The CH-
         stretching mode at 3020 cm^{-1} of chloroform is investigated.

2970 cm^{-1}. ν_2 is an A_1 state with frequency $\nu_2 = 667$ cm^{-1} and ν_5 is
an E-state with frequency $\nu_5 = 762$ cm^{-1}. The combination $2\nu_3 + 3\nu_5$ may
be related to the high-frequency shoulder. Other combinations like
$2\nu_4 + 2\nu_6$ or even $10\nu_6 + \nu_3$ may explain the peaks in the 2900 cm^{-1} regime. All these intensities must be due to Fermi resonance mixing
with ν_1.

     For the dominant resonance at 2970 cm^{-1} we estimate from the
infrared spectrum an intensity ratio R = 0.35. The frequency diffe-

rence between the initially prepared state and this final state is
$E_i - E_f = \hbar\omega = 46$ cm^{-1}.

    As our second example we take dichloromethane (CH_2Cl_2). In this
case we have two CH-stretching modes at the frequencies $\nu_1 = 2985$ cm^{-1}
and $\nu_6 = 3048$ cm^{-1} (Fig. 3). The experiment shows that the lifetime
is the same regardless which CH-stretching mode is excited. The observed lifetime of 50 ps (Fig. 4) is considerably longer than the
lifetime of chloroform. One factor contributing to the lengthening
can be easily understood. Since the two stretching modes exchange
their energy within the time of excitation, both have to be depopulated during the relaxation process. The final state is probably the
overtone of the CH-bending mode $\nu_2 = 1423$ cm^{-1}. The energy of the

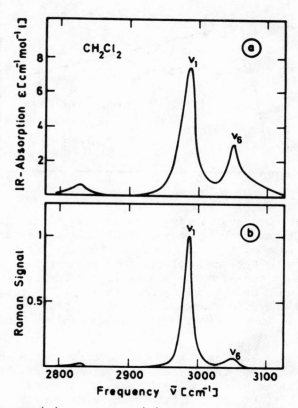

Fig. 3.   Infrared (a) and Raman (b) spectrum of dichloromethane
          depicting the symmetric (ν_1) and asymmetric (ν_2) CH_2-
          stretching modes. The band at 2839 cm^{-1} represents the
          overtone of the CH-bending mode in small Fermi resonance
          with the ν_1 mode.

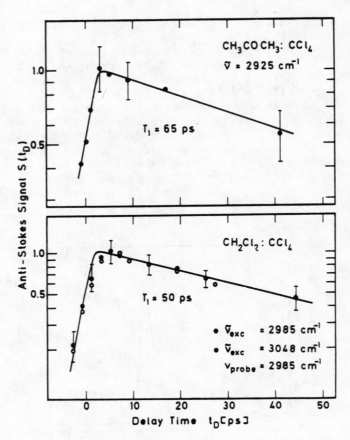

Fig. 4. Anti-Stokes scattering signal versus time. Top: Decay of
the CH_3-stretching mode at 2925 cm^{-1} of acetone (see Fig.
5b). Bottom: Rise and decay of the CH_2-stretching mode
with ν_1 = 2985 cm^{-1} of dichloromethane. The time dependence is the same for excitation of the ν_1 or ν_2 mode at
2985 cm^{-1} or 3048 cm^{-1}, respectively. The result suggests
rapid energy exchange between the two CH_2-stretching modes,
but slow decay via the overtone of the CH-bending modes.

overtone is around 2840 cm^{-1} very close to twice the energy of ν_2.
The intensity ratio of the infrared spectrum gives us b^2 = 0.07.
The energy difference is $E_i - E_f$ = 170 cm^{-1}. This small value and the
small energy shift of $2\nu_2$ indicate weak Fermi resonance interaction,
which correlates well with the long lifetime.

A similar situation as for dichloromethane exists for acetone
(CH_3COCH_3). There are three double degenerate CH-stretching modes
(see Fig. 5). We only need to discuss one CH_3-group, since the two

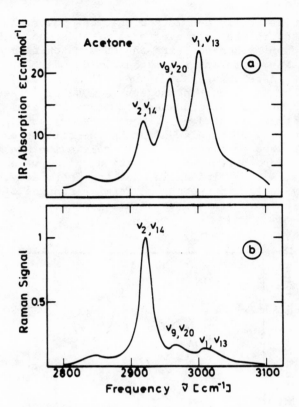

Fig. 5. Infrared (a) and Raman (b) spectrum of acetone around 3000 cm^{-1}. The band at 2843 cm^{-1} corresponds to an overtone of the CH-bending mode. The overtone is in small Fermi resonance with the ν_2, ν_{14} modes of A_1 symmetry.

are equivalent and do not interfere. The three CH_3-stretching modes communicate fast and lengthen the lifetime by a factor of three. there are three pairs of CH-bending modes. The overtones of these can interact with the lowest CH-stretching mode which is of A_1-symmetry. The combination tones may also interact with the higher CH-stretching modes. Since the spectrum does not resolve all contributions, we cannot pick the final state in a unique manner. We relate the small infrared (and Raman) band at 2843±2 cm^{-1} to the overtone of the almost degenerate bending modes ν_4 or ν_{10} with frequency 1423 cm^{-1}. So we get again a very small energy shift. These overtones are relatively far displaced from the A_1 CH-stretching mode ν_2 (83 cm^{-1}). The intensity ratio is about 0.04.

The molecules $CHCl_3$, CH_2Cl_2 and CH_3COCH_3 have relatively low CH-bending modes, so that the overtones of these modes have relatively little Fermi resonance with the CH-stretching modes. The situation is quite different for methanol (CH_3OH). Here, the CH-bending mode ν_4 is relatively high (ν_4 = 1448 cm^{-1}). Its overtone is well above the CH-stretching mode ν_2 = 2943 cm^{-1} (see Fig. 6). The overtone of ν_4 and the fundamental ν_2 are in Fermi resonance with an intensity ratio of 1 to 2. The transfer from ν_2 to $2\nu_4$ is so fast that $2\nu_4$ may be populated during the time of excitation. There is probably no equilibrium before relaxation and there are several decay channels. For the transition ν_2 to $2\nu_4$ one gets R = 0.5 and $E_i - E_f$ = 30 cm^{-1}. We find again that the fast decay of 1.5 ps correlates very well with the strong Fermi resonance of the stretching mode and the overtone of the bending mode.

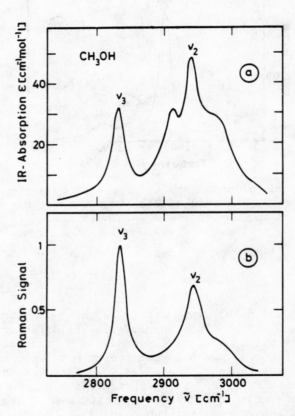

Fig. 6.   Infrared (a) and Raman (b) spectrum of methanol. The overtone of a CH-bending mode is located between the two CH-stretching modes ν_2 and ν_3. The large intensity of the overtone points to large Fermi resonance.

Comparing methanol and ethanol (CH_3-CH_2-OH) it is difficult to understand that the lifetime is so much longer (20 ps) for the latter molecule. We have seen that the strong Fermi resonance is largely responsible for the fast decay. In ethanol the spectrum does not show such Fermi resonances (Fig. 7). We interpret this result as follows:

The presence of the high bending mode frequency in methanol is due to the strong electronegativity of the OH-group. If an CH_2-group is entered between the CH_3- and the OH-group, this effect is reduced. Thus the CH-bending mode of the CH_3-group in ethyl alcohol is lowered compared to methanol and the overtone no longer makes good Fermi resonance with the stretching mode. Within the CH_2-group the anharmonic coupling is smaller anyhow. There is an additional factor which lengthens the lifetime. The number of CH-stretching modes in ethanol

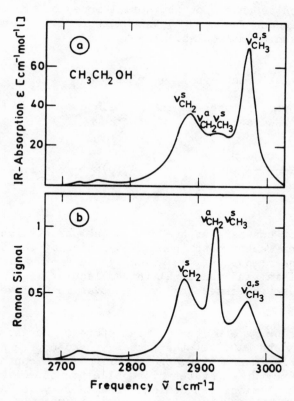

Fig. 7. Infrared (a) and Raman (b) spectrum of ethanol. There is small Fermi resonance with overtones of CH-bending modes located around 2950 cm^{-1}.

is 5. These can equilibrate since the decay into the overtones takes now longer. They lengthen the lifetime by a factor of 5. If one assigns the shoulder at 2750 cm^{-1} as final state, one gets R = 0.15 and $E_i - E_f$ = 60 cm^{-1}.

A SIMPLIFIED MODEL

So far we have seen that there exists a qualitative correlation between the Fermi resonance mixing and the presence of a bottleneck effect on the one hand and the lifetime on the other. We want to try to come to a quantitative description of this observation. Our goal is to relate the lifetime to other observables such as the intensity ratio which can be determined empirically.

We start with a semiclassical model for the transition rates. The interaction may be represented by a product consisting of an operator V which depends upon the normal modes of the molecule Q_j and a time-dependent factor U(t) which describes the interaction with the surrounding molecules.

$$H_{int} = V(Q_1 \ldots Q_n)U(t) . \tag{1}$$

The transition rate between an initial state ψ_i and a final state ψ_f may be expressed as:

$$k_{if} = \left| \frac{\langle \psi_i V(Q_1 \ldots Q_n) \psi_f \rangle}{\hbar^2} \right|^2 \int_{-\infty}^{+\infty} dt \ll U(t)U(0) \gg \cdot \exp(i\omega t) . \tag{2}$$

The first factor consists of the vibrational transition matrix element. The second factor describes the dynamics of the time-dependent interaction. The frequency ω is given by the energy difference between the initial and the final intramolecular eigenstates $\hbar\omega = E_i - E_f$. The bracket $\ll \gg$ indicates the average over initial velocities and angular momenta.

The dephasing rate of the initial state ψ_i may be obtained as the limit $\psi_i \to \psi_f$ and $\omega \to 0$. If the states ψ_i and ψ_f are strongly mixed due to anharmonic interaction, the matrix element contains in the basis of the unperturbed states diagonal elements. We will show now that these contributions can dominate and thereby we derive an interrelation between the depopulation rate T_1^{-1} and the dephasing rate T_2^{-1}.

We have already mentioned that the eigenstates denoted as ψ_i and ψ_f are eigenstates of the full vibrational Hamiltonian of the molecule under investigation. In particular the anharmonic inter-action is included. If we consider only interaction between two states, we can represent the exact eigenstates ψ_i and ψ_f in terms of the zero-order states ψ_i^o and ψ_f^o which may be the harmonic normal states:

$$\psi_i = a\psi_i^o - b\psi_f^o$$
$$\psi_f = b\psi_i^o + a\psi_f^o \quad . \tag{3}$$

The mixing coefficients a and b must fulfill the condition $a^2 + b^2 = 1$. b may be expressed in terms of the anharmonic coupling matrix ele-ment $W_{if} = \langle \psi_i^o \, W_{anh} \psi_f^o \rangle$ and the energy difference $\delta = E_i^o - E_f^o$ of the unperturbed energy levels

$$b^2 = \frac{(4W_{if}^2 + \delta^2)^{1/2} - \delta}{2(4W_{if}^2 + \delta^2)^{1/2}} \quad . \tag{4}$$

The perturbed eigenvalues are given as

$$E_{i,f} = \frac{E_i^o + E_f^o}{2} \pm \frac{1}{2} \left[(E_i^o - E_f^o)^2 + 4W_{if}^2 \right]^{1/2} \quad . \tag{5}$$

If we introduce the zero-order normal-mode function from Eq.(3) in the matrix element of (2) we obtain

$$\langle \psi_i \, V \, \psi_f \rangle = (a^2 - b^2) \, \langle \psi_i^o \, V \, \psi_f^o \rangle + ab \langle \psi_i^o \, V \, \psi_i^o \rangle - ab \langle \psi_f^o \, V \, \psi_f^o \rangle \quad . \tag{6}$$

For our systems we can identify ψ_i^o with the first excited state of a CH-stretching mode. ψ_f^o is an overtone, a combination tone of CH-bending modes, or in special cases even a higher combination tone. The matrix element $\langle \psi_i^o \, V \, \psi_f^o \rangle$ on the r.h.s. of (6) becomes very small if many quanta are changed in the transition from ψ_i to ψ_f. For in-stance, for a transition of the CH-fundamental into an overtone three quanta are changed. Such matrix elements are at least an order of magnitude smaller than the diagonal matrix elements. Which of the two diagonal matrix elements of Eq. (6) dominates, has been analyzed for the exponential repulsive interaction on the basis of a binary collision model.[6] A dependence upon the direction of the trajectory is predicted. The second term can dominate if the trajectory points in the direction of the stretching mode of the H-atom. However, the last term is more efficient for most other collision directions. This has to do with the normal mode displacement of a bending mode. It is also due to the fact that the diagonal matrix element of V for $V \propto Q^2$ formed with the second excited harmonic oscillator wave function (overtone) is larger by a factor of two compared to the first, the fundamental. In addition, an empirical fact speaks in favor of the final state matrix element. The linewidth which relates to the dephasing time T_2 is usually larger for the final state. Holding only the last term from Eq. (6) we get

$$k_{if} = a^2 b^2 \left| \frac{<\psi_f^o \ V \ \psi_f^o>}{\hbar} \right|^2 \int_{-\infty}^{+\infty} dt \ll U(t)U(0) \gg \cdot \exp(i\omega t) \ . \qquad (7)$$

Now we see the close relation to the dephasing rate

$$T_2^{-1}(f) = \left| \frac{<\psi_f \ V \ \psi_f>}{\hbar} \right|^2 \int_{-\infty}^{+\infty} dt \ll U(t)U(0) \gg \ . \qquad (8)$$

We approximate in Eq. (8) ψ_f by ψ_f^o and try to estimate the ω-dependence of the integral in Eq. (7) in order to introduce T_2 in Eq. (7). Again we take the binary collision model as guidance. The main ω-dependence is of the form[6,7]

$$k_{if} \propto \exp\{-(\frac{\omega}{\Omega})^{2/3}\}$$

with

$$\Omega = \frac{\alpha}{2\pi} \cdot (\frac{2}{3})^{3/2} \cdot (\frac{kT}{m_{eff}})^{1/2} \ . \qquad (10)$$

α is the reciprocal interaction length and m_{eff} the effective collision mass which also includes the effect of rotation.[8] In principle, Ω can be determined from Eq. (10). This equation applies well only if $\hbar\omega$ is larger than kT. For smaller values of ω we performed more detailed calculations on small molecules[6] and obtained for Ω typical values close to 100 cm^{-1}. Experimentally we can estimate T_2 from the Raman linewidth of the final state, or if this is not available, from the infrared linewidth of the final state. The mixing coefficient a or b may be determined empirically.

Assuming that only the zero-order state of the CH-stretching mode carries oscillator strength in the infrared excitation process, the ratio of the intensities of the two states in Fermi resonance is given as[9]

$$\frac{I_f}{I_i} = R = \frac{b^2}{a^2} .$$ (11)

Using the property $a^2 + b^2 = 1$, one obtains for the transition rate between two selected states:

$$k_{if} = \frac{R}{(1+R)^2} \cdot \exp \{-(\frac{\omega}{\Omega})^{2/3}\} T_2^{-1}(f) .$$ (12)

If N states are initially populated via fast energy exchange and if there is only one effective final channel, the rate is shortened by the factor N. Our final result for the lifetime reads:

$$T_1 = N \cdot (1+R)^2 R^{-1} \exp(\frac{\omega}{\Omega})^{2/3} \cdot T_2(f).$$ (13)

Table 1. Theoretical (Eq. 13) and experimental lifetimes. T_2 is estimated from the Raman linewidth $\Delta \nu$ as $T_2 = (2\pi c \Delta \nu)^{-1}$

Molecule	N	R	$E_i - E_f$	$T_2(f)$	T_1(Eq.13)	T_1(exp)
$CHCl_3$	1	0.35	46 cm^{-1}	$\cong$ 0.3 ps	$\cong$ 2.8 ps	1.6 ps
CH_2Cl_2	2	0.07	140 cm^{-1}	$\cong$ 0.4 ps	$\cong$ 46 ps	50 ps
CH_3COCH_3	3	0.04	83 cm^{-1}	$\cong$ 0.3 ps	$\cong$ 59 ps	60 ps
CH_3OH	1	0.5	30 cm^{-1}	$\cong$ 0.2 ps	$\cong$ 1.4 ps	1.5 ps
CH_3CH_2OH	5	0.15	60 cm^{-1}	$\cong$ 0.2 ps	$\cong$ 18 ps	20 ps

There are certainly large uncertainties in the determination of $T_2(f)$ and in R. For very small Fermi resonances one tends to overestimate the value of R, since overtones can have a small infrared intensity due to diagonal anharmonicities in the absence of Fermi resonance with the CH-stretching mode. For very strong Fermi resonance inter- action it becomes difficult to determine the decay channels, since they may get excited during the excitation process (ethanol). Even more difficult is the accurate determination of $T_2(f)$, since there are often overlapping bands in this energy regime. Still it seems interesting that the main differences in the observed lifetimes can be understood on the basis of Eq. (13).

MODEL STUDIES FOR THE ANHARMONIC COUPLINGS

We have seen that the empirical determination of the Fermi resonance mixing is limited. One would like to predict the mixing coefficients from the force field. Unfortunately one finds in the literature only information about a 'General Harmonic Force Field' (GHFF)[10] which is determined empirically from the observation of the energy states for selectively deuterated molecules. In some cases ab initio calculations are known, but they are not in good agreement with the experimentally observed energies and intensities for overtones and combination bands.

We want to introduce here a semi-empirical method[6] to determine the cubic anharmonicity constants. We start from a GHFF. Let us summarize the valence coordinates as $\vec{r}$. Then we write the potential as

$$V = \frac{1}{2} \vec{r}^t \, K \, \vec{r} \tag{14}$$

with the harmonic force field K. We want to extend this by cubic terms to get

$$V = \frac{1}{2} \sum_{ij} K_{ij} \, r_i \, r_j + \sum_{i,j,l} K_{ijl} \, r_i \, r_j \, r_l \, . \tag{15}$$

The parameters K_{ijl} are unknown. We introduce for the local stretching modes a Morse potential. We can then relate the cubic term K_{iii} to the dissociation energy.

Interestingly, we don't need more empirical data in order to estimate many more nondiagonal cubic coupling terms. In particular, the coupling between the CH-stretching modes and the overtones of the bending modes can be calculated because the transformation between the valence coordinates and the normal modes is non-linear. These contributions to the coupling between normal modes are particularly large for the bending modes.

If $\vec{X}$ denotes the Cartesian coordinates of an atom, we have

$$r_i = \sum_{m=1}^{3M} B_{im} \, X_m + \sum_{m,n=1}^{3M} X_m \, C_{mn}^i \, X_n \tag{16}$$

with

$$B_{im} = \frac{\partial r_i}{\partial X_m} \quad \text{and} \quad C_{mn}^i = \frac{1}{2} \frac{\partial^2 r_i}{\partial X_m \partial X_n} \, . \tag{17}$$

The coordinate r_i stands here either for a valence length or a valence angle. The matrices B_{im} and C_{mn}^i depend only upon the geometry of the molecule and can be calculated. If the matrix A gives us the transformation between the cartesian coordinates $\vec{X}$ and the normal mode coordinates $\vec{Q}$, we can express the $\vec{r}$ in terms of the $\vec{Q}$:

$$r_i = \sum_{l=1}^{3M-6} \frac{\partial r_i}{\partial Q_l} Q_l + \sum_{l,l'=1}^{3M-6} \frac{1}{2} \frac{\partial^2 r_i}{\partial Q_l \partial Q_{l'}} Q_l Q_{l'}, + \ldots \qquad (18)$$

with

$$\frac{\partial r_i}{\partial Q_l} = \sum_{m=1}^{3M} B_{im} A_{ml}$$

and

$$\frac{1}{2} \frac{\partial r_i}{\partial Q_l \partial Q_l} = \sum_{m,n=1}^{3M} C_{mn}^i A_{ml} A_{nl'} .$$

If we introduce these into our expression for the potential Eq. (15) we get

$$V = \frac{1}{2} \sum_{l,p=1}^{3M-6} k_{lp} Q_l Q_p + \sum_{l,p,q=1}^{3M-6} k_{lpq} Q_l Q_p Q_q + \ldots \qquad (19)$$

with

$$k_{lp} = \sum_{ij} K_{ij} \frac{\partial r_i}{\partial Q_l} \frac{\partial r_j}{\partial Q_p} \qquad (20)$$

and

$$k_{lpq} = \frac{1}{2} \sum_{ij} K_{ij} \left(\frac{\partial r_i}{\partial Q_l} \cdot \frac{1}{2} \frac{\partial^2 r_j}{\partial Q_p \partial Q_q} + \frac{\partial r_j}{\partial Q_l} \cdot \frac{1}{2} \frac{\partial^2 r_i}{\partial Q_p \partial Q_q} \right)$$

$$+ \sum_i K_{iii} \frac{\partial r_i}{\partial Q_l} \frac{\partial r_i}{\partial Q_p} \frac{\partial r_i}{\partial Q_q} . \qquad (21)$$

Usually the convention $l \leqq p \leqq q$ is taken in expression (19). To arrive at these normal coordinate force constants the constants (21) and (20) have to be summed up over all permutations of the indices (l,p,q).

APPLICATION TO METHYLIODID

We have chosen methyliodide as an example because there exists a lot of information about the molecule which can help to test the model. Earlier we analyzed already transitions from the lowest states ν_6 and ν_3[8,11]. We found that vibrational energy is effectively transferred into rotation. Here we are mainly interested in the transitions between the high-frequency CH-stretching modes ν_1 and ν_4 and the overtones of the CH-bending modes ν_2 and ν_5. The molecule is in some respect typical for the interactions of a CH_3-group. It is special in the sense that the modes ν_4, ν_5 and ν_6 are degenerate. The other three ν_1, ν_2 and ν_3 are totally symmetric. We find for the diagonal anharmonic force constant of the CH-coordinates from the dissociation energy $K_{111} = K_{222} = K_{333} = -5.47$ mdyn/$\overset{\circ}{A}{}^2$. The calculations show that there are two types of coupling constants which can become large. First there are those which get their main contribution from the local anharmonicities of the stretching motion. These are k_{111}, k_{144} and k_{444}. The others get most of their contribution from the nonlinear part of the transformation Eq. (18). The coupling constants are given in Table 2.

The wave function for the totally symmetric CH-stretching vibration gets its main admixture from the overtone of ν_2 and the totally symmetric component of the overtone from ν_5.

$$\psi_1 = 0.92\ \psi^{\circ}_{\nu_1} + 0.367\ \psi^{\circ}_{2\nu_5} + 0.137\ \psi^{\circ}_{2\nu_2}\ . \tag{22}$$

The final state $\psi_{2\nu_5}$ is strongly influenced by another Fermi resonance with the combination tone $\nu_3 + \nu_5 + \nu_6$. This level, on the other hand, is strongly coupled to $2\nu_3 + 2\nu_6$. In both cases the relevant coupling constant is k_{356}. The wave function $\psi_{2\nu_5}$ is approximately given by

$$\psi_{2\nu_5} = -0.298\ \psi^{\circ}_{\nu_1} + 0.387\ \psi^{\circ}_{\nu_3\nu_5\nu_6} + 0.795\ \psi^{\circ}_{2\nu_5} - 0.352\ \psi^{\circ}_{2\nu_3 2\nu_6}$$
$$- 0.074\ \psi^{\circ}_{2\nu_2}\ . \tag{23}$$

The wave functions of the combination tones read:

$$\psi_{\nu_3\nu_5\nu_6} = 0.196\ \psi^{\circ}_{\nu_1} + 0.802\ \psi^{\circ}_{\nu_3\nu_5\nu_6} - 0.458\ \psi^{\circ}_{2\nu_5} - 0.327\ \psi^{\circ}_{2\nu_3 2\nu_6}$$
$$+ 0.046\ \psi^{\circ}_{2\nu_2} \tag{24}$$

and

$$\psi_{2\nu_3 2\nu_6} = -0.048\ \psi^0_{\nu_1} + 0.454\ \psi^0_{\nu_3 \nu_5 \nu_6} + 0.148\ \psi^0_{2\nu_5}$$
$$+ 0.877\ \psi^0_{2\nu_3 2\nu_6} - 0.013\ \psi^0_{2\nu_2} . \tag{25}$$

Table 2. Anharmonic Force Constants for CH_3I

-185	8	0.4	170	69	13	-69	-17	-24	-37
-587	-7	33	110	-22	245				
34	-80	-374	73	-118	3		CH_3I		
6	-17	-74	10	-26	5				
-143	-29	16	-62	162	124	-7	-4	-66	15
k_{111}	k_{112}	k_{113}	k_{122}	k_{123}	k_{133}	k_{222}	k_{223}	k_{233}	k_{333}
k_{144}	k_{145}	k_{146}	k_{155}	k_{156}	k_{166}				
k_{244}	k_{245}	k_{246}	k_{255}	k_{256}	k_{266}		k_{lpq}		
k_{344}	k_{345}	k_{346}	k_{355}	k_{356}	k_{366}				
k_{444}	k_{445}	k_{446}	k_{455}	k_{456}	k_{466}	k_{555}	k_{556}	k_{566}	k_{666}

This result shows already how complex the mixing between the different states becomes. The effect on the relaxation rates is for the detailed calculation even stronger than predicted by the simplified model of chapter 3.

We do not want to present the details of the calculations. The model has been discussed elsewhere[11]. The results are summarized in Table 3. For comparison the predictions in the absence of anharmonic coupling are also included. One finds that the highest rate from ν_1 is predicted for the transition to $\nu_3 + \nu_5 + \nu_6$. The value is 18 msec^{-1}torr^{-1} for Xenon as a collision partner. In the absence of the anharmonic interaction the rate would be 0.0036 msec^{-1}torr^{-1}. That means the Fermi resonance changes the rate by more than two orders of magnitude even though the final state is separated in energy by 130 cm^{-1}. This prediction clearly demonstrates the importance of anharmonic interactions. It is again predicted that the CH-stretching modes equilibrate rapidly followed by a slower decay into overtones of the bending modes or higher combination tones.

Table 3. Rate constants k_{if} in $msec^{-1}torr^{-1}$ in brackets are the rates without the anharmonic interaction

$\psi_f=$	$2\nu_5(A_1)$	$\nu_3+\nu_5+\nu_6(A_1)$	$2\nu_3+2\nu_6(A_1)$
$\psi_i=\nu_1$	6.4 ($1.7\cdot10^{-1}$)	18 ($3.6\cdot10^{-3}$)	1.0 ($3.3\cdot10^{-4}$)

$\psi_f=$	ν_1	$2\nu_5(E)$	$\nu_3+\nu_5+\nu_6(E)$
$\psi_i=\nu_4$	$2.7\cdot10^2$ ($3.1\cdot10$)	6.5 ($6.5\cdot10^{-2}$)	2.2 ($1.2\cdot10^{-3}$)

This is in qualitative agreement with experiments.[12] For quantitative comparison one needs a good estimate for the collision number in the liquid.

Acknowledgment:
The authors like to thank Dr. A. Fendt for important experimental contributions.

1. A. Laubereau and W. Kaiser, Rev. of Mod. Phys. 50, 607 (1978).
2. A. Laubereau, G. Kehl and W. Kaiser, Optics Commun. 11, 74 (1974); R. R. Alfano and S.L. Shapiro, Phys. Rev. Lett. 29 1655 (1972).
3. A. Laubereau, S.F. Fischer, K. Spanner and W. Kaiser, Chem. Phys. 3, 335 (1978).
4. A. Fendt, S.F. Fischer and W. Kaiser, Chem. Phys. Letters submitted.
5. D. W. Oxtoby, Adv. Chem. Phys. 40, 1 (1979).
6. R. Zygan-Maus, Dissertation, Technische Universität München, Physik-Department (1980).
7. R. N. Schwarz, Z.I. Slawsky and K. Herzfeld, J. Chem. Phys. 20, 1591 (1952).
8. A. Miklavc and S. F. Fischer, J. Chem. Phys. 69, 69 (1978).
9. D. C. McKean, Spectrochim. Acta 29A, 1559 (1973).
10. J. L. Duncan, D. C. McKean, G. K. Speirs, Mol. Phys. 24, 553 (1972).
11. R. Zygan-Maus and S. F. Fischer, Chem. Phys. 41, 319 (1979).
12. K. Spanner, A. Laubereau and W. Kaiser, Chem. Phys. Lett. 44, 88 (1976).

IRREVERSIBILITY QUESTIONS IN CHEMISTRY, QUANTUM-COUNTING, AND TIME-DELAY

Karl Gustafson

Department of Mathematics
University of Colorado
Boulder, Colorado 80309

INTRODUCTION

I would like to bring together here three questions, each of independent and fundamental interest, bound together by a common underlying theme: how do we distinguish irreversibility effects in microscopic processes? The three questions come from our models for molecules, counters, and particles. Although somewhat idealized here, they lie at the heart of the foundations of theoretical chemistry, measurement theory, and quantum mechanics.

This exposition will be of necessity brief. However a number of references are given.

There seems to be little doubt that chemists, at least many chemists, think of atoms, molecules, and collisions in a time dependent way. Perhaps more correctly, the experimentalists among them think, of necessity, of frequency shifts but even then still mix in doppler and other time delay effects. The theorists even in their time-independent number crunching have developed methods in terms of "how long the molecule stays around."

The three problems we will discuss are not unrelated to the above three sentences and deal, respectively, with irreversibility in chemical processes at the molecular level, in experimental measurement at the quantum level, and in time delay at the scattering level.

1. CHEMISTRY

Much recent work in chemistry is based upon the view that regions of the phase plane in classical dynamics that exhibit ergodic behavior have their analogues in the quantum dynamics of molecules. Roughly speaking, the idea is that a randomization of internal energy precedes bond breaking.

Most chemists know the amusing folklore about how bond breaking came about historically, and most know the less amusing still remaining mysteries about it. Some even question whether molecules really exist. We refer to the Proceedings of two recent conferences [1, 2] for more about these important subjects.

In George, Henin, Mayne, and Prigogine [3] and in other physics and chemistry papers, attention is focused on reconciling a dynamical reversible evolution U_t and a related thermodynamic irreversible evolution W_t. Following [3], one may attempt to establish a change of representation Λ so that the given deterministic evolution is transformed into an evolution displaying thermodynamic properties. In particular, in [3] such Λ and corresponding Ljapunow entropy functionals are shown to exist in a Friedrichs field theoretical model.

In Misra, Prigogine, and Courbage [4], the above problem is studied further, with emphasis on the change from deterministic systems to probabilistic descriptions. By requiring Λ to be invertible rather than, say, a projection, there is no drastic "contraction of description" or "coarse graining" involved in going to the stochastic description.

In the more general setting of [4] Λ was to be (i) bounded selfadjoint on a measure space $L^2(\Omega, S, \mu)$, (ii) positivity preserving, (iii) trace preserving, (iv) information preserving, and (v) equilibrium preserving. In [4] it was conjectured that Λ^{-1}, even when unbounded, could not be of the same type. On a physical basis, this amounts to concluding that Λ^{-1} cannot be positivity preserving, for otherwise $W_t^* = \Lambda U_t \Lambda^{-1}$ would not be a truly probabilistic description to be arrived at from a deterministic evolution U_t.

In Goodrich, Gustafson, and Misra [5] the above conjecture is answered affirmatively, and for a larger class of transformations Λ. Let us give the precise result of [5] here.

Theorem. [5]. Let (i) Λ be a closed not necessarily selfadjoint operator with domain $D(\Lambda)$ containing all characteristic functions of Borel sets E in S, (ii) $\Lambda(\rho) \geqq 0$ for $\rho \geqq 0$ and $\rho \in D(\Lambda)$, (iii) $\int \Lambda(\rho) d\mu = \int \rho d\mu$ for $\rho \in D(\Lambda)$, (iv) range $R(\Lambda)$ containing all characteristic functions E, (v) $\Lambda(1) = 1$, and

(vi) $\Lambda^{-1}(\rho) \geqq 0$ for $\rho \geqq 0$ and $\rho \in R(\Lambda)$: then Λ is induced from a point transformation of the phase space.

By the above result of [5] , given a reversible evolution U_t and change of representation Λ satisfying (i)-(vi) , the corresponding $W_t^* = \Lambda U_t \Lambda^{-1}$ must also be reversible. Thus by reversing this argument, beginning with a forward moving W_t^* that loses information, should there exist a way to convert it to a reversible U_t via a change of representation Λ within the class allowed above, then one could come back to W_t^* now no longer information losing.

This establishes on a rigorous basis the principle often used intuitively in chemistry that a forward moving chemical (e.g., Markov) process that loses information cannot be reversed.

At least, it cannot be reversed under the rules of the game as described above. How far one wants to push the class of Λ (e.g., nonlinear?) is not yet clear and would depend on physical considerations.

There are related problems in models for particle decay. In Gustafson and Misra [6, 7] a connection between certain stochastic processes of a "shift" type and quantum mechanical decay laws was investigated. There one ends up considering reduced evolutions of the form $Z_t = PU_t P$ in models for unstable particles such as mesons. In other words, "Λ" is a projection P , and if interpreted as in the present paper, Λ loses information by restricting the overall evolution to functions in a subspace or more visually to functions of restricted support in the case that P is given by a spatial characteristic function.

We will not attempt here to cover all of these related aspects of dealing with dissipative systems but leave them with the comment that in the next section you will see $PU_t P$ again in the context of measurement theory and in the third section you will see irreversibility and decay considerations entered into the context of the delay-time for a particle acted upon by another.

2. QUANTUM-COUNTING

The following problem was pointed out to me by J.M. Jauch [8] . Versions of it had no doubt been around for some time at least among those concerned with the measurement theory of quantum mechanics. See for background the article by Wigner [9] . Most of the models and results given here were presented at the Bozeman Conference five years ago [10] . Some parts were later worked out with B. Misra and have appeared in Misra and Sudershan [11] .

Let C be a subset of R^3 , and P_c the projection given by multiplication of any $L^2(R^3)$ function by χ_c , the characteristic function of C . Let φ_0 be an initial completely prepared state function in $L^2(R^3)$, e.g., with compact support away from C , and let $\varphi_t = U_t\varphi_0$ be the wave packet evolving under some unitary evolution U_t . We assume an opaque counter and proceed to count as follows.

Since counting should always take place in a finite time interval we look at such an interval $[0,t]$. First we ask if we have counted φ_t at $t = 0$, $t/2$, or at t . The probability of so counting is given by (let $\varphi_0 = \varphi$ and $P_c = P$ for simplicity of notation)

$$\mathbb{P} = \|P\varphi\|^2 + \|PU(t/2)P^\perp\varphi\|^2 + \|PU(t/2)P^\perp U(t/2)P^\perp\varphi\|^2 .$$

That is, we may count at $t = 0$, or not then but at $t/2$, or at neither of those but at t . More generally the $n^{\underline{th}}$ approximation to the probability of counting is

$$\mathbb{P} = \langle P\varphi,\varphi\rangle + \langle (U(t/n)P^\perp)^* P(U(t/n)P^\perp)\varphi,\varphi\rangle +$$

$$\cdots + \langle (U(t/n)P^\perp)^{*n} P(U(t/n)P^\perp)^n\varphi,\varphi\rangle$$

$$= \langle T_n\varphi,\varphi\rangle$$

where T_n is noted to be a positive, bounded selfadjoint operator. One may then define the probability of counting to be $\lim \mathbb{P}_n$ as $n \to \infty$ and the counting operator T_t to be $\lim T_n$ as $n \to \infty$.

The only problem is that it is not clear when these operators exist. Partial results and a discussion may be found in [12] . These problems will lead to a new theory of nonmonotone operator limits.

More generally one should be able to consider various physically interesting evolutions U_t and more general measurors such as P_c the E_λ of the position operator Q and other observables of physical interest. Further considerations from the point of view of quantum logic may be found for example in Srinivas [13] .

As an example suppose the range of P , $P\mathcal{H} = \mathcal{M}$, is a subspace without nonregeneration (see, e.g. [7]). By this we mean that relative to the evolving wave packet φ_t under U_t we have the condition

$$PU_s P^\perp U_t P = 0 \qquad \forall t \geq 0 , \quad s \geq 0 .$$

Then easily $\lim T_n$ exists and $T_t = \lim T_n = PU_t P$. It follows that the probabilities T_t form a semigroup.

To see this, suppose there is no regeneration. Then

$$PU(t/n)[I - P]U(t/n)P = 0$$

and

$$PU(2t/n)P = PU(t/n)PU(t/n)P .$$

Similarly

$$PU(3t/n)P = PU(2t/n)PU(t/n)P$$

and

$$(PU(t/n)P)^n = PU(t)P .$$

This is the simplest case.

Let us recall some measurement theory. Measurements of the first kind are those in which one does not destroy the particle, the experiment can be repeated. Measurements of the second kind destroy the particle. Ideal measurements are those that can be performed continuously. Nonideal measurements are the others.

Most measurement models (e.g., Feynmann, Piron treatments) are essentially ideal measurements of the first kind. The above model is that of an ideal measurement of the second kind.

It has the disadvantage that if there is not decay initially there is none thereafter. That is, the decay observability does not depend on the amount of Δt . To avoid a paradox one needs to go to a theory including instantaneous response.

It is hoped that a proper putting in of irreversibility considerations will avoid the paradoxes and still yield an understanding of nonimmediate response so that most existing counters need not be regarded as invalid.

3. TIME DELAY

Here one considers two situations. In the first, one has a free particle. In the second one has the particle interacting with another.

This is a typical situation in simple scattering theory. None-theless there remain questions about the time delays involved in the second situation as contrasted to the first. These considerations seem to be fundamental to a thorough understanding of resonances and other features of interacting particles.

Let us first define time delay. We use the counter notation for analogy and simplicity. A dense set of φ may be considered but they need certain compact energy support (away from 0) and regularity properties for technical reasons.

Def: Time delay = Sojourn time difference

$\qquad\qquad$ = Mean time difference spent by particle φ in space C

$$= \int_{-\infty}^{\infty} \|\chi_C e^{-iHt} \Omega_- \varphi\|^2 dt$$

$$- \int_{-\infty}^{\infty} \|\chi_C e^{-iH_0 t} \varphi\|^2 dt \quad ,$$

where $H = H_0 + V$, H_0 the Hamiltonian for the free particle and V representing the effect of the interaction with the other particle.

If we let $C = B_r$ the r ball of radius r in R^3, then the above time delay we may denote by T_r, and we are also interested in $T(\varphi) = \lim T_r$ as $r \to \infty$ as the total time delay caused by the inter-action, and in particular, when it is given by the Eisenbud–Wigner formula

$$T_\lambda(\varphi) = -iS^*(\lambda)dS(\lambda)/d\lambda = \delta'(H_0)$$

where

$$S(\lambda) = S \text{ matrix at energy } \lambda = e^{i2\delta(\lambda)}$$

where

$\qquad\qquad \delta(\lambda)$ = the phase shift

$\qquad\qquad \delta(H_0)$ = the phase shift operator.

In Martin [14] and Amrein, Jauch, and Sinha [15] the above relation between time-delay and the S-matrix was derived under cer-tain hypotheses which in the context of potential scattering amounted to assuming that the potential V is spherically symmetric and $0(r^{-4-\varepsilon})$, $\varepsilon > 0$, at $r \to \infty$. In Jauch, Sinha, and Misra [16] and

Martin and Misra [17] a time-independent method, utilizing a trace class condition, was employed, and although no spherical symmetry was assumed, the relevant decrease of $V(r)$ at infinity was $O(r^{-3-\varepsilon})$.

Recently Tee [18] investigated a sharpening of the approach of [14, 15] . It may be seen that the approach of [18] , with some modification, may be pushed through to obtain the Eisenbud-Wigner relation for V that are $O(r^{-3-\varepsilon})$.

For earlier work on time-delay see Jauch and Marchand [19] and Smith [20] . For the existence of a weighted time-delay operator for V which are, roughly, $O(r^{-3-\varepsilon})$ in R^3 , but without connection to the S - matrix, see Lavine [21] . For recent work on a time-delay operator as a dressed limit in the context of hyperbolic equations see Lax and Phillips [22] and Amrein and Wollenberg [23] . For original papers on the Eisenbud-Wigner relation, see Eisenbud [24] and Wigner [25] .

In a recent paper [26] we show that the Eisenbud-Wigner relation for time-delay holds for potentials $V(r)$ that are $O(r^{-5/2-\varepsilon})$ at ∞ . This improves previous results in which V was required to be $O(r^{-4-\varepsilon})$ and $O(r^{-3-\varepsilon})$, respectively. No doubt the sharp condition is $V(r)$ that are $O(r^{-2-\varepsilon})$ but a technical difficulty blocks the proof.

Probably irreversibility considerations play a role in a full understanding of time delay. This contention may be supported both physically and mathematically. On physical grounds, although we don't really know what mediates the retardation effects in the slowing down of the particle, they must exist and must correspond to some loss of freedom. Mathematically, in perturbation theory one is always struck by the lack of reciprocity in going from H_0 to $H_0 + V$ and then back to H_0 .

There is the old anecdote of a German on an Austrian train that was running three hours late. "Why do you bother with timetables in this country?" he asked the conductor. The latter replied "If we had no timetables, we wouldn't know how late we are!"

4. ADDITIONAL REMARKS

The following comments relate to those of other speakers and common themes of this conference.

Let us consider for example K flows. Recall the situation:

$K-flow \Rightarrow L|_{N_0^{\perp}}$ is absolutely continuous, unif. spectral mult.

$\Rightarrow$ There exists an increasing Lyapunov Entropy Functional $\langle U_t \rho, MU_t \rho \rangle$

$\Rightarrow L|_{N_0^{\perp}}$ is absolutely continuous

$\Rightarrow U_t = e^{iLt}$ is mixing

$\Rightarrow$ Ergodicity $(N_0 = N(L))$

where N_0 in general is the subspace of functions constant on the energy surface under consideration. K-flows are "intrinsically unstable" in the dynamical sense and in fact exhibit exponential divergence of initially close points. A K-flow may be thought of as a generalized regular representation and as such is equivalent (essentially, in the same manner as shown in [7]) to a quantum mechanical momentum evolution or direct sum thereof. See [6] for an interpretation of these situations in terms of correlation functions.

Λ then introduces (if properly chosen) an "intrinsic randomness", and then $M = \Lambda^* \Lambda$ introduces an "intrinsic irreversibility" when applied to a basic flow U_t .

The Λ change of representation operator especially in its delocalizing forms, may be regarded as a "scrambler". This observation is not only in tune with the theme of the present conference but also illustrates in a general way the serious problem of "basis dependence" discussed numerous times at this conference. By such a Λ , or more stringently by a projection P or by some less stringent extension of its notion, one scrambles a deterministic flow into a flow exhibiting stochastic properties. Both flows are from the same master flow and are not really very different. But depending on which basis or mixed states you stand on, things look different.

From this just stated viewpoint, then, one should recast the irreversibility questions as questions dealing with "what is lost". In other words, in the intramolecular line broadening theoretical models, when one projects (let us call the projection P) the flow onto a reduced flow, the original group, which according to conservation of energy and momentum principles should contain all information, has been stripped of some information. The same situation occurs in coarse graining. But nothing has really been lost or made irreversible if you admit that in the complementary subspace under the projection, the other part of the flow is still going on. The word "lost" should be changed to "made inaccessible" in the chemical context of section 1 and to "converted" in the measurement context of section 2 and to "dissipated" in the interaction picture of section 3.

For example in the counter model discussed in section 2 the assumption that the projection was onto a nonregenerative subspace was made precise by the condition $PU_sP^\perp U_tP = 0$ for all $t,s \geq 0$. It is easy to see that this condition is exactly equivalent to $Z_t = PU_tP$ being a semigroup for all $t \geq 0$. The reduced process Z_t is no longer reversible (under its own action) but nonetheless may move forward, albeit with "dissipation".

For P commuting with the group, Z_t remains a group, whereas for P not satisfying the regeneration condition, Z_t is not even a semigroup. Many of these possibilities and others such as when the Z_t remain isometries, are worked out in [12]. For simplicity one could consider the following heirarchy of possible interest in the chemical reaction frame:

Z_t a Markovian semigroup;

Z_t some other (e.g. Martingalian) semigroup;

Z_t a semigroup without stochastic classification;

Z_t no longer a semigroup but stochastic in some sense;

Z_t no longer a semigroup and no longer classifiable.

As one proceeds down the list, one encounters situations in which increasing amounts of "information" have been "lost". Better: "converted", into other forms from which the corresponding "information' is no longer accessible to the eye of the beholder. Admittedly, the latter information may be really inaccessible, as heat or noise for example. But the degree of inaccessibility is no change in concept and amounts to a separate question.

Under the invertible Λ transformations considered in section 1, one loses less: Z_t is always a semigroup and in many cases a Markovian one. It is not clear however that by just preserving positivity (i.e. populations) one has accounted for all the information that one desires preserved.

REFERENCES

1. R.G. Woolley, ed., Quantum Dynamics of Molecules: The New Experimental Challenge to Theorists, Proc. Cambridge NATO Advance Study Institute, Sept. 15-29, 1979, Plenum Press.

2. K. Gustafson and W. Reinhardt, eds., Classical, Semiclassical, and Quantum Mechanical Problems in Mathematics, Chemistry, and Physics, Proc. Boulder Conference on Mathematical Physics, Mar. 27-29, 1980, Plenum Press.

3. C.I. George, F. Henin, F. Mayne, and I. Prigogine, New quantum rules for dissipative systems, Hadronic Journal 1 (1978), 520-573.

4. B. Misra, I. Prigogine, and M. Courbage, From deterministic dynamics to probabilistic descriptions, Physica 98A (1979), 1-26.

5. K. Goodrich, K. Gustafson, and B. Misra, On a converse to Koopman's Lemma, Physica 102A (1980), 379-388.

6. K. Gustafson and B. Misra, Correlations and evolution equations, Proc. III Mexico-United States Symp. on Diff. Eqns., Mexico City, January, 1975, Boletin de la Sociedad Matematica Mexicana, 1975.

7. K. Gustafson and B. Misra, Canonical commutation relations of quantum mechanics and stochastic regularity, Letters in Mathematical Physics 1 (1976), 275-280.

8. J.M Jauch, private communication, Boulder, April, 1974.

9. B.d'Espanat, ed., E.P. Wigner, in Foundations of Quantum Mechanics, Academic Press, 1971.

10. K. Gustafson, Some Open Operator Theory Problems in Quantum Mechanics, lecture given at the Bozeman C^* Algebra Conference, August, 1975.

11. B. Misra and E. Sudershan, The Zenós paradox in quantum theory, J. Math. Phys, 18 (1977), 756-763.

12. K. Gustafson, The Counter Problem, to appear.

13. M. Srinivas, Foundations of a quantum probability theory, J. Math. Phys. 16 (1975), 1672-1685.

14. Ph. Martin, On the time-delay of simple scattering systems, Comm. Math. Phys. 47 (1976), 221-227.

15. W.O. Amrein, J.M. Jauch, and K.B. Sinha, Scattering Theory in Quantum Mechanics, W.A. Benjamin, Inc., Reading, Mass., 1977.

16. J.M. Jauch, K. Sinha, and B. Misra, Time-delay in scattering processes, Helv. Phys. Acta 45 (1972), 398-426.

17. Ph. Martin and B. Misra, On the trace-class operators of scattering theory and the asymptotic behavior of scattering cross section at high energy, J. Math. Physics 14 (1973), 997-1005.

18. R. Tee, Time delay in quantum scattering, dissertation, University of Colorado, 1978.

19. J.M. Jauch and J.P. Marchand, The delay time operator for simple scattering systems, Helv. Phys. Acta 40 (1967), 217-229.

20. F. Smith, Lifetime matrix in collision theory, Phys. Rev. 118 (1960), 349-356.

21. R. Lavine, Commutators and local decay, Scattering Theory in Mathematical Physics, Eds. J.A. LaVita and J.P. Marchand, Reidel, Dordrecht, Holland (1974), 141-156.

22. P.D. Lax and R.S. Phillips, The time delay operator and a related trace formula, Topics in Functional Analysis, Advances in Mathematics, Supplementary Studies, 3, Academic Press, New York

(1978), 197-215.
23. W.O. Amrein and M. Wollenberg, On the Lax-Phillips scattering theory, to appear.
24. L. Eisenbud, dissertation, Princeton University, 1948.
25. E.P. Wigner, Lower limit for the energy derivative of the scattering phase shift, Phys. Rev. 98 (1955), 145-147.
26. K. Gustafson and K. Sinha, On the Eisenbud-Wigner Formula for Time-Delay, Letters of Math. Physics, to appear.

ENERGY SCRAMBLING OUT IN THE CONTINUUM:

WEYL'S THEORY AND PREDISSOCIATION

Erkki Brändas   Nils Elander

Quantum Chemistry Group   Research Institute of Physics
Uppsala University   Fack
751 2o Uppsala, Sweden   1o4o5 Stockholm 5o, Sweden

INTRODUCTION

In a recent review[1] on high resolution techniques in spectroscopy on small molecules "time resolved molecular spectroscopy" is
defined as molecular studies based upon the decay properties of
excited levels including separated rotational levels and their fine-
structure components. Since these studies include dynamical properties
of a molecule, careful consideration must be given to its environmental interactions. Systematic effects like spectral blends, cascades
from upper levels, resonant transfer, collisional deactivation,
radiation trapping and various escape processes need to be estimated
in order to give realistic molecular lifetimes. According to Erman[1]
normal accuracy of absolute lifetime recordings of resolved levels
are in the range of 3-1o% while careful considerations of possible
systematic errors may yield recordings with an accuracy of 1% or
better. The molecular lines and f-values that are obtained can be
used to give direct information on molecular and atomic abundances in
interstallar medium as well as for determining isotope ratios in
cosmic objects.

It is generally observed in modern time resolved spectroscopy
that predissociations are very common in nature even in small fundamental molecules. Furthermore, since the tunneleffect in predissociation and preassociation can be used to determine dissociation energies,
molecular formation and chemical reactions at low temperature, it is
necessary to have a mathematically rigorous theoretical treatment
that describes the appropriate dynamical features.

In this review we will employ Weyl's theory[2] for second-order,
ordinary linear differential equations (or coupled set of equations)

543

to study predissociation phenomena in some diatomic molecular systems.
It is important to note that we are particularly interested in the
analytical structure of the differential equation and its spectral
properties in relation to the interaction potentials. Although the
present study requires accurate numerical determination of potential
curves, our formulation may be considered as a first step towards a
more general treatment where the Born-Oppenheimer approximation may
no longer be valid. General analyticity requirements based on
dilatation analytic Hamiltonians[3,4] may offer a possibility for
further development of this method as has been indicated by some
preliminary studies.[5]

     In the following sections we will review and stress some of the
essential features of the Weyl-Titmarsh[2,6] m-function theory as well
as of the inverse spectral problem.[7] We will also discuss barrier
penetration problems, curve crossings between states of the same
symmetry, couplings between adiabatic curves of the Coriolis type,
spin interactions etc. in view of the analytic structure given by
Weyl's theory. Some earlier applications to HgH,[8] CH[9] SiO[5] as well
as recent experimental results on SiH,[10] NH,[11] OH[12] will be discussed.

II WEYL'S THEORY

     In addition to the classic 1910 thesis of Herman Weyl[2] and the
work of Titchmarsh[6] in which the m-coefficient was first explicitly
introduced, elementary formulations based on the circle method can
be found in Coddington and Levinson[13] chapter 9. For a more detailed
exposition of the relationship between the spectrum of the differential operator and analytic properties of the m-coefficient, see
Chaudhuri and Everitt.[14] For an excellent review of the present
status of the Weyl-Titmarsh theory see Everitt and Bennewitz.[15]

     We will briefly consider the symmetric, second-order equation

$$H\chi = -(p\chi')' + q\chi = \lambda\chi \tag{1}$$

in the interval $I = [a,\infty)$, $a \geq 0$. We assume p and q to be continuous
on I and $p > 0$ for $x\epsilon I$ and $\lambda = E+i\epsilon$, $\epsilon \neq 0$. Given the initial conditions

$$\varphi(a) = \sin\alpha \qquad p(a)\,\varphi'(a) = -\cos\alpha;$$

$$\psi(a) = \cos\alpha; \qquad p(a)\,\psi'(a) = \sin\alpha \tag{2}$$

where α could be chosen so that ψ is regular at the origin ($\psi(0)=0$),
a general solution χ, in particular $\chi\epsilon L^2(a,\infty)$, can be written as

$$\chi(x,\lambda) = \varphi(x,\lambda) + m(\lambda)\psi(x,\lambda) \tag{3}$$

The following theorem regarding the m-coefficient introduced above now holds:[15] m-theorem: Let the interval I, the coefficients p and q and the initial solutions $\varphi(\lambda)$ and $\psi(\lambda)$ be given as above; then there exists at least one pair of analytic functions (m_+, m_-) with the following properties

(i) m_+ maps the complex number fields C_+ onto itself,

i.e. $m_+: C_+ \to C_+$

(ii) m_+ belong to the class of Cauchy analytic functions

holomorphic in C_+, i.e. $m_+ \in H(C_+)$

(iii) $\chi(\lambda) = \varphi(\lambda) + m_+(\lambda)\psi(\lambda) \in L^2(a,\infty)$ $(\lambda \in C_+)$

in fact $\int_\infty^\infty |\chi(x,\lambda)|^2 dx = \dfrac{im[m_+(\lambda)]}{im[\lambda]}$

(iv) $[m_+(\lambda)]^* = m_+(\lambda^*)$ $(\lambda \in C_+)$

(v) either

(a) limit-circle (LC) at ∞, $\varphi(\lambda) \in L^2(a,\infty), \psi(\lambda) \in L^2(a,\infty)$

in which case (m_+, m_-) is not unique

or

(b) limit point (LP) at ∞, in which case m_+ is unique.

$\varphi(\lambda) \notin L^2(a,\infty)$ and $\psi(\lambda) \notin L^2(a,\infty)$ $(\lambda \in C_+ \cup C_-)$

The connection between the spectrum $\sigma(H)$ and $m(\lambda)$ follows by considering its associated Nevanlinna representation. The main result in the theory of Nevanlinna functions required for our purpose is: ρ-theorem: Let $m_+: C_+ \to C_+$ and $m_+ \in H(C_+)$. There exists then a uniquely determined function ρ, called the spectral function of m, with the properties

(i) $\rho: R \to R$ (R=real line) is monotonically increasing on R.

(ii) $\rho(E) = 1/2\,[\rho(E-0) + \rho(E+0)]$; $E \in R$ and $\rho(o) = 0$

(iii) $\int_{-\infty}^{+\infty} \dfrac{1}{E^2+1}\, d\rho(E) < \infty$

and uniquely determined real numbers β and γ with $\gamma > 0$ such that

$$m(\lambda) = \beta + \gamma\lambda + \int\limits_{-\infty}^{+\infty} [\frac{1}{E-\lambda} - \frac{E}{E^2+1}] \, d\rho(E) \qquad (\lambda\epsilon C_+) \qquad (4a)$$

where the integral is absolutely convergent. Eq (4a) reduces to

$$m(\lambda) = K + \int\limits_{-\infty}^{+\infty} \frac{d\rho(E)}{E-\lambda}, \quad \lambda\epsilon C_+ \qquad (4b)$$

where K is a constant (independent on λ), if

$$\int\limits_{1}^{\infty} \frac{im \, m_+(i\nu)}{\nu} \, d\nu < \infty \qquad (4c)$$

In particular, when m is derived from the differential equation (1) and satisfies the m-theorem, the Nevanlinna representation takes the form

$$m(\lambda) = -tg\alpha + \int\limits_{-\infty}^{+\infty} \frac{d\rho(E,\alpha)}{E-\lambda} \qquad (\lambda\epsilon C_{\pm}) \qquad (5)$$

We have here explicitly incorporated the dependence on the initial angle . The form (5) is given by Hille;[16] see also Wray.[17] From (5) it follows that $m(\lambda)$ is directly related to the resolvent

$$R(z) = (\lambda I - H)^{-1} \qquad (6)$$

see, e.g. reference 18.

Instead of working with the Nevanlinna form one may directly evaluate m from, see e.g. Titchmarsh[6] and also ref. 18, i.e.

$$m = \frac{\sin\alpha \, p(a)\chi'(a) + \cos\alpha \, \chi(a)}{\sin\alpha \, \chi(\alpha) - \cos\alpha \, p(a)\chi'(a)} \qquad (7)$$

or in terms of logarithmic derivatives

$$z(a) = p(a)\chi'(a)/\chi(a)$$
$$m = \frac{z(a)\sin\alpha + \cos\alpha}{\sin\alpha - z(a)\cos\alpha} \qquad (8)$$

In fact (7) and (8) are very convenient for numerical purposes, since the potentials used in our application are of such a simple form that the asymptotic behaviour of χ is known. Consequently, according to Poincare's theorem, $m_{\pm}(\lambda)$ must have a meromorphic continuation

into C+, but with the provision that the two functions m+ in C+ may
not be continuations of each other. The complex poles obtained from
one of the branches, let us say $m_+(\lambda)$, corresponds to a quasibound
state, i.e.

$$\varepsilon_r = E - i\, \frac{\Gamma}{2} \tag{a}$$

where the position is given by $E = \mathrm{Re}(\varepsilon_r)$ and the lifetime of the
state is $\tau = h/\Gamma$, with $\Gamma = 2\mathrm{Im}(\varepsilon)$. For more details on the numerical
determination of $m(\lambda)$ and its analytic structure we refer to referen-
ces 19, 8, 9b and 2o. For some preliminary studies on the interpretation of complex eigenvalues and time evolution with particular reference to Weyl's theory see ref. 21.

In the following sections, we will interpret experimental results from time resolved spectroscopy[1] by means of the analytic
behaviour of m_+ in C_-. In particular we will concentrate on the
appearance of complex poles of m_+ (in C_-) and its interpretation
in terms of associated quasistationary states.

III SOME COMMENTS ON THE INVERSE PROBLEM

The inverse problem is the problem of finding the potential q
from spectroscopic information. This is one of the key problems in
scattering theory, see e.g. the excellent monograph by Chadan and
Sabatier.[23] Rather than commenting upon the various procedures and
their improvements we will give the main theorem of inverse spectral
theory.[1] The problem can be formulated as follows: Do there exist
restrictions to be placed on the function ρ, defined in the ρ-theorem
or equivalently on the Nevanlinna function $m(\lambda)$ such that $m(\lambda)$ is
generated as the m-coefficient of eqs. (1), (2) and (3)? It turns
out that this question has a positive answer[7] and we will formulate
it as a theorem.

iρ-theorem: Let the non-decreasing function ρ: R→R satisfy the
conditions:

  (i) the set of points of R where ρ is strictly increasing has
      at least one finite limit-point

  (ii) $\int\limits_{-\infty}^{o} e^{x\sqrt{|E|}}\, d\rho(E)$ is convergent for all $x \varepsilon I$

  (iii) if τ: R→R is defined by

          $\tau(E) = \rho(E) - 2\sqrt{E/\pi}$ $E\varepsilon[o,\infty)$

                  $= \rho(E)$ $E\varepsilon(-\infty, o]$

then for all $x \epsilon I$ the integral

$$\int_{1}^{\infty} \frac{\cos x \sqrt{E}}{E} \, d\rho(E)$$

is convergent and represents a function which has four continuous derivatives on I;

then ρ is the spectral function of a uniquely determined boundary value problem given by (1) (with p=1) and the boundary condition

(a) $\sin -\chi'(a)\cos = o$;   (1o)

with a uniquely given potential q, real valued and continuous on I and $\alpha \epsilon(o,\pi)$.

Hence the function $m(\lambda)$ is generated as a m-coefficient of the differential equation (1) and the boundary condition (1o), if the spectral function $\rho(E)$ satisfies the conditions (i)-(iii).

In practice one constructs the potential from spectroscopic information. The actual connections have been reviewed by Wheeler.[24] For more details see also ref. 22 and 9. In the following we will illustrate the importance of the preceeding theory to the description of predissociating rovibronic levels in diatomic molecules.

IV THE RESULTS OF TIME RESOLVED SPECTROSCOPY IN THE LIGHT OF SPECTRAL THEORY

The theoretical description of the decay of excited levels in a diatomic molecule, as obtained in time resolved spectroscopy, can be divided into three categories, (i) the pure radiative decay in a free isolated molecule, (ii) the competition of the radiative decay and the nonradiative decay in a free molecule and (iii), the decay processes induced by the environment, such as electric and magnetic fields, collisions, stimulated emission etc. Presently we concentrate on the two first aspects of the problem. We use the nomenclature given in ref. 5.

The theoretically simplest cases of nonradiative decay are the processes that can be described as barrier penetrations. The predissociation by rotation in the $X^2\Sigma^+$ groundstate of, HgH[25] was orifinally reported and analysed by Hulthén.

Using massreduced quantum numbers and the spectroscopic information from HgH, HgD and HgT as well as scattering data, Stwalley[26] has constructed a very accurate potential energy curve for this ground state(see fig.1). He used considerations based on phase shift theory t

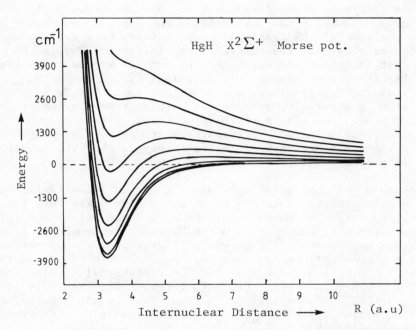

Fig. 1. Effective potentials for N = 0,5,10,15,20,25,30,35,40, between R = 2.0 and R = 11.0 a.u.

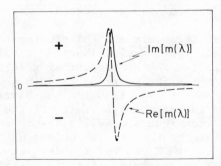

Fig. 2. Shape of complex m-function around the quasi-bound state E(o.3o), $5311 \leq E \leq 5315.5$ cm^{-1}. Note the peak of Im m(λ) and the sign change of Re m(λ) at the resonance energy.

determine the positions and widths of the quasi-stationary rovibrational levels. Hehenberger et al[8],[27] used Stwalley's potential
and Weyl's theory to accurately describe the continuous part of
the energy spectrum. In particular the complex poles associated
with the position of the quasi-stationaly states were determined.
These studies show that the numerical application of Weyl's theory
is both fast, accurate and reliable. The complex valued m-function
is calculated for a series of real energies. Fig. 2 shows the
Re $(m(E))$ and Im $(m(E))$ close to a pole. The fact that Re (m)
changes sign from positive to negative values as the real part of
the energy passes the pole position from below gives a computationally very reliable tool for finding sharp as well as very broad
resonances. Another strength of Weyl's theory is that the spectral
density is a directly computed quantity.

It turned out that Stwalley's HgH $X^2\Sigma^+$ potential actually contained more information than its constructor had looked for. Through
a careful analytic continuation with Padé approximants of the full
m-function from the real axis into the "second Riemannsheet" it
actually had more structure than was earlier known. The asymmetric
shape of the spectral density (see fig. 3) was shown to be due to the
presence of the very broad $v = 4$, $N = 9$ rovibrational levels. Phase
shift data could not resolve this structure. Weyl's theory is an
excellent method for studying the spectral density close to and above
the barrier maximum. The Weyl's theoretical results were later confirmed experimentally.

Since time resolved spectroscopy, in particular the High Frequency Deflection technique [1] can be used to determine predissociation
rates four orders of magnitude smaller than was possible with earlier
methods, it was of importance to compare the exact Weyl theoretical
treatment with currently used semiclassical method of LeRoy. The
$B^2\Sigma^-$ state of the CH radical was considered a good candidate.[9a]

It was found that the semiclassical treatment was accurate
enough for intermediate type levels. That is (within the relative
experimental error limits) the semiclassical results agree with the
Weyl theoretical estimates for energies clearly above the asymptotic
energy limits and of course clearly below the barrier maximum. The
resonances just above the asymptotic limit are the ones that can be
studied with the HFD technique and Weyl's theory is thus of importance for correctly interpreting the experimental data. An extensive
study of the $B^2\Sigma^-$ state of CH was later undertaken.[9b] The scope of
this investigation was to determine if it is possible to describe
the experimentally observed predissociations within the adiabatic
approximation. In other words, if there is a spectral function $\rho(E)$,
which includes the experimentally found eigenvalues, satisfying the
Gelfand-Levitan-Gasymov conditions[7],[28] (see above) then it is possible
to find an appropriate unique, potential. Fig. 4.

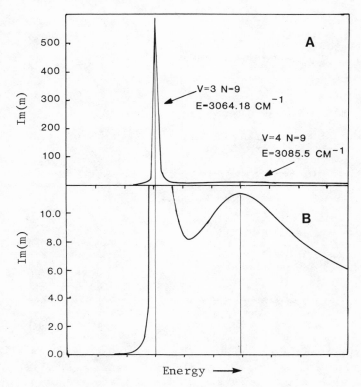

Fig. 3.   Im(m) for N = 9 associated with v = 3,4.
          The second plot is a magnification of the first

The results can be summarized as: Two sets of potential energy curves, one set for each spin component, are capable of describing the experimentally derived

   (i) energy levels and (ii) spin splitting for both bound and
        quasibound levels and
  (iii) the absolute as well as the relative nonradiative decay
        rates for the quasibound levels.

     All computed quantities were within the experimental error limits. The levels of $A^2\Delta$ v = 1, N = 17,18,19 of CH have a larger nonradiative lifetime than the adjacent rovibronic levels. These $A^2\Delta$ levels have the appropriate symmetry to interact with the $B^2\Sigma$ continuum. These adiabatic semiclassically constructed potential energy curves could however not be used to extrapolate $\rho(E)$ outside its region of construction. The calculated v = 1, N = 17,18,19 levels of the $B^2\Sigma$ states were about 3oo-4oo cm^{-1} too low to produce the observed $A^2\Delta$ -$X^2\Pi$ -$B^2\Sigma$ predissociation channel. Ab initio calculations by Lie, Hinze and Liu[29] indicate that the $B^2\Sigma$ potential barrier is a result of a crossing between the $C(^3P)+H^2(S)$ limit configuration

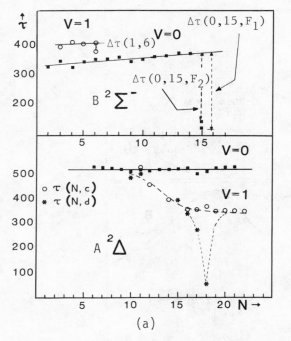
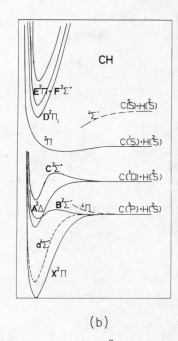

(a)   (b)

Figs. 4.   Experimentally observed (9a) lifetimes of the $B^2\Sigma^-$ (top)
           and the $A^2\Delta$ (bottom) rovibronic levels of the CH radical.
           The $v = 0, N = 15$ levels of the B state are predissociated.
           The difference between the assumed radiative lifetimes
           (unfilled circle) and the observed lifetimes (full square)
           are indicated as $\tau(0,15,F_1)$ and $\tau(0,15,F_2)$ for the two
           spin components, F_1 and F_2, respectively. The appropriate
           potential energy curves are found in Fig. 4b.

electronic state and an ionic electronic state dissociating into
$C^-(1s^2 2s^2 sp^3 \ ^2D)+H^+$. The observed $A^2\Delta-X^2\Pi-B^2\Sigma^-$ interaction is thus
better described within the diabatic framework.[30] In view of this
and several other problems the one-dimensional Weyl-Titchmarsh theory
has been generalized to a system of N coupled singular differential
equations.[31] A molecular system can be described with a system of N
coupled equations. This system has the general structure in matrix
form:

$$y'' = (V-\lambda \cdot I)y \tag{11}$$

Such a system of equations posses 2N linearly independent solutions
for any given parameter λ. In particular a set of linearly independent
initial-value solutions at a point $x = x_0$ can be defined by

$$\begin{bmatrix} \varphi \\ \varphi' \end{bmatrix}_{x=x_o} = \begin{bmatrix} I \\ 0 \end{bmatrix} \text{ and } \begin{bmatrix} \psi \\ \psi' \end{bmatrix}_{x=x_o} = \begin{bmatrix} 0 \\ I \end{bmatrix} \tag{12}$$

In analogy with the one-dimensional theory, the 2N x2N Z matrix can be constructed

$$Z(x) = \begin{bmatrix} \varphi & \psi \\ \varphi' & \psi' \end{bmatrix} (x) \tag{13}$$

In particular

$$Z(x_o) = \begin{bmatrix} I & 0 \\ 0 & I \end{bmatrix} \tag{14}$$

Continuing the generalization of the one-dimensional theory, the N general solutions are of the form

$$\chi = \varphi + \sum_{k=1}^{N} \psi M_{jk} \tag{15}$$

or in matrix form

$$\begin{bmatrix} \chi \\ \chi' \end{bmatrix} = Z \begin{bmatrix} I \\ M \end{bmatrix} \tag{16}$$

The eigenvalue problem can be evaluated by defining the logderivative matrix

$$R = y' \cdot y^{-1} \tag{17}$$

and solving the related Riccati equation

$$R' = (V - \lambda \cdot I) - R^2 \tag{18}$$

Assuming the equations to be uncoupled at the origin and at infinity (18) can be integrated from some point a, sufficiently close to the origin, to x_o to give the logderivative matrix R_a and from infinity to x_o to give R_b. The eigenvalues of (18) are given by the zeros of

$$R_a(x_o) - R_b(x_o) = M_a - M_b \tag{19}$$

The full $A^2\Delta - \chi^2\Pi - B^2\Sigma^-$ interaction describing the $A^2\Delta$ predissociation can in this fashion be represented by a set of coupled differential equations. Predissociations experimentally found in SiH;[12] NH;[11] OH,[12a] OD[12b] as well as the photophragment spectroscopy results on O_2^{+32} can be analysed with the same approach.

Outlook for the future

The present work is carried on along two parallel lines. The form of the spectral density for N coupled equations is currently not known even though the positions of the resonances are given by (19). The ansatz of analytic continuation is not theoretically satisfactory, since it involves an additional approximation. By using complex rotation of Riccati's equation with complex boundary conditions the integration can be carried out right through the complex pole.[33] Numerical work along these lines is currently going on.

REFERENCES

1.   P. Erman, Specialist Periodic Reports, Chem. Soc. London 6, 5o1 (1979).
2.   H. Weyl, Math. Ann. 68, 22o (191o).
3.   E. Balslev and J. Combes, Commun. Math. Phys. 22, 28o (1971).
4.   B. Simon, Ann. Math., 97, 247 (1973).
5.   E. Brändas, N. Elander and P. Froelich, Int. J. Quant. Chem. 14, 443 (1978).
6.   E.C. Titchmarsh, Eigenfunction Expansions Associated with Second Order Differential Equations (Clarendon, Oxford 1946; 1962) Vol. I, (1958), Vol. II.
7.   I.M. Gelfand and B.M. Levitan, Izv. Akad Nauk SSSR 15, 3o9 (1951) (Transl. Amer. Math. Soc. (2) 1, 253).
8.   M. Hehenberger, B. Laskowski and E. Brändas, J. Chem. Phys. 65, 4559 (1976).
9a.   J. Brzozowski, P. Bunker, N. Elander and P. Erman, Astrophys. J. 2o4, 414 (1976).
9b.   N. Elander, M. Hehenberger and P.R. Bunker, Physica Scripta 2o, 631 (1979).
1o.   T.A. Carlson, N. Durić, P. Erman and M. Larsson, J. Phys. B 11, 3667 (1978).
11.   W. Hayden-Smith, J. Brzozowski and P. Erman, J. Chem. Phys. 64, 4628 (1976).
12.   J. Brzozowski, P. Erman and M. Lyyra, Physica Scripta 17, 5o7 (1978).
13.   E.A. Coddington and N. Levinson, Theory of ordinary differential equations (McGraw-Hill, New York, 1955).
14.   J. Chaudhuri and W.N. Everitt, Proc. Royal Soc. Edinburgh (A) 68, 95 (1969).
15.   W.N. Everitt and G. Bennewitz, in Tribute to Ake Pleijel, Proceedings of the conference, September 1979, Dept. of Mathematics, University of Uppsala, Uppsala Sweden, pg.49.
16.   E. Hille, Lectures on ordinary differential equations (Addison-Wesley, London, 1969).
17.   S.D. Wray, Proc. Royal Soc. Edinburgh (A) 14, 41 (1974/75).
18.   E. Brändas, M. Hehenberger and H.V. McIntosh, Int. J. Quant. Chem. 9, 1o3 (1975).

19. E. Brändas and M. Hehenberger, in Lecture Notes in Math. 415, 316 (Springer-Verlag, Heidelberg, 1974).

2o. E. Brändas, Proceedings of the Bielefeld Symposium on Numerical Integration of Differential Equations, April 198o (to be published, Springer Verlag).

21. E. Brändas, Physica 82A 97 (1976).

22. N. Elander, Dissertation, University of Stockholm, Faculty of Mathematics and Natural Science (1977).

23. K. Chadan and P.C. Sabatier, Inverse Problems in Quantum Scattering Theory, (Springer Verlag, 1977).

24. J.A. Wheeler, in Studies in Math. Physics. Essays in Honor of Valentine Bergmann (ed. E.H. Lieb, B. Simon and A.S. Weightman), Princeton Series in Physics, Princeton University Press, Princeton, N.J., 1976.

25. E. Hulthén, Z. Physik. 37, 32 (1925).

26. W.J. Stwalley, J. Chem. Phys. 63, 3o62 (1975).

27. M. Hehenberger, P. Froelich and E. Brändas, J. Chem. Phys. 65, 4571 (1976).

28. M.G. Gasimov and B.M. Levitan, Uspekhi Math. Math. Nauk 19, 3 (1964).

29. G.C. Lie, J. Hinze and B.J. Liu, J. Chem. Phys. 59, 1872 (1973).

3o. T.F. O'Malley, Adv. in Atomic and Molecular Phys., eds. D.R. Bates and I. Esterman, Vol. 7, p. 223 (Acad. Press, New York, 1971).

31. M. Hehenberger, E. Brändas and N. Elander, Int. J. Quant. Chem. Symp. S 12, 67 (1978).

32. A. Carrington, P.G. Roberts and P. Sarre, Mol. Phys. 35, 1523 (1978).

33. M. Rittby, N. Elander and E. Brändas (to be published).

DYNAMICAL EQUATIONS FOR THE WIGNER FUNCTIONS

Jens Peder Dahl

Department of Chemical Physics
Technical University of Denmark
DTH 301, DK-2800 Lyngby, Denmark

1. INTRODUCTION

The motion of a classical particle is conveniently described
as taking place in phase space, i.e., the direct product of configuration space and momentum space. The wavefunction associated
with a quantum mechanical particle is, on the other hand, a quantity
in configuration space *or* momentum space, and phase space might only
seem to be of interest in cases where the uncertainty relation between position and momentum may be ignored.

It is, nevertheless, possible to introduce an exact phase space
description of a quantum mechanical particle, and such a description
has some attractive characteristics. As a consequence, there exists
a considerable literature on the subject. A fairly comprehensible
introduction has been given by de Groot and Suttorp,[1] the first
formulations being due to Weyl[2] and Wigner.[3]

A distinctive feature of the phase space formulation of
quantum mechanics, as it has been developed over the years, is that
it treats states and transitions in an equivalent manner. Thus,
there is a phase space function associated with every quantum state,
and with every quantum transition as well. This function is the
celebrated Wigner function.

The Wigner function for a given state or transition is usually
generated from wavefunctions in configuration space or momentum
space. It is, however, important to realize that the phase space
description is a description in its own right. As discussed in
detail elsewhere[4] it has a strong group theoretical basis. The
quantum mechanical significance of a point in phase space is *not*

that it defines a simultaneous position and momentum of a particle – this would of course be inconsistent with the laws of quantum mechanics – it is instead that each such point defines an inversion operator which is a bona-fide observable in Dirac's sense.

In the present contribution we shall focus the attention upon the dynamical equations which Wigner functions must satisfy for pure states and transitions between such states. These equations are differential equations in phase space and can hardly be claimed to be new. With a single and incomplete exception, however, it does not seem to have been realized that one may sometimes solve these equations and thus obtain the Wigner functions in a direct and in- dependent manner. This aspect will, accordingly, be the main theme of our discussion.

To make the presentation reasonably self-contained we present some of the main features of the phase space representation of quantum mechanics in the following section. Section 3 is devoted to the formulation of the dynamical equations. These are then solved for three simple cases in Sections 4-6. Section 7 contains our conclusions and reference to further work.

2.   THE WEYL-WIGNER CORRESPONDENCE

The phase space formulation of quantum mechanics is based on the Weyl-Wigner correspondence between operators in Hilbert space and ordinary functions on phase space. Let $\vec{q} = (q_1, q_2, \ldots, q_f)$ and $\vec{p} = (p_1, p_2, \ldots, p_f)$ be the cartesian coordinates and momenta, respectively, associated with a system of particles. Similarly, let $\vec{Q} = (Q_1, Q_2, \ldots, Q_f)$ and $\vec{P} = (P_1, P_2, \ldots, P_f)$ be the corresponding quantum mechanical operators, satisfying Heisenberg's commutation relations

$$[Q_i, P_j] = i \hbar \, \delta_{ij} \quad , \tag{1}$$

where, as usual, $\hbar = h/2\pi$. With each point $(\vec{p}, \vec{q})$ of phase space we may then associate an operator,

$$\Pi(\vec{p}, \vec{q}) = \left(\frac{1}{2h}\right)^f \iint d\vec{u} \; d\vec{v} \; \exp[\frac{i}{\hbar} (\vec{q} \cdot \vec{u} + \vec{p} \cdot \vec{v})]$$

$$\times \exp[-\frac{i}{\hbar} (\vec{Q} \cdot \vec{u} + \vec{P} \cdot \vec{v})] \quad , \tag{2}$$

which is hermitian and unitary, i.e.,

$$\Pi(\vec{p}, \vec{q})^\dagger = \Pi(\vec{p}, \vec{q}) \quad , \tag{3}$$

$$\Pi(\vec{p}, \vec{q})^2 = 1 \quad . \tag{4}$$

The operator $\Pi(\vec{p}, \vec{q})$ is, as first noted by Grossmann[5] and Royer,[6] the parity operator with respect to the phase space point $(\vec{p},\vec{q})$. As pointed out by the present author[4] it is an observable in Dirac's sense and may be taken as the basic operator in the phase space formulation of quantum mechanics. The main features are the following (for references and further discussion, see[4]).

Let $\mathcal{H}_1$ be the Hilbert space associated with our system of particles and $\mathcal{H}_2$ the space of linear operators acting on the vectors of $\mathcal{H}_1$. An operator $A \in \mathcal{H}_2$ may then be expressed in the form

$$A = \left(\frac{2}{h}\right)^f \iint d\vec{p} \; d\vec{q} \; a(\vec{p},\vec{q}) \; \Pi(\vec{p},\vec{q}) \quad . \tag{5}$$

The function $a(\vec{p},\vec{q})$ is, per definition, the Weyl transform of the operator A.

If a, b and c are the Weyl transforms of A, B and C respectively, and $C = AB$, then we have the following expression for c[1]:

$$c(\vec{p},\vec{q}) = \exp\left[\frac{i\hbar}{2}\left(\frac{\partial}{\partial\vec{q}_1} \cdot \frac{\partial}{\partial\vec{p}_2} - \frac{\partial}{\partial\vec{p}_1} \cdot \frac{\partial}{\partial\vec{q}_2}\right)\right] a(\vec{p},\vec{q}) \; b(\vec{p},\vec{q}) \; . \tag{6}$$

The subscript 1 on a differential operator indicates that this operator only acts on the first function in the product $a(\vec{p},\vec{q}) \cdot b(\vec{p},\vec{q})$. Similarly, the subscript 2 is used with operators which only act on the second function in the product.

The functions c_+ and c_- corresponding to the operators

$$C_+ = \tfrac{1}{2}(AB + BA) \tag{7}$$

and

$$C_- = \frac{1}{i\hbar}(AB - BA) \tag{8}$$

are clearly given by the expressions

$$c_+(\vec{p},\vec{q}) = \cos\left[\frac{\hbar}{2}\left(\frac{\partial}{\partial\vec{q}_1} \cdot \frac{\partial}{\partial\vec{p}_2} - \frac{\partial}{\partial\vec{p}_1} \cdot \frac{\partial}{\partial\vec{q}_2}\right)\right] a(\vec{p},\vec{q}) \; b(\vec{p},\vec{q}) \tag{9}$$

and

$$c_-(\vec{p},\vec{q}) = \frac{2}{\hbar}\sin\left[\frac{\hbar}{2}\left(\frac{\partial}{\partial\vec{q}_1} \cdot \frac{\partial}{\partial\vec{p}_2} - \frac{\partial}{\partial\vec{p}_1} \cdot \frac{\partial}{\partial\vec{q}_2}\right)\right] a(\vec{p},\vec{q})b(\vec{p},\vec{q}). \tag{10}$$

With each pair of vectors, $|\psi_i\rangle$ and $|\psi_j\rangle$, in $\mathcal{H}_1$ we may associate an operator $|\psi_j\rangle\langle\psi_i|$ which for $i=j$ is a projection operator, for $i \neq j$ a transition operator. The Weyl transform of this operator is a constant, h^f, times the Wigner function $\Gamma_{ij}(\vec{p},\vec{q})$, which in terms of coordinate wavefunctions $\langle\vec{q}|\psi\rangle$ or momentum wavefunctions $\langle \vec{p}|\psi\rangle$ is given by the well-known expressions

$$\Gamma_{ij}(\vec{p},\vec{q}) = \left(\frac{2}{h}\right)^f \int d\vec{q}{}' \; \langle \vec{q}+\vec{q}{}'|\psi_j\rangle\langle\psi_i|\vec{q}-\vec{q}{}'\rangle \; \exp\left(-\frac{2i}{\hbar}\,\vec{p}\cdot\vec{q}{}'\right) \quad (11)$$

and

$$\Gamma_{ij}(\vec{p},\vec{q}) = \left(\frac{2}{h}\right)^f \int d\vec{p}{}' \; \langle \vec{p}+\vec{p}{}'|\psi_j\rangle\langle\psi_i|\vec{p}-\vec{p}{}'\rangle \; \exp\left(\frac{2i}{\hbar}\,\vec{p}{}'\cdot\vec{q}\right) \quad , \quad (12)$$

For the sake of simplicity we have suppressed the time dependence of Γ_{ij}, stemming from the time dependence of $|\psi_i\rangle$ and $|\psi_j\rangle$.

We note the following properties of the Wigner functions:

$$\Gamma_{ji}(\vec{p},\vec{q}) = \Gamma_{ij}(\vec{p},\vec{q})^* \quad , \tag{13}$$

$$\iint d\vec{p}\; d\vec{q}\; \Gamma_{ij}(\vec{p},\vec{q}) = \langle\psi_i|\psi_j\rangle \quad , \tag{14}$$

$$\int d\vec{p}\; \Gamma_{ij}(\vec{p},\vec{q}) = \langle\psi_i|\vec{q}\rangle\langle\vec{q}|\psi_j\rangle \quad , \tag{15}$$

$$\int d\vec{q}\; \Gamma_{ij}(\vec{p},\vec{q}) = \langle\psi_i|\vec{p}\rangle\langle\vec{p}|\psi_j\rangle \quad , \tag{16}$$

and further that

$$\langle\psi_i|A|\psi_j\rangle = \iint d\vec{p}\; d\vec{q}\; \Gamma_{ij}(\vec{p},\vec{q})\; a(\vec{p},\vec{q}) \quad , \tag{17}$$

with $a(\vec{p},\vec{q})$ being the Weyl transform of the operator A.

These properties show that the Wigner functions may be treated as distributions. These distributions are in general complex valued when $i \neq j$, but real when $i=j$. But although $\Gamma_{ii}(\vec{p},\vec{q})$ is a real function it is generally not a non-negative function, and hence it is not a probability density. This is in accordance with the fact that the uncertainty principle precludes the existence of a proper probability density on phase space. The number $\Gamma_{ij}(\vec{p},\vec{q})$ associated with the phase space point $(\vec{p},\vec{q})$ has, nevertheless, a well defined meaning. It is the matrix element of the parity operator $\Pi(\vec{p},\vec{q})$:

$$\Gamma_{ij}(\vec{p},\vec{q}) = \left(\frac{2}{h}\right)^f \langle\psi_i|\Pi(\vec{p},\vec{q})|\psi_j\rangle \quad . \tag{18}$$

The number $\Gamma_{ii}(\vec{p},\vec{q})$ is, in particular, proportional to the expectation value of the "observable" $\Pi(\vec{p},\vec{q})$ in the quantum state $|\psi_i\rangle$.

3.   DYNAMICAL EQUATIONS FOR THE WIGNER FUNCTIONS

Let us now assume that the time development of our system is governed by the time-independent Hamiltonian H, with the Weyl transform $h(\vec{p},\vec{q})$. The time-dependent Schrödinger equation gives immediately that

$$i\hbar \frac{\partial}{\partial t} \left(|\psi_j\rangle\langle\psi_i|\right) = H|\psi_j\rangle\langle\psi_i| - |\psi_j\rangle\langle\psi_i|H \quad , \tag{19}$$

and hence, according to (10), and with the time dependence explicitly indicated:

$$\frac{\partial}{\partial t} \Gamma_{ij}(\vec{p},\vec{q};t) = \frac{2}{\hbar} \sin\left[\frac{\hbar}{2}\left(\frac{\partial}{\partial \vec{q}_1} \cdot \frac{\partial}{\partial \vec{p}_2} - \frac{\partial}{\partial \vec{p}_1} \cdot \frac{\partial}{\partial \vec{q}_2}\right)\right]$$

$$\cdot h(\vec{p},\vec{q})\, \Gamma_{ij}(\vec{p},\vec{q};t) \quad .$$

(20)

This is the well-known equation of motion for the Wigner function $\Gamma_{ij}(\vec{p},\vec{q};t)$, first discussed by Wigner[3] and Moyal.[7]

Let us further assume that $|\psi_i\rangle$ and $|\psi_j\rangle$ are eigenvectors of the hermitian operator A with the (real) eigenvalues α_i and α_j respectively. We then have the equations

$$A|\psi_j\rangle\langle\psi_i| + |\psi_j\rangle\langle\psi_i|A = (\alpha_i+\alpha_j)|\psi_j\rangle\langle\psi_i|$$

(21)

and

$$A|\psi_j\rangle\langle\psi_i| - |\psi_j\rangle\langle\psi_i|A = (\alpha_j-\alpha_i)|\psi_j\rangle\langle\psi_i| \quad ,$$

(22)

and hence, by observing (9) and (10):

$$\cos\left[\frac{\hbar}{2}\left(\frac{\partial}{\partial \vec{q}_1} \cdot \frac{\partial}{\partial \vec{p}_2} - \frac{\partial}{\partial \vec{p}_1} \cdot \frac{\partial}{\partial \vec{q}_2}\right)\right] a(\vec{p},\vec{q})\, \Gamma_{ij}(\vec{p},\vec{q};t)$$

$$= \frac{\alpha_i+\alpha_j}{2}\, \Gamma_{ij}(\vec{p},\vec{q};t)$$

(23)

and

$$\sin\left[\frac{\hbar}{2}\left(\frac{\partial}{\partial \vec{q}_1} \cdot \frac{\partial}{\partial \vec{p}_2} - \frac{\partial}{\partial \vec{p}_1} \cdot \frac{\partial}{\partial \vec{q}_2}\right)\right] a(\vec{p},\vec{q})\, \Gamma_{ij}(\vec{p},\vec{q};t)$$

$$= \frac{\alpha_j-\alpha_i}{2i}\, \Gamma_{ij}(\vec{p},\vec{q};t) \quad .$$

(24)

Taken together, these two equations define an eigenvalue problem for the determination of $\Gamma_{ij}(\vec{p},\vec{q};t)$. Although this may have been fairly obvious to several workers in the field it does not seem to have been explicitly discussed in the literature. A partial exception is the work by Uhlhorn[8], to be mentioned later.

A situation of particular importance arises when $|\psi_i\rangle$ and $|\psi_j\rangle$ are eigenstates of the hamiltonian, with eigenvalues E_i and E_j respectively. We get then, according to (19):

$$i\hbar \frac{\partial}{\partial t} \Gamma_{ij}(\vec{p},\vec{q};t) = (E_j-E_i)\, \Gamma_{ij}(\vec{p},\vec{q};t) \quad ,$$

(25)

and hence:

$$\Gamma_{ij}(\vec{p},\vec{q};t) = f_{ij}(\vec{p},\vec{q}) \exp[\frac{i}{\hbar}(E_i-E_j)t] \quad . \tag{26}$$

This time dependence also follows directly from (11) and (12). It is obvious that $f_{ij}(\vec{p},\vec{q})$ is the Wigner function at t=0 or, equivalently, that f_{ij} is obtained by substituting the solutions of the time independent Schrödinger equation in (11) and (12).

We may now replace $\Gamma_{ij}(\vec{p},\vec{q};t)$ in (23) and (24) with $f_{ij}(\vec{p},\vec{q})$, provided that the operator A commutes with the hamiltonian.

In what follows we shall limit ourselves to systems with a single degree of freedom (f=1). Thus we consider the one-dimensional motion of a single particle with mass m and the quantum mechanical hamiltonian

$$H = \frac{P^2}{2m} + V(Q) \quad . \tag{27}$$

The Weyl transform of H is the corresponding classical hamiltonian

$$h(p,q) = \frac{p^2}{2m} + V(q) \quad . \tag{28}$$

Choosing A in (21) and (22) as the hamiltonian we obtain, from (23) and (24):

$$\left\{\frac{p^2}{2m} + V(q) - \frac{\hbar^2}{2m}\frac{\partial^2}{\partial q^2} - \frac{\hbar^2}{2}\frac{d^2V}{dq^2}\frac{\partial^2}{\partial p^2}\right\} f_{ij}(p,q)$$

$$\tag{29}$$

$$+ \sum_{r=4,6,\ldots} \frac{1}{r!}\left(\frac{i\hbar}{2}\right)^r \frac{d^rV}{dq^r}\frac{\partial^r}{\partial p^r} f_{ij}(p,q) = \tfrac{1}{2}(E_i+E_j) f_{ij}(p,q)$$

and

$$\left(-\frac{p}{m}\frac{\partial}{\partial q} + \frac{dV}{dq}\frac{\partial}{\partial p}\right) f_{ij}(p,q) + \sum_{r=3,5,\ldots} \frac{1}{r!}\left(\frac{i\hbar}{2}\right)^{r-1}\frac{d^rV}{dq^r}\frac{\partial^r}{\partial p^r} f_{ij}(p,q)$$

$$\tag{30}$$

$$= \frac{i}{\hbar}(E_i-E_j) f_{ij}(p,q) \quad .$$

By solving these two simultaneous equations we can, at least in principle, determine $f_{ij}(p,q)$. We shall here determine the solutions for a potential of the form

$$V(q) = A + Bq + Cq^2 \quad , \tag{31}$$

with A, B and C being constants. This is in a certain sense the "classical" case, since (20), the equation of motion, now becomes the Liouville equation of classical statistical mechanics.[3]

We must distinguish between three cases, corresponding to a harmonic, a linear and a constant potential, respectively.

4. THE HARMONIC POTENTIAL

With a suitable choice of origin and zero of energy we may eliminate A and B whenever $C \neq 0$. Thus we consider the potential

$$V(q) = \tfrac{1}{2} m \omega^2 q^2 \quad , \tag{32}$$

which is the potential for a harmonic oscillator with the angular frequency ω. Eqns. (29) and (30) become:

$$\left\{ \frac{p^2}{2m} + \tfrac{1}{2} m \omega^2 q^2 - \frac{\hbar^2}{2m} \frac{\partial^2}{\partial q^2} - \frac{\hbar^2}{2} m \omega^2 \frac{\partial^2}{\partial p^2} \right\} f_{ij}(p,q)$$

$$= \tfrac{1}{2} (E_i + E_j) f_{ij}(p,q) \tag{33}$$

and

$$\left(- \frac{p}{m} \frac{\partial}{\partial q} + m \omega^2 q \frac{\partial}{\partial p} \right) f_{ij}(p,q) = \frac{i}{\hbar} (E_i - E_j) f_{ij}(p,q) \quad . \tag{34}$$

By suitably scaling p and q, Eqn. (33) becomes the Schrödinger equation for a two-dimensional, isotropic harmonic oscillator, with its angular momentum determined by (34). The corresponding eigenfunctions are well known.[9] It is, however, illuminating to obtain the solutions as follows.

Imagine phase space filled with a fluid of virtual particles moving in accordance with the laws of classical mechanics. The velocity of a particle at (p,q) is then

$$(\dot{p}, \dot{q}) = \left(- \frac{dV}{dq} , \frac{p}{m} \right) \quad . \tag{35}$$

This velocity is independent of time, i.e., the flow is steady, the stream-lines being the classical phase space trajectories. The points on a trajectory may be distinguished by giving the dynamical time $\tau(p,q)$, i.e., the time at which a chosen virtual particle on the trajectory through (p,q) is found at that point.

We may now perform a transformation to new canonical variables

$$\sigma = \frac{p^2}{2m} + V(q) \quad , \tag{36}$$

$$\tau = \tau(p,q) \quad , \tag{37}$$

with the new momentum σ being identical with the hamiltonian $h(p,q)$ and the new coordinate being the dynamical time. We have then, according to (35):

$$\frac{\partial}{\partial \tau} = - \frac{dV}{dq} \frac{\partial}{\partial p} + \frac{p}{m} \frac{\partial}{\partial q} \quad , \tag{38}$$

and hence (34) takes the form

$$\frac{\partial}{\partial \tau} f_{ij}(\sigma,\tau) = - \frac{i}{\hbar} (E_i - E_j) f_{ij}(\sigma,\tau) \quad . \tag{39}$$

Solving this equation gives immediately

$$f_{ij}(\sigma,\tau) = \exp[- \frac{i}{\hbar}(E_i - E_j)\tau] g_{ij}(\sigma) \quad . \tag{40}$$

The phase space trajectories corresponding to the potential (32) are ellipses, and each virtual particle executes a periodic motion with the angular frequency ω. Adding $2\pi/\omega$ to the dynamical time must accordingly amount to transforming each phase space point into itself. The requirement that $f_{ij}(p,q)$ be single-valued gives, therefore, the following condition:

$$E_i - E_j = n_{ij} \hbar \omega , \quad n_{ij} = 0, \pm 1, \pm 2, \ldots \tag{41}$$

This condition shows that the harmonic oscillator is quantized such that its energy levels form an equidistant sequence, the distance between two consecutive levels being $\hbar\omega$.

We shall return to Eqn. (41) and its implications in Section 7. Here we proceed by inserting (41) into (40) to obtain:

$$f_{ij}(\sigma,\tau) = \exp(- i n_{ij} \omega \tau) g_{ij}(\sigma) \quad . \tag{42}$$

It follows from the solution to the classical problem that

$$\tau(p,q) = - \frac{1}{\omega} \tan^{-1} \left(\frac{p}{m\omega q}\right) \quad , \tag{43}$$

for a convenient choice of zero dynamical time. By using this result together with (36) and (32) it is readily seen that

$$\frac{1}{m} \frac{\partial^2}{\partial q^2} + m\omega^2 \frac{\partial^2}{\partial p^2} = 2\omega^2 \left(\frac{\partial}{\partial \sigma} + \sigma \frac{\partial^2}{\partial \sigma^2}\right) + \frac{1}{2\sigma} \frac{\partial^2}{\partial \tau^2} \quad . \tag{44}$$

When this, together with (42), is inserted into (33) we find that $g_{ij}(\sigma)$ must satisfy the following equation:

$$[- \frac{\hbar^2}{8} (2\omega^2 \frac{d}{d\sigma} + 2\omega^2 \sigma \frac{d^2}{d\sigma^2} - \frac{1}{2\sigma} \omega^2 n_{ij}^2)$$
$$+ \sigma - \frac{1}{2}(E_i + E_j)] g_{ij}(\sigma) = 0 \quad . \tag{45}$$

In addition, $g_{ij}(\sigma)$ must be finite everywhere, as (18) shows.

From the discussion following Eqn. (41) it is clear that we may write:

$$E_i = E_o + n_i \hbar \omega \quad , \quad E_j = E_o + n_j \hbar \omega \quad , \tag{46}$$

where n_i and n_j are positive integers or zero, and

$$n_{ij} = n_i - n_j \quad . \tag{47}$$

It is then straightforward to solve Eqn. (45). We find:

$$E_o = \frac{1}{2} \hbar \omega \quad , \tag{48}$$

and

$$g_{ij}(\sigma) = N_{ij} \exp(- \frac{s}{2}) s^{\frac{1}{2}|n_i - n_j|} L_n^{|n_i - n_j|}(s) \quad , \tag{49}$$

where n is the larger of n_i and n_j. N_{ij} is a constant, and

$$s = \frac{4\sigma}{\hbar\omega} \quad . \tag{50}$$

$L_n^m(s)$ is the associated Laguerre polynomial:

$$L_n^m(s) = \frac{d^m}{ds^m} L_n(s) \quad , \tag{51}$$

$$L_n(s) = e^s \frac{d^n}{ds^n} (s^n e^{-s}) \quad . \tag{52}$$

Combining (26), (40) and (49) gives finally:

$$\Gamma_{ij}(p,q;t) = N_{ij} \exp(- \frac{s}{2}) s^{\frac{1}{2}|n_i - n_j|} L_n^{|n_i - n_j|}(s)$$
$$\times \exp[i(n_i - n_j) \omega (t-\tau)] \quad , \tag{53}$$

with τ and s defined by (43) and (50), respectively.

To determine the constant N_{ij} we recall the discussion between Eqns. (10) and (17), and assume that the state vectors are normalized to unity so that

$$\langle \psi_i | \psi_j \rangle = \delta_{ij} \quad . \tag{54}$$

N_{ii} may then be determined from the requirement (14), which gives:

$$\iint dp \, dq \, \Gamma_{ii}(p,q) = 1 \quad , \tag{55}$$

and N_{ij} may be determined by using (17), with $A = |\psi_i\rangle\langle\psi_j|$:

$$\iint dp \, dq \, |\Gamma_{ij}(p,q)|^2 = h^{-1} \quad . \tag{56}$$

N_{ij} contains, for $i \neq j$, an arbitrary phase factor, whose value must be assigned by some consistent phase convention. Thus we get:

$$N_{ij} = \frac{2}{h} \left[\frac{n_j!}{(n_i!)^3} \right]^{\frac{1}{2}} e^{i\delta_{ij}} \quad , \tag{57}$$

where δ_{ij} is real.

This completes our determination of the Wigner functions from the dynamical equations (33) and (34). The result is in accordance with that obtained by other methods, based on a direct application of (11)[10] or (18).[4] With the usual phase choice of the eigenstates of the harmonic oscillator we have

$$N_{ij} = (-1)^{n_i} \frac{2}{h} \left[\frac{n_j!}{(n_i!)^3} \right]^{\frac{1}{2}} \quad . \tag{58}$$

The full method of the present section has, as far as we know, not been discussed or pursued earlier. It must, however, be pointed out that Uhlhorn,[8] in a very interesting paper, did present the diagonal ($i=j$) forms of Eqns. (33) and (45), and from them determined the solution $f_{ii}(p,q)$.

The energy level of the harmonic oscillator are discrete. In the next two sections we shall consider situations in which the energy eigenvalues are continuously distributed.

5.   THE LINEAR POTENTIAL

With C in Eqn. (31) being zero, while $B \neq 0$, we may eliminate A by a suitable choice of origin or zero of energy. We are thus led to consider the potential

$$V(q) = Fq \quad , \tag{59}$$

where F is a constant. This is, e.g., the potential for a charged particle in a constant electric field.

The classical phase space trajectories are now parabolas, and the new canonical variables become

$$\sigma = \frac{p^2}{2m} + Fq \quad , \tag{60}$$

$$\tau = -\frac{p}{F} \quad . \tag{61}$$

Eqns. (38)-(40) retain their form, but there is no quantisation condition corresponding to (41), since the domain of τ is from $-\infty$ to $+\infty$. The equivalent of (45) becomes

$$\left[-\frac{\hbar^2 F^2}{8m} \frac{d^2}{d\sigma^2} + \sigma - \tfrac{1}{2}(E_i + E_j) \right] g_{ij}(\sigma) = 0 \quad . \tag{62}$$

The solution which is everywhere finite is

$$g_{ij}(\sigma) = N_{ij} \; Ai \left[\left(\frac{8m}{\hbar^2 F^2} \right)^{1/3} \left(\sigma - \frac{E_i + E_j}{2} \right) \right] \quad , \tag{63}$$

where Ai (z) is the Airy function.[11]

Hence we get:

$$\Gamma_{ij}(p,q;t) = N_{ij} \; Ai \left[\left(\frac{8m}{\hbar^2 F^2} \right)^{1/3} \left(\sigma - \frac{E_i + E_j}{2} \right) \right]$$
$$\times \exp\left[\frac{i}{\hbar} (E_i - E_j)(t - \tau) \right] \quad , \tag{64}$$

with σ and τ as defined by (60) and (61), respectively.

To determine N_{ij} we assume that the vectors $|\psi_i\rangle$ and $|\psi_j\rangle$ are normalized such that

$$\langle \psi_i | \psi_j \rangle = \delta(E_i - E_j) \quad . \tag{65}$$

This gives, according to (14), the requirement

$$\iint dp \; dq \; \Gamma_{ij}(p,q;t) = \delta(E_i - E_j) \quad . \tag{66}$$

Since the transformation defined by (60) and (61) is canonical, we have that $dpdq = d\sigma d\tau$. Using that

$$\int_{-\infty}^{\infty} Ai(x)dx = 1 \quad , \tag{67}$$

it is then easily found that

$$N_{ij} = \frac{1}{h} \left(\frac{8m}{\hbar^2 F^2} \right)^{1/3} e^{i\delta_{ij}} \quad , \tag{68}$$

where δ_{ij} is real and tends to zero as $E_i \to E_j$. It is evidently possible to choose the phase factor of $|\psi_i\rangle$, considered as a function of E_i, such that $\delta_{ij} = 0$ always. This choice corresponds, in fact, to the phase factors usually encountered in the literature for the vectors $|\psi_i\rangle$, when these are expressed in the coordinate[12] or the momentum[12,13] representation. This may be seen by actually evaluating the integrals in (11) and (12).

The general expression (64) is apparently new, but the diagonal ($i=j$) form has previously been derived by Heller[13] by means of (12). His expression agrees with ours, except that the argument of the Airy function has the opposite sign. This seems to be due to a spurious minus sign that has slipped in between his Eqns. (A5) and (A6).

It is apparent from the present section, that the linear potential represents a case where the solution of the dynamical equations is the simpler and more transparent way to obtain the Wigner functions.

6.   THE CONSTANT POTENTIAL

When both B and C in Eqn. (31) are zero we may eliminate A by a suitable choice for the zero of energy, to obtain

$$V(q) = 0 \quad . \tag{69}$$

The classical phase space trajectories are straight lines parallel to the q-axis, and the canonical variables corresponding to (36) and (37) become:

$$\sigma = \frac{p^2}{2m} \quad , \tag{70}$$

$$\tau = \frac{mq}{p} \quad . \tag{71}$$

We see, however, that each (σ, τ) set determines two phase space points, and hence the use of σ and τ as basic variables is not straightforward. The complication arises, of course, because there are two trajectories for each non-zero value of the energy, one above and one below the q-axis.

To deal with this degeneracy we must construct the dynamical

equations for an operator which commutes with the hamiltonian. The
operator to choose is clearly the momentum operator P.

We assume, then, that $|\psi_i\rangle$ is an eigenvector for P with the
eigenvalue π_i, and for H with the eigenvalue

$$E_i = \frac{\pi_i^2}{2m} \quad . \tag{72}$$

A similar assumption is made for $|\psi_j\rangle$. Eqn. (26) is then still
valid, and Eqns. (23) and (24) become, with $a(p,q) = p$:

$$(p - \frac{\pi_i + \pi_j}{2})\, f_{ij}(p,q) = 0 \quad , \tag{73}$$

$$\frac{\partial}{\partial q}\, f_{ij}(p,q) = \frac{i}{\hbar}\, (\pi_j - \pi_i)\, f_{ij}(p,q) \quad . \tag{74}$$

These equations are readily solved, and we get that

$$\Gamma_{ij}(p,q;t) = \frac{1}{h}\, \delta(p - \frac{\pi_i + \pi_j}{2})\, \exp[\frac{i}{\hbar}\, (\pi_j - \pi_i)q]$$
$$\times \exp[\frac{i}{\hbar}\, (E_i - E_j)t] \quad . \tag{75}$$

The normalization constant has here been chosen such that

$$\iint dp\; dq\; \Gamma_{ij}(p,q;t) = \delta(\pi_i - \pi_j) \quad , \tag{76}$$

which, according to (14), imply that $|\psi_i\rangle$ and $|\psi_j\rangle$ are normalized by
the delta function of momentum, i.e.

$$\langle\psi_i|\psi_j\rangle = \delta(\pi_i - \pi_j) \quad . \tag{77}$$

The usual phase factor has in (75) been put equal to 1. This
amounts to working with coordinate wavefunctions of the form (at
t=0):

$$\langle q|\psi_i\rangle = h^{-\frac{1}{2}}\, \exp(\frac{i}{\hbar}\, \pi_i\, q + i\gamma) \quad , \tag{78}$$

where γ is an arbitrary, real number which is the same for all
$|\psi_i\rangle$.

The function (75) will of course also satisfy the dynamical
equations (29) and (30), which now become

$$[\frac{p^2}{2m} - \frac{\hbar^2}{2m}\frac{\partial^2}{\partial q^2} - \frac{1}{2}(E_i+E_j)] \, f_{ij}(p,q) = 0 \quad , \tag{79}$$

$$-\frac{p}{m}\frac{\partial}{\partial q} f_{ij}(p,q) = \frac{i}{\hbar} (E_i-E_j) \, f_{ij}(p,q) \quad . \tag{80}$$

But since (73) and (74) already determine f_{ij} completely, these equations give nothing new.

In closing this section we make the obvious remark, that the Wigner functions corresponding to eigenvectors of H which are not eigenvectors of P are linear combinations of functions of the type (75).

7.   CONCLUSIONS AND DISCUSSION

The point we have wanted to stress in the present contribution is, that the phase space representation of quantum mechanics is a representation in its own right. We have illustrated this by presenting a set of dynamical equations for the Wigner function and by showing that these equations may actually be solved in some simple cases.

The classical phase space trajectories appear in these cases as the natural carriers of the Wigner function, not only for stationary states, but also for the transitions between such states. This is a result of great interest in the semiclassical limit, as discussed elsewhere.[14] The point may be illustrated by Eqn. (41), which was derived, not by considering the stationary states, but by studying the transitions directly. The equation states, that the transition frequency $(E_i-E_j)/\hbar$ must be a multiple of the frequency associated with the classical orbit. This condition is, in the semiclassical limit, retained for more general potentials, and thus the study of Wigner functions leads to a direct derivation of the correspondence principle.[14] The trajectory carrying the transition function corresponds to the energy $(E_i+E_j)/2$.

The dynamical equations can probably not be solved exactly where more general potentials than (31) are involved. The equation of motion, as given by (20), can now no longer be associated with the flow of virtual particles in a precise sense, and hence the role played by the classical trajectories becomes more unclear. It is, nevertheless, well known that the Wigner function associated with states generally tend to favorize the classical trajectories.[13,15] The situation is, however, rather complex when more degrees of freedom are involved,[16] and there seems to be a clear need for a closer study of Wigner functions for actual systems. Such work is now being carried out at this laboratory, with a focus on the description of genuine atomic and molecular states.[17]

The great importance which trajectory calculations have for our present understanding of molecular dynamics is well illustrated in accompanying papers by Heller, Marcus and Rice.

REFERENCES

1. S.R. DeGroot and L.G. Suttorp, "Foundations of Electrodynamics," North-Holland Publishing Company, Amsterdam (1972).
2. H. Weyl, "The Theory of Groups and Quantum Mechanics," Dover, New York (1931).
3. E. Wigner, Phys.Rev. 40, 749 (1932).
4. J.P. Dahl, Physica Scripta, in press.
5. A. Grossmann, Commun.math.Phys. 48, 191 (1976).
6. A. Royer, Phys.Rev. A 15, 449 (1977).
7. E.J. Moyal, Proc.Cambridge Phil.Soc. 45, 99 (1949).
8: U. Uhlhorn, Arkiv för Fysik 11, 87 (1956).
9. L. Pauling and E.B. Wilson, "Introduction to Quantum Mechanics," McGraw-Hill, New York (1935).
10. H.J. Groenewold, Physica 12, 405 (1946).
11. M. Abramowitz and I.A. Stegun, "Handbook of Mathematical Functions," U.S. National Bureau of Standards, Washington D.C. (1964).
12. L.D. Landau and E.M. Lifshitz, "Quantum Mechanics," Pergamon Press, London (1958).
13. E.J. Heller, J.Chem.Phys. 67, 3339 (1977).
14. J.P. Dahl, to be published.
15. M.V. Berry, Phil.Trans.Roy.Soc. London 287, 237 (1977).
16. S.A. Rice, in "Quantum Dynamics of Molecules," ed. R.G. Woolley, Plenum Press, New York (1980).
17. J.P. Dahl and M. Springborg, to be published.

VIBRATIONAL MOTION IN THE REGULAR AND CHAOTIC REGIMES, CLASSICAL
AND QUANTUM MECHANICS

R. A. Marcus

Department of Chemistry
California Institute of Technology
Pasadena, California 91125

INTRODUCTION

There are many experiments which are related to intramolecular
energy transfer. Initially, they involved primarily thermal unimolecular reactions[1] and chemical activation.[2] More recently,
experiments which are more state-selective have also been performed
in which one zeroth order mode (e.g., some bond or normal mode) has
been excited. These experiments include infrared multiphoton
absorption,[3] reactions induced by excitation of high CH overtones,[4]
dissociation of vibrational state-excited van der Waals' complexes,[5]
and fluorescence spectra from mode-selected vibrationally-excited
substituted aromatic molecules.[6]

Paralleling the experiments has been considerable theoretical
interest in the subject of intramolecular energy transfer
('randomization'). In the present article some recent studies of
our group on semiclassical and quantum behavior of states of simple
systems are described. These studies were initially for systems
with few degrees of freedom, but related concepts are expected to
apply to larger systems. A more detailed review of this work and
of that in other laboratories will appear elsewhere.[7]

CLASSICAL MECHANICS

There have been major developments in the classical mechanics
of coupled anharmonic oscillators (and other coordinates) and so
applicable to molecular vibrations. In particular, Poincaré specu-
lated around the turn of the century that there were three regimes
for the motion of classical mechanical system--one where the
classical motion is largely 'regular', one where the motion is both

573

regular, or for different initial conditions, chaotic, and the third
where the motion is largely chaotic.[8]

More recently, in the middle 1950's and early 1960's the famous
Kolmogorov-Arnold-Moser theorem (KAM) was developed.[9] It was shown
there that at very small values of the perturbation parameter (small
perturbation from an 'integrable' system), the motion is 'regular'
for almost all initial conditions: i.e., for almost all initial
conditions it has m good action variables for a system with m
coordinates. Numerically calculated trajectories for coupled
anharmonic oscillators have revealed that most initial conditions
yield this regular motion even at larger perturbations, but that at
still larger ones (or at still higher energies) the classical motion
becomes chaotic (or, as it is sometimes called, stochastic, ergodic,
irregular) for an increasing fraction of initial conditions.[10]

As a consequence of having these m good action variables,
regular motion is expressible in a Fourier series having at most m
fundamental frequencies, plus combinations and overtones, and is,
thereby, 'quasi-periodic'.[9] A test for finding if the motion is
quasi-periodic or if it is, instead, chaotic is given by numerical
(computer) experiments in a variety of ways, such as linear instead
of exponential separation (in time) of two neighboring trajectories,[11]
the pattern of the Poincaré surfaces of section discussed later,[10]
and the power spectrum for the trajectory,[12] which has lines corresponding to the above fundamentals, overtones and combinations for
a quasi-periodic motion, or has, instead, 'broadened' bands (many
lines) for chaotic motion.[12]

The implications of a quasi-periodic behavior for chemistry,
in those cases where classical mechanics is a useful first approximation, are several fold. For example, a quasi-periodic trajectory
winds itself around a manifold of m-dimensions in a 2m dimensional
phase space, namely around a 'torus'. A chaotic trajectory tends,
instead, to occupy all of phase space that is not preempted by the
tori and that is consistent with the given energy and angular
momentum. Thus, only in the chaotic case does the time average of
any property calculated from the trajectory tend to become approximately equal to the microcanonical phase space average for the
given angular momentum. The restriction, instead, to an m-
dimensional torus in a 2m-dimensional phase space involves a considerable confinement of the quasi-periodic trajectory in this
phase space. Different theories of unimolecular reactions have
been based on these two extremes.[7] Again, the regularity of the
vibrational spectrum is expected to more readily permit a coherent
absorption of infrared photons than the chaotic case, with its
more complicated spectrum. Distinguishing a chaotic spectrum from
a quasi-periodic one in the case of molecules with many coordinates
may nevertheless be difficult.

Classical quasi-periodic states of motion are of two kinds:[7] those which do not extensively mix the energies of some choice of zeroth order modes and those which do, though in a nonchaotic way. These two types of quasi-periodic trajectories are obtained using, for example, a Hamiltonian frequently employed in nonlinear dynamics studies:

$$H = \tfrac{1}{2}(p_x^2 + p_y^2 + \omega_x^2 x^2 + \omega_y^2 y^2) + \lambda x(xy + \eta x^2) \quad . \tag{1}$$

Hamilton's equations of motion are integrated numerically. In Figure 1 y(t) is plotted vs. x(t) for the case where ω_x and ω_y are incommensurate.[13,14] An example is given in Figure 2 for the case[15] of $\omega_x = \omega_y$ and in Figure 3 for the case[16] of $\omega_x = 2\omega_y$. All of these trajectories are quasi-periodic.

There is clearly a considerable difference between Figure 1 on the one hand and Figures 2 and 3 on the other. In Figure 1 the amplitude of the y-motion is approximately independent of the x-coordinate and vice versa. I.e., in this quasi-periodic state of the system, there is relatively little mixing of the energy of the zeroth order modes (the x and y modes) during the motion, because the unperturbed frequencies are incommensurate. In contrast, in Figure 2, which has a 1:1 resonance, the motion is sometimes along the x-direction, sometimes along the y-direction, and sometimes in between. I.e., even in this quasi-periodic state there is extensive modal energy mixing. Figure 3 is intermediate; there is a higher order resonance, and so the system has some but a more limited modal energy mixing than that in Figure 2. Thus, modal energy mixing does not itself imply classical chaos, since it does occur in quasi-periodic states--those with zeroth order internal resonances--as well as in chaotic states. Indeed, for the system in Figure 2 the distribution of the internal angular momentum

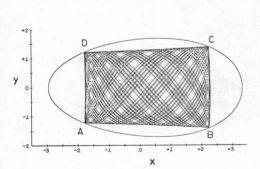

Fig. 1. A quasi-periodic trajectory for Eq. (1) with ω_x
   and ω_y incommensurate.

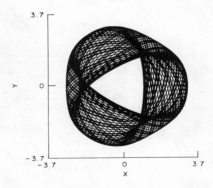

Fig. 2. A quasi-periodic trajectory for Eq. (1) with $\omega_y =$
   ω_y.

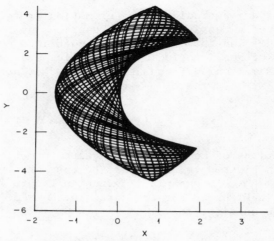

Fig. 3. A quasi-periodic trajectory for Eq. (1)
with $\omega_x = 2\omega_y$. Plot of y(t) vs. x(t).

$(yp_x - xp_y)$ is quite nonchaotic. This point has frequently been
overlooked in the literature. An example of a chaotic trajectory[16]
for a 2:1 resonant system is given in Figure 4. The trajectory is
trying to occupy all of the energetically accessible phase space.
There is none of the structure of caustics evident in Figures 1-3.

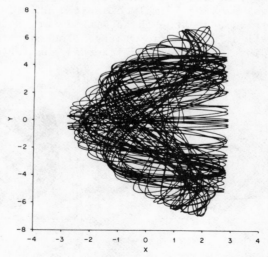

Fig. 4. A chaotic trajectory for Eq. (1)
with $\omega_x = 2\omega_y$. Plot of y(t) vs. x(t).

SEMICLASSICAL MECHANICS

　　We turn next to the corresponding semiclassical version of the
quantum mechanical behavior. Using semiclassical arguments it has
been possible to calculate from classical mechanics the quantum
mechanical energy eigenvalues. The first successful technique for
smooth potentials[13] employed exact classical trajectories. The
integrals in the left hand side of the semiclassical Eq. (2) were
evaluated for coupled anharmonic oscillators.

$$\oint_{C_i} \sum_j p_j \, dq_j = (n_i + \tfrac{1}{2})h \qquad (i = 1 \text{ to } m) \; . \qquad (2)$$

$\oint$ is a cyclic integral (an integral over a closed path); j goes from
1 to m. For a system of m oscillators there are m topologically
independent closed paths C_i, and there are m integers n_i. The left
hand side of Eq. (2) is the i'th action variable. We evaluated each
of these action variables for m = 2 by choosing the C_i to be the
caustics (e.g., the boundaries AB and back, and BC and back, in
Figure 1). The initial conditions, including the energy, were then
adjusted until the two n_i's calculated in Eq. (2) after evaluation of
the left hand sides (m = 2) were integers. The method gave good
agreement with the quantum mechanically calculated eigenvalues[13] of
the Hamiltonian in Eq. (1).

　　This method was then extended by choosing each C_i to be a curve
(a series of points) appearing in a Poincaré surface of section.[14]
In a system of two coordinates (x,y) one plots p_y vs. y for a trajectory each time a trajectory passes some value of x, e.g., x = 0,
with $p_x > 0$. A typical plot for several trajectories is given in
Figure 5. The area under a curve for one of them gives one of the

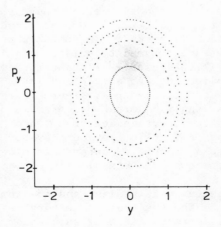

Fig. 5. Example of a surface of section at x = 0 for several
　　trajectories, each typically of the form in Figure 1.

integrals over C_i, while the plot of p_x vs. x at some y gives another
topologically independent integral. Once again, good agreement was
obtained for the eigenvalues both with the first method and with the
quantum mechanical ones.

The integral in Eq. (2) for Figure 4 is independent of the
plane (x = a constant) chosen for the surface of section, as long
as this plane cuts the same caustics (e.g., AB and DC in Figure 1),
and an analogous remark applies to the integral for the other
Poincaré surface of section. The method has since been extended to
treat systems with zeroth order internal resonances $\omega_1 : \omega_2$ = 1:1, 1:2,
by introducing curvilinear coordinates.[17] The method cannot be used
in the classically chaotic regime: the surface of section has a
'shotgun' pattern there. Indeed, the good action variables in the
left hand side of Eq. (2) do not appear to exist there.

The method has been supplemented by a variety of approximate
classical techniques, perturbative, perturbative-iterative and
variational (reviewed in Ref. 7). In general these methods and the
exact trajectory method are complimentary: the former (at least
methods involving iteration) require more complicated computer programs but, at least for polynomial potentials, are computationally
faster than the trajectory method. On the other hand, the small
divisor problem causes perturbation methods ultimately to diverge.[18]
The trajectory method provides direct information on the shape of
the tori, information which can also be used to introduce appropriate coordinate systems for use with the approximate methods.
Approximate techniques are essential for treating large systems,
but problems of convergence may appear.

These methods are, like the exact trajectory method, strictly
applicable for calculating quantum mechanical eigenvalues only when
good action variables, the left hand sides of Eq. (2), actually
exist. In the classically chaotic regime they appear not to exist,
except for isolated residual families of tori, and some additional
method or approximation is needed. In the perturbative-type methods
good action variables are assumed to exist, even in the chaotic
regime, and Eq. (2) is then used.[19] Thus far, however, this method
has not yet been applied to what are called later in this paper
quantum mechanically chaotic systems. There is reason to believe
that these approximations will break down for such systems,[7]
although they may capture certain features. Another possible
method, one involving families of periodic systems,[20] has not yet
been applied to systems of coupled anharmonic oscillators in either
the classically or the quantum mechanically chaotic regimes.

Comparisons between semiclassical and quantum mechanical results
have also been made by the spectral trajectory method, and in the
classically quasi-periodic regime they show good agreement for both
the positions of the spectral lines and their intensities

(fundamentals and overtones) when use is made of the correspondence principle.[12] Once again, much less is understood about the relationship between exact quantum and classical spectra in the classically chaotic regime. Examples of the spectrum for the quasi-periodic and the chaotic trajectories in Figures 3 and 4 are given in Figures 6 and 7, respectively.[21]

QUANTUM MECHANICS

We have compared above the eigenvalues and the spectra, semi-classical with quantum. One can also compare a wavefunction with the trajectory which, by obeying Eq. (2), corresponds semi-classically to it. The quantum mechanical wavefunction ψ corresponding to Figure 3 is given, in the form of $|\psi|^2$, in Figure 8.[16] The base of the former has the same shape as the region occupied by the trajectory for this 2:1 Fermi resonant system.

The relation between classical and quantum behavior in the classically chaotic regime is only beginning to be understood. For example, we studied, for a particular value of λ, the 'Henon-Héiles' system (Eq. (1), with $\omega_x = \omega_y = 1$ and $\eta = -1/3$). We found that the quantum mechanical spectrum reflected regular spacings in the sequences of eigenvalues, even when the classical spectrum was 'chaotic', in the classically chaotic energy regime.[22] Thus, classical chaos is not a sufficient condition for quantum chaos (cf. also Ref. 23). The occurrence of classical chaos arises, according to Chirikov's and Ford's theory,[24] when there are

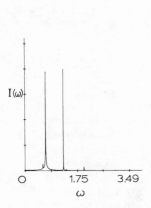

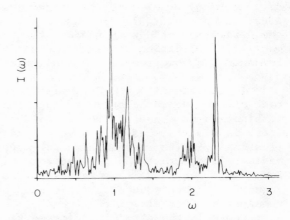

Fig. 6. Spectrum of a trajectory similar to that in Figure 1.

Fig. 7. Spectrum for the trajectory in Figure 4.

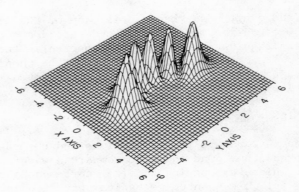

Fig. 8.   A wavefunction in the quasiperiodic region,
corresponding to the trajectory in Figure 3.

overlapping internal classical resonances. Each resonance has a
center and a 'width'. When these widths overlap classical chaos
begins.[24] However, when each width contains essentially no quantum
states, each classical resonance has no quantum mechanical counter-
part, and so the overlap has none either.[25,26] Only when the
spacing of the quantum states in the region near the resonant centers
is small enough will, by this argument, classical chaos imply quantum
chaos for a quantum state. Conditions which can be tested have been
suggested.[25,26] Some test of Ref. 25 has been made by Kay.[23]

   Information about quantum mechanical resonances, which are
expected to cause quantum mechanical chaos (e.g., quite uneven
spacings in eigenvalue sequences), particularly when the resonances
'overlap', is obtained from plots[22] of eigenvalues versus perturbation parameter λ. When two such eigenvalue curves undergo an
avoided crossing near some λ, a frequency (an energy difference
divided by h) has become nearly zero, and one has an analog of a classical resonance.[26] In a classical resonance, some frequency, i.e.,
the derivative of H with respect to some action variable, nearly
vanishes, reflecting a near commensurability of the frequencies of
fundamentals, overtones, or combinations.

   In the vicinity of the avoided crossing of two eigenvalue plots
each of the two associated wavefunctions is approximately a linear
combination of those of the 'uncrossed' curves. When a particular
state is simultaneously involved in many such 'overlapping avoided
crossings',[27] its wavefunction has a very complicated set of
maxima, in contrast to the high regular set depicted in Figure 8.
Further, the eigenvalue spacings can no longer be resolved into
regular sets of sequences, i.e., the spectrum is now chaotic, even
though the eigenvalue differences which constitute the spectrum are
near those in the earlier regime where the spacings were more even.
The state can be termed quantum mechanically chaotic.[26]

For some systems quantum chaos, described as above, and classical chaos begin at about the same energy,[28],[29] i.e., the additional
conditions needed for quantum chaos are apparently satisfied, and for
others, as in the Henon-Héiles system mentioned earlier, only
classical chaos was observed at the perturbation parameter studied.[22]

One approximate method for detecting the onset of quantum chaos
was given recently and involves the use of perturbation theory to
predict avoided crossings.[29] Plots were made of eigenvalues, calculated perturbatively, versus a perturbation parameter λ. Such plots
yield crossings instead of avoided crossings. To obtain the avoided
crossings one can use degenerate perturbation theory in the vicinity
of each crossing. When the avoided crossings near some λ become
extensive at some energy, that energy is the predicted value for the
onset of quantum chaos.

We turn finally to the eigenvalue spectrum expected for quantum
mechanically 'integrable' systems and for systems near them in some
sense.[30] A classical dynamical system of m coordinates is defined
as integrable,[31] i.e., as having m good action variables, if there
exist m functions F_i in phase space for which the Poisson bracket
$\{F_i, F_j\}$ vanishes, with $F_1 = H$. The quantum mechanical analog of
this definition is obtained, it has been suggested,[30] using the well-
known correspondence[32] between classical and quantum mechanics in
which the classical Poisson brackets are replaced by commutators and
dynamical variables by Hermetian operators. Thus, one might define
an m-coordinate quantum mechanical system as being 'integrable' if
for it there exist m operators $\underline{F}_i$ with $[\underline{F}_i, \underline{F}_j] = 0$ and $\underline{F}_1 = \underline{H}$.[30]
Thus, for such systems, a complete set of commuting observables
actually exists, whereas for most quantum mechanical systems such a
set does not exist, according to this analogy. (In almost all
classical systems there do not exist F_i's for which $\{F_i, F_j\} = 0$ and
$F_1 = H$ for all initial conditions.)

Examples of these integrable systems are those which permit
separation of variables and another[33] for which variables may not
yet have been separated but for which the F_i's have been found with
$\{F_i, F_j\} = 0$. Indeed, Liouville has apparently shown[34] that given
these F_i's with the property that $\{F_i, F_j\} = 0$, and $F_1 = H$, the
classical problem can be reduced to quadratures, i.e., is
'integrable'. (Nowadays, this property has been taken as a definition of integrability.[31])

A consequence of this quantum mechanical integrability, i.e.,
of there being m commuting operators, is that the latter yield a
set of m quantum numbers. They, in turn, give rise to a regular
series of spectral lines and nodal patterns of wavefunctions.
Similar conclusions on the wavefunctions were made earlier from
actual pictures of them.[35] The latter picture is further

supported by the semiclassically corresponding trajectory, which fills the same spatial region in which the eigenfunction is concentrated (e.g., cf. Figures 3 and 8). Conjectures on spectra had also been made earlier by analogy with the classical behavior.[36]

In Hamiltonian systems which are close to these integrable systems, i.e., for which the perturbation parameter (or because of scaling the energy) is small enough, it was suggested that qualitatively similar properties of the spectra and wavefunctions (regularly spaced sequences of eigenvalues and more or less regular nodal patterns) might occur.[30] Deviations from these patterns of spectra and nodal surfaces will arrise when avoided crossings (absent in integrable systems) occur. When a quantum state is involved simultaneously in many 'overlapping avoided crossings' it has a 'statistical' (highly delocalized) wavefunction.[27]

We have differentiated between modal and nonmodal energy mixing quasi-periodic states. For some observables it should not matter if instead modal energy mixing has occurred as a result of chaotic behavior. For other observables, those directly connected with the study of sufficiently highly resolved spectra, there would be a difference between the two types of behavior. Some implications of classical and quantum results in this paper for experiments are discussed elsewhere.[7]

This research was supported in part by the National Science Foundation. I am particularly indebted to my co-workers, Don Noid and Michael Koszykowski, for many helpful discussions. This article is Contribution No. 6381 from the California Institute of Technology.

REFERENCES

1. P. J. Robinson and K. A. Holbrook, "Unimolecular Reactions," John Wiley, New York (1972); W. Forst, "Unimolecular Reactions," Academic Press, New York (1973).
2. A. N. Ko, B. S. Rabinovitch, and K. J. Chao, J. Chem. Phys. 66, 1374 (1977); J. M. Farrar and Y. T. Lee, ACS Symp. Ser. 66, 191 (1978).
3. P. A.Schulz, Aa. S. Sudbø, D. J. Krajnovich, H. S. Kwok, Y. R. Shen, and Y. T. Lee, Ann. Rev. Phys. Chem. 30, 379 (1979).
4. K. V. Reddy and M. J. Berry, Faraday Disc. Chem. Soc. 67, 188 (1979); B. D. Cannon and F. F. Crim, J. Chem. Phys. 73, 3013 (1980).
5. D. H. Levy, Ann. Rev. Phys. Chem. 31, 197 (1980).
6. J. B. Hopkins, D. W. Powers, and R. E. Smalley, J. Chem. Phys. 73 683 (1980) and references cited therein; R. A. Coveleskie, D. A. Dolson, and C. S. Parameter, J. Chem. Phys. 72, 5774 (1980).
7. D. W. Noid, M. L. Koszykowski, and R. A. Marcus, "Quasi-Periodic and Stochastic Behavior in Molecules," Ann. Rev. Phys. Chem. 32, 000 (1981).

8. H. Poincare, "New Methods of Celestial Mechanics," (Transl.
    NASA, Washington, D.C. 1957) (1897).

9. V. I. Arnol'd, "Mathematical Methods of Classical Mechanics,"
    Springer-Verlag, New York (1978).

10. J. Ford, "Fundamental Problems in Statistical Mechanics,"
    E. G. D. Cohen, editor, Vol. 3, North Holland, Amsterdam (1975).

11. G. Benettin, L. Galgani, and J. M. Strelcyn, Phys. Rev. A14,
    2338 (1976).

12. D. W. Noid, M. L. Koszykowski, and R. A. Marcus, J. Chem. Phys.
    67, 404 (1977); M. L. Koszykowski, D. W. Noid, and R. A.
    Marcus, to be published.

13. W. Eastes and R. A. Marcus, J. Chem. Phys. 61, 4301 (1974).

14. D. W. Noid and R. A. Marcus, J. Chem. Phys. 62, 2119 (1975).

15. D. W. Noid and R. A. Marcus, J. Chem. Phys. 67, 559 (1977).

16. D. W. Noid, M. L. Koszykowski, and R. A. Marcus, J. Chem.
    Phys. 71, 2864 (1979).

17. D. W. Noid, M. L. Koszykowski, and R. A. Marcus, J. Chem.
    Phys. 73, 391 (1980).

18. C. L. Segal, Ann. Math, 42, 806 (1941); C. L. Segal, Math.
    Ann. 128, 144 (1945); C. L. Segal and J. K. Moser, "Lectures
    on Celestial Mechanics," Springer-Verlag, Berlin (1971).

19. R. T. Swimm and J. B. Delos, J. Chem. Phys. 71, 1706 (1979);
    C. Jaffe, Ph.D. Thesis, University of Colorado, Boulder (1979).

20. M. C. Gutzwiller, Phys. Rev. Lett. 45, 150 (1980).

21. D. W. Noid, M. L. Koszykowski, and R. A. Marcus, J. Chem. Ed.
    57, 624 (1980).

22. D. W. Noid, M. L. Koszykowski, M. Tabor, and R. A. Marcus,
    J. Chem. Phys. 72, 6169 (1980).

23. K. G. Kay, J. Chem. Phys. 72, 5955 (1980).

24. B. V. Chirikov, E. Keil, and A. M. Sessler, J. Stat. Phys.
    3, 307 (1971); G. H. Walker and J. Ford, Phys. Rev. 188, 416
    (1969); J. Ford and G. H. Lunsford, Phys. Rev. A1, 59 (1970).

25. R. A. Marcus, Ann. N.Y. Acad. Sci. 357, 159 (1980).

26. E. V. Shuryak, Sov. Phys. JETP 44, 1070 (1977).

27. R. A. Marcus, "Horizons in Quantum Chemistry," K. Fukui and
    B. Pullman, editors, Reidel, Dordrecht (1980) p. 127.

28. D. W. Noid, M. L. Koszykowski, and R. A. Marcus, Chem. Phys.
    Lett. 73, 269 (1980).

29. R. Ramaswamy and R. A. Marcus, "Perturbative Examination of
    Avoided Crossings," J. Chem. Phys. 74, 000 (1981).

30. R. Ramaswamy and R. A. Marcus, unpublished.

31. J. Moser, "Topics in Nonlinear Dynamics," AIP Conf. Proc.,
    No. 46, S. Jorna, editor, Am. Inst. Phys., New York (1978), p. 1.

32. P. A. M. Dirac, "The Principles of Quantum Mechanics," Oxford
    University Press, London, 4th ed. (1958).

33. M. C. Gutzwiller, Ann. Phys. N. Y. 124, 347 (1980).

34. Ref. 9, p. 272.

35. D. W. Noid, M. L. Koszykowski, J. D. McDonald, and R. A.
    Marcus, unpublished.

36. I. C. Percival, J. Phys. B. 6, 1229 (1973).

TRAJECTORY CALCULATIONS AND COMPLEX COLLISIONS

Christoph G. Schlier

Fakultät für Physik der Universität Freiburg

D 7800 Freiburg, FRG

INTRODUCTION

The theoretical treatment of energy scrambling in molecules is generally one of bound systems, even if many of the experiments by which these theories can be tested, are experiments on states which undergo unimolecular decay. The argument, that a small leakage in phase space will not change the overall behaviour of the molecular motion, be it quasiperiodic or stochastic*, is certainly a sound one if only this decay is improbable on a time scale, which is typical for the rest of the motion.

In this talk I want to discuss the situation when this approximation can no longer be taken for granted. Specifically, I will discuss energy randomization in a colliding system A+BC, which is further specialized in our computations to the simple reactive system $H^{+}+D_2$ and its isotopic variants. Many general statements are, however, not limited to triatomic systems. The overall situation is pictured schematically in Fig. 1. One should, however, be aware of the true situation which is multidimensional. At very low energies the KAM theorem[1,2] assures us that the motion is quasiperiodic. Then, near some critical energy E_{cl}, a "stochastic transition" occurs, and energy begins to flow freely between the degrees of freedom. At the dissociation limit nothing special happens, except that dissociation out of the random motion takes place with some (small) probability.

---

* We use "stochastic" as a general term to comprise "ergodic", "mixing" and the like.[2,3] Many references, e.g. ref. 1, call this "ergodic" in a loose sense.

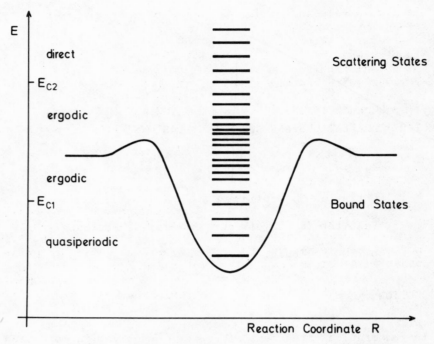

Fig. 1: Schematic diagram of quasiperiodic, stochastic, and direct
         scattering regimes. The wall may be the centrifugal wall if
         no other exists.

If one prepares this situation in a low energy scattering experiment,
the projectile will (with some probability) be trapped by the target,
forming a long-lived scattering complex, in which one expects energy
randomization to take place.

    At still higher energies, however, the stochastic regime dies
out in our computations exponentially and one is left with simple
direct scattering events, where energy is transferred but in no way
randomized. One can argue that the lifetime of the collision complex
has become too short for efficient energy scrambling.

    In our work we try to shed some light on what happens in this
transition of complex to direct scattering. The computational method
is classical trajectory calculation, and we choose the system H_3^+ not
only because accurate energy surfaces exist,[4-6] but also because
there are precise experiments.[7,8] These experiments show that in low-
energy collisions every channel is populated, as soon as it is open,
and suggest that the product distributions can be described by some
statistical theory.

The plan of the talk is as follows. In the first part I will discuss some problems, which arise if one tries to transfer the discussion of energy scrambling from bound to unbound systems. Then I will discuss, what we actually can calculate by trajectories. The main conceptual problem here is to define what one counts as a long-lived complex while the practical one is to keep integration errors low. In the final part I will sketch how, nevertheless, trajectory calculations can help us to define what one may characterize as the "most dynamically biased" statistical model for a complex forming collision.

ENERGY SCRAMBLING IN COLLISIONS

The usual way of talking about molecular motion is to consider the motion as a flow of trajectories in phase space. A group of trajectories is characterized at any fixed time by a subset of phase space points, say A, and the flow is depicted as a measure preserving transformation $A \to U_t A$. Since energy conservation holds, phase space must be restricted to the energy shell and made measurable thereon. Other conservation laws (e.g. of total angular momentum J) must be treated similarly. If the system cannot dissociate, the available phase space is compact and has finite Liouville measure. Mixing motion can then be defined by saying that in the limit of infinite time the flow $U_t A$ distributes "randomly", e.g. in an uncorrelated and equal manner into any given allowed subset of phase space, B, i.e.

$$\lim_{t \to \infty} \mu(U_t A \cap B) = \mu(A) \cdot \mu(B) \tag{1}$$

Trying to apply this to collisions one hits upon several problems:

(1) Phase space is no longer finite, i.e. measurable by the Liouville measure, and it is not clear, whether other measures preserved unter U_t exist, which could replace it.

(2) One might think of restricting phase space further, say, by restricting eq. (1) to a sphere of large radius in configuration space. This sphere is then measurable but mixing certainly will not hold, because there are trajectories corresponding to glancing collisions, which do not mix. We will show below that this is even true, when you take the centrifugal wall of the colliding ion-molecule system as a boundary: There exist non-mixing collisions transgressing the centrifugal barrier.[9]

(3) The trajectories taken within such a boundary are not infinitely long, such that the limit in eq. (1) cannot be taken meaningfully. Even weaker definitions will be restricted by the finite number of recurrences of a trajectory to the neighborhood of any point within the complex. As a consequence we can never talk of

"mixing" or "ergodicity" in the mathematical sense, where it is allowed to make the "coarse-graining", which salvages from determinism, as fine as one wishes.

I am mentioning these points, not because I have solutions to offer in the mathematical sense. As far as I can see few papers treat "ergodicity" for systems with a non-compact phase space, in which escape is possible. A notable exception is the restricted 3-body problem[10], which has all kinds of trajectories: bound periodic, bound ergodic, "scattering" ones, and even sticking trajectories in the strict sense, fortunately in an amount of zero measure. What I want to do instead is to look onto the situation from the other end: What do we expect from a system with "random" or energy-"scrambling" collisions? We look for a flow of trajectories starting from a large sphere, S, around the scattering center that will be randomized on that same sphere after the collision. Symbolically, for $A \subset S$ and $B \subset S$, and allowing t to depend on the specific trajectory

$$\mu(U_t A \cap B) = \mu(A) \cdot \mu(B) \tag{2}$$

In a more familiar language, where we use (quantum) state indices rather than sets of trajectories initial conditions, this reads

$$P(f|i) = P(f) \cdot P(i) \tag{3}$$

or $P_{i \to f} = P_{i \to c} \cdot P_{c \to f}$

i.e. the probability of scattering from i to f bears no correlation between i and f, therefore can be factored into one factor describing "complex formation" and another one describing "complex decay". This is then a (weak) analogue of the mixing property, eq. (1); and it "solves" problems (1) and (3) above, though not problem (2). To remedy this, one has to add a non-random probability, $P^o_{i \to f}$, for "direct" scattering from i to f, by which we end up with the well-known expression[11]

$$|S_{if}|^2 = P^o_{i \to f} + P_{i \to c} P_{c \to f} \tag{4}$$

with normalization

$$\sum_i (P^o_{i \to f} + P_{i \to c}) = \sum_f (P^o_{i \to f} + P_{c \to f}) = 1 \tag{5}$$

A stochastic* theory of molecular collisions has to disregard the

---

*One should distinguish between "stochastic" behaviour, where eq.(3) holds at almost every energy, and a "statistical" theory, which allows only approximate "coarse-grained" averages over an energy interval. Note that the number theoretical examples cited in ref. 12 to support the second view belong mathematically to the first catagory.

direct probability $P_{i \to f}^o$. It will then assume non-correlation between initial and final states, or "loss of memory" (within the limits set by conservation laws). But, in general, it will make a second assumption: equiprobability. That means, an a priori probability measure in phase space (or an a priori counting procedure for quantum states) must be postulated, then the $P(i)$ and $P(f)$ of eq. (3) have to be taken proportional to the phase space volume with respect to this measure (or proportional to the count of accessible elementary quantum states). While eq. (1) presupposes that the measure μ is preserved under the trajectory flow (i.e. obeys the Liouville theorem), this is no longer necessary in the later equations. That opens the possibility of different statistical theories.[13] In practice also velocity weighted measures ("flux measures") are used, moreover phase space is restricted either to the E shell or to the E, J-shell, finally the selection of that part of phase space which leads to direct collisions is treated in different ways. All these theories are compatible with eq. (4).

The question which is the "true" statistical theory, i.e. which is the true measure in phase space applicable to a certain type of collision, is, in principle, open to test by experiments. There is moreover, no reason, why it should be the same for all scattering systems. In practice, the test is a bit difficult because the predictions by different theories are not too different, and because there are few systems where one can safely disregard errors of the potential energy surface, which one uses as input to the theory (except in the "thermodynamic approach"[14] which does not need such information).

Here is, where classical trajectories enter the picture. What we propose, and have partially achieved, is to use trajectories to compute not only $P_{i \to f}^o$ but also $P_{i \to c}$, then use microscopic reversibility to get $P_{c \to f}$, and construct statistical cross-sections from these data. This approach has also been used in ref. 15.

TRAJECTORY CALCULATIONS

The method of classical trajectories as applied to molecular collisions is well known.[16] If one has a realistic potential energy surface their use is a true though classical simulation of the collision process. In general the deviation of this simulation from quantum calculations is small if one avoids situations heavily dependent on tunnel effects, interferences, and the like.

On applying this technique to randomizing collisions one might of course try to compute the trajectories through the complex, and forget all statistical theories. This will become more and more difficult, when one approaches collision complexes which deserve the description as "mixing". For, if the probability of complex decay becomes independent of the way by which it was formed, i.e. if the complex "forgets" its initial conditions, the integrating routine must "forget" the initial conditions too, in contradiction to what it is supposed to do.

The ease with which one handles this situation, which is mathematically characterized by instability and stiffness[17] of the systems of differential equations, depends on system parameters. H_3^+ with its deep and strongly anharmonic well seems to be especially difficult, and we were not without undue effort able to compute the trajectories for more than the first 5oo fs. However, - and that is the important thing - one does not need to compute more than a short piece of the trajectory in order to make a statistical theory. This piece must only be long enough to define whether a trajectory is to be counted as direct or as "complex" i.e. "stochastic".

There are two obvious ways of doing this, and three less obvious ones:

(a) The first obvious, though unpractical way to define complex collisions is to look for the dependence of some final property of the trajectories (e.g. scattering delay) on some initial condition (e.g. vibrational phase) and call that fraction of initial phase space leading to complex collisions, where this dependence is "erratic".[18] Apart from the question whether all properties or initial conditions give the same answer, I call this procedure unpractical, because it does not save you from computing the trajectories to their end, which obviates the need for a statistical theory.

(b) A trajectory is counted as complex forming if it spends some time in a certain part of configuration space. The idea is that, when A+BC stay together as ABC for long enough time, energy and other observables will randomize. This idea is sound, but be aware: the time needed to randomize different observables must not be the same. So people often demand that $\tau > \tau_{rot}$, the rotation time of the complex, which is rather long, especially if taken as the rotation time at the orbiting distance. Of course, this is necessary to randomize the decay angle, but the time necessary to mix the vibrotational states may be much shorter. As another example I may mention $H^+ + H_2$, where we find energy scrambling within about loo fs, but experiments with ortho- and para-H_2 show[19] that the nuclear spin is not even randomized after the average complex lifetime of more than 1 ps.

Another word of caution is necessary: it may well be that the time needed for energy scrambling depends on the shape of the complex which has been formed. As an example, in $H^+ + D_2$ we find orbiting trajectories, which lead to very little interaction of the orbiting motion of H^+ with the vibration-rotation of D_2, but nevertheless are long-lived on the time scale discussed.

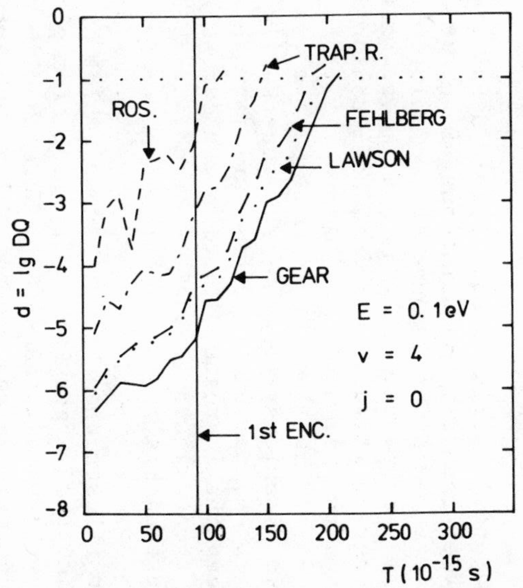

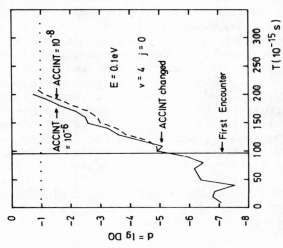

Fig. 3 Same as Fig. 2. Gear Routine with accuracy parameter ACCINT changed after first diatomic encounter; cf. also text.

Fig. 2 Logarithmic derivative of phase space distance of one trajectory in single vs. double precision; different integrating routines.

(c)   A third criterion to tell whether a trajectory is mixing or not
has been proposed by Duff and Brumer.[20] They monitored the distance
in phase space, $D(t)$, of two trajectories started near each other,
which is expected to increase exponentially under mixing conditions.
They found that for a model system an increase $D(t)/D(o)$ of about
1000 guarantees randomization of the collision complex. Similar
behaviour has been observed by us comparing single trajectories calculated twice, once in single precision, and once in double precision.
Again the aberration is roughly exponential and its logarithmic derivative $d \log D(t)/dt$ is independent of other circumstances, particularly
of the type of integrating routine used, while the pre-exponential
factor depends on the latter (see Fig. 2). We found, moreover, that
the difference between the approximate and the "exact" trajectory
develops at the first "binary encounter" (turning point of the
smallest interatomic distance R_{ik}) and can be influenced little
thereafter (see Fig. 3). The great advantage of the method of Duff
and Brumer is that it checks a property of the trajectories which
we will see is directly connected with "mixing", while its disadvantage is the necessity to compute two trajectories for one.

(d)   This is avoided if one uses not $D(t)$ itself but looks for the
deeper cause of the exponential increase. Let us write the system
of ordinary differential equations, which we want to integrate, as

$$\dot{y} = \underline{f} \, (\underline{y}) \quad . \tag{6}$$

This equation can be linearized to yield

$$\dot{y} = \underline{\underline{J}} \, \underline{y} \text{ with } J_{ik} = \partial f_i / \partial y_k \tag{7}$$

the Jacobian (or Lyapunov) matrix of eq. (6). If J has real eigenvalues (the Lyapunov exponents), the solutions of eq. (6) are unstable,[21] and the distance in phase space of two solutions started
near each other increases approximately as $\exp \lambda_{max} t$, where λ_{max} is
the largest eigenvalue (positive, since in a Hamiltonian system all
eigenvalues come in pairs $\pm\lambda$).

Now, an unstable dynamic system must not necessarily be mixing,
but special kinds of them, called C-systems, are.[3,22] This renders
the connection between the increase $D(t)$ of neighbour solutions, the
value of λ_{max} along the trajectory, and the stochastic behaviour of
the trajectory understandable.

To give another illustration, Fig. 4 shows a plot of the largest
real root (dashed) and the largest imaginary root of a piece of HD_2^+
trajectory together with the distances R_{ik}. Note that in the 3-
dimentional case these eigenvalues are not only a property of the
potential but also of the angular motion of the intermediate complex.
One observes that the eigenvalues of the Jacobian peak whenever a
"binary encounter" of a pair of nuclei takes place. One may,

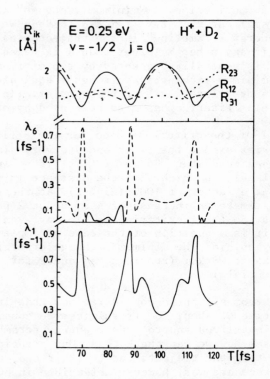

Fig. 4: Internuclear distances and their relation to the largest real and largest imaginary roots (λ_6 resp. λ_1) of Jacobian. Three diatomic encounters give peaks of λ's.

therefore, say that these near encounters are the "source" of mixing in the collisions. As a consequence of this behaviour of λ_6 one can expect that a good criterion for "stochastic trajectories" is what has been called the "stochastic impulse" by Hutchinson and Wyatt,[23] but was used before by Benettini et al.[24]:

$$\Lambda = \int Re\lambda_1 \, dt \quad . \tag{8}$$

The former authors found that $\Lambda \geq 7$ assures stochastic behaviour of a trajectory.

(e) Finally, in our calculations[7,9] we used the count of "minimum exchanges", M, as a criterion. A "minimum exchange" is the change of identity of the smallest R_{ik}, say from R_{12} to R_{23}. We found that $M \geq 8$ insures "energy scrambling" in the trajectories, with not too much difference if any other M larger than 4 is taken. Since one binary encounter is generally bracketed by two minimum exchanges (cf. Fig. 4), 8 minimum exchanges are roughly equivalent to 4 binary encounters or "vibrations" of the complex. (In a somewhat different form this criterion has also been used by others.[25,26])

Actually all of the criterions used for the definition of complex trajectories except the first one are very similar. In HD_2^+ 8 minimum exchanges need about 8o fs of time (if no intermediate orbiting takes place), in which time the distance in phase space increases by about a factor of 1000 (Fig. 2). At the same time from Fig. 4 one can estimate an increase of Λ of about 2 per pair of minimum exchanges. So the 4 criteria are similar even on a quantitative basis, and it is a question of the ease of implementation which one to use. In my opinion the minimum exchange criterion is the easiest as long as it works. (Problems may occur for very asymmetric or very soft potentials).

We conclude from all these considerations that in the H_3^+ system a collision lifetime of about 1oo fs suffices to consider the collision complex as long-lived and to expect energy scrambling. Of course, the question arises how we can check this. One principal possibility consists in computing the trajectories to their end, and see how they behave.[15] This procedure was, however unfeasible in our 3-D calculations on H_3^+, because of the exponentiating inaccuracy of the trajectories. But we can take advantage of the fact that for H_3^+ we have an accurate potential so that comparisons with experiments are meaningful. We want to compare the computed complex forming cross section

$$\sigma_{i \to c} = \pi b_{max}^2 \cdot \frac{N(M \geq 8)}{N_{total}} \tag{9}$$

with an "experimental" value. Now $\sigma_{i \to c}$ is not strictly an experimental observable, at least not for lifetimes as short as here. If we assume, however, that the branching ratio between

$$H^+ + D_2 \to HD_2^+ \to HD + D^+ \text{ and } \to D_2 + H^+ \tag{1o}$$

can be taken from statistical theory, and that direct collisions play a minor role, which can be taken into account by using the computed direct cross sections then

$$\sigma_{\to complex, \, exp} = (\sigma_{D^+_{exp}} - \sigma_{D^+_{direct}})(1 + \frac{P_{c \to D^+}}{P_{c \to H^+}}) \tag{11}$$

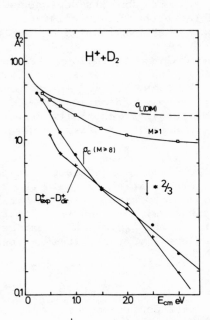

Fig. 5: Cross sections for H^+ and D_2 vs. energy. σ_c is the computed complex formation cross section, $2/3$ ($D_{exp} - D_{dir}$) the "experimental" one, cf. text. The Langevin cross section σ_L is far off.

Taking $3/2$ for the last bracket we arrive at Fig. 5, where the computed and the experimental complex forming cross sections have been plotted vs. collision energy together with the Langevin "capture" cross section, which is ordinarily assumed to determine complex formation.

Having thus gained some confidence in our definition we have computed a few other properties of HD$_2^+$ complexes. Among other things we find:

(a)   Internal excitation of the target, D_2, reduces the complex formation cross section somewhat but its dependence on internal energy is in no way comparable to that on translational energy. Vibrational and rotational excitation act similarly.

(b)   From dimensional analysis it follows that only mass ratios not the absolute mass can influence cross sections, assuming that energies and distances remain the same. We find large isotope effects (up to a factor of two) for the complex formation from D^+ + HD compared to D^+ + H_2. Probably the trapping probability depends on the mass ratio of the first binary encounter, which determines the energy taken away from the projectile.

(c)   We also calculated lifetimes of the complexes under the assumptions that the linear decay, which we could observe, can be interpreted as the beginning of an exponential one. Fig. 7 of ref. 9 shows a plot of lifetimes vs. total energy. It is gratifying that the total energy, which is so meaningless as a parameter for complexformation, shows up here as the important parameter determining τ.

STATISTICAL MODELS

As I said in the introduction we want to use detailed complex formation cross sections to formulate the "most dynamically biased statistical model", which still obeys the non-correlation axiom (eq. (3)), microscopic reversibility, and angular momentum conservation. The idea is that the detailed cross section $\sigma_{i \to f}$ can be taken to be a direct part (easily computable from trajectories) plus a complex part onto which we focus from now on. The latter is given by

$$\sigma_{i \to f}^E = \sum_{J,M} \delta_{i \to c}^{EJM} \cdot P_{c \to f}^{EJM} \tag{12}$$

(where no real M-dependence exists, such that $\sum_M$ gives only a factor 2J+1). $\sigma_{i \to c}^{EJM}$ is calculable from the trajectories without calculating them through the complex, as we have discussed. One gets

$$\sigma_{\alpha v j \to c}^{EJM} = \frac{\pi}{k^2} \frac{1}{2j+1} \sum_l P_{\alpha v j l}^{EJM} \tag{13}$$

where

$$P_{\alpha v j l}^{EJM} = \left(\frac{N_{complex}}{N_{total}} \right)_{\alpha v j l}^{EJ} \tag{14}$$

I.e. trajectories must be started from a given channel α with internal state v, j of BC, total energy E, total angular momentum J, and orbital angular momentum l (or corresponding impact parameter).

Computing trajectories for fixed J is somewhat different from the usual procedure, where J is taken at random between $|1+j|$, but poses no problems. Since our calculations up to now are oldfashioned in this respect, we checked that σ_c does not depend on the angle between $\underline{j}$ and $\underline{l}$, which allows some approximations on the old data. For new calculations one would do better to take l at random between $|J+j|$ leading directly to

$$P_{\alpha v j}^{EJM} = P_{\alpha v j}^{EJ} = (\frac{N_{complex}}{N_{total}})\, _{\alpha v j}^{EJ} \tag{15}$$

Experience shows that these P's depend in a very smooth manner on E and v,j, such that efficient interpolation is possible.

Now to the decay probabilities, $P_{c \to f}$ in eq. (12): Microscopic reversibility tells us that

$$P_{c \to \alpha v j(1)}^{EJM} \propto P_{\alpha v j(1) \to c}^{EJM}$$

as computed before, where normalization of the P's provides an equation for the proportionality constant. Thus one has finally

$$\sigma_{\alpha v j \to \alpha' v' j'}^{EJM} = \frac{\pi}{k^2} \frac{1}{2j+1} (\Sigma_l P_{\alpha v j 1}^{EJ})(\Sigma_{1'} P_{\alpha' v' j' 1'}^{EJ}) \cdot \frac{1}{N^{JE}} \tag{16a}$$

with

$$N^{JE} = \sum_{\alpha''v''j''1''} P_{\alpha''v''j''1''}^{EJ} \tag{16b}$$

for normalization. Again, there is no real M-dependence, and we drop the index M on the P's.

Similar, though more complicated formulae can be written for differential cross sections.

Calculations along these lines are proceeding in our laboratory. The effort is greater than for the computation of direct cross sections, since at a given total energy all open channels have to be considered. We found that interpolation can profitably be used, especially with respect to j.

The results will be compared with other statistical theories, a first example was already given in Fig. 8 of ref. 9. The difference between statistical theories seem big enough to be tested by experiments, since the relative accuracy of our energy loss spectra is better than lo%.

To conclude, let me emphasize that trajectory calculations seem to be a convenient means to discuss the scrambling of energy and of other properties in colliding molecules, at least if the systems are small. It is possible to study different influences (mass ratios, potential shapes), and to visualize different mechanisms (i.e. the fact that orbiting collisions do not contribute much to scrambling). Moreover they provide us with the means to improve on current statistical theories of collision.

ACKNOWLEDGEMENTS

This work is the result of the common effort of E. Teloy (who proposed to count minimum exchanges), D. Gerlich (who did most of the experiments), U. Nowotny (who did most trajectories), D. Brass (who looked into the failure of integration routines) and other members of the laboratory. Computations were done at the University Computing Center. Financial aid of the Deutsche Forschungsgemeinschaft is gratefully acknowledged.

REFERENCES

1. S.A. Rice, Internal Energy Transfer in Isolated Molecules: Ergodic and Nonergodic Behaviour in: "Advances in Laser Chemistry", A.H. Zewail, ed., Springer, Berlin (1978).
2. O. Penrose, Foundations of Statistical Mechanics, Rep. Progr. Phys. 42: 1937 (1979).
3. V.I. Arnold and A. Avez, "Ergodic Problems of Classical Mechanics", W.A. Benjamin, New York (1968).
4. R.K. Preston and J.C. Tully, Effects of Surface Crossings in Chemical Reactions: The H_3^+ System, J. Chem. Phys. 54: 4297 (1971).
5. C.F. Giese and W.R. Gentry, Classical Trajectory Treatment of Inelastic Scattering of H^+ with H_2, HD, and D_2, Phys. Rev. A lo: 2156 (1974).
6. R. Schinke, M. Dupuis and W.A. Lester Jr., Proton-H_2 Scattering on an ab initio CI potential Energy Surface, J. Chem. Phys. 72: 39o9 (198o).
7. D. Gerlich, Reaktionen von Protonen mit Wasserstoff bei Stoß-energien von 0.4 bis 10 eV, Dissertation, Freiburg (1977).
8. E. Teloy, Proton-Hydrogen Differential Inelastic and Reactive Scattering at low Energies, in: "Electronic and Atomic Collisions", G. Watel ed., North Holland, Amsterdam (1978).
9. D. Gerlich, U. Nowotny, Ch. Schlier, and E. Teloy, Complex Formation in Proton-D_2 Collisions, Chem. Phys. 47:245 (198o).

1o. J. Moser, "Stable and Random Motions in Dynamical Systems",
     Princeton University Press, Princeton (1973).
11. W.H. Miller, Study of the Statistical Model for Molecular
     Collisions, J. Chem. Phys. 52: 543 (197o).
12. M. Quack and J. Troe, Statistical Methods in Scattering, in:
     "Theoretical Chemistry, Advances and Perspectives" Vol. 6,
     H. Eyring and D. Henderson eds., Academic Press, New York
     (to appear).
13. A Ben-Shaul, On Statistical Models and Prior Distributions in
     the Theory of Chemical Reactions, Chem. Phys. 22: 341 (1977).
14. R.D. Levine and R.B. Bernstein, Thermodynamic Approach to
     Collision Processes, in: "Modern Theoretical Chemistry" Vol.2.
     W.H. Miller ed., Plenum Press, New York (1976).
15. J.W. Duff and P. Brumer, Exponentiating Trajectories and
     Statistical Behaviour, J. Chem. Phys. 71: 2693 (1979), and:
     Statistical Behaviour and the Detailed Dynamics of Collinear
     F+H$_2$ Trajectories, J. Chem. Phys. 71: 3895 (1979).
16. See e.g. D.G. Truhlar and J.T. Muckermann, Quasiclassical and
     Semiclassical Methods, in: "Atom-Molecule Collision Theory",
     R.B. Bernstein ed., Plenum Press, New York (1979).
17. See e.g. K.E. Atkinson, "An Introduction to Numerical Analysis",
     Wiley, New York (1978).
18. D.E. Fitz and P. Brumer, Geometric Effects on Complex Formation
     in Collinear Atom-Diato Collisions, J. Chem. Phys. 7o: 5527
     (1979).
19. D. Gerlich (to be published from his laboratory).
2o. J.W. Duff and P. Brumer, Exponentiating Trajectories and
     Statistical Behaviour in Collinear Atom-Diatom Collisions,
     J. Chem. Phys. 67: 4898 (1977).
21. See e.g. A. Halanay, "Differential Equations", Academic Press,
     New York (1966).
22. J. Ford, The Transition from Analytical Dynamics to Statistical
     Mechanics, Adv. Chem. Phys. 24: 155 (1973).
23. J.S. Hutchinson and R.E. Wyatt, Detailed Dynamics of Collinear
     F+H$_2$ Trajectories, J. Chem. Phys. 7o: 35o9 (1979).
24. G. Benettini, R. Brambilla, and L. Galgani, A Comment on the
     reliability of the Toda Criterion for the Existence of a
     Stochastic Transition, Physica 87A: 381 (1977).
25. E. Pollak and P. Pechukas, Unified Statistical Model for Complex
     and Direct Reaction Mechanisms: A test on the Collinear H+H$_2$
     Exchange Reaction, J. Chem. Phys. 7o: 325 (1979).
26. J.P. Davis, a Combined Dynamical-Statistical Approach to
     Calculating Rates of Complex Bimolecular Exchange Reactions,
     J. Chem. Phys. 71: 52o6 (1979) and 73: 2o1o (198o).

NEW PRINCIPLES IN CONDENSED PHASE CHEMISTRY

S. A. Adelman

Department of Chemistry
Purdue University
West Lafayette, IN 47907

Today I will discuss new theoretical methods[1] for describing and simulating, via classical trajectories, condensed phase vibrational energy transfer and chemical reaction processes. These methods are based on a new <u>exact</u> formulation of irreversible dynamics[2] which I call the molecular timescale generalized Langevin equation (MTGLE) theory. The MTGLE theory provides a natural framework for many-body dynamics on ultrashort timescales, e.g. the timescale of a chemically effective molecular encounter. The MTGLE theory is thus naturally suited as a framework for condensed phase chemical reaction dynamics.

The MTGLE theory is also, in principle, applicable to other short timescale relaxation phenomena. Examples of interest to people at this meeting include: energy scrambling and non-linear dynamics in large molecules, optical dephasing in molecular crystals.[3]

In this talk I will try to:

1. Explain the physical simplifications which make condensed phase reaction dynamics a tractable theoretical problem.

2. Show how the MTGLE theory provides a formal framework within which models which exploit these physical simplifications may be developed.

3. Show how the MTGLE models permit one to reduce classical mechanical condensed phase chemical problems to numerically tractable but physically realistic effective few-body trajectory calculations.

To focus our attention let us consider, as an example, the following aqueous SN_2 reaction

$$Cl^- + CH_3Cl \xrightarrow{\ H_2O\ } ClCH_3 + Cl^-.$$

This reaction may become the "H + H_2" of theoretical liquid state chemical kinetics. The reason is that it is one of the simplest reactions which displays the basic features shared by most solution reactions. These features include: an initial encounter leading to the formation of a caged <u>ionic</u> complex, repeated collisions in the common cage, vibrational energy transfer, a barrier crossing, final breakup of the common cage to yield separated products.

Classical trajectory simulations could, in principle, provide detailed information about the molecular mechanism of this reaction. While such simulations would be relatively straightforward in the gas phase, a liquid state simulation presents a complex many-body problem.

Fortunately the main features of the many-body aspects of the reaction problem may be understood in physical terms. These features are:

(i)   Solute-solvent energy transfer.
(ii) Caging of the solute by the solvent.

The key point is that these features are mainly due to inter-action of the solute with the <u>first</u> solvation shell. This immedi-ately suggests the following strategy. Treat the influence of the first (and perhaps also the second, third etc.) solvation shell on the reaction realistically. Model the bulk water as a simple energy reservoir.

The MTGLE theory provides a rigorous framework which permits one to conveniently implement this rough strategy.

Solution reactions, despite simplifications, are, however, physically complex. Thus as a first illustration of how the MTGLE theory works, I will discuss a far simpler many-body molecular collision problem. This is the problem of gas atom scattering off a one-dimensional nearest neighbor harmonic chain (Fig. 1). The gas atom strikes atom 0 colinearly, exchanges energy through atom 0 with the chain, and then departs. The atom-chain model played a key role in the early development of the concepts of the MTGLE theory.

   The following physical points concerning atom-chain collisions
are of general importance for condensed phase chemical problems.

   (i) The many-body collision reduces to a two-body problem if
atom 1 is clamped at equilibrium or equivalently if the force constant $\omega_{c_1}^2$ (Fig. 1) coupling chain atom 0 and 1 vanishes. This two-
body problem is the collision of a gas atom with an isolated or
<u>Einstein</u> harmonic oscillator (Fig. 1). For this reason we call ω_{e_0}
the Einstein frequency of chain atom 0. The Einstein limit is realized physically for collisions of high energy gas atoms off cold
(T=0K) chains. For such collisions, the scattering is completed
before the impact energy can displace atom 1 (Fig. 1) from its
initial equilibrium position. At lower gas atom energies the
Einstein limit breaks down due to energy flow into the chain. The
collision dynamics, as is apparent from Fig. 1, are still, however,
strongly influenced by the Einstein frequency ω_{e_0}. Thus the

Einstein limit must be realistically treated in any full theory.

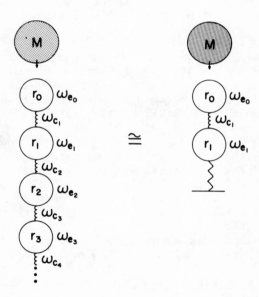

Fig. 1. Reduction of Many-Body Dynamics for Atom-Chain Collisions

(ii) The many-body aspects of the problem, in this case energy
flow into the bulk chain, must also be included in any qualitatively
satisfactory theory. A trapped gas atom, for example, will desorb
from an Einstein model solid in less than, say, 1 psec. Gas atoms
can stick to a real solid, however, for macroscopic times ($\gtrsim$ 1 sec.).

The many-body aspects of the problem may fortunately be modelled.
The physical idea behind the modelling is simple. Initial energy
transfer to the chain must be treated realistically; i.e. the short
timescale or Einstein limit must be correctly described. Once the
transferred energy has left the impact zone and entered the bulk
chain, however, its detailed flow becomes irrelevant to the collision dynamics and thus may be modelled. The one restriction on
modelling is that the energy of impact must be properly dissipated
(no Poincaré recurrences).

The effects of the $\sim 10^{23}$ bulk chain atoms may therefore be
modelled by a special spring. This spring is purely absorptive for
T = 0K chains. For chains at finite temperature, the special spring
emits as well as absorbs energy. The emission and absorption balance so as to maintain the temperature of the chain. (This balance
is the content of the so-called fluctuation-dissipation theorem.)

Since 1, 2, 3, ... chain atoms may be treated explicitly, the
procedure just described leads to a convergent sequence of models
for the full many-body system. Figure 1 illustrates the modelling
of the full chain by a two-atom model.

Now let us return to liquids. Before considering reactions
let us look at the dynamics of a single solute atom of mass m. We
will qualitatively discuss the dynamics from heuristic chemical
viewpoint. We imagine that the atom is surrounded by a concentric
sequence of nested solvation shells. This picture is illustrated
in Fig. 2.

We next discuss the many-body dynamics using the nested solvation shell picture. For conceptual simplicity, we first imagine
that the solvation shells are initially "frozen" in their most
probable positions (Fig. 2). Thus initially there is no net force
on the atom. Next displace the atom slightly, say to the right,
by an amount Δx_0. A restoring force on the atom directed to the
left is set up by this displacement. The restoring force is pro-
portional to Δx_0 and may be written as $F = -m \, \omega_{e_0}^2 \, \Delta x_0$. We will

refer to ω_{e_0} as the Einstein frequency of the atom. The reason is

that if the solvation shells remained frozen, the atom would vibrate
like an isolated or Einstein harmonic oscillator with frequency
ω_{e_0}. This vibratory motion is often referred to as the cage effect.

NESTED–SHELL PICTURE

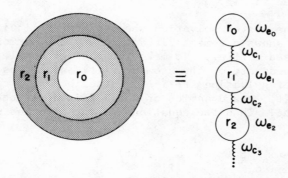

Fig. 2.   MTGLE Picture of Many–Body Dynamics

The solvation shells, of course, do not remain frozen since the first solvation shell is directly coupled to the atom, the second shell is directly coupled to the first, etc. Assume the strength of the successive couplings is measured by a sequence of force constants $\omega_{c_1}^2$, $\omega_{c_2}^2$, $\omega_{c_3}^2$. Thus for example, $\omega_{c_1}^2$ is the coupling constant linking the atom and the first shell.

Using the nested shell picture, the many–body dynamics will be described in the following simple manner. The displaced atom ini-tially oscillates with its Einstein frequency ω_{e_o}. This, in turn, stimulates the second shell to oscillate with an Einstein frequency ω_{e_2} etc.

The net effect of this sequence of stimulated oscillations is that the energy of displacement, $\frac{1}{2} m \omega_{e_o}^2 \Delta x_o$, is dissipated into the bulk water and that the solute atom executes damped oscillatory motion. The details of this motion are determined by the set of Einstein frequencies ω_{e_o}, ω_{e_1}, ... and coupling constants $\omega_{c_1}^2$, $\omega_{c_2}^2$,...

This picture is, of course, incomplete since atomic motion in liquids is perpetual rather than damped. The prediction of damping comes because of our assumption that the shells were initially frozen at their most probable configuration. The shells, of course,

actually execute random thermal motion about the most probable configuration.   The result is that the damped oscillatory motion describes the average atomic trajectory.   A typical trajectory fluc-
tuates about this average.   The atomic velocity fluctuations after
the initial displacement energy has relaxed away are those expected
from the Maxwell-Boltzmann velocity distribution.   These points
are illustrated in Fig. 3.

Within the nested solvation shell picture, liquid state many-
body dynamics shows a remarkable analogy to the harmonic chain dynamics discussed earlier.   To make the analogy complete, let us
denote the coordinate of the atom by $r_o(t)$ and let us imagine that
the dynamics of solvation shell 1 is described by a <u>collective</u>
<u>coordinate</u> $r_1(t)$, the dynamics of shell 2 is described by a collec-
tive $r_2(t)$, etc.   Then the equations of the atom and nested shells
may be recast as

$$\ddot{r}_o(t) = -\omega_{e_o}^2 \, r_o(t) + \omega_{c_1}^2 \, r_1(t)$$

$$\ddot{r}_1(t) = -\omega_{e_1}^2 \, r_1(t) + \omega_{c_1}^2 \, r_o(t) + \omega_{c_2}^2 \, r_2(t) \tag{1}$$
$$\vdots$$

where ω_{e_o}, ω_{e_1}, ... and ω_{c_1}, ω_{c_2}, ... are respectively, the Einstein
frequencies of the shells and the coupling constants linking the

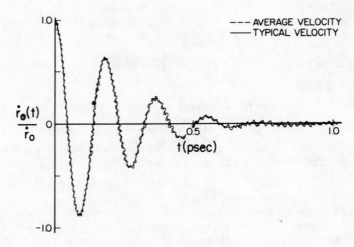

Fig. 3.   Velocity of a Displaced Atom

shells. Notice that the above equations are identical in form to
the harmonic chain equations of motion illustrated in Fig. 1. We
will, therefore, refer to Eq. (1) as the equivalent harmonic chain
equations.

Our "derivation" of the equivalent chain equations leaves room
for skepticism. A statistical mechanical analysis,[2] namely the
MTGLE theory, however, shows that the classical or quantum dynamics
of an arbitrary dynamical variable $r_o(t)$ may be rigorously recast
in the harmonic chain form of Eq. (1). The Einstein frequencies
and coupling constants may be calculated from the underlying many-
body dynamics via an algorithm prescribed by the theory. The coordinates $r_1(t)$, $r_2(t)$, ... are mathematical constructs defined by
the theory.

For atomic dynamics in a liquid the variables $r_1(t)$, $r_2(t)$, ...
may be loosely interpreted as collective coordinates describing
solvation shells. This interpretation is clearly a rough one. The
shell correspondence, for example, breaks down completely as one
goes down the chain since high order solvation shells do not exist.
The equivalent chain equations are apparently as close as rigorous
theory can get to the chemists intuitive picture of solvation shells.

Equation (1) describes an infinite chain. Thus, so far, all
we have done is to transform one many-body problem (liquids) in
another many-body problem (chains). The equivalent chain, however,
may be truncated in the same manner as the real harmonic chain discussed earlier (see Fig. 1). A two-atom truncated equivalent chain
thus roughly corresponds to the treating the solute atom and the
first solvation shell in detail and to modelling the solvent outside
the shell as an energy reservoir. A three atom chain roughly corresponds to treating two shells explicitly, etc.

The following points are crucial. Solute-solvent energy transfer and solute caging, as stressed above, are the main ways that
many-body effects influence reaction dynamics. These effects,
however, depend largely on solute interaction with the first solvation shell. Thus the simplest MTGLE approximant, the truncated two-
atom chain therefore provides a physically realistic reduction of
the liquid state many-body problem to an effective two-body problem.
The implications of this result for theoretical liquid state chemical kinetics are very far reaching. As a simple example, the problem
of atomic recombination in liquids is reduced, via the two-atom chain
model, to an effective four body problem.

While the two-atom chain model gives a qualitatively correct
description of many-body dynamics, its quantitative accuracy is not
immediately clear. Thus the problem of truncated chain convergence
arises.

One convenient method for investigating convergence is based on the following remarkable property of the MTGLE equivalent chain. Suppose we have an <u>infinite</u> chain which is initially quiescent; i.e. all chain atoms are initially located at their equilibrium positions and have zero initial velocities. Next the terminal chain atom 0 is given a unit velocity at time t = 0. Then the theory shows that $\dot{r}_0(t)$, the velocity of the terminal chain atom, is rigorously equal to the velocity autocorrelation function of the physical atom. The velocity of the terminal atom of a truncated rather than an infinite chain may be similarily calculated. The result is equal to the velocity autocorrelation function of the chain model for the full many-body system. This approximant may be compared with the exact velocity autocorrelation function. This yields a check of the accuracy of the chain model.

The procedure just described was applied in an analysis of the velocity autocorrelation function of the Na^+ ion dissolved in pyridine. The far infrared absorption due to cage oscillations of the Na^+ was measured by W. Edgell and M. Balk (J. Chem. Phys. to be published). The velocity autocorrelation function was determined by Fourier transforming the absorption spectrum to the time domain. A two-atom chain model was also constructed from the spectrum using the prescriptions of the MTGLE theory. The time correlation function of the two-atom chain model was also determined. This is shown in Fig. 4. Agreement between the experimental (points) and two-atom chain (line) correlation functions is excellent. Notice that the Na^+ velocity autocorrelation function is highly oscillatory. This is characteristic of the strong caging expected for a well solvated ionic species. The two-atom chain model, as expected, faithfully reproduces the cage oscillations.

We next show the results of a model classical trajectory study of atomic recombination in a solvent. What we do is monitor the dynamics of a pair of truncated two-atom chains which are initially separated by 7Å but which recombine due to the presence of a strong attractive interatomic potential. Chain and potential parameters are chosen so as to crudely model $I\cdot + \cdot I$ recombination in hydrocarbon solvents.

The point of this study is to illustrate that short-timescale solvent response can influence recombination dynamics. Such dynamics are usually thought of as being diffusion controlled. This means that the recombination rate depends only on the diffusion coefficients of the recombining species. The diffusion coefficients, in turn, depend only on the <u>zero frequency</u> Fourier transform of the atomic velocity autocorrelation functions. We show in Fig. 5 that recombination dynamics varies as the Einstein frequency ω_{e_c} and coupling constant ω_{c_1} are varied with the diffusion constants fixed.

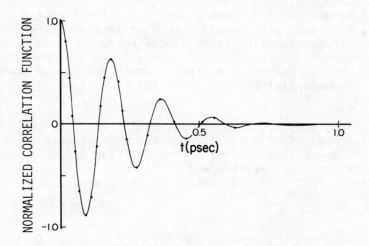

Fig. 4. Velocity Autocorrelation Function of Na^+ in the Solvent Pyridine

PROBABILITY P(t) OF AN UNREACTED PAIR

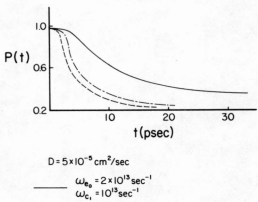

$D = 5 \times 10^{-5} cm^2/sec$

$\omega_{e_0} = 2 \times 10^{13} sec^{-1}$
$\omega_{c_1} = 10^{13} sec^{-1}$

$\omega_{e_0} = 10^{13} sec^{-1}$
$\omega_{c_1} = 10^{13} sec^{-1}$

$\omega_{e_0} = 10^{13} sec^{-1}$
$\omega_{e_1} = 2 \times 10^{13} sec^{-1}$

Fig. 5. Recombination Rate as a Function of Time

The model studies just presented suffer from a qualitative flaw. The reason is that the chain parameters used are appropriate for isolated atoms. As the radicals recombine, the solvation shell structure changes due to overlap of the individual atomic cages and eventual formation of a common cage. Thus the Einstein frequency, coupling constant, etc. should depend on interatomic separation. For more general reactions than simple recombination, these parameters should, in fact, depend on the full solute configuration. Thus one expects parameter surfaces, analogous to gas phase Born-Oppenheimer potential surfaces, to play a key role in liquid state reactions.

A generalized MTGLE theory based on these ideas has recently been developed.[1] The detailed application of this theory to realistic simulations of butane isomerization and iodine recombination in carbon tetrachloride and vibrational energy relaxation of nitrogen in liquid argon is currently in progress.

REFERENCES

1. S. A. Adelman, J. Chem. Phys. 73, 3145 (1981).
2. S. A. Adelman, J. Chem. Phys. 74, 4646 (1981).
3. A. Zewail, Acc. Chem. Research 13, 360 (1980).

INDEX